HISTORY OF MATHEMATICS ❖ VOLUME 32

Episodes in the History of Modern Algebra (1800–1950)

Jeremy J. Gray
Karen Hunger Parshall
Editors

Editorial Board

American Mathematical Society
Joseph W. Dauben
Peter Duren
Karen Parshall, Chair
Michael I. Rosen

London Mathematical Society
Alex D. D. Craik
Jeremy J. Gray
Robin Wilson, Chair

2000 *Mathematics Subject Classification*. Primary 01A55, 01A60, 01A70, 01A72, 01A73, 01A74, 01A80.

For additional information and updates on this book, visit
www.ams.org/bookpages/hmath-32

Library of Congress Cataloging-in-Publication Data
Episodes in the history of modern algebra (1800–1950) / Jeremy J. Gray and Karen Hunger Parshall, editors.
   p. cm.
   Includes bibliographical references and index.
   ISBN-13: 978-0-8218-4343-7 (alk. paper)
   ISBN-10: 0-8218-4343-5 (alk. paper)
   1. Algebra—History. I. Gray, Jeremy, 1947– II. Parshall, Karen Hunger, 1955–

QA151.E65 2007
512.009—dc22
                                                                                              2007060683

Copying and reprinting. Individual readers of this publication, and nonprofit libraries acting for them, are permitted to make fair use of the material, such as to copy a chapter for use in teaching or research. Permission is granted to quote brief passages from this publication in reviews, provided the customary acknowledgment of the source is given.

Republication, systematic copying, or multiple reproduction of any material in this publication is permitted only under license from the American Mathematical Society. Requests for such permission should be addressed to the Acquisitions Department, American Mathematical Society, 201 Charles Street, Providence, Rhode Island 02904-2294, USA. Requests can also be made by e-mail to reprint-permission@ams.org.

© 2007 by the American Mathematical Society. All rights reserved.
Printed in the United States of America.

Copyright of individual articles may revert to the public domain 28 years after publication. Contact the AMS for copyright status of individual articles.
The American Mathematical Society retains all rights
except those granted to the United States Government.
∞ The paper used in this book is acid-free and falls within the guidelines
established to ensure permanence and durability.
The London Mathematical Society is incorporated under Royal Charter
and is registered with the Charity Commissioners.
Visit the AMS home page at http://www.ams.org/

10 9 8 7 6 5 4 3 2 1   12 11 10 09 08 07

Episodes in the History of Modern Algebra (1800–1950)

Contents

Acknowledgments   1

Chapter 1.   Introduction
                 JEREMY J. GRAY AND KAREN HUNGER PARSHALL   3
  Algebra: What? When? Where?   3
  Episodes in the History of Modern Algebra   5
  Concluding Remarks   10
  References   10

Chapter 2.   Babbage and French *Idéologie*: Functional Equations, Language, and the Analytical Method
                 EDUARDO L. ORTIZ   13
  Introduction   13
  Speculation on the Origin of Languages   14
  Senses, Languages, and the Elaboration of a Theory of Signs   16
  The Position of Grammar   17
  On the Language of Calculation   18
  Babbage and a "Language" for the Solution of Functional Equations   19
  Babbage's Notation   21
  Babbage's Treatment of Functional Equations   23
  Babbage and First-Order Functional Equations in One Variable   26
  The Aftermath of Condillac in France: The Beginning of a Discussion   28
  De Gérando's Critique of Condillac: A Turning Point in *Idéologie*   31
  De Gérando's Theory of Signs and Its Mathematical Implications   33
  Babbage and de Gérando's Views on Signs   36
  Destutt de Tracy's *Élémens d'idéologie*   37
  Destutt de Tracy's Views on Artificial Languages   39
  Final Remarks   40
  References   42

Chapter 3.   "Very Full of Symbols": Duncan F. Gregory, the Calculus of Operations, and the *Cambridge Mathematical Journal*
                 SLOAN EVANS DESPEAUX   49
  Introduction   49
  The Establishment of a "Proper Channel" for the Research of Junior British Mathematicians   50
  The Calculus of Operations before the *Cambridge Mathematical Journal*   52
  The Revival: The Introduction of the Calculus of Operations into the *Cambridge Mathematical Journal*   54

The Conversation Begins: The Adoption of the Calculus of Operations by
Contributors to the *Cambridge Mathematical Journal*   60
Conclusion   67
References   69

Chapter 4. Divisibility Theories in the Early History of Commutative Algebra and the Foundations of Algebraic Geometry
OLAF NEUMANN   73
On Some Developments Rooted in the Eighteenth Century   74
Developments Inspired by Gauss   77
From Kummer to Zolotarev   84
Complex Analytic and Algebraic Functions   87
Kronecker's Modular Systems   89
The Work of David Hilbert and Emanuel Lasker   94
The Work of Emmy Noether and Her Successors   97
Concluding Remarks   99
References   100

Chapter 5. Kronecker's Fundamental Theorem of General Arithmetic
HAROLD M. EDWARDS   107
Introduction   107
Statement and Proof of Kronecker's Theorem   109
Conclusions   114
References   115

Chapter 6. Developments in the Theory of Algebras over Number Fields: A New Foundation for the Hasse Norm Residue Symbol and New Approaches to Both the Artin Reciprocity Law and Class Field Theory
GÜNTHER FREI   117
Introduction   117
The Beginnings of Structure Theory   118
Wedderburn's General Structure Theorems   121
Hurwitz and the Arithmetic of Quaternions   122
The Structure of Skew Fields: Connections with Algebraic Number Theory   124
The Theory of Semisimple Algebras   127
The Local Theory and the Theorem of Brauer-Hasse-Noether   133
The New Norm Residue Symbol and New Approaches to Both the Reciprocity Law and Class Field Theory   135
Summary and Conclusions   141
References   143

Chapter 7. Minkowski, Hensel, and Hasse: On The Beginnings of the Local-Global Principle
JOACHIM SCHWERMER   153
Introduction   153
Toward an Arithmetic Theory of Quadratic Forms   155
Mathematical Digression: Quadratic Forms over Rings   158
Hermann Minkowski's Early Work   159
Hermann Minkowski's Letter to Adolf Hurwitz in 1890   161

Hensel's p-adic Numbers: Series, Expansions, or Numbers as Functions	162
Mathematical Digression: Valuations and p-adic Fields	164
Hasse's Thesis and *Habilitationsschrift*	165
From the Small to the Large, or a Local-Global Principle	167
A Talk by Hasse in Königsberg in 1936	168
Conclusion	170
Appendix: Helmut Hasse to Hermann Weyl, 15 December, 1931	171
References	173

Chapter 8. Research in Algebra at the University of Chicago: Leonard Eugene Dickson and A. Adrian Albert
         DELLA DUMBAUGH FENSTER   179

Introduction	179
Leonard Dickson: Student	179
Leonard Dickson: University of Chicago Faculty Member	180
A. Adrian Albert and the Classification of Division Algebras	181
A. Adrian Albert: Professional Overview	188
Conclusions	192
References	194

Chapter 9. Emmy Noether's 1932 ICM Lecture on Noncommutative Methods in Algebraic Number Theory
         CHARLES W. CURTIS   199

Introduction	199
Brauer's Factor Sets, the Brauer Group, and Crossed Products	202
Cyclic Algebras and the Albert-Brauer-Hasse-Noether Theorem	206
The Principal Genus Theorem	210
Applications to Algebraic Number Theory by Hasse and Chevalley	212
Conclusion	215
References	217

Chapter 10. From *Algebra* (1895) to *Moderne Algebra* (1930): Changing Conceptions of a Discipline–A Guided Tour Using the *Jahrbuch über die Fortschritte der Mathematik*
         LEO CORRY   221

Introduction	221
The *Jahrbuch über die Fortschritte der Mathematik*	225
Algebra by the Turn of the Century: The *Jahrbuch* in 1900	227
Some Tentative Changes: 1905–1915	229
Transactions of the American Mathematical Society: 1910	232
The *Jahrbuch* after 1916	233
Concluding Remarks	237
References	239

Chapter 11. A Historical Sketch of B. L. van der Waerden's Work in Algebraic Geometry: 1926–1946
         NORBERT SCHAPPACHER   245

Introduction	245
1925: Algebraizing Algebraic Geometry *à la* Emmy Noether	250
1927–1932: Forays into Intersection Theory	256

1933–1939: When in Rome ... ?	264
1933–1946: The Construction Site of Algebraic Geometry	272
Appendix: Extract from a Letter from Hasse to Severi	276
References	278

Chapter 12. On the Arithmetization of Algebraic Geometry
            SILKE SLEMBEK   285

Introduction	285
Earlier Rewritings of Algebraic Geometry	286
The Mathematical Situation: Why Another Rewriting Seemed Necessary	289
Imitating Geometry with Modern Algebra and Arithmetic Ideal Theory	292
Leaving the Beaten Path	294
The Arithmetic Proof	296
Conclusion	296
References	298

Chapter 13. The Rising Sea: Grothendieck on Simplicity and Generality
             COLIN MCLARTY   301

The Weil Conjectures	303
Abelian Categories	305
The Larger Vision	311
Anticipations of Schemes	313
Schemes in Paris	315
Schemes in Grothendieck	317
Toward the Séminaire de Géométrie Algébrique	321
References	322

Index	327

Acknowledgments

In the spring of 2003, the Mathematical Sciences Research Institute (MSRI) in Berkeley, California hosted a semester in commutative algebra. MSRI's Director, David Eisenbud, and its then Deputy Director, Michael Singer, wanted to supplement this purely mathematical agenda with a related program in the history of mathematics and asked us if we would consider organizing such an historical component. The outcome of this collaboration was a week-long workshop in the history of nineteenth- and twentieth-century algebra held at MSRI in April of 2003.

The workshop was made possible financially thanks to support from both MSRI and the National Science Foundation (SES-0216982). It brought together some twenty speakers from Austria, Brazil, France, Germany, Israel, Switzerland, the United Kingdom, and the United States for a series of hour-long lectures delivered over a five-day period. Each day closed with a moderated discussion in which the participating historians of mathematics and the mathematicians visiting MSRI engaged in a dialog about what they viewed as the critical issues in the history of modern algebra and why. The present volume brings together suitably revised, chapter-length versions of twelve of those lectures.

Many people collaborated to make this volume a reality. At MSRI, David Eisenbud and Michael Singer welcomed all of us and provided an intellectual environment in Berkeley that allowed free and open discussion as well as the interchange of a wide variety of ideas. In the MSRI office, Loa Nowina-Sapinski smoothed the way for our intellectually intense week by taking care not only of all of the visa and other paperwork but also of all of the local arrangements in Berkeley. With the end of our spring week together in California, however, work of a different kind began.

As editors, we thank not only all of the authors whose research appears here but also all of the referees who read and provided extensive critical comments on the submissions. The authors first prepared drafts of their chapters in light of the discussions that took place in Berkeley, then responded to the critique of the (anonymous) referees, and finally worked with us in answering the numerous queries that inevitably arise in the final editing stages of any project. At the offices of the American Mathematical Society in Providence, Barbara Beeton then masterfully solved all of the typesetting problems that came up as we were preparing the final version of the manuscript for publication. The result, we hope, is a relatively consistent volume of historical scholarship that will shed new light on some of the historical complexities associated with trying to understand the development of modern algebra in the nineteenth and twentieth centuries.

London, United Kingdom   Jeremy J. Gray
Charlottesville, Virginia, United States   Karen Hunger Parshall
15 October, 2006

CHAPTER 1

Introduction

Jeremy J. Gray and Karen Hunger Parshall
Open University, United Kingdom and University of Virginia, United States

Algebra: What? When? Where?

It is an interesting, and difficult, problem even to define algebra; to begin to give it a history presents yet another challenge. Some historians of mathematics identify the abstract and general features of Mesopotamian mathematics with algebra, and so consider it the oldest branch of written mathematics, dating from at least 2000 B.C.E.[1] Others see its origins around 150 C.E. in Diophantus's *Arithmetica* with its solutions of general, indeterminate problems [Klein, 1968]. Still others find its more modern roots in medieval Islam, noting that the very word originated only in the ninth century in al-Khwārizmī's text entitled *al-Kitāb al-mukhtasar fī hisāb al-jabr wa'l-muqābala* or *The Compendious Book on Calculation by Completion and Balancing*. There, "al-jabr"—which was ultimately Latinized into "algebra"—translates as "completion" or "restoration." It referred to adding the same thing to both sides of an equation or restoring all of the terms to a standard form, and al-Khwārizmī himself regarded it as different from—although not unrelated to—the kind of geometrical procedures that Euclid had utilized in the *Elements* (300 B.C.E.) and that had held sway over the course of the intervening millennium [van der Waerden, 1985]. Those historians who insist on algebra as something distinct from geometry, however, see its emergence in the sixteenth-century, European texts of mathematicians such as Girolamo Cardano, Niccolò Tartaglia, and Rafael Bombelli on cubic and quartic equations as well as on the introduction of complex numbers,[2] while others require the presence of the characteristic features of high school algebra today, namely, the familiar signs for the arithmetic operations and the use of letters for constants and unknowns. For the latter, algebra came about in the late sixteenth and seventeenth centuries in the work of François Viète, Thomas Harriot, Pierre de Fermat, and René Descartes.[3] Finally, mathematicians tend to share yet another view of algebra's origins. For

---

[1]For a very sophisticated and highly nuanced interpretation of Babylonian algebra, see [Høyrup, 2002]. Many histories of algebra begin their story in ancient Mesopotamia. See, for example, [van der Waerden, 1983], [Scholz, 1990], [Sesiano, 1999], and [Bashmakova and Smirnova, 2000].

[2]See, for example, [Cardano, 1545/1968] for a sense of this sixteenth-century Italian work.

[3]For a sense of the seventeenth-century scene, see [Bos, 2001]. In [Nový, 1973], the history of algebra "begins" some time after 1770.

them, algebra is modern algebra, namely, the subject that was put into book form and given its name by Bartel van der Waerden beginning in 1930.[4]

All of these definitions have something to recommend them, and, indeed, there is a degree of continuity between these types of algebra that suggests a coherence to the topic across the centuries. The ancient Mesopotamians possessed systematic methods for solving numerical questions about numbers and their reciprocals and about numbers and squares. Interest in this topic passed to medieval Islamic mathematicians and eventually from them to the Renaissance Italians. When, in the late eighteenth and early nineteenth centuries, the quintic equation proved a stumbling block to the production of algorithms or formulas for solving polynomial equations of successive degrees, a series of mathematicians took up structural investigations of the problem. Although the list of structurally minded algebraists could justifiably begin with Harriot,[5] it most certainly includes the names of the eighteenth- and early nineteenth-century mathematicians, Alexandre Vandermonde, Joseph-Louis Lagrange, Niels Henrik Abel, and Évariste Galois.[6]

If the solution of equations has long engaged mathematical minds, so too has the problem of providing integer—and, indeed, natural number—solutions to mathematical questions. Whether this began and continued in a culture that found it natural to restrict the range of solutions in this fashion—so natural it went without comment—or whether it was a deliberately imposed restriction, a long-standing tradition connects algebra with number theory, and specifically with the branch today called algebraic number theory. Fermat may have labored in vain to interest his contemporaries in the subject, and such was his melancholy opinion, but with Leonhard Euler and Lagrange the subject burgeoned in the eighteenth century and became one of the key areas of interest to the German mathematical community in the nineteenth.[7]

Algebra has also established a close connection with geometry. While some have argued that Euclid was doing "geometrical algebra" as early as 300 B.C.E., the modern connection between the two fields is much more recent, not extending back chronologically much before the time of Fermat and Descartes.[8] This, however, became a vital connection, with each discipline at times turning to the other for central insights and methods. Euler was the first to *define* the conic sections algebraically, rendering it obvious that the same $b^2 - 4ac$ test that distinguished ellipses from hyperbolas distinguished the real and complex roots of a polynomial equation. In the nineteenth century, mathematicians from George Boole, Arthur Cayley, and James Joseph Sylvester in Great Britain to Charles Hermite, Paul Gordan, and David Hilbert on the Continent pursued this clue energetically and, in so doing, developed the new field of invariant theory.[9] Algebraic and, especially, projective geometry was one half of a subject, invariant theory the other, and so

---

[4]On the rise of the structural approach to algebra that culminated, in some sense, in van der Waerden's text, see [Corry, 1996].

[5]On Harriot's mathematical work, see [Stedall, 2003].

[6]For an analysis of these developments, see, for example, [Edwards, 1984].

[7]See [Euler, 1770-1771/1984] with its additions by Lagrange.

[8]The debate over "geometrical algebra" flared famously in an exchange between Bartel van der Waerden and Sabatei Unguru on the pages of the *Archive for History of Exact Sciences*. See [Unguru, 1975] and [van der Waerden, 1976].

[9]On these developments, see, for example, [Parshall, 1989].

began an amicable and long-running family dispute between those who stressed geometry and those who emphasized algebra.

Algebra even has a claim to being if not the core of mathematics then at least its language. Most papers in any branch of mathematics have an algebraic cast to them determined by the very symbolism used, even if the intellectual difficulties are in another field. This fact led some eighteenth-century philosophers to regard mathematics—that is, algebraic mathematics—as the best language because it generated arguments having the highest degree of certainty. Two hundred years later, mathematicians turned to syntax and formal languages to seek assurances about the nature of mathematics, while linguists such as Noam Chomsky defined language itself as a purely syntactic object expressed in mathematical terms. Algebras, languages, complexity, and issues in computation and the theory of algorithms are central to such topics as the P versus NP question.

Algebra, then, has not only an ever-evolving and distinct identity but also close connections to arithmetic and number theory, to geometry, and to topics in linguistics and computer science. It is a vast and vastly complicated area to try to pin down from an historical point of view. As noted in the acknowledgments, this book grew out of a conference hosted by the Mathematical Sciences Research Institute, Berkeley in 2003. It thus explores a number of *episodes* in the modern period of algebra's development, specifically episodes that stretch from the time of the French Revolution in the late eighteenth century to the arrival of modern algebra in the early twentieth century to the advent of category theory as an algebraic tool in the mid-twentieth century.

Episodes in the History of Modern Algebra

The volume opens with a chapter by Eduardo Ortiz on "Babbage and French *Idéologie*: Functional Equations, Language, and the Analytical Method" that explores eighteenth- and early nineteenth-century conceptions of mathematics as a language. As Ortiz shows, this was a particular concern of the Abbé de Condillac, who advocated inventing and adapting languages in the light of the experience provided by various sciences, taking algebra as a model. This revived a debate that had flourished in the seventeenth century about the relationship between logic and grammar, but Condillac reversed the old view and argued that mastery of grammar should come before the study of logic. As this may suggest, Condillac was interested in the idea that mental processes involved in understanding, such as analogy and association of ideas, should feature more prominently in the language in which thought is expressed. Charles Babbage read Condillac's work as a young man and applied it to his own research on functional equations. The idea of analogy in a context that, for Babbage, was much more algebraic than analytic, struck him forcefully, and he devised a method for constructing general solutions when special cases are given. The method, Ortiz argues, seems capable of further use even today, but has never before been adequately described in the historical literature.

The themes of language, the operational calculus, and broader forces at play in the history of mathematics are continued in the chapter by Sloan Despeaux on Duncan Gregory and the *Cambridge Mathematical Journal*. The emerging generation of British mathematicians in the 1820s and 1830s was largely indifferent to the strictures of Augustin-Louis Cauchy and other continental Europeans about the lack of rigor in the formal methods of the operational calculus. They applied it cavalierly, but with some success, to various kinds of differential equations, and while they had to admit they could not make it work for equations with variable coefficients, they were nonetheless led to interesting reflections about algebraic processes. Moreover, although they did not succeed in their goal of creating a single, unified field that embraced polynomial equations, differential equations, and functional equations, they did create and sustain a journal for mathematics in Britain that served as a key line of communication for a developing community of British mathematicians. Among the journal's notable publications was George Boole's work on linear differential equations with constant coefficients, a paper which highlighted not only the potential strength of the Cambridge-style, formal methods but also the journal's effectiveness in conveying new mathematical ideas to a self-taught Boole with no real Cambridge connections other than access to its pages.

The episodes in the history of modern algebra that Ortiz and Despeaux detail seem far from the modern algebra that has often been seen as having sprung with all the inevitability of a great invention in the early decades of the twentieth century from the group around Emmy Noether. It is as hard for mathematicians to imagine the twentieth century without it as it is to imagine the world without Hollywood, and this has hindered historians' ability to describe the transformation from the older algebra to the new. It has thus seemed easier to look at the precursors in the Noether line, such as Richard Dedekind, than to explore the work of those who stand outside it. The next two chapters here, however, take on that harder task.

Olaf Neumann traces the concept of divisibility from its origins in (and even before) the eighteenth century to the twentieth-century work of Emmy Noether and Wolfgang Krull. The idea was to establish, over a variety of domains, theories of divisibility that mimicked that of elementary number theory. The development of this idea, as Neumann shows, was not unilateral. Following Ernst Kummer's invention of ideal complex numbers in the mid-nineteenth century, two rather distinct approaches evolved. One—championed by Karl Weierstrass, Egor Zolotarev, Kurt Hensel, Krull, and others—centered on the idea of embedding the various domains either in principal ideal domains or in direct products of principle ideal domains. The other grew out of Dedekind's theory of ideals and engaged a string of mathematicians from David Hilbert through Adolf Hurwitz and Francis Macaulay to Emmy Noether. Thus, while the "direct" line to Noether is explored, so is another, equally important, intertwining line. The two lines taken together played a key role in the development of the modern concepts of commutative ring theory.

Another key nineteenth-century figure who stands largely off of the "direct" line to Noether is Leopold Kronecker. In his chapter, "Kronecker's Fundamental Theorem of General Arithmetic," Harold Edwards explores just what Kronecker meant by this theorem, which is a replacement for the fundamental theorem of algebra. The question is how to compute with roots of a polynomial equation, or, equivalently, how to factor polynomials with coefficients in an algebraic number field. Kronecker, in keeping with his constructivist approach to mathematics, advocated not regarding these roots as real numbers, and therefore had to develop a method for working with them. He succeeded in his task and saw his approach as lying at the heart of what he called the "precious" Galois principle. This led Kronecker to his new fundamental theorem, and Edwards here not only gives his own account of it but also uses this account to shed light on his own understanding of some of Kronecker's murkier arguments.

The next four chapters in this volume deal with the way in which the structural theory of algebras and hypercomplex number systems was brought to bear on the theory of numbers. Key figures here are Leonard Eugene Dickson and A. Adrian Albert on the American side, as Della Fenster discusses in her chapter, and Richard Brauer, Helmut Hasse, and Emmy Noether on the German side, as Günther Frei, Joachim Schwermer, and Charles Curtis discuss in theirs. The picture that emerges is one of fruitful and competitive interaction between the Americans and the Germans.

Günther Frei takes us back to the mid-nineteenth-century origins of the structure theory of algebras in the British ideas of William Rowan Hamilton, Charles Graves, and Arthur Cayley and traces the development of that theory through the work of Joseph H. M. Wedderburn on the general structure theory and Dickson's research on associative and cyclic algebras in the early decades of the twentieth century. In his 1923 book, *Algebras and Their Arithmetics*, Dickson laid out many of these ideas as well as work at the interface of the theory of algebras and algebraic number theory on the so-called arithmetic of algebras. The German translation of his book, which appeared four years later in 1927, sparked rapid developments in the arithmetic-algebraic school of research that centered on the work of Brauer, Hasse, and Noether. In particular, Frei traces in this broader context the development and extension of the Reciprocity Law from Hilbert's interpretation of it as an infinite product over all primes of norm residue symbols, to Hasse's success in providing explicit expression of these symbols in the quadratic case, to later breakthroughs in treating higher-order cases. As he demonstrates, Dickson's book was critical in linking the results of the American and German schools and, thereby, in giving rise to powerful, new techniques in modern algebra.

Another key technique in the theory of algebras *per se* is the local-global principle for algebras. Its origins are associated with Hasse's work in number theory, the subject of Joachim Schwermer's chapter. Schwermer grounds his analysis in the work of Gauss on quadratic forms, takes it through the novel ideas Hermann Minkowski introduced in his study of quadratic forms in several variables, and culminates in Hasse's introduction of p-adic methods in the formulation of a local-global principle for number fields. As Schwermer shows, quoting a letter from Hasse to Hermann Weyl, Hasse seems to have gotten this idea from his reading of Minkowski, and Schwermer traces how Hasse brought together ideas from Minkowski, Kurt Hensel, and Emmy Noether in formulating his own theory.

Della Fenster's chapter also treats an aspect of Hasse's work, focusing on the history of the often-so-called Brauer-Hasse-Noether Theorem but highlighting Adrian Albert's work in that direction. She carefully unpacks the story of how Brauer, Hasse, and Noether's joint paper of 1931 showed that, indeed, a central division algebra is cyclic, only to be followed two months later by a joint paper of Albert and Hasse that gave a new proof of the same result and an account of what had preceded it. Fenster shows that Albert's work had its origins in a vigorous, American research effort that had stemmed from Wedderburn's reduction in 1907 of the study of associative algebras over a field to the classification of division algebras (also called skew fields) and from his suggestion that these algebras are best regarded as algebras over their centers. Wedderburn had also noted that the dimension of a central division algebra over its base field must be a perfect square, while Dickson had introduced (in 1906) the concept of a cyclic division algebra. The question thus naturally arose whether a central division algebra is necessarily cyclic, since those of small dimension were shown to be, although not without some effort. Albert, a student of Dickson, not surprisingly approached the Brauer-Hasse-Noether Theorem from this direction.

Although as Frei's and Fenster's chapters show, the work of the Americans and the Germans intersected in the solution of key, open questions, their chapters as well as Schwermer's equally underscore the fact that mathematical concerns were different on the two sides of the Atlantic. In his chapter, Charles Curtis explores a fundamental German interest, namely, the use of noncommutative methods in number theory as animated by Emmy Noether's confidence in the fruitfulness of such an approach. Curtis concentrates on Noether's address to the International Congress of Mathematicians in Zürich in 1932, where she spoke on the application of noncommutative methods to commutative algebra and, in particular, to algebraic number theory. As Fenster also discussed, it was the profound immersion of the Germans in number theory—and especially in the theory of p-adic numbers—that had given them the edge over Albert, while the Americans had the greater experience in dealing with algebras. Curtis looks at the different techniques of Brauer and Noether, then at Hasse's involvement and the vindication of Noether's hopes for the approach, and finally at Max Deuring's contributions to it.

In different ways and in different contexts what is at issue in the chapters by Frei, Schwermer, Fenster, and Curtis—chapters which deal chronologically with the decades immediately preceding and immediately following the turn of the twentieth century—is the emergence of the mathematician's notion of "modern algebra." What, though, it makes sense to ask, was algebra like immediately before and during the transformation into this "modern" phase? This is precisely the historical question that Leo Corry addresses in his chapter, "From *Algebra* (1895) to *Moderne Algebra* (1930)," by closely examining the changing categorization of algebraic research in the principal abstracting journal of the period, the *Jahrbuch über die Fortschritte der Mathematik*. In the period from the 1890s to 1930, a vast amount of new algebra was created, and this shows up in the content of books by authors such as Dickson. These books nevertheless adhered, to varying degrees, to the overview of algebra that had been presented in 1895 by Heinrich Weber in his definitive three-volume *Lehrbuch der Algebra*. The decisive novelty in Bartel van der Waerden's two-volume *Moderne Algebra* of 1930–1931 was the new organization or direction of the subject. Weber and his immediate successors saw algebra in general and Galois theory in particular as the study of the solution of equations illuminated by the concepts of group and field. Polynomials in several variables form a closely related subject; algebraic number theory lies a little farther away and closer to analytic number theory. Later developments (many of them discussed in the present volume) involving hypercomplex number systems, the structure of algebras, and abstract field theory complicated this picture, as Corry shows with a shrewd use of the *Jahrbuch*. Van der Waerden and the "Noether boys" inverted the picture, so that structural features (groups, rings, fields) became central to what algebra was about, and topics such as the solution of equations and the real numbers became consequences or applications of them.

While van der Waerden's book was critical in the process that defined "modern algebra," his work in algebraic geometry was also important in redefining that field in the first half of the twentieth century. The final three chapters in this volume look precisely at the overlap between the history of algebra and the history of algebraic geometry. In the first, Norbert Schappacher examines and contextualizes van der Waerden's research in algebraic geometry. As he shows, van der Waerden's ideas arose at precisely the historical moment when the standing of contemporary Italian work on algebraic geometry was in decline and the modern structural algebra was on the rise. It is tempting to suppose, and commonly held, that it was van der Waerden's strong identification with the program to bring in modern algebra that drove him to rewrite algebraic geometry. A variant of this unduly simple thesis suggests that it was a lack of rigor among the Italians that impelled van der Waerden to bring in algebra. As Schappacher convincingly argues, neither of these simplistic readings is, in fact, correct. Indeed, many mathematicians found it hard both to grasp the work of Francesco Severi and others and to formulate it in other, perhaps more algebraic, ways, but other factors—publishing practices and politics, among them—also played a role. In any event, André Weil and Oscar Zariski shifted algebraic geometry into a modern algebraic mode, and the subtleties of van der Waerden's evolving position on the shift underscore not only the complexity of the process but also the range of possibilities involved in it. For example, van der Waerden was by no means devoted to rewriting all geometric concepts in terms of ideal theory. He was much more concerned with introducing and exploiting his concept of a generic point on an algebraic variety and with defining intersection multiplicity. He thus stayed closer to the geometric subject matter than did, for example, Zariski.

Silke Slembek's chapter focuses precisely on the evolution of Zariski's algebraic geometric ideas. As Zariski himself famously observed, his book on *Algebraic Surfaces* (1935) forever diverted him from the Italian approach in which he had first matured as a mathematician. Slembek argues that Zariski was, indeed, concerned in the 1930s with creating, through the medium of commutative algebra, a way of making rigorous the Italian insights into geometry. He called his approach the "arithmetization" of algebraic geometry because of his use of Krull's arithmetical ideal theory, but as Slembek shows, he was also profoundly innovative. For example, Zariski concentrated on the treatment—and, if possible, the resolution—of singularities of an algebraic surface. He wanted a proof that was not only rigorous (Robert Walker had provided one in 1935 that met Zariski's standards) but that could claim to be geometric rather than analytic. He succeeded in this in 1939 by exploiting his notion of the arithmetically normal variety. Zariski showed that an algebraic surface can be reduced to an arithmetically normal algebraic surface and that the resolution of singularities for such surfaces can be rigorously established. As Schappacher's and Slembek's chapters underscore, van der Waerden and Zariski may be seen to have taken different paths, the one more geometric and the other more algebraic.

The volume's final chapter moves the story forward in time another two decades to Alexandre Grothendieck and his metaphor of the rising sea. As Colin McLarty explains, the remarkably fruitful collaboration between Grothendieck and Jean-Pierre Serre from the mid-1950s to about 1970 was based on a striking contrast of approaches. Serre was a master of the elegant incisive strike that cracks a nut precisely with a single blow; Grothendieck sought to dissolve the shell slowly in a body of liquid, the "rising sea." One source of their shared interests was the Weil conjectures, which Serre explained to Grothendieck in cohomological terms in

1955. McLarty traces the origins of cohomological studies in France in the preceding decade and shows how, from this, Grothendieck generated an imposing amount of new category-theoretic ideas. In this context, McLarty gives a short history of sheaves and schemes as they were created in the Bourbaki milieu and shows how Grothendieck's breadth of vision and capacity for working with very many simple ideas led him to recreate algebraic geometry.

Concluding Remarks

The history of a topic as large as algebra is no more finished than the topic itself. There are developments in algebra that followed from the work described here, and there are developments that took place at the same time that are not discussed here at all. That is as it should be. Any readable book offers only a selection of what could be said. But the absence of a terminus for this book, welcome as that is on many grounds, makes the drawing of conclusions rather tentative.

Three aspects stand out and might suggest avenues for further research. One is the tendency exhibited by the mathematicians considered here to look for concepts with which to organize their work. This is, of course, one of the characteristic features of modern algebra, but it can be refined in at least two ways. On the one hand, as Edwards and Neumann have shown here and elsewhere, Kronecker, while noted for his "constructive" mathematics, was also a highly conceptual thinker with a developed program. On the other hand, the craft of mathematics—the sheer ability to find a result through hard calculation—does not disappear. It is enough to note, with Curtis, that Noether's arithmetical considerations are today's cohomological arguments. The balance between concept and craft is an endlessly fascinating one.

The second aspect is the somewhat cooperative, somewhat competitive dimension of these algebraic researches. A mathematical presentation of the topics here would lead cleanly and directly to the main results. These chapters, however, highlight the significant role such social factors as community, communication, and education play in the evolution of mathematics. Mathematics, we suggest, cannot fully be understood without acknowledging these factors.

The third, and the oldest observation of the three, is how amply the mathematical work here described vindicates the opinion of Gauss that the study of number theory leads to deep and hidden connections between superficially different branches of mathematics. Such, too, was the opinion of Kronecker, and ultimately the opinion of Grothendieck. We can surely be confident that the on-going work of mathematicians will continue to make work for historians to do.

References

Bashmakova, Isabella and Smirnova, Galina. 2000. *The Beginnings and Evolution of Algebra*. Trans. Abe Shenitzer. N.p.: The Mathematical Association of America.

Bos, Henk J. M. 2001. *Redefining Geometrical Exactness: Descartes' Transformation of the Early Modern Concept of Construction*. New York: Springer-Verlag.

Cardano, Girolamo. 1545/1968. *The Great Art or the Rules of Algebra*. Trans. T. Richard Witmer. Cambridge, MA: The MIT Press.

REFERENCES

Corry, Leo. 1996. *Modern Algebra and the Rise of Mathematical Structures.* Science Networks. Vol. 17. Basel: Birkhäuser Verlag.

Edwards, Harold M. 1984. *Galois Theory.* New York: Springer-Verlag.

Euler, Leonhard. 1770–1771/1984. *Elements of Algebra.* Trans. John Hewlett. New York: Springer-Verlag.

Høyrup, Jens. 2002. *Lengths, Widths, Surfaces: A Portrait of Old Babylonian Algebra and Its Kin.* New York: Springer-Verlag.

Klein, Jacob. 1968. *Greek Mathematical Thought and the Origin of Algebra.* Trans. Eva Brann. Cambridge, MA: The MIT Press.

Nový, Luboš. 1973. *Origins of Modern Algebra.* Leiden: Noordhoff International Publishing.

Parshall, Karen Hunger. 1989. "Toward a History of Nineteenth-Century Invariant Theory." In *The History of Modern Mathematics.* Ed. David E. Rowe and John McCleary. 2 Vols. Boston: Academic Press, Inc., 1:157-206.

Scholz, Erhard, Ed. 1990. *Geschichte der Algebra: Eine Einführung.* Mannheim: Bibliographisches Institut Wissenschaftsverlag.

Sesiano, Jacques. 1999. *Une Introduction à l'histoire de l'algèbre: Résolution des équations des Mésopotamiens à la Renaissance.* Lausanne: Presses polytechniques et universitaires romandes.

Stedall, Jacqueline A. 2003. *The Greate Invention of Algebra: Thomas Harriot's Treatise on Equations.* Oxford: Oxford University Press.

Unguru, Sabatei. 1975. "On the Need to Rewrite the History of Greek Mathematics." *Archive for History of Exact Sciences* 15, 67-114.

Van der Waerden, Bartel. 1976. "Defence of a 'Shocking' Point of View." *Archive for History of Exact Sciences* 15, 199-210.

———. 1983. *Geometry and Algebra in Ancient Civilizations.* New York: Springer-Verlag.

———. 1985. *A History of Algebra from al-Khwārizmī to Emmy Noether.* New York: Springer-Verlag.

CHAPTER 2

Babbage and French *Idéologie*: Functional Equations, Language, and the Analytical Method

Eduardo L. Ortiz
Imperial College London, United Kingdom

> "C'est bien là ce qu'est l'algèbre: aussi l'algèbre est-elle une langue, et les langues ne sont elles-mêmes que des espèces d'algèbres."[1]

Introduction

This chapter focuses on the interface between mathematics and some specific trends in French philosophy in the late eighteenth and early nineteenth century. In particular, it explores the extent to which these philosophical ideas may have influenced the efforts of Charles Babbage (1791–1871) to construct a specific mathematical language in England. His goal was to develop a notation capable of reducing the solution of problems in the difficult field of functional equations to some form of algebraic manipulation. In that, he displayed a fairly sophisticated and modern conception of algebra.

In 1815–1816, Babbage published a two-part "Essay," in which he proposed a new nomenclature applicable to the solution of functional equations. Here, I show that the ingenious and often complex "language" proposed by Babbage had an unsuspected power; with its help it is possible, for example, to treat functional differential equations, the solution of which would not be achieved until the mid-twentieth century.

I suggest in this chapter that Babbage's conception of the power of language as an analytical tool in science, a conception that dictates his theoretical approach to functional equations, belongs to the constellation of ideas proposed by Étienne Bonnot (1714–1780), better known as the Abbé de Condillac. However, by the time Babbage published his "Essay," these views were widely disputed in French philosophy of science circles, where there was intense debate on the role and structure of language and on its impact on the development of science. These new ideas came from the school of the *Idéologues*, which was broadly regarded as the inheritor of Condillac's tradition.

In a paper written in 1821, Babbage quoted a critique of Condillac's views published by Joseph Marie de Gérando (1772–1842) in 1800. Babbage seems to have accepted some of de Gérando's views; he claimed, in a paper written in 1821

---

[1][Destutt de Tracy, 1801c, p. 239].

but only published in 1827, that he had independently arrived at similar conclusions [Babbage, 1827]. This means that in 1815–1816, Babbage was operating with a delay of at least ten to fifteen years relative to the critical analysis taking place in France.

The conception of the role of language in science adopted by Babbage in his 1815–1816 "Essay," however, was not supported, but was rather strongly criticized, by de Gérando in his extensive treatise on the theory of signs. Given this, Babbage's 1821 views on the subject must have differed considerably from those he had supported in 1815–1816. It was in the 1820s, in fact, that Babbage abandoned the apparently promising line of thought that had combined mathematical research and an abstract approach to language and that he had initiated with his "Essay."

Here, a discussion of Babbage's mathematical work follows a brief overview of late eighteenth-century ideas on language and precedes an account of the *Idéologues*' ideas on language, analysis, method, and mathematics around the opening decades of the nineteenth century. This background will support the argument that by the time Babbage became acquainted with de Gérando's work, that is, when he published his 1821 paper, there was an even wider divergence between Babbage's views and those then currently held in scientific philosophical circles in France.

De Gérando's 1800 views had already been superseded by more advanced research published between 1801 and 1817 by Antoine-Louis-Claude Destutt de Tracy (1754–1836), the acknowledged leader of the *Idéologues*, and by other members of his group. They were actively working on the formulation of a general theory of ideas from a philosophical perspective. In parallel and in full contact, a second group led by Pierre Cabanis (1757–1808) explored the possibility of laying a physiological foundation for the philosophers' work.

This discussion has a further point of interest. Babbage belonged to the minority of English mathematicians generally regarded as deeply interested in new developments in France.[2] It thus seems reasonable to question the strength of the communication channels through which exchanges in areas of philosophy that affected mathematical research circulated between France and England in the final years of the Napoleonic period and early in the Restoration. I briefly consider this topic by way of conclusion.

Speculation on the Origin of Languages

For traditionalists, speculation on the origin of languages[3] presumed a break with biblical tradition that held that language was a part of Creation; for this reason, they had little interest in the subject. The rationalists shared this indifference, although for a different reason; they saw language and reason as inherent in the human condition. In the eighteenth century, however, a number of factors—historical, philosophical, and scientific—contributed to spark a concern for languages, their origins, development, structure, and later the comparative study of them.

By the end of the eighteenth century, while Latin and modern French were the languages favored by scholars and scientists throughout the Western world, their limitations in giving the precision required to communicate in certain areas of

---

[2]For a biography of Babbage and a discussion of his connections with France, see [Hyman, 1981] and the references given therein.

[3]See [Condillac, 1746] and, for a general overview, [Harnois, 1929], [Kuehner, 1944], and the references given therein.

philosophy and science began to be acknowledged. Since, in an inquiry, the accuracy of thought was held to be in direct relation to the sensitivity and reliability of the tools used to conduct it, and since mathematics was then viewed as the canon of reliable thought, a tendency began to develop to assimilate other disciplines to the methodology or, at least, to some external forms of the language of the exact sciences. One way of attempting this pairing was through the formulation of specific languages designed, as closely as possible, like those used in mathematics [Ortiz, 1999]. Although the general tendency was to mimic the language of algebra, translations into the language of geometry, as well as the relations between algebraic and geometric language, were also discussed.

Beginning in the late eighteenth and early nineteenth centuries, scholars began to consider specific languages equipped to deal with the needs of highly specialized fields, such as classification, three-dimensional graphical representation, chemistry, mechanical machinery [Ortiz and Bret, 1997; Quere, 2005], and others. These new languages occasionally intersected with mathematics; for example, classification and probability theory, dimension reduction in the representation of solids and descriptive geometry, chemistry and algebraic equations, early thought on abstract machine theory and algebra all posed specific problems that attracted the attention of mathematicians.

Among language theoreticians, Condillac was one of the early champions of a line of thought that relied on mimicking algebraic language.[4] He discussed these matters in a number of influential books, but particularly in his *Essai sur l'origine des connoissances humaines: Ouvrage où l'on réduit à un seul principe tout ce qui concerne l'entendement humain* [Condillac, 1746]; *Traité des sensations* [Condillac, 1754]; *Grammaire*, part of the *Cours d'études pour l'instruction du Prince de Parme* [Condillac, 1775];[5] and *La Logique* [Condillac, 1780].[6] In particular, the *Cours d'études* was an influential sixteen-volume work published between 1767 and 1775 that Condillac had written while he was tutor (from 1758 to 1768) to the young Bourbon Duke Ferdinand of Parma, a grandson of Louis XV. Before the French Revolution, it was not unusual for an extensive educational treatise like this to arise in such a teaching context. Condillac returned to the subject of language in mathematics at the end of his life in his unfinished book, *La Langue des calculs* [Condillac, 1798b] (hereinafter referred to as *La Langue*).[7]

In his philosophical work, Condillac embarked on a project of understanding language by adopting a number of new mid-eighteenth-century ideas in which the shift from the object to the mechanics of thought played a part. Accordingly, he became interested in the development of theories of language and thought that he based on the analysis of sensations. As [Réthoré, 1864] and later [Knight, 1968] have pointed out, Condillac approached the study of the role of the senses from an empiricist viewpoint using a rationalist methodology. Condillac stated that if

---

[4]Condillac's biography remains poorly understood, but see [Baguenault de Puchesse, 1910].
[5][Condillac, 1775] also contains other works of interest in connection with this discussion, such as *De l'art de penser*, *De l'art d'écrire*, and *De l'art de raisonner*.
[6]The latter was published in 1780 but probably finished in 1778.
[7]This work was published posthumously in 1798 as the last volume of the first edition of his complete works. The manuscript is preserved at the *Bibliothèque Nationale de France*, Paris, n. a. fr. 6344, 6345, 6346, 11098–11117. The edition [Condillac, 1798b] incorporates modifications made to the original manuscript. Sylvain Auroux and Anne-Marie Chuillet have recently published a scholarly edition of *La Langue* [Condillac, 1981], with references to different editions of this work.

man is man through experience, then his language is his language through experience as well. He also rooted logic in experience, viewing it as a mirror of the rationality expressed by nature through its regularities and its fundamental mechanical-mathematical laws.

Senses, Languages, and the Elaboration of a Theory of Signs

In their search for clues to the role of the senses in the construction of a theory of language, French philosophers developed an interest in a number of new topics that revealed the emergence of a wider and more relative view of language. They expected that the topics they considered would help them find the elements of a genetic and structural analysis generally applicable to all forms of expression.

One of their sources was accounts of the development of individuals—like feral children—whose normal communication with other human beings had been, in some way, disturbed. Another was discussions of systems of signs designed specifically for and by people—such as the blind and deaf-mutes—who had experienced a serious diminution of their sensorial facilities either at birth or in early infancy. The question, for example, of the acquisition of the notions of length and volume by the blind was of considerable interest both to philosophers and to mathematicians interested in the possible origins of mathematical concepts.[8] The emergence of mental faculties in small children, and the role of language in their development, was also regarded as a valuable source of information. For those interested in a genetic approach to the emergence of communication systems in human beings and in the role of sensations in their establishment, these common concerns also helped to strengthen the bond between experts on education and philosophers interested in the development of a theory of signs, who usually modeled their work on algebra. Education was increasingly becoming a point of subtle philosophical speculation.

In order to analyze the role of experience in the development of cognitive abilities, a problem to which the perception of space was not unrelated, Condillac conducted a thought experiment on a "statue" of a human being, that is, one incapable of sentiment or movement [Condillac, 1754].[9] He then progressively attributed the different human senses, or selective combinations of them, to the statue, effectively introducing a technique of breaking cognitive powers of the mind into special units that later mathematicians would apply in discussions of the impact of different axioms or groups of axioms particularly in geometry.

Another line of inquiry employed by Condillac involved discussing which categories of ideas were associated with a given sense, and how information gathered by one sense might add to the function of another. Yet another focused on the new interest in the "primitive" and on then recent anthropological discoveries.[10] In this connection, Condillac wrote that "[w]e who believe ourselves to be educated should go often among the most ignorant of people in order to learn from them the beginnings of our discoveries: because it has been a long time since we were

---

[8] A remarkable case of the latter line of inquiry is [Suzanne, 1806, 1807, and 1809].

[9] On the use of the example of the statue by other eighteenth-century authors, see Georges Lyon's remarks in the "Introduction" to the Paris 1886 edition of Condillac's *Traité des sensations* [Lyon, 1886].

[10] The word "primitive" was used to designate those people who were deemed closer to nature or to the origins of knowledge.

nature's disciples" [Condillac, 1798b, p. 38].[11] In particular, philosophers began to take into account reports of voyagers who had visited remote lands and encountered societies that, in the past or in the present (see, for example, [De Gérando, 1800a]), used languages and concepts structured differently from those then currently in use in Europe.[12] All of this contributed to a heightened awareness of language and its philosophical implications and interpretations.

The Position of Grammar

A shift began in the seventeenth century away from the old conception of grammar as prescriptive and as the underpinning of logic and rhetoric in the old *Trivium* (traditionally including the three language-related subjects of logic, grammar, and rhetoric). This shift also contributed to support change in the conception of language. Claude Favre de Vaugelas (1585–1650) formulated grammar [Vaugelas, 1647] as a convention and not as a rule of nature,[13] while Jansenist philosophers of Port Royal attempted to subordinate grammar to innate rational principles following the methodology Descartes gave in his *Discours de la méthode* [Descartes, 1637].[14] Their claims that grammar matched the logical structure of human reason paved the way for the possibility of a universal grammar common to all languages.

To this panoply of ideas, Condillac added a different, genetic meaning of grammar, in which he suggested that the relationship between the emergence of language and the development of human understanding be reversed. Although his ideas clearly had roots in the work of other philosophers—among others, Ramon Llull, Gottfried Leibniz, and John Locke—they nevertheless had a significant impact. Contrary to the Jansenists of Port Royal, Condillac believed that the rules of thought did not emerge and should not be learned before grammar: grammar should be used to master logic. As a result of his influence, grammar once again began to be positioned before logic; for Condillac, grammar was *the* analytic method.[15]

In his view, an analysis of sensations allowed man to begin mastering mental processes through the invention of language. Organization of pure sensations began to take form as clusters of simultaneous ideas were broken into pieces and put into a sequence; these clusters came from a language of action, that is, a basic language made up of gestures and cries. Language was thus the product of an individual, expressive, and not necessarily communicative experience, processed from the senses and made collective through the universality of the human body and mind. It was expanded and enriched through the association of ideas, while errors in mental

---

[11] "Nous qui nous croyons instruits, nous aurions donc souvent besoin d'aller chez les peuples les plus ignorans, pour apprendre d'eux le commencement de nos découvertes: car c'est sur-tout ce commencement dont nous aurions besoin; nous l'ignorons, parce qu'il y long-temps que nous ne sommes plus les disciples de la nature."

[12] Condillac specifically referred to a voyage made by Charles-Marie de la Condamine [de la Condamine, 1745], in which he learned both of the numerical systems employed by South American Indians and of the lack of proper notions of substance, essence, and being in their languages. See [Knight, 1968, pp. 157-158 (note)] as well as [Benedict, 2001].

[13] Condillac commented that "grammar theoreticians have often given us their whims as laws [les grammaires nous ont donné ses caprices pour des lois]" [Condillac, 1798b, p. 5].

[14] See [Arnauld and Nicole, 1662] (revised in 1664, 1668, 1671 and 1683) and [Arnauld and Lancelot, 1660] (revised in 1662).

[15] Condillac defined grammar in a more restricted way than was usual at his time. For him, it preceded vocal language. See [Knowlson, 1975] and [Condillac, 1798b, p. 10], where the language of action in mathematics is associated with finger calculation.

processing were attributed to the imperfect translations of language and to its forays from what nature actually intended. The use of the important tool of "analogy," aided by the related notion of "association of ideas," made possible an extension of these basic components into more abstract and organized forms of language and thought.[16]

Once the language of action was analyzed and symbols replaced natural expressions, an analysis of discourse, that is, the construction of a general abstract grammar, could begin. Condillac devoted a full chapter of his *Traité de l'art de penser* to the discussion, advancing some provocative ideas on analysis. He discussed knowledge and errors through perceptions coming from the senses, ideas and their causes, the generation of complex ideas, links between ideas and their effects, and the concepts of synthesis, substance, and even infinity. He perceived analysis as decomposition, comparison, and evaluation of relations [Condillac, 1798a, 6:221]. Condillac also imposed definite conditions on a process to be regarded as an "analysis," emphasizing the importance of a methodical decomposition capable of abstracting all qualities of an object and of presenting the sequence of abstract ideas genetically, that is, by making it possible to grasp the totality of what was being analyzed. Condillac also implied that analysis was the true secret of the art of discovery [Condillac, 1798a, pp. 222-223].

The idea of this analysis of discourse and of its construction of a general grammar suggested the possible realization of the dream of a perfect, final language. It both stimulated considerable discussion and was met with reservations even by some of Condillac's more enthusiastic scientific followers and admirers, such as Antoine Lavoisier (see below).

On the Language of Calculation

Accepting that to construct the art of reasoning is to develop a well-made language, grammar ascends to a primordial position in the art of thinking. Considering language as an analytic method, that is, identifying it with analytical logic, brings it closer to its most elaborate model: algebra. In turn, a thorough analysis of the language of algebra becomes a necessary requirement. Condillac set himself the task of finding the grammar of algebra. In his view, this was a topic fit for metaphysicians, rather than for ordinary mathematicians. Finding such a grammar was the main objective of *La Langue des calculs*, and perhaps a reason that it remained unfinished. Condillac declared in it that mathematics was not his object, but rather the vehicle for a much larger objective: to show how all sciences could be given the accuracy often assumed to be exclusive to mathematics [Condillac, 1798b, p. 8].

According to Condillac, algebra operated by analogy. It involved constructing long lists of identical propositions or equivalent statements; from these, useful connections were selected and extracted in order to advance the process of an algebraic calculation (see [Nový, 1973]). Condillac suggested that the same machinery as that used in algebra could be applied to the manipulation of more general ideas, using sensations as atomic elements instead of the concept of quantity and discussing their translations and transformations.

---

[16]Condillac defined "analogy" as "un rapport de resemblance," the expression of one thing in many different manners [Condillac, 1798b, pp. 2-3]. This notion later provided a link with algebra.

The possibility of a transcendental system of calculation with ideas lay behind Condillac's expectation that metaphysical analysis and mathematical analysis are equivalent and that, consequently, one and the same analysis would apply to them both [Condillac, 1798b, p. 218]. According to Condillac, "[c]ertainly, to calculate is to reason, and to reason is to calculate: if there are two names, there are not two operations" [Condillac, 1798b, pp. 226–227].[17] He warned, however, that because of the nature of the ideas involved, analysis in metaphysics was definitely more difficult than in mathematics. Condillac closed the first book of *La Langue*, stating that, since algebra is nothing but a language, he had played in algebra the role that a grammarian plays in language. He also averred that he had achieved the goal of discussing the grammar of algebra and reflected on the fact that languages are just more or less perfect analytical methods. If the language associated with a given science could be taken to a greater perfection, then anyone "speaking" that language would be able to understand that particular science with much greater precision. Therefore, to create a science was nothing more than to create a language, and to study a science was to learn "a language well made [une langue bien fait]" [Condillac, 1798b, p. 228]. This well-known statement needed reconsideration, and, as noted above, some of Condillac's followers later attempted to do just that.

Babbage and a "Language" for the Solution of Functional Equations

In 1815–1816, Babbage published in the *Philosophical Transactions of the Royal Society* a two-part "Essay," in which he introduced what he called a "calculus of functions," that is, a set of rules and procedures for the treatment of functional equations, including functional differential and functional integro-differential equations of different orders in one or several variables.[18] Independent of their intrinsic mathematical relevance, these equations appeared in the solution of interesting questions in geometry and physics. In addition, the topic was directly linked to issues concerning the design and calculation of mathematical tables. Babbage's work thus pertained to an important area bordering on algebra and analysis. It supported, and in some ways heralded in England, an interest in the use of operational techniques in mathematical analysis.

When Babbage approached this difficult field, work on functional equations had already been published by some of Europe's most distinguished mathematicians: Jean d'Alembert, Leonhard Euler, Joseph-Louis Lagrange, and Louis Arbogast.[19] Babbage, however, attributed his main source of inspiration to two papers by Gaspard Monge on the determination of two functions from given conditions.[20] In the first of these, Monge had discussed the solution of functional equations by means of curves of double curvature[21] and curved surfaces; in the second, he had remarked that the determination of the general solution of equations in partial finite differences depends on being able to integrate one or more equations in finite differences,

---

[17]"Certainement calculer c'est raisonner, et raisonner c'est calculer: si ce sont-là deux noms, ce ne sont pas deux opérations."

[18]See [Babbage, 1815] and [Babbage, 1816].

[19]Babbage indicated that, on account of communication restrictions at the time, he did not have access to Arbogast's work, which had won the prize of the Petersburg Academy in 1790.

[20]See [Monge, 1770–1773] and [Monge, 1773].

[21]Alexis Clairaut had introduced this name for curves in space [Clairaut, 1731]; Charles Babbage and John Herschel had defined them as curves "formed by the successive intersection of a curved surface, whose parameter varies" [Babbage and Herschel, 1813, p. ix].

for which the ratio of the independent variable to its finite differences, whether variable or constant, is given. At roughly the same time, Laplace had also discussed functional equations—particularly $F[x, \varphi(x), \varphi(\alpha(x))] = 0$, where F is a given functional expression, x is the independent variable, $\alpha(x)$ is a prescribed function, and $\varphi(x)$ is the function to be determined—and had tried to relate their solution, in Babbage's words, to a "peculiarly elegant" approach to problems in finite differences [Babbage, 1815, p. 394]. Laplace was not the only mathematician who saw in this connection a possible route to the solution of functional equations. In his pioneering researches of 1813, John Herschel (1792–1871) had followed the same road, showing that Laplace's method, which was essentially a process of reduction to finite difference equations, could be applied to all functional equations of the first order [Herschel, 1813].[22] However, as Babbage subtly observed, Herschel's "solution is equally elegant and general to that of Laplace's; it leaves nothing to be regretted, but the narrow limits of our knowledge respecting the integration of equations of finite differences" [Babbage, 1815, p. 394]. The field of finite difference partial differential equations was still in need of further research. In Babbage's view, what was wanted was specific methods for the solution of functional equations [Babbage, 1815, p. 395].

Babbage's interest in these ideas apparently dated to 1809, the year before he entered Cambridge University as a student.[23] At the time, he was considering some curious generalizations of classical geometrical problems. Among them was a problem of Pappus: to inscribe a sequence of progressively smaller tangent circles inside a given semi-circle, and then to find the relationship between the area of that semicircle and the sum of the areas of the inscribed circles. Although Babbage had wanted to take up the same question for more general systems of curves, he stopped working on these problems for some two years until John Herschel, with whom he discussed them, suggested that all these questions, in their most general form, were related to the solution of functional equations of arbitrary order [Babbage, 1815, p. 391].

Babbage had moved to Cambridge in October 1810; by the end of 1811, he had bought, for the princely sum of seven pounds [Hyman, 1982, p. 21], the first eight volumes of the *Journal de l'École polytechnique*, a repository of the most innovative research in mathematics (and also in other subjects) done at that school. The Analytical Society—which counted among its members Babbage, Herschel, and George Peacock—was constituted at Cambridge in the spring of 1812 and had begun publishing its *Memoirs* in 1813. In the preface to these *Memoirs*, a document of over twenty pages possibly written by Babbage and amended by Herschel [Hyman, 1982, p. 25 (note 10)], a number of ideas on analytical methodology, notation, and

---

[22]Not everyone would agree with the characterization here of Herschel's work as "pioneering." In the bibliography of Herschel's work that he compiled for the *Mathematical Monthly*, John Runkle expressed quite different views: "[t]he latter portion of this Part III, on Functional Equations, is all wrong" [Runkle, 1861, p. 220]. On Herschel's scientific personality, see Silvan Schweber's prefatory essay to [Herschel, 1981, pp. 1–58]. This work reproduces Herschel's memoir on the equations of differences in volume one.

[23]An unpublished 289-page manuscript by Babbage on "The History of the Origin and Progress of the Calculus of Functions during the Years 1809, 1810 ... 1817" is preserved in the archive of the Museum for the History of Science, Oxford. References to this material are given in [Dubbey, 1978, Chap. 4].

symbolic analysis appear that are to be found in later works of Babbage, especially in his work on functional equations.

Babbage did not approach his research on functional equations by formulating a general theoretical framework; rather, he approached it through a series of difficult and progressively more general problems, to which his tools were applied and gradually developed and extended. His genetic approach was based on the formulation and development of a notation—a form of "language"—which was, in many ways, similar to that of algebra. In this language, functions performed a role analogous to that of quantity in algebra. These techniques allowed him to tackle some difficult problems in the field of functional equations. In what follows, I describe briefly and then test Babbage's analytical approach. I show that his "language," despite certain shortcomings, has a largely unsuspected power. Using his techniques, it is possible—and with simplicity—to solve difficult functional and functional differential equations, including one solved only in the middle of the twentieth century.

Babbage's Notation

Babbage began his "Essay" [Babbage, 1815] with a discussion of the term "function," which he defined as "designating the result of every operation that can be performed on quantity" [Babbage, 1815, p. 389].[24] He regretted that, although its use had become central in mathematics, "the various applications of which it admits, and the questions to which it gives rise, do not appear to have met with sufficient attention" [Babbage, 1815, p. 389]. In the "Essay," he proposed to outline a "new calculus" [Babbage, 1815, p. 389] that, as he saw it, followed naturally from the concept of function he had given earlier. In his view, this new calculus "comprehends questions of the greatest generality and difficulty, and would require the invention of new methods for its improvements" [Babbage, 1815, p. 389].

Babbage remarked that calculations involving functions consist of two "parts"—one direct and one the inverse of the other—giving examples taken from algebra and from the calculus of finite differences and infinitesimal calculus. In all cases, the inverse method "is by far the most difficult" [Babbage, 1815, p. 389], and the first part of the "Essay" aimed to consider it "with respect to functions" [Babbage, 1815, p. 390]. Babbage wanted to generate a language with functions as its fundamental elements, where these functions were implicitly defined by functional equations. His main tool would be a systematic use of analogy with algebraic notation, namely, a process of analysis and recomposition of functions. Although there is explicit reference to Condillac's work neither in the "Essay" nor in the comprehensive index of Babbage's works composed for [Babbage, 1989], there are clear similarities between the two of thought and even of expression.

Using analogy, Babbage remarked that finding an unknown quantity x—implicitly defined as the root of a given equation (or by a given condition)—and finding the "form" of a function—implicitly defined by a given functional equation—appear to be related problems. In one case, the unknown is the quantity x; in the second, it is "the form assumed by quantity" [Babbage, 1815, p. 389]. In the first case, "the various powers of the unknown quantity enter into the equation; in the other, the different orders of the function are concerned" [Babbage, 1815, p. 390].

---

[24]A detailed and careful analysis of the impact of work on functional equations on the definition of function is given in [Dhombres, 1986].

At this point, and in close analogy with algebra, Babbage introduced new notation, rules, and symbols to develop his "language."[25] In the context of functional equations, the atomic element in his language was, as noted, that of function. If a function has "known functional characteristics" [Babbage, 1815, p. 390], that is, if it is specified, it was denoted by one of the first letters of the Greek alphabet—$\alpha, \beta, \gamma, \ldots$—while if the function is unspecified, it was denoted by $\varphi, \chi, \psi, \ldots$. He next introduced for his elements the notions of "order" to which he had already alluded. For clarity, I use parentheses, which Babbage omitted, to write, for example, $\psi(\psi(x))$ or $\psi^2(x)$ instead of his $\psi\psi x$ or ψ^2. He called $\psi^2(x)$ the "second function" of $\psi(x)$; if the function is iterated n times, the "n^{th}-function" is $\psi^n(x)$. For example, if
$$\psi(x) := a + x, \text{ then } \psi^n(x) := na + x.$$
Having defined an "order" for iterated functions, he then defined an order for functional relations: a functional equation is of the first order, if it only contains the "first function," that is, the original function $\psi(x)$ (or $\psi^1(x)$). Therefore,
$$\left[\psi(x) + \psi\left(\frac{1}{x}\right)\right]^n ax + x^2 = 0$$
is a functional equation of the first order, while
$$\left[\psi^2(x) + \psi\left(\frac{1}{x}\right)\right]^n ax + x^2 = 0$$
is a functional equation of the second order because a "second function" enters into its expression.

Similarly, Babbage defined iterated functions associated with functions of several variables. In this case, however, the situation was far more complex, since combinatorial problems began to play a more visible role in his definitions, in particular, for n^{th}-functions. The case of just two variables and $n = 2$ is indicative: $\psi(x,y)$ admits two second functions, $\psi((x,y),y)$ and $\psi(x,\psi(x,y))$, for which Babbage introduced the notations $\psi^{2,1}(x,y)$ and $\psi^{1,2}(x,y)$, respectively. Yet, as Babbage remarked, the notation encounters difficulties even in this simple case: "besides these two there is another, which arises from taking the second function simultaneously relative to x, and y; it is $\psi(\psi(x,y),\psi(x,y))$. This ought not to be written $\psi^{2,2}(x,y)$ for it is not the second function first taken relative to x and then to y, nor is it the converse of this" [Babbage, 1815, p. 391].[26]

Symmetric functions of two or more variables also played an important role in the multidimensional extension of Babbage's approach, when he considered functional equations of several variables. If $\varphi(x,y)$ is symmetric in its variables, he added a bar to indicate it. If more than two variables are present, this notation was not sufficiently informative, since a function $\varphi(x, y, v, w)$ may have more than one

---

[25]Babbage also referred to the use of analogy in later works; he indicated his desire "to point out some of the more prominent points in which the calculus of functions resembles common algebra or the integral calculus" [Babbage, 1817, p. 215]. He also expressed his determination to use analogy as a tool to "transfer the methods and artifices employed in the latter calculus, to the cultivation and improvement of the former" [Babbage, 1817, p. 216], emphasizing again his interesting, unitary view of mathematics and his perception of the importance of formal relations in the hypothetical synthesis he was proposing.

[26]I return below to this particular type of function, which Babbage reconsidered on its own merits.

symmetry defined for its variables. To increase the information value of his notation, Babbage was forced to add a second index (he used a perpendicular bar for it) to indicate that, say, x and y are symmetric in one sense and v and w in another. He gave the example of this function in four variables— $(v+z+axy)/(v^3z^3ax^2y^2)$— "which is symmetrical in one sense relative to x and y, and in a different sense with respect to v and z" [Babbage, 1815, p. 396]. It should also be mentioned that Babbage and Herschel had already pioneered some interesting ideas on notation two years earlier in 1813 [Babbage and Herschel, 1813, pp. i-iii].

The fact that notation for iterated functions could be quite complex became even more apparent in a paper Babbage wrote in 1822 on notation for the calculus of functions (see, in particular, [Babbage, 1822, p. 74]). Babbage was fascinated by the power that the notation he was developing had to store information in a very compact form. As he remarked, "[t]he surprising condensation of meaning comprised in [a] small space and yet exempt from even the slightest tinge of obscurity is nowhere more conspicuous than it is throughout the calculus of functions" [Babbage, 1822, p. 75]. As an example of compacting information through the use of suitable notation, he offered the equation $\psi^{10,10}(x,y) = \psi(x,y)$, where $\psi^{2,2}(x,y)$ is defined by the simultaneous replacement $\psi(\psi(x,y),\psi(x,y))$. If written in extenso, x and y would be repeated 512 times and ψ would occur 1023 times, as he had shown by induction [Babbage, 1822, p. 68]; the whole expression would consist of 2047 letters. Babbage returned to these questions in [Babbage, 1827, pp. 331-332], considering the equation $\psi^{9,9}(x,y) = (x,y)$, where $\psi^{2,2}(x,y)$ is defined as before.

The notational advances in the natural sciences introduced, for example, by Linnaeus in botany and by Lavoisier in chemistry, supported outstanding achievements in the field of scientific classification. The impact of these accomplishments across the frontiers of the sciences can be associated with Babbage's efforts to develop mathematical notation and, later, with his remarkable applications of it. Babbage drew attention to the fact that suitable notation could be used to characterize—and then to compress—large sets of mathematical information. He also suggested the possibility of using intelligently chosen notation to detect regularities in the large mathematical data-sets it defined. On a more abstract level, he remarked that, by exploiting symmetries and combinatorial properties, it was possible to quantify the incidence of given functional symbols, or of given non-trivial algebraic expressions in these data sets. As an example, he discussed the large data-sets defined by the equations $\psi^{10,10}(x,y) = \psi(x,y)$ or $\psi^{9,9}(x,y) = (x,y)$, which he had introduced earlier in his algebraic work. These ideas suggest that Babbage had a remarkably modern conception of algebra.

Babbage's Treatment of Functional Equations

Babbage's approach to functional equations essentially depended on the simultaneous use of his special notation and, more technically, of the assumption that a particular solution to the problem in question is known. He claimed that, progressing from these elements, a general solution to a given problem on functional equations could be constructed. For Babbage, a particular solution was one that contained arbitrary constants, while a general one was one that contained one or more arbitrary functions [Babbage, 1815, p. 394]. However, the number of arbitrary functions entering into the complete solution of functional equations of orders higher than the first is not obvious from his discussion. On that point, Babbage admitted that "I have little at present to offer; the difficulty of the subject, and the wide extent of the enquiries to which it would lead, induce me to postpone it until I have more time for the consideration" [Babbage, 1815, pp. 420-421].

Babbage was aware that his system of signs was not complete and that in considering functional equations of several variables, he would confront even greater difficulties. He was forced to acknowledge that "the first step in that direction must be an improvement in the notation" [Babbage, 1815, p. 423]. The same theme reappeared in the second part of his "Essay," when he considered equations where functions of various orders were present; he tried to reduce these to the first order. He allowed that "great difficulties still remain" [Babbage, 1816, p. 241], as it was not always easy to find particular solutions on which to support his structure. Even if found, the first-order equations are "of considerable difficulty" [Babbage, 1816, p. 241]. He also accepted that the development of this tool was "in its present state unequal" [Babbage, 1816, p. 180] and conceded that "the notation I have used should only be considered as of a temporary nature; it may be employed until some more convenient one be devised" [Babbage, 1816, p. 180]. He argued, however, that his notation should be kept "until our acquaintance with this subject becomes more intimate" [Babbage, 1816, p. 180].

Babbage remarked that functional equations of the first order had long been known and discussed, but he believed that the method he used to treat them was "entirely new" [Babbage, 1815, p. 391]. He warned that equations of higher order than the first "have never been even mentioned; it is these which present the most interesting speculations, and which are involved in the greatest difficulties" [Babbage, 1815, p. 391].

In the first part of his "Essay," Babbage had not attempted to give a complete solution of functional equations in its greater generality, but to propose "a new method of solving all functional equations of the first order" [Babbage, 1815, p. 392]. Later, he aimed to reduce equations of a higher order to that case. Equations in several variables were left to be considered in the second part of the "Essay," communicated on 14 March, 1816 and published in the same year. In both parts of his paper, he ignored questions relating to uniqueness, existence of solution, and existence of the inverse operator, as when he encountered multiplicity in the definition of the inverse [Babbage, 1816, pp. 190-191]. His discussion showed that he was well aware that he did not possess the tools necessary even to attempt a discussion of these fundamental topics.

In his approach, Babbage used, with some freedom, Condillac's conception of language as an analytical tool. However, in his line of attack, he did not take into account distinctions clearly made within Condillac's school well before his own papers were written in 1815–1816. For example, Destutt de Tracy—anticipated by Lavoisier—had already indicated the importance of clarifying the main theoretical points in the area under discussion as a step to be taken *before* the introduction of any new notation. Questions of existence and uniqueness were among them. (I return to these general prescriptions in the next section in connection with a brief discussion of Destutt de Tracy's ideas.)

In his Problem VIII, for example, Babbage introduced an interesting method of reduction to deal with functional equations of the form

$$F[x, \varphi(x), \varphi(\alpha(x)), \ldots, \varphi(\nu(x))] = 0,$$

where $\alpha(x), \ldots, \nu(x)$ are prescribed functions, when a particular solution containing one or more arbitrary constants was known [Babbage, 1815, p. 406]. He accepted the fact that the question of the number of arbitrary constants and functions required to find a general solution using his methods had not yet been made clear [Babbage, 1815, p. 408]. However, in the final lines of the first part of his "Essay," he indicated that he was "in possession of methods which give the general solution of equations of all orders, and even to those which contain symmetrical functions" [Babbage, 1815, p. 423]. He also provided general approaches to treat first-order functional equations with several variables, including those of differential and partial differential type.

In the second part of the "Essay," Babbage reiterated his confidence in the existence of a calculus that could be employed as an "instrument of discovery in the most difficult branches of analysis," that is, in a language that would serve as an analytic method [Babbage, 1816, p. 179]. The prescription applied not only to science abstractly but also to specific problems. Since, in Babbage's view, it was "peculiarly adapted to the discovery of those laws of action by which one particle of matter attracts or repels another of the same or of a different species," it should be applicable to all branches of natural philosophy, such as "the hidden laws which govern the phenomena of magnetic, electric, or even chemical action" [Babbage, 1816, p. 179].

Still, in the problems he considered, how to characterize a general solution often remained an open question. That was the case, for example, in Problems IX, XII, and XXXVI.[27] In the second part of his "Essay," Babbage did use, however, the reduction results he had developed in the first part to deal with functional equations of the form $F[x, \varphi(x), \varphi(\alpha(x)), \varphi(\alpha^2(x)), \ldots, \varphi(\alpha^n(x))] = 0$.[28] He also considered cases where not only derivatives but also integral expressions of the unknown form were present. Returning to a possible motivation for these studies, he mentioned an interesting problem that had been discussed by d'Alembert: given a sphere of particles of matter, determine the law of attraction among these particles so that the force of attraction of the whole sphere acting on particles at a distance is the same as that between the particles themselves.[29]

In the second part of his "Essay," Babbage dealt as well with fractions and negative numbers in his calculus. For example, in Problem XXXIX, he referred to fractional or negative-order functional equations and to functional differential equations of a fractional order in its derivatives [Babbage, 1816, pp. 248-250]. The question of negative or fractional orders had already attracted the attention of Lacroix, who had regarded the topic as both important and difficult, and who had discussed it briefly in his treatise on calculus.[30] For Babbage, Lacroix's works in analysis ([Lacroix, 1797–1798] and [Lacroix, 1800]) were a main mathematical—and philosophical—source on the new methods used in France. However, by the

---

[27]See [Babbage, 1816, pp. 199, 206, and 241-244, respectively].

[28]See, for example, Problem XXXII in [Babbage, 1816, p. 230].

[29]Babbage was probably alluding here to the *Huitième mémoire, remarques sur quelques questions concernant l'attraction* in [D'Alembert, 1761–1780, I (1761), pp. 246-264], to *Sur la loi de l'attraction* in [D'Alembert, 1761–1780, VI (1773), pp. 404-407], and to the paper immediately before it on algebraic functions [D'Alembert, 1761–1780, VI (1773), pp. 393-404]. Taking into account Herschel's interests in physics, and his acquaintance later with Ottaviano Mossotti (1791–1863), these questions may have attracted the attention of Herschel and Babbage in connection with problems in molecular physics.

[30]Babbage contributed to the translation into English of this work. See [Lacroix, 1816].

time Babbage had become an adherent of them, Lacroix's philosophical views on mathematics had already been seriously challenged by distinguished members of the *Idéologie* movement, in particular, by Maine de Biran, a young and brilliant student of Destutt de Tracy.

Babbage closed his two-part "Essay" with a call to extend his methods to equations with two or more functional characteristics, to discuss more thoroughly elimination as well as extremal problems, and to introduce in this area the methods of the calculus of variations. He regretted, however, perhaps exaggeratedly, that he had made "little progress in each of them" [Babbage, 1816, p. 256.], remarking that his goal had been "to direct the attention of the analyst to a new branch of the science" that "seems eminently qualified to reduce into one regular and uniform system the diversified methods and scattered artifices of the modern analysis" [Babbage, 1816, p. 256]. Babbage had already showed his concern for the unification of methods arising in different areas of mathematics in the preface to the *Memoirs of the Analytical Society*. There, he had expressed his admiration for the work done by "[o]ur continental neighbours" in "digesting various points [of analysis] into a systematic form" with the idea of "reducing into a reasonable compass the whole essential part of analysis, with its applications" [Babbage and Herschel, 1813, p. xxii].

Babbage and First-Order Functional Equations in One Variable

With the above contextualization of Babbage's mathematical ideas, consider now the first two problems that Babbage treated in the first part of his "Essay." A technical look at these problems illustrates the basic workings of his technique, while consideration of a third problem (XXXIII) highlights that technique's remarkable power. It goes without saying that arguments for problems involving several unknowns are far more elaborate and complex; however, they are not essentially different in approach from those considered here. It should also be noted that although Babbage often introduced interesting views and ideas in his "Essay," they do not essentially modify his overarching approach: the use of particular solutions in association with his special "notation" and the reduction to cases already solved.

Problem I: Consider the functional equation $\psi(x) = \psi(\alpha(x))$, and assume that a particular solution of it is known: $f(x) = f(\alpha(x))$. Let φ be an arbitrary function. To create a general solution, set $\psi := \varphi[f(x)]$. Then, $\psi := \varphi[f(x)] = \varphi[f(\alpha(x))]$ solves the given equation [Babbage, 1815, p. 394].

Example 1: Let $\psi(x) = \psi(-x)$. In this case, $\alpha(x) = -x$. Take as a particular solution $f(x) = x^2$. Then $f(x) = f(-x)$. Defining $\psi(x) = \varphi(f(x)) = \varphi(x^2)$, the general solution ψ is of the form $\varphi(x^2)$ [Babbage, 1815, pp. 394-395].

Problem II: Reconsider now the problem, $\psi(x) = \psi(\alpha(x))$, with the more general involution condition $\alpha^2 = I$, that is, $\alpha^2(x) = x$. Assume that a particular solution is known such that $f(x) = f(\alpha(x))$, and construct the general solution in terms of a symmetric function φ with two arguments, one of them, f, is specified by a particular solution, and the other, f_1, is unspecified and has to be determined: $\psi(x) := \varphi[f(x), f_1(x)]$. Since $\psi(x) = \varphi[f(x), f_1(x)] = \varphi[f(\alpha(x)), f_1(\alpha(x))]$, it follows that $f(x) = f_1(\alpha(x))$, and $f_1(x) = f(\alpha(x))$.

Set $x := \alpha(x)$ in the former of these expressions, then, $f(\alpha(x)) = f_1(\alpha^2(x)) = f_1(x)$, which fixes f_1 in terms of f. Therefore, $\psi(x) = \varphi[f(x), f(\alpha(x))]$ [Babbage, 1815, pp. 396-397].

Example 2: Consider the functional equation $\psi(x) = \psi\left(\frac{x}{ax-1}\right)$. Set $f(x) := x$ and $\alpha(x) := x/(ax-1)$; the latter satisfies the involution condition. Then $\psi(x) = \varphi[x, x/(ax-1)]$, where, as before, φ is symmetric. In particular, if $a = 0$, $\psi(x) = \psi(\alpha(x)) = \varphi[x, -x] = \varphi(x^2)$, as in Example 1 ([Babbage, 1815, p. 397]).

Problem XXXIII: Consider the functional differential equation

$$\psi(\alpha(x)) = \frac{d\psi(x)}{dx},$$

with $\alpha : \alpha^2 = I$, and set $x := \alpha(x)$ to get

$$\psi(x) = \frac{d\psi(\alpha(x))}{d\alpha(x)}.$$

Applying d/dx and taking into account that

$$\frac{d}{dx}\left\{\frac{d\psi(\alpha(x))}{d\alpha(x)}\right\} = \frac{d}{dx}\left\{\left[\frac{d\psi(\alpha(x))}{dx}\right]\left[\frac{d\alpha(x)}{dx}\right]^{-1}\right\},$$

yields

$$\psi(\alpha(x)) = \frac{d}{dx}\left\{\left[\frac{d\psi(\alpha(x))}{dx}\right]\left[\frac{d\alpha(x)}{dx}\right]^{-1}\right\}.$$

Call the latter expression condition (1) [Babbage, 1816, pp. 235-237].

Example 3: Consider

$$\psi\left(\frac{1}{x}\right) = \frac{d\psi(x)}{dx}$$

where $\alpha^2 = I$. Then, setting $\psi(\alpha(x)) := z$ in condition (1) gives

$$z = \frac{d}{dx}\left\{\left[\frac{dz}{dx}\right]\left[\frac{d\alpha(x)}{dx}\right]^{-1}\right\}.$$

Since $d\alpha(x)/dx = -1/x^2$, we have $x^2 z'' + 2xz' + z = 0$, which reduces to the form $y'' + y = 0$. Therefore, $y(x)$ is a sum of sines and cosines. Changing back to the old variable yields:

$$\psi(x) = Ax^{\frac{1}{2}} \cos\left[\left(\frac{3^{\frac{1}{2}}}{2}\right) \log x \left(\frac{\pi}{6}\right)\right],$$

(see [Babbage, 1816, pp. 235-237]).

The solution of this functional differential equation, which arises in photography and other areas of physics, was given by Ludwick Silberstein in the *Philosophical Magazine* in 1940. In his well-known repository of solutions of differential equations, Erich Kamke gave Silberstein's equation as the penultimate example of known solutions of functional differential equations, making no reference to Babbage's technique (see [Kamke, 1944, p. 660]). As Example 3 shows, however, Babbage had already offered the means for solving it as early as 1816. Thus, even given the limitations that owed to Babbage's lack of concern for questions of existence and uniqueness, his notation would seem to have wider applicability than has been acknowledged heretofore.[31] In a limited way and starting from a very classical standpoint, Babbage may be regarded among the early enthusiastic precursors of functional analysis. Although following a similar algebraic line, his work preceded that of Salvatore Pincherle by decades.[32]

The Aftermath of Condillac in France: The Beginning of a Discussion

The results that Babbage published in his "Essay" of 1815–1816 reflected both mathematical and philosophical ideas informed by Babbage's reading of Condillac's work. As noted above, however, that work had been the subject of intense discussion and debate in the opening years of the nineteenth century, a discussion and debate of which Babbage seemed unaware until 1821. An understanding of the evolution of Babbage's thought after the publication of the "Essay" thus requires a sense of some of the philosophical work that originated in France in response to Condillac—particularly within the *Idéologie* school—and that paved the way for both a deeper understanding of signs and the laying of the foundations of a theory of signs.

Toward the end of the 1790s, conceptually deep fractures had begun to appear between Condillac and a new generation of his disciples. These distant followers of Condillac were members of a French philosophical school known commonly as *Idéologie*, the undisputed leader of which was Antoine Destutt de Tracy.[33] François Picavet, the author of a monograph that remains essential reading on the *Idéologie*

---

[31]Kamke was not the only one who ignored Babbage's contribution. Salvatore Pincherle considered the historical development of the theory of functional equations in his articles for the German [Pincherle, 1904–1916, pp. 789-791] and French [Pincherle, 1912, p. 46] editions of the encyclopedia of mathematics. In them, he made very brief and indirect references to Babbage's work, quoting the translation by Joseph Gergonne of a note by Babbage [Gergonne, 1821]. Babbage's original piece had been included in [Herschel, 1820].

[32]Pincherle's contributions to the foundations of functional analysis were reviewed in [Birkhoff and Kreyszig, 1984, p. 265].

The history of functional equations has also been discussed in [Aczél, 1984] and [Dhombres, 1984]. No references to Babbage's work are given in Aczél's work, while Dhombres acknowledged the existence of his first "Essay" in a note. In a more recent and extensive paper, probably the most reliable study on the history of functional equations to date, Dhombres accurately placed Babbage in the chronology of the subject, but he again used Gergonne's note as a reference without entering into a discussion of the mathematical nature of Babbage's contribution [Dhombres, 1986, p. 139].

Finally, John Dubbey devoted an entire chapter of his pioneering book on Babbage to his subject's work on the calculus of functions, without entering into its technical aspects or its applicability [Dubbey, 1978]. Since that publication, [Boyle, 1986], [Panteki, 1991] (see also [Panteki, 2003]), [Grattan-Guinness, 1992], and [Friedelmeyer, 1993], among others, have reconsidered Babbage's contributions to the calculus of functions. Babbage's original memoirs are now also available in Martin Campbell-Kelly's excellent scholarly edition of Babbage's complete works [Babbage, 1989], where a good number of the numerous small misprints in Babbage's papers have been spotted and corrected.

[33]Antoine-Louis-Claude, Comte Destutt de Tracy, was born to an old aristocratic French family, twice ennobled and endowed with a very considerable fortune and influence in Court. He received training in the humanities at the progressive University of Strasbourg. In the second half of the 1790s, Destutt de Tracy began to publish philosophical work; initially he followed, critically, Condillac and Locke, but he soon began to develop original ideas. On Destutt de Tracy, see [Arnault, Jay, de Jouy, and Norvins, 1822, 5:434-435] and [Newton de Tracy, 1852, 1:338-351]. [Kennedy, 1978, pp. 1-37] gives important new information on Destutt de Tracy's life and ideas. The ideas of Destutt de Tracy were initially discussed in [Picavet, 1891]; see also [Moravia, 1968], and [Gusdorff, 1978]. The latter contains interesting insights on the intellectual life of the period. Destutt de Tracy's views on the philosophy of language are discussed in [Goetz, 1993].

school, regards Destutt de Tracy and his followers as members of a *second generation* of *Idéologues*, placing Condillac, Condorcet and other contemporaneous French philosophers in a first generation of the same school [Picavet, 1891].[34] From different angles, all of these *Idéologues* were attempting to construct a science of ideas. Babbage, even before writing his "Essay" on functional equations, had expressed an interest in an improved version of the old Baconian "philosophical theory of invention" [Babbage and Herschel, 1813, p. xxi]. It is perhaps not surprising then that he would have been attracted by the work of the *Idéologues* of both the first and the second generations and by their efforts in the field of the analysis of ideas.

Although *Idéologie* stemmed from Condillac's ideas and, more distantly, from a methodological background in which the controversies surrounding Cartesianism are still visible,[35] members of the second generation of *Idéologues* showed a fairly wide diversity of views both among themselves and when directly compared to those of Condillac. These generations, and subcultures within generations, were nevertheless united by a deep interest in the connections between science and language.

If an understanding of the origin of language had been the subject of a substantial literature in the late eighteenth century centered around Condillac's work,[36] the beginning of the nineteenth century was marked by a much closer interest in a systematic development of a theory of signs. This "general linguistics" was seen as the key to a better understanding not only of ordinary language (see [Martinet, 1960] and [Prieto, 1966]) but also of science and the mind's processes. To understand more precisely the impact of the second generation of *Idéologues* on mathematics per se, however, requires a careful differentiation of levels within the thought of the *Idéologues*.

Soon after the *Institut national de France* was established in 1795, a section directly inspired by the work of Condillac on the *Analyse des sensations* was created in its Second Class, the philosophy department.[37] It was, however, around this time that the validity of Condillac's ideas had begun to be reconsidered critically. The most conspicuous members of this new section—and two of the critics—were Corresponding Members Destutt de Tracy and Pierre Cabanis, the former a philosopher and the latter a natural scientist whose work influenced medical science.[38] In 1798, the section's membership discussed and approved an "open prize" on the subject: "To determine what the influence of signs has been on the formation of ideas."[39]

---

[34] Although they do, indeed, have, features in common, the two successive generations of *Idéologues* by no means shared a monolithic structure. Instead, they exhibited rich, interesting, and different internal structures, to which brief reference will be necessary in both this and the next section.

[35] The *Idéologues* were critical of Descartes's philosophy, but they adopted the new methodological formulations on reasoning he gave in [Descartes, 1637]. See [Picavet, 1891, Chap. 1].

[36] See [Harnois, 1929], [Kuehner, 1944], and the references given therein.

[37] On the organization of the *Institut*, and other French institutions of the time, see [Crosland, 1967]. On social attitudes at the time, see [Terrall, 2002]. The department of philosophy was suppressed by Napoleon in 1803.

[38] [Cabanis, 1802] opened a modern discussion of the relations between the physical organization of human beings and their thinking process. In this work, Cabanis tried to reconstruct Condillac's theory of sensations on a more solidly scientific foundation.

[39] "Déterminer quelle a été l'influence des signes dans la formation des idées." Details about the organization of this prize can be found in [de Gérando, 1800b, 1:Introduction]. According to [Tisserand, 1930–1939, I (1930), p. ix], the *Institut* had already announced a prize on a similar topic two years earlier and had renewed its call in 1798.

The work of Condillac and, more recently that of Destutt de Tracy and Cabanis, had put this *à l'ordre du jour* in Parisian philosophical circles, and several brilliant, young philosophers, among them de Gérando and Maine de Biran, naturally considered entering the competition.

Condillac's complete works had appeared in print in 1798, and, in it, his work *La Langue des calculs* was made publicly available for the first time. The fact that the publication of Condillac's works involved a number of the *Idéologues*[40] may have been a catalytic agent for the *Institut* to offer a prize in the theory of signs. This was an implicit way of inviting a critical review of Condillac's early attempts in the field of analytical philosophy using linguistic tools.

On 2 April, 1799, the prize was awarded to de Gérando, a junior collaborator of Destutt de Tracy. A year after receiving the prize, de Gérando published in Paris an extensive book on signs and the art of thinking considered in their mutual relations [De Gérando, 1800b]. The text of this book was not, however, the piece he had submitted for the prize, but rather a much rewritten and elaborated version of it [De Gérando, 1800b]. In its new format, de Gérando's work became a large book, extending over four volumes and totalling nearly two thousand pages; in two parts, each of these is divided into two sections with each section occupying one whole volume. This work reveals a new stage in the development of the ideas of the French *Idéologues*; when compared with the previous scholarly work of Destutt de Tracy, de Gérando's writing is clearly aimed at an audience much wider than that of professional philosophers. One of the remarkable and explicitly declared objectives of his book is that it is an exercise in public relations with the increasingly influential lobby of scientists. De Gérando was not indifferent to the members of that group who had an interest in philosophy. In fact, he counted several scientists among his personal friends, among them, André-Marie Ampère and Gergonne. De Gérando declared his wish to establish better channels of communication between those living in the world of metaphysics and those working in the positive sciences [De Gérando, 1800b, 1:xli]. His work did, in fact, reach scientists, even those outside France: one of the latter was, rather belatedly, Charles Babbage.

To achieve his objectives, de Gérando avoided using the technicalities of philosophical language and limited considerably the scope of his discussion. He also lacked Destutt de Tracy's originality and finesse of thought. De Gérando stated that he was aware of parallel developments in philosophical circles in Germany, but decided to omit them from his work. These developments, however, were too important to be left out. Two years later they were discussed by Destutt de Tracy in *De la métaphysique de Kant* [Destutt de Tracy, 1802], an important work that showed the increasing impact of Kant's ideas on French scientifico-philosophical circles in the first decade of the nineteenth century. This influence would only deepen over the next two decades.[41] De Gérando also decided to omit a discussion of what he called "other systems of metaphysics" that were then competing with the *Idéologues*'s views, although this was not because he lacked the proper philosophical background to attempt it.[42]

---

[40]In particular, it involved Pierre Laromiguière (1756–1837).

[41]On the impact of Kant's ideas on mathematics, see [Pierobon, 2003].

[42]The use of the word metaphysics here is important. Initially, *Idéologie* was called a "new system of metaphysics." This word was also used in the sense of "foundations" by, for example, Carnot and Lagrange. De Gérando did write his account of the "other systems of metaphysics"

De Gérando's Critique of Condillac: A Turning Point in *Idéologie*

The views expressed in de Gérando's book mark an important split in method between members of the two successive generations of French *Idéologues*. Condillac's optimistic views on the power of language as an analytical method were challenged in the works of de Gérando and especially in the more refined work of Destutt de Tracy, Maine de Biran, and others in the same school (see below). De Gérando was particularly critical of traditional studies on languages, believing that they had inappropriately explored the connections between the art of signs and the art of speaking or writing with far greater attention than they had its connections with the art of thinking [De Gérando, 1800b, 1:ix-x], that is, with the subject of logic. However, while he saw room for further and more refined improvement, he did not wish the new studies he advocated to support a break in the connections between ideas and sensations [De Gérando, 1800b, 1:x], which in his view had been so carefully and intelligently developed by Condillac.

Although there were differences—and sometimes deep differences— between the two philosophers, de Gérando did not intend to demolish Condillac; indeed, much of his work remained an integral part of the thought of the school's second generation. This permanence at a time of deep philosophical debate and intense social and political change was, in fact, one of the elements that gave coherence to the school of *Idéologie*, the identity and unity of which are sometimes difficult to grasp.

In his book, de Gérando credited Leibniz and John Locke with playing foundational roles in the systematic analysis of language, but he recognized Locke as having paved the way, and he believed Condillac had followed in Locke's footsteps.[43] In full command of Locke's doctrine, in de Gérando's view, Condillac had paid more attention than any previous philosopher to the elaboration of a theory of signs. He had also illuminated some of the more obscure areas of the sciences of ideas, by showing the role of language in decomposing thought and by arriving at what de Gérando now disqualified as a "sort of analytical method" [De Gérando, 1800b, 1:xiii-xiv].[44] De Gérando also credited Condillac both with having perceived that the mechanism of abstract thought involves a succession of translations and with having started an exploration of the relationship between signs and ideas [De Gérando, 1800b, 1:xiv]. However, de Gérando criticized Condillac for the excessive emphasis he had placed on the interplay between languages and analysis, a point that could have benefitted Babbage had he known of it earlier. De Gérando stressed that Condillac had stopped short of considering the role played by signs as instruments in certain operations of the mind. For example, he referred to the statue-man, indicating that Condillac must have assumed that it had a certain degree of intelligence since it was able to use artificial signs. In de Gérando's view, Condillac had also omitted discussion of the equally difficult question of the nature of the association—via the notion of analogy—of ideas. Other authors had even attempted to discuss this complex concept in terms of Newtonian attraction.

---

four years later in [de Gérando, 1804]. Although Destutt de Tracy was critical of this work, it was reprinted twice over the next four decades. See [De Gérando, 1822–1823] and [De Gérando, 1847].

[43]Voltaire was instrumental in bringing Locke's ideas to France [Picavet, 1891, p. 2]. On Condillac's critical views on Locke, see [Duchesnau, 1974]. [Zimmer, 2004] discusses the influence of John Wallis on Locke, who was, briefly, his pupil.

[44]De Gérando based this assessment on his reading of Condillac's *Traité de l'art de penser*.

Although de Gérando again acknowledged Condillac's emphasis on the impact that well-designed languages have on the progress of our understanding, he nevertheless sharply criticized his predecessor on matters of signs, language, and method. Relative to signs, de Gérando faulted Condillac for having attempted to consider them from a general point of view without tracing either the history of their introduction or the properties different species have in common, for not having studied the indirect influence that signs have on the development of our mind, and for having considered the role signs play in our intelligence but not how our intelligence relates to signs. With respect to language, he criticized him for proposing a perfectly analogical language without making proper references to its rules, and for having discussed the advantages of these languages without taking into account their drawbacks[45] or even their possible existence [De Gérando, 1800b, 1:xix]. He also accused Condillac of methodological errors: he had limited himself to indicating the road to truth without exploring it more fully, and he had formulated general principles without properly exploring their applications [De Gérando, 1800b, 1:xviii]. However, in a gesture of respect to the master after the damage had been done, de Gérando conceded that these flaws were, perhaps, a consequence of Condillac's potent creativity, and that he may have left to others the detailed organization of what his mind had imagined [De Gérando, 1800b, 1:xviii].

De Gérando next moved to the important question of the relationship between language and analysis. Here, he was very specific, launching a frontal attack on Condillac's grand ideas on language as an analytical tool, rather than leaving open the possibility of correcting the limitations and shortcomings in his conceptions. He also censured Condillac for making statements that he viewed as too absolute, for example, the study of a science is limited to learning a language, and a well developed science is nothing but a well-made language. Nor did de Gérando agree with Condillac's propositions that the only privilege mathematics had over the natural sciences was that it possessed a better language and that this difference could easily be eroded by providing sciences with a similarly efficient sign facility [De Gérando, 1800b, 1:xxi].[46] More importantly, perhaps, he criticized Condillac for inspiring what he called the "seductive hope [l'espérance séduisante]" that a process as simple as the reformation of language could end all disputes, prevent all future errors, and reveal all truths [De Gérando, 1800b, 1:xxi]. In fact, he accused Condillac of encouraging the belief that all the secrets on which the perfection of the human spirit depends may be hidden in the mysterious depths of an art of signs [De Gérando, 1800b, 1:xxii].[47] De Gérando attributed the wide and uncritical acceptance of some of Condillac's assertions to the high authority and respect the latter enjoyed as a philosopher.

The position of language in science clearly changed substantially in the twenty years from 1780 to 1800, years that both separated and united Condillac and his new critics. These changes of perception on the role of language in science, deeply influenced by Condillac's *La Logique* of 1780, were perhaps more perceptible in the writings of subsequent philosophers than in the more reserved views expressed by

---

[45]De Gérando used the word "inconvéniens."

[46]De Gérando drew here from 7, 8, and 218 of [Condillac, 1798b, pp. 7, 8, and 218]. Except for the quotation on p. 8, de Gérando's quotations of Condillac are not literal. Similar statements can be found in [Condillac, 1780].

[47]He was probably referring here to efforts in line with Llull's work.

scientists such as Lavoisier.[48] Although de Gérando's critique of Condillac referred mainly to ideas expressed in *La Langue des calculs*, the issues the former wished to explore were by no means confined to that treatise. This may owe in part to the fact that Condillac often explored the same ideas in successive works. *La Langue* was no exception to this rule; in it are clear traces of several of his previous works, particularly his influential *Grammaire* and *Logique*.

The new critical attitude adopted by de Gérando in his work of 1800, even though imperfect, heralded a profound shift in the young French philosopher's views on the arguments offered by his venerable predecessor, Condillac. Until 1800, the latter had been the undisputed master of the important group of philosophers that later became known in France as the *Idéologues*. De Gérando's work marked the tentative beginnings of a new grouping within Condillac's school. While still accepting some of his basic ideas on sensations, these younger philosophers began to distance themselves from their elder's attractive, although now seemingly simplistic, ideas on the overwhelming power of language to dictate the path of science and knowledge. In particular, they took exception to Condillac's implicit view that removing conceptual obstacles was not a necessary precondition for the construction of a scientific language.[49]

De Gérando's Theory of Signs and Its Mathematical Implications

While the first part of de Gérando's book, that is volumes I and II, was a direct answer to the *Institut*'s prize question, the second part, volumes III and IV, dealt more specifically with the influence the *perfection* of systems of signs could have on the art of thinking and on the progress of human knowledge. Specifically, the third volume concerned the question of our understanding of facts and how improvements in the art of signs can affect them, while the fourth dealt with the same problems for abstract statements as well as with a number of interesting miscellaneous questions. As de Gérando pointed out, the first half of his treatise concerned experience, and principles, the second half rules or deductions.

A first clash with Condillac can already be detected toward the end of the sixth chapter in the first volume, where de Gérando agreed that language was to be regarded as a method, but qualified this explaining that the word "method" is, in this context, improperly used. In his view, languages are "facilities, means of analysis"; they are auxiliaries to a method but not the "method itself."[50] Having said this and using what, in fact, had become a strategy in his book, de Gérando then respectfully returned to Condillac, crediting his predecessor with the original ideas which he was now developing.

---

[48]See [Lavoisier et al., 1787].

[49]Contradictions in the field of algebra arose in [Condillac, 1798b], for example, in connection with the possible enlargement of the realm of quantity with complex numbers, a concept Condillac was naturally forced to reject.

[50]De Gérando wrote that "languages are *facilities* (or tools), *means of analysis* that, in our hands are auxiliaries to follow a methodic inquiry, but they are not the *method* itself [les langues sont des *occasions*, des *moyens d'analyse*, c'est-à-dire qu'elles sont entre nos mains un secours pour mieux suivre la méthode, mais elles ne sont pas la *méthode* même]" [De Gérando, 1800b, 1:158-159, note (his emphasis)].

From a mathematical standpoint, the ninth chapter in the second volume is perhaps the most interesting. In it, de Gérando meditated on the nature of abstract reasoning in relation to the idea of quantity. He discussed algebraic calculation, ideas of quantity,[51] and the practical application of mathematical truths [De Gérando, 1800b, 1:211 and 215]. For him, the creative power of algebra resided in the use of letters as both known and unknown quantities. He also considered the important question of analogy in the algebraic language.

In the closing years of the eighteenth century, driven in part by an unprecedented expansion of advanced mathematics teaching and the need to simplify its presentation, the foundations of calculus began to attract considerable attention from mathematicians. The publication in 1797 of [Carnot, 1797] and [Lagrange 1797], the latter based on his lectures on the theory of functions,[52] suggests genuine interest in the matter. De Gérando claimed in his book that an analysis of the principles on which the metaphysics—that is, the foundations—of calculus is based was of considerable interest in philosophy [De Gérando, 1800b, 1:212]. Unfortunately, he did not develop this idea further, moving on instead to the less fundamental area of calculation, where he claimed that prodigious advances in the science of calculation raised the hope that abstract science, as well as other sciences, would progress as long as they mirrored the format of mathematics. Questions relating to infinity, however, commanded the attention they deserved neither from Condillac nor from the second generation of *Idéologues*.

Relative to algebra per se, De Gérando stressed that it was there, where signs carry the notion of quantity, that the process of analysis showed all of its power [De Gérando, 1800b, 1:219]. Like other members of his school, he perceived mathematics as an area of singular prominence in science and, returning to the idea of the mathematization of scientific knowledge, did not hesitate to state his belief that mathematics would ultimately become the scene upon which the prodigies of the art of thinking would be properly displayed [De Gérando, 1800b, 1:236].

He hinted, however, that language alone was not the cause of the prominence of mathematics; it also rested on a special foundation. In de Gérando's view, two very different ingredients coexisted in algebraic language, one dependent on the nature of the signs used and the other on the laws controlling their composition. As "grammairians" would put it, these were the elements of discourse and syntaxis [De Gérando, 1800b, 1:263]. He returned to these questions in the fifth chapter of his treatise's third volume to offer a rather limited definition of what he understood the language of a branch of science to be. There, he described it as the set of names used to designate the facts that compose that science [De Gérando, 1800b, 3:153]. While his discussion had initially been on elements and on laws of composition for them or, in the language of linguistics, between the "dictionary" and the "grammar" that rules the combinations of the elements in that dictionary, his second, more limited understanding of language as a catalog placed too much stress on the elements and too little on the rules. This regression in his viewpoint on the definition of language emphasizes how unstable the understanding of an abstract formulation of language still was in the period.

---

[51] De Gérando used the word "grandeur."

[52] Dominique Joseph Garat, in parallel with Lagrange, lectured at the *École Normale* on "l'analyse de l'entendement," a new formulation of metaphysics that was a forerunner of *Idéologie*.

Some of these ideas reappeared independently in Babbage's work of 1815–1816, thereby situating him among the pioneers both of functional analysis and of primitive notions of abstract space from which his techniques for functional equations seem to have stemmed. The writings of 1800–1820 of the French *Idéologues* deserve no less credit, however; they sketched a conceptual philosophico-linguistic framework for these ideas.

De Gérando accepted that scientific languages cannot be viewed as permanent, and that not even Lavoisier's nomenclature could be expected to remain unreformed for long [De Gérando, 1800b, 3:174-175]. Lavoisier would not have disagreed with these remarks, as he himself expressed the same notion. While in agreement with Condillac's ideas on language and method, Lavoisier had displayed a more measured attitude toward the role of language in science. In an interesting note submitted to the *Académie royale des Sciences* by Lavoisier, Louis Guyton de Morveau, Claude-Louis Bertholet, and Antoine de Fourcroy on 17 April, 1787 [Lavoisier *et al.*, 1787], the authors expressed the view that the creation of a scientific language was an open-ended process, making it even more difficult to implement when the science in question was undergoing fundamental changes. They stated in unambiguous terms that they were aware that they were trying to construct a language for a science without being in possession of all its scientific facts; they thus regarded their new nomenclature as far from perfect [Lavoisier, 1787, 5:360]. These authors expected, however, that the language they were proposing would be flexible enough to adapt to new discoveries. Like Babbage relative to his notation for functional equations, the chemists anticipated the need to persevere with confidence in the use of the language they were formulating until more experience and better suggestions became available. There are, however, fundamental differences in the methodology followed by the French chemists in the formulation of their language and an approach that might have been suitable for use in mathematics. (See the discussion of Destutt de Tracy's views below.)

In his fourth and final volume, de Gérando returned to a discussion of abstraction and the impact the perfection of a system of signs might have on the progress of human understanding. He felt that the latter might allow for new and more elaborate operations on ideas. Before discussing the rules and possibilities of a new methodical language, however, he analyzed the restrictions imposed by the particular nature of the language's elements. It was clear to him that a new methodical language would not be an "algebra" in the traditional sense of the word, since its structure and rules would not be those of ordinary algebra. For example, methodical languages would not share with the algebraic language the facility of having "mobile" signs capable of representing with equal validity known and unknown quantities. In a methodical language, de Gérando held, "unknowns" in the algebraic sense could not exist, since the formation of signs was controlled by the formation of the corresponding (and therefore known) ideas. He concluded that the signs and methods of algebra were inapplicable to metaphysics because of the very nature of the topics discussed there [De Gérando, 1800b, 4:170-171]. He distanced himself even more clearly from Condillac when he remarked that the signs and methods of one could not easily be applied to the other due to the different nature of the questions addressed in algebra and metaphysics.

Babbage and de Gérando's Views on Signs

Even if effective calculation was not central to *La Langue*, several critics of this book censured this aspect of it when it appeared in 1798. This was a time when, mainly for the purpose of educational reform but also for reasons unrelated to it, considerable attention was paid in France to the question of computation. Although de Gérando conducted a much deeper and thorough critique of Condillac's work, he also returned to the question of his predecessor's treatment of practical computation in *La Langue*. In his discussion of the advantages of algebraic calculation, de Gérando expressed his own views on the praxis of calculation.[53] He emphasized the simplicity that ensued in calculation from the use of algebraic signs, the swiftness these signs impart to reasoning on quantities, and the question of the capacity of a suitable symbolism to store information and to help focus thought.

The latter points naturally attracted Babbage. He quoted de Gérando's views, making them his own, in a paper on the influence of signs in mathematical reasoning read to the Cambridge Philosophical Society in December of 1821 but only printed in its *Transactions* in 1827 [Babbage, 1827].[54] In this paper, as well as in [Babbage and Herschel, 1813, p. iii], Babbage commented on symbolic manipulation and operational techniques that may have influenced those British mathematicians who later followed the operational road.[55]

In his 1821 paper, Babbage, generally following de Gérando's work on signs of 1800, gave a more subtle evaluation of the power of language than he had in 1815–1816. His quotations thus reflected a delay of over twenty years in his intellectual discourse with French scientifico-philosophical circles. He argued in 1821 that he had independently arrived at the views de Gérando had expressed in 1800; this independent discovery must have occurred in the second half of the 1810s, after he had completed his "Essay" on functional equations. However, by the time Babbage wrote quoting de Gérando's ideas, these had been largely superseded in France by more advanced and refined philosophical research.

Throughout his lifetime, Babbage continued to explore the ramifications and use of language in different forms and contexts, although not always with success. Clearly, the views on the power of language he had acquired in his youth did not remain untouched by time, but he never returned to a more detailed discussion of the interplay between language, mathematics, and philosophy. In connection with his mechanical inventions, for example, language became more of a tool, a key to defining both special paths in a system of elements and operating rules with

---

[53] These concerns should be seen against the backdrop of intense work on tabulation, for example, in Gaspard Riche de Prony's decimal tabulation project.

[54] Through the family of his future wife, Georgiana Whitmore, Babbage became acquainted with Napoleon's younger brother, Lucien Bonaparte, who lived in exile in Worcestershire in the period when Babbage was a student at Cambridge University. Before Lucien's relations with his brother deteriorated, leading to his exile, he enjoyed an important patronage position in France; de Gérando was among his favorites. The latter's name may have been mentioned to Babbage by Lucien.

[55] [Peacock, 1834] reflected influences different from those acting on Babbage's works of the decade from 1810 to 1820. Peacock seemed unaware of the writings of Destutt de Tracy and his followers. However, by the end of the 1820s, English philosophers like Jeremy Bentham had begun to take notice of the existence of Destutt de Tracy. See, for example, Bentham's correspondence from 1818 in [Bentham, 1989, pp. 291-292].

perfectly well-defined characteristics.[56] No longer was language a guiding analytical principle as it had been in his 1815–1816 "Essay" on functional equations.

Destutt de Tracy's *Élémens d'idéologie*

If de Gérando's book on the theory of signs heralded a revision of Condillac's ideas, it was no substitute for them. His work, in many ways significant, lacked Condillac's scope, depth, originality, and systematic treatment. It must be seen as a preliminary critique and, as the author stated in the introduction to his book, not as a definite formulation or new theory. The task of systematically reconstructing Condillac's wide views was begun by Destutt de Tracy and, later, by some of his disciples.

In several academic papers, and later in books published in the early years of the nineteenth century, Destutt de Tracy reconsidered language and scientific method, introducing a new personal perspective. His five-volume *Élémens d'idéologie*, which appeared between 1801 and 1817, proved instrumental in the diffusion of his ideas [Destutt de Tracy, 1801c; 1803; 1804; 1817]. It attracted considerable attention in France as well as abroad, owing to its rapid translation into several foreign languages.

This series directly paralleled Condillac's *Cours d'études* and some of its related work (such as his treatise on logic). It was written as a possible textbook for students of the newly created *Écoles Centrales*, a most interesting experiment in secondary education started in the revolutionary period. Destutt de Tracy's series was thus addressed to a much less privileged audience than Condillac's course of studies for the Prince of Parma and so formed part of a broad educational movement, in the promotion of which the school of *Idéologie* played a central role by the turn of the nineteenth century.

Although Destutt de Tracy had used the name *Idéologie* before, it was only after the publication of the first volume of his *Élémens*, when his philosophical work became more widely known, that the name became extensively used. Initially, members of the group in which Destutt de Tracy was the most prominent figure were addressed as *Idéologistes*, but later the name *Idéologues* prevailed. They were perceived as a coherent movement with sufficient common traits and consistent chains of new ideas to be designated as a philosophical school. Destutt de Tracy's group accepted Condorcet, d'Alembert, Condillac, and other French as well as foreign philosophers, Locke among them, as the true initiators of their movement.

Destutt de Tracy divided the field of *Idéologie*, or more properly rational *Idéologie*,[57] into three sections. In the first, he gave a history of means of knowledge divided into three parts: a general introduction to the theory—called *Idéologie proprement dit*—in which he tried to bring an order to human understanding; a discussion of *grammaire générale* or the medium through which ideas and knowledge are structured for communication; and an outline of *logique* that explored the ways in which ideas are and can be combined [Destutt de Tracy, 1801c, pp. 4-5].

Idéologie, like other philosophical schools before and since, attempted a totalitarian approach to knowledge. It could not limit itself to a doctrinal corpus

---

[56][Grattan-Guinness, 1992] argued, correctly, that Babbage's thought was dominated by the notion of algorithm.

[57]This name indicated that he had left aside the psycho-physiological approach to the theory of ideas of his close friend, the physiologist Cabanis.

of theory and leave aside problems in the fields of morals, history, practice, and the sciences. If there are no volumes on history in Destutt de Tracy's works that compare with Condillac's comprehensive *Cours d'études*, the second section of the *Élémens d'idéologie*, entitled "Traité de la Volonté," was a methodological attempt to apply human means of knowledge to the study of *volonté* and its effects. This second section was, in turn, divided into three parts: the study of actions in general, referred to as "economics"; the study of feelings or "morals," where he attempted to produce a rationally deduced code of norms; and the study of how to direct the two through political sciences or "legislation."[58]

The third section of the *Élémens* was to be devoted to the application of human means of knowledge to the study of the environment. The topics were to have been: bodies and their properties, that is, physics; properties of extension, or geometry; and properties of quantity, or calculus [Destutt de Tracy, 1801c, p. iv]. This last section remained entirely unwritten; Destutt de Tracy expected others, perhaps with more specific knowledge, to complete it. It could be argued that he left this open to allow scientists to become more directly involved in the theoretical development of *Idéologie*. However, changes in the political situation in France during the Napoleonic era ultimately mitigated against this kind of involvement.[59]

With this overview of the structure of the *Élemens*, consider now the arguments Destutt de Tracy laid out specifically in his section on logic [Destutt de Tracy, 1804, 1:32-36]. There, he pointed to the difficulties arising outside the realm of algebraic reasoning and algebraic symbols and engaged in a comparative discussion of algebraic and ordinary language, a topic on which he had advanced some ideas in [Destutt de Tracy, 1801a]. In his discussion, he warned of serious limitations inherent in Condillac's dream. In 1798, he had stated that it was wrong to equate reasoning with calculation, since the latter only concerned the limited notion of quantity.[60]

Destutt de Tracy returned to the question, already discussed by de Gérando and quoted by Babbage, concerning the limited appeal algebraic language makes to memory. In his view, algebraic language owed its operational power to the fact that, in the course of a long calculation, it is not necessary to return constantly to reconsider the meaning of the symbols being used. He attributed this facility to algebra's internal simplicity and to the fact that algebra only deals with a very definite genre of ideas: quantity. This pointed to a further point of divergence with the possible formulation of a language adapted to metaphysics.

In their attempts to formulate general models of mind and human reasoning, early nineteenth-century philosophers like Destutt de Tracy clearly outlined a division of functions in which processing, accumulation, and in-and-out communication appear as distinct units. Later, similar models were adapted to the design of automatic calculating machinery, for example, by Babbage. Discussions on the language of algebra had, in fact, never been far from the notion of automatic calculation.

---

[58] Destutt de Tracy later called this last part "government," a topic he considered specifically in his studies on Montesquieu [Destutt de Tracy, 1798a]. (Thomas Jefferson translated the latter book into English.) Destutt de Tracy took up education in [Destutt de Tracy, 1801a].

[59] Although this third section was never undertaken, there were interesting efforts to link some of the topics Destutt de Tracy proposed for it to the general philosophy supported by the *Idéologues*, either in agreement or in dissonance with it. Works of Gergonne, Ampère, Maine de Biran, and others, not discussed here, can be regarded as part of this legacy.

[60] See [Destutt de Tracy, 1798b, p. 123 (note)].

Destutt de Tracy's Views on Artificial Languages

Condillac had opened his book *La Langue des calculs* with the well-known statement: "All language is analytic method, and all analytic method is language. These two truths, as simple as they are new, have been demonstrated; the first, in my grammar; the second, in my logic; and one could convince oneself of the light that they shed on the art of speaking and on the art of reasoning, [since] they reduce to one and the same art."[61] Destutt de Tracy had expressed his disagreement with the idea, often derived from Condillac's dictum, that to renovate a particular chapter of science and send it down a new avenue of progress only required a renewal of its nomenclature and the introduction of a language *méthodique*. He wrote that there were those who believed that, in order to effect exceptional advances in a science, it was only necessary to renew its nomenclature by more greatly systematizing it. In Destutt de Tracy's view, however, "this is by no means what it is about."[62] In 1816, Babbage remained one of those distant disciples of Condillac who still upheld the outdated ideas Destutt de Tracy decried.

For Destutt de Tracy, the construction of a language for an area of science was a more complex task, far beyond a formal definition. For him, it was essential to identify and clarify the obscure points of the science, and to do it in such a way that the words used did not express any more than the right ideas introduced, that is, that they remained true to the accepted facts [Destutt de Tracy, 1826, p. 27 (note)]. In addition, a language was "made" and "fixed" when those using it applied its terms in a sense that was both accurate and specific.

Returning to his critique of Condillac's prescription, Destutt de Tracy stated clearly that, even accepting the advantages of a good nomenclature and of a well designed language, it was not words that create science [Destutt de Tracy, 1826, p. 27 (note)]. In a direct reference to Lavoisier's work, he used phlogiston theory to show that until chemists fixed the precise sense of the words *phlogistique*, *combustion*, and *combustible*, little advance was possible.[63] Had they not disposed of the first of these enigmatic words, the idea of the role of oxygen would not have become as clear in their minds. In Destutt de Tracy's view, without understanding the significance of these facts, it would have been impossible both to rectify ideas in chemistry to the extent Lavoisier and his colleagues did and to construct a language capable of expressing these new ideas. He had turned Condillac's propositions upside down.

In his work, Destutt de Tracy considered some of Condillac's over-simplifications of the question of languages specifically designed for use in science and discussed the possibility that they had a structure far more complex than that of ordinary algebra. He did not believe that scientific languages (with the exception of that of algebra—or perhaps those of areas of mathematics, in general—due to its structural simplicity) could be viewed as potential keys to the formulation of the path

---

[61]"Toute langue est une méthode analytique, et toute méthode analytique est une langue. Ces deux vérités, aussi simples que neuves, ont été démontrées; la première, dans ma grammaire; la seconde, dans ma logique; et on a pu se convaincre de la lumière qu'elles répandent sur l'art de parler et sur l'art de raisonner, qu'elles réduisent à un seul et même art" [Condillac, 1798b, p. 1].

[62]"[C]e n'est point du tout cela dont il s'agit" [Destutt de Tracy, 1826, pp. 26-27].

[63]Phlogiston theory held that different inflammable materials have in them, in different proportions, a fluid and volatile a substance (called *phlogistique* from the Greek *phlox* or flame) responsible for their different ability to burn. This theory was developed by the German chemist Georg Ernst Stahl (1660–1734) and supported by the French chemist Guillaume François Rouelle (1703–1770), who had been Lavoisier's teacher.

forward in a specific field of science for which the laws are as yet unknown. In his view, languages were representations of an actual contingent scientific structure. He did not rule out, however, the existence of formal analogies between algebra and the various scientific languages. He expressed the equation between algebra and specific scientific languages in the following terms: "That is what algebra is: algebra is also a language, and languages are themselves only kinds of algebras."[64]

This was clearly not *ordinary* algebra as Condillac had postulated. Finding the algebra of algebras ascends to a paradigmatic position and becomes a question parallel to that of finding the theory of theories. Clearly, the *Idéologues* were neither the first nor the only philosophical school looking for a theory of theories. D'Alembert, one of the heralds of *Idéologie*, had been a contributor to this topic. He had considered this question in the preliminary discourse to the *Encyclopédie*, where he wrote that he wished to make of the universe a unique event and a great truth.[65] In his view, this was to be achieved through a dynamic process of successive approximation, perfecting erroneous theories through the development of science, passing from "partial" error to "partial" truth. Destutt de Tracy proposed the same process in the discussion of language.

Final Remarks

As we have seen, in his "Essay" published in 1815–1816, Babbage developed a formal notation for the treatment of functional equations capable of solving interesting examples. The solution of one of them, the functional differential equation discussed above, was only found well over a century after the publication of Babbage's "Essay" using an independent ad hoc approach. It could have been easily derived using Babbage's rules of language, if his ideas had been more seriously considered. However, his elegant formal process gave satisfactory results, provided that some fundamental questions, such as the existence of the inverse operator and the uniqueness of the solution, did not interpose themselves with the course of his procedure. His language also suffered from serious problems relating to the complexity of formulating an unambiguous notation.

I have argued here that, in his ambition to resolve the problems of an area of science through the construction of an analytical method which, in essence, was an operative "language," Babbage followed a conception of language that situated him firmly within Condillac's circle of ideas. Condillac's approach was developed in the 1760s and 1770s when his *Grammaire*, his *Logique*, and possibly *La Langue des calculs* were thought out and sketched; its direct influence began to be felt in different fields of science, among them chemistry, beginning in the 1780s.

By the time Babbage's "Essay" was published, however, French philosophers interested in questions related to the philosophy of science and mathematics had become highly critical of Condillac's definitive statements and were constructing new formulations of the concept of language and its relationship with science. Babbage did not seem to have been aware of these critiques until well after their publication.

---

[64]"C'est bien là ce qu'est l'algèbre: aussi l'algèbre est-elle une langue, et les langues ne sont elles-mêmes que des espèces d'algèbres" [Destutt de Tracy, 1801c, p. 239].

[65]"The universe, for those who wish to embrace it from a unified viewpoint, will be—if it is permissible to say so—but a unique event, a great truth [L'Univers pour qui fauroit l'embraffer d'un feul point de vûe, ne feroit, s'il eft permis de le dire, q'un fait unique une grande vérité]" [D'Alembert, 1751, p. ix]. See also [D'Alembert, 1894].

According to his biographers, Babbage made his first visit to Paris with John Herschel probably in 1819; this was the same year the Cambridge Philosophical Society was founded.[66] During that trip, Babbage met with a number of scientists who were in direct contact with leading *Idéologues* and who shared with them positions in many official bodies including the *Académie des Sciences*. As shown above, Babbage finally quoted de Gérando's critique of Condillac's theories on language and analysis in a paper written in 1821. It is significant that after that date he stopped pursuing mathematical work based, like his work on functional equations of 1815–1816, on the development of analytical scientific languages.[67]

However, by 1817 the views expressed in de Gérando's book of 1800, which had served to announce an important methodological rift between members of successive generations of the French school of *Idéologues*, had been largely superseded in France. The views sketched in de Gérando's critical assessment had crystallized in Destutt de Tracy's well-designed and original formulations inserted in his *Élémens d'idéologie* and were being reconsidered by some of his then disciples, among them, Maine de Biran. By 1821, Babbage seemed not to have been seriously influenced by this more contemporaneous work. He had thus been some ten to fifteen years behind critical philosophical advances in France. Since these later ideas and refinements would have constituted a philosophical export of considerable mathematical significance, and since they were clearly not transmitted to or through Babbage, it seems reasonable to question the extent and efficacy of the channels of communication for philosophical ideas that existed between France and Britain in the first two decades of the nineteenth century.[68]

---

[66]See, for example, [Hyman, 1982, pp. 40-44].

[67]Much later, in 1836, De Morgan expressed a renewed interest in Babbage's "calculus of functions" in [De Morgan, 1836].

[68]Even if Babbage's contacts with France, and continental Europe in general, became stronger after his 1819 visit to Paris (in 1819 he offered the names of "Lacroix, Biot and Laplace" as references when he unsuccessfully applied for a chair in Edinburgh [Babbage, 1864, p. 473]), the question about the early part of the century remains. Reference was made above to the difficulties Babbage encountered in England in gaining access to foreign, and in particular to French, scientific literature in 1815. The same view was expressed two years earlier in [Babbage and Herschel, 1813, p. iii]: "Recent French publications are not easily procured, nor is it surprising that to obtain those of the German Analysts is almost impossible, when Delambre regrets their scarcity even in France." In Babbage's case, these remarks are fully supported by his limited quotations—in that period—of contemporary French mathematical and philosophy of science literature. The "Index" in Campbell-Kelly's edition of Babbage's works [Babbage, 1989, 11:377-390] clearly shows this.

The general question of scientific communication in times of conflict has an extensive literature (for the case of mathematics, see [Booss-Bavnbek and Høyrup, 2003] and the references given therein). In the specific case of relations between France at the time of the Napoleonic Wars, Gavin De Beer subscribed to the view that the sciences have never been at war, which he supported in his influential book [De Beer, 1960]. (For a more comprehensive analysis of these events, from a modern perspective, see [Crosland, 2005].) De Beer's views are based on the permission given by the French government to Humphry Davy (who, previously, had been honored for his scientific achievements by Napoleon) to visit and inspect the volcanoes of the Aubergne and to extend that visit to the *Institut* in Paris in 1813. De Beer also makes reference to interesting correspondence between British (particularly Joseph Banks) and French scientists at the time, which he reproduced in his documented book.

Arguments given in this chapter support the view that communication in mathematics and in the philosophy of science had a rather exceptional character in that period.

References

Aczél, Janòs. Ed. 1984. *Functional Equations: History, Applications, and Theory*. Dordrecht-Boston: D. Reidel Publishing Co.

Arnauld, Antoine and Lancelot, Claude. 1660. *Grammaire générale et raisonnée*. Paris: Chez Pierre Le Petit.

Arnauld, Antoine and Nicole, Pierre. 1662. *La Logique, ou l'art de penser*. Paris: Chez Charles Savreux.

Arnault, Antoine; Jay, Antoine; Étienne de Jouy, Victor Joseph; and [Marquet de Montbreton de] Norvins, Baron Jacques. 1822. *Bibliographie nouvelle des contemporains, ou dictionnaire historique et raisonné* . . . . Paris: Librarie historique.

Babbage, Charles. 1809–1817. "History of the Origin and Progress of the Calculus of Functions during the Years 1809, 1810. . .1817." Manuscript. Oxford: Museum for the History of Science.

——————. 1815. "An Essay Towards the Calculus of Functions, Part I." *Philosophical Transactions of the Royal Society of London* 105, 389-423.

——————. 1816. "An Essay Towards the Calculus of Functions, Part II." *Philosophical Transactions of the Royal Society of London* 106, 179-256.

——————. 1817. "Observations on the Analogy Which Subsists between the Calculus of Functions and Other Branches of Analysis." *Philosophical Transactions of the Royal Society of London* 107, 197-216.

——————. 1822. "Observations on the Notation Employed in the Calculus of Functions." *Transactions of the Cambridge Philosophical Society* 1, 63-76.

——————. 1827. "On the Influence of Signs in Mathematical Reasoning." *Transactions of the Cambridge Philosophical Society* 2, 325-377.

——————. 1864. *Passages from the Life of a Philosopher*. London: Longman, Green.

——————. 1989. *The Works of Charles Babbage*. Ed. Martin Campbell-Kelly. 11 Vols. London: William Pickering.

Babbage, Charles and Herschel, John F. W. 1813. "Preface." *Memoirs of the Analytical Society* 1, i-xxii.

Baguenault de Puchesse, Count Gustave. 1910. *Condillac: Sa vie, sa philosophie, son influence*. Paris: Plon-Nourrit et Cie.

Benedict, Barbara M. 2001. *Curiosity: A Cultural History of Early Modern Inquiry*. Chicago: University of Chicago Press.

Bentham, Jeremy. 1989. *The Correspondence of Jeremy Bentham*. Vol. 9 (January 1817 to June 1820). Ed. Stephen Conway. Oxford: Clarendon.

Birkhoff, Garrett and Kreyszig, Erwin. 1984. "The Establishment of Functional Analysis." *Historia Mathematica* 11, 258-321.

Booss-Bavnbek, Bernhelm and Høyrup, Jens. Ed. 2003. *Mathematics and War*. Basel: Birkhäuser Verlag.

Boyle, Stephen. 1986. "Babbage's Work on the Calculus of Functions and Its Historical Roots." M. Sc. Thesis: Imperial College, London.

Cabanis, Pierre. 1802. *Rapports du physique et du moral de l'homme*. Paris: Crapart, Caille et Ravier.

Carnot, Lazare. 1797. *Réflexions sur la métaphysique du calcul infinitésimal*. Paris: Duprat.

Clairaut, Alexis-Claude. 1731. *Recherches sur les courbes à double courbure*. Paris: Nyon.

Condamine, Charles-Marie de la. 1745. *Relation abrégée d'un voyage fait dans l'intérieur de l'Amérique méridionale*. Paris: Vve. Poissot.

Condillac, Étienne Bonnot de. 1746. *Essai sur l'origine des connaisances humaines: Ouvrage où l'on réduit à un seul principe tout ce qui concerne l'entendement humain*. Amsterdam: Chez Pierre Mortier.

_____. 1754. *Traité des sensations*. 2 Vols. London-Paris: de Buré l'aîné.

_____. 1775. *Cours d'études pour l'instruction du Prince de Parme*. 16 Vol. Parma: Imprimerie royale.

_____. 1780. *La Logique, ou les premiers développements de l'art de penser*. Paris: L'Esprit.

_____. 1798. *Oeuvres de Condillac*. 23 Vols. Paris: Imprimerie de Ch. Houel.

_____. 1798a [Ann VI]. "De l'art de penser." In [Condillac, 1798, 6:1-255].

_____. 1798b [Ann VI]. "La Langue des calculs." In [Condillac, 1798, 23:1-484].

_____. 1981. *La Langue des calculs*. Ed. Sylvain Auroux and Anne-Marie Chouillet. Lille: Presses universitaires de Lille.

Crosland, Maurice. 1967. *The Society of Arcueil: A View of French Science at the Time of Napoleon I*. London: Heinemann.

_____. 2005. "Relationships between the Royal Society and the Académie des Sciences in the Late Eighteenth Century." *Notes and Records of the Royal Society of London* 59(1), 25-34

D'Alembert, Jean Le Rond. 1751. *Discourse préliminaire de l'Encyclopédie*. Paris: Chez Briasson/David/Le Breton/Durand (Quotations from the 1772 Ed. Geneva: Chez Cramer l'ainé Cie.).

_____. 1761–1780. *Opuscules mathématiques, ou mémoires sur différens sujets de la géométrie, de mécanique, d'optique, d'astronomie & c*. 8 Vols. Paris: David.

_____. 1894. *Discours préliminaire de l'encyclopédie, publié intégralement d'après l'édition de 1763 avec les avertissements de 1759 et 1763, la dédicace de 1751, des variantes, des notes, une analyse et une introduction par François Picavet*. Paris: A. Colin et Cie.

De Beer, Gavin. 1960. *The Sciences Were Never at War*. London: Nelson.

De Gérando, Joseph Marie. 1800a [An VIII]. *Considérations sur les diverses méthodes à suivre dans l'observation des peuples sauvages*. Paris: Société d'observateurs de l'homme.

_____. 1800b [An VIII]. *Des signes et de l'art de penser considérés dans les rapports mutuels*. 4 Vols. Paris: Chez Goujon fils.

_____. 1822–1823. *Philosophie moderne: Histoire comparée des systèmes de philosophie*. 4 Vols. Paris: A. Eymery.

_____. 1847. *Philosophie moderne: Histoire comparée des systèmes de philosophie*. 4 Vols. Paris: Ladrange.

De Morgan, Augustus. 1836. *Treatise on the Calculus of Functions: Extracted from the Encyclopedia Metropolitana*. London: W. Clowes & Son.

Descartes, René. 1637. *Discours de la méthode pour bien conduire sa raison et chercher la verité dans les sciences*. Leiden: Imprimerie de Ian Marie.

Destutt de Tracy, Antoine. 1798a [Ann VI]. *Commentaire sur l'esprit des lois de Montesquieu.* Paris: Delaunay.

———. 1798b [Ann VI]. "Mémoire sur la faculté de penser." *Mémoires de l'Institut national des sciences et des arts* 1, 283-450. (Quotations from [Destutt de Tracy, 1992, 35-177]).

———. 1801a [Ann IX]. *Projet d'éléments d'idéologie à l'usage des Écoles Centrales de la République française.* Paris: P. Didot.

———. 1801c [Ann IX]. *Élémens d'idéologie.* Vol. I. *Idéologie proprement dite.* Paris: Courcier (Quotations from [Destutt de Tracy, 1826a]).

———. 1802 [Ann X]. "De la métaphysique de Kant." *Mémoires de l'Institut national des sciences et des arts* 4, 544-606 (Quotations from [Destutt de Tracy, 1992, 243-306]).

———. 1803. *Élémens d'idéologie.* Vol. 2. Pt. 3. *De la grammaire.* Paris: Courcier (Quotations from [Destutt de Tracy, 1826b]).

———. 1804. *'Élémens d'idéologie.* Vol. 3. Pt. 3. *De la logique.* 2 Vols. Paris: Courcier (Quotations from [Destutt de Tracy, 1826c]).

———. 1817. *Élémens d'idéologie.* Vols. 4-5. *Traité de la volonté et de ses effets.* Paris: Vve. Courcier (Quotations from [Destutt de Tracy, 1826d]).

———. 1826a [Ann IX]. *Élémens d'idéologie.* Vol. 1. *Idéologie proprement dite.* Brussels: August Wahlen.

———. 1826b. *Élémens d'idéologie.* Vol. 2. Pt. 3. *De la grammaire.* Brussels: August Wahlen.

———. 1826c. *Élémens d'idéologie.* Vol. 3. Pt. 3 *De la logique.* Brussels: August Wahlen.

———. 1826d. *Élémens d'idéologie.* Vols. 4-5. *Traité de la volonté et de ses effets.* Brussels: August Wahlen.

———. 1992. *Mémoire sur la faculté de penser, De la métaphysique de Kant, et autres textes.* Corpus des oeuvres de philosophie en langue française. Paris: Fayard.

Dhombres, Jean. 1984. "On the Historical Role of Functional Equations." In *Functional Equations: History, Applications, and Theory.* Ed. Janòs Aczél. Dordrecht-Boston: D. Reidel Publishing Co., 17-31.

———. 1986. "Quelques aspects de l'histoire des équations fonctionnelles liés à l'évolution du concept de fonction." *Archive for the History of Exact Sciences* 36, 91-181.

Dubbey, John M. 1978. *The Mathematical Works of Charles Babbage.* Cambridge: Cambridge University Press.

Duchesnau, François. 1974. "Condillac, Critique de Locke." *Studi internationali di filosofia* 6, 77-98.

Friedelmeyer, Jean-Pierre. 1993. "Le Calcul des dérivatives d'Arbogast dans le projet d'algébrisation de l'analyse à la fin du XVIIIe siècle." Thèse de Doctorat: Université de Nantes.

Gergonne, Joseph D. 1821. "Des équations fonctionnelles; par M. Charles Babbage." *Annales de mathématiques pures et appliquées* 12, 73-103.

Goetz, Rose. 1993. *Destutt de Tracy: Philosophe du langage et science de l'homme.* Geneva: Droz.

Grattan-Guinness, Ivor. 1992. "Charles Babbage As an Algorithmic Thinker." *IEEE Annals of the History of Computing* 14 (3), 34-48.

Gusdorf, Georges. 1978. *Les Sciences humaines et la pensée occidentale.* 8 Vols. Vol. 8. *La Conscience révolutionaire: Les Idéologues.* Paris: Payot.

Harnois, Guy. 1929. *Les Théories du langage en France de 1660 à 1821.* Paris: Société d'édition "Les Belles Lettres."

Herschel, John F. W. 1813. "On Equations of Differences and Their Application to the Determination of Functions from Given Conditions." *Memoirs of the Analytical Society* 1, 65-114.

_____. 1820. *Collection of Examples of the Applications of the Calculus of Finite Differences.* Cambridge: J. Deighton and Sons.

_____. 1981. *Aspects of the Life and Thought of Sir John Frederick Herschel.* Ed. Sylvan S. Schweber. 2 Vols. New York: Arno Press.

Hyman, Anthony. 1982. *Charles Babbage: Pioneer of the Computer.* Princeton: Princeton University Press.

Kamke, Erich. 1944. *Differentialgleichungen: Lösungsmethoden und Lösungen.* Leipzig: Akademische Verlagsgesellschaft.

Kennedy, Emmet. 1978. *Destutt de Tracy and the Origins of "Ideology".* Philadelphia: American Philosophical Society.

Knight, Isabel. 1968. *The Geometric Spirit: The Abbé de Condillac and the French Enlightenment.* New Haven: Yale University Press.

Knowlson, James. 1975. *Universal Language Schemes in England and France, 1600–1800.* Toronto: University of Toronto Press.

Kuehner, Paul. 1944. *Theories on the Origin and Formation of Languages in the Eighteenth Century in France.* Philadelphia: University of Pennsylvania Press.

Lacroix, Silvestre François. 1797–1798 [Ann V–VI]. *Traité du calcul différentiel et du calcul intégral.* 2 Vols. Paris: Chez J. B. H. Duprate.

_____. 1800 [Ann VIII]. *Traité des différences et des séries.* Paris: Chez J. B. H. Duprate.

_____. 1816. *An Elementary Treatise on the Differential and Integral Calculus, Translated from the French.* Pt. 1 by Charles Babbage. Pt. 2 by George Peacock. Pt. 3 by John F. W. Herschel. With an Appendix [by Herschel] and Notes [by Peacock and Herschel]. Cambridge: J. Deighton & Sons.

Lagrange, Joseph Louis. 1797. *Théorie des fonctions analytiques.* Paris: Imprimerie de la République.

Lavoisier, Antoine. 1787. *Sur la nécessité de réformer et the perfectionner la langue de la chimie.* Paris: Chez Couchot (Quotations from [Lavoisier, 1864–1893, 5:354-364]).

_____. 1864–1893. *Oeuvres complètes de A. Lavoisier.* 6 Vols. Paris: Imprimerie nationale.

Lavoisier, Antoine; de Morveau, Louis Guyton; Bertholet, Claude-Louis; and de Fourcory, Antoine. 1787. *Méthode de nomenclature chimique.* Paris: Chez Couchet.

Lyon, Georges. Ed. 1886. *Condillac: Traité des sensations.* Paris: F. Alcan.

Martinet, André. 1960. *Éléments de linguistique général.* Paris: A. Colin.

Monge, Gaspard. 1770–1773. "Mémoire sur la détermination des fonctions arbitraires dans les intégrales de quelques équations aux différences partielles." And "Second mémoire sur le calcul intégral de quelques équations aux différences partielles." *Mémoires de l'Académie des Sciences de Turin* 5, 16-79.

──────────. 1773. "Mémoire sur la construction des fonctions arbitraires qui entrent dans les intégrales des équations aux différences partielles." *Mémoires des savants étrangers de l'Académie des Sciences de Paris* 6, 267-304.

Moravia, Sergio. 1968. *Il Tramonto dell'Illuminismo: Filosofia e politica nella società francese (1770–1810)*. Bari: Editori Laterza.

Newton de Tracy, Sarah. 1852. *Essais divers, lettres et pensées*. 2 Vols. Paris: Typographie Plon frères.

Nový, Luboš. 1973. *Origins of Modern Algebra*. Prague: Academia.

Ortiz, Eduardo L. 1999. "Geometría, lógica y teoría de las máquinas: El Ensayo de Lanz y Betancourt de 1808 sobre la teoría de máquinas." *Fórmula* 5, 261-272.

Ortiz, Eduardo L. and Bret, Patrice. 1997. "José María de Lanz and the Paris-Cadiz Axis." In *Naissance d'une communauté internationale d'ingenieurs*. Ed. Irina Gouzvitch and Patrice Bret. Paris: Musée de La Villette, 56-77.

Panteki, Maria. 1991. "Relationships Between Algebra, Differential Equations, and Logic in England, 1800–1860." 2 Vols. Ph.D. Thesis: Middlesex Polytechnic University.

──────────. 2003. "French 'logique' and British 'logic': On the Origins of Augustus De Morgan's Early Logical Enquiries: 1805–1835." *Historia Mathematica* 30, 278-340.

Peacock, George. 1834. "Report on the Recent Progress and Present State of Certain Branches of Analysis." In *Report of the Third Meeting of the British Association for the Advancement of Science Held at Cambridge in 1833*. London: John Murray, 185-352.

Picavet, François. 1891. *Les Idéologues: Essai sur l'histoire des idées et des théories scientifiques, philosophiques, religieuses, etc. en France depuis 1789*. Paris: F. Alcan.

Pierobon, Frank. 2003. *Kant et les mathématiques: La Conception kantienne des mathématiques*. Paris: J. Vrin.

Pincherle, Salvatore. 1904–1916. "Funktionaloperationen und Gleichungen." In *Encyklopädie der mathematischen Wissenschaften mit Einschluss ihrer Anwendungen*. 6 Vols. Vol. 2.1. *Analysis*. Leipzig: B. G. Teubner Verlag, 761-817.

──────────. 1912. "Équations et opérations fonctionnelles." In *Encyclopédie des sciences mathématiques pures et appliquées*. 6 Vols. Vol. 5. *Analyse*. Paris: Gauthier-Villars, 1-78.

Prieto, Luis J., 1966. *Messages et signaux*. Paris: Presses universitaires de France.

Quere, Bernard. 2005. "La communication scientifique et technique par les utiles graphiques de 1750 à 1850 dans le contexte de la Bretagne." Unpublished Doctoral Dissertation: École des Hautes Études en Sciences Sociales.

Réthoré, François. 1864. *Condillac ou l'empirisme et le rationalisme*. Paris: A. Durand.

Runkle, John Daniel. 1861. "Complete Catalogue of the Writings of Sir John Herschel." *The Mathematical Monthly* 3, 220-227.

Silberstein, Ludwick. 1940. "On a Functional Differential Equation." *Philosophical Magazine* 30, 186-188.

Suzanne, Pierre-Henri, 1806–1809. *De la manière d'étudier les mathématiques: Ouvrage destiné à servir de guide aux jeunes gens, à ceux sur-tout qui veulent approfondir cette science, ou qui aspirent à l'École impériale polytechnique.* 3 Vols. Paris: De l'auteur.

Terrall, Mary. 2002. *The Man Who Flattened the Earth: Maupertuis and the Sciences in the Enlightenment.* Chicago: University of Chicago Press.

Tisserand, Pierre. Ed. 1920–1949. *Oeuvres de Maine de Biran.* 14 Vols. "Introduction." Vol. 1 Paris: F. Alcan and Presses universitaires de France.

Vaugelas, Claude Favre de, Baron de Peroges. 1647. *Remarques sur la langue française vtiles à ceux qui veulent bien parler et bien escrire.* Paris: Camusat.

Zimmer, Carl. 2004. *Soul Made Flesh: Thomas Wallis, the Discovery of the Brain and How it Changed the World.* London: Free Press.

CHAPTER 3

"Very Full of Symbols": Duncan F. Gregory, the Calculus of Operations, and the *Cambridge Mathematical Journal*

Sloan Evans Despeaux
Western Carolina University, United States

Introduction

John F. W. Herschel focused the first part of his 1845 presidential address to the British Association for the Advancement of Science on recent mathematical accomplishments at his *alma mater*, the Cambridge University. Augustus De Morgan, a fellow Cambridge graduate and professor of mathematics at University College, London, advised Herschel that "[y]ou should not forget the Cambridge 'Mathematical Journal.' It is done by the younger men .... It is full of very original contributions. It is, as is natural in the doings of young mathematicians, very full of symbols. The late Dr [sic] F. Gregory, whom you must notice most honourably ... gave his extensions of the Calculus of Operations ... in it. He was the first editor. He was the most rising man among the juniors" [De Morgan, 1845]. Herschel concurred with this appraisal in his address, citing Gregory and the *Journal* as "[a]nother instance of the efficacy of the course of study in this University, in producing not merely expert algebraists, but sound and original mathematical *thinkers* (and, perhaps, a more striking one, from the generality of its contributors being men of comparatively junior standing)" [Herschel, 1845, p. xxix].

Gregory's mathematical reputation has not diminished; indeed, today he is recognized as a central figure in the development of the British approach to algebra in the 1830s through the 1840s. Along with a group of British mathematicians including George Peacock, Augustus De Morgan, George Boole, Arthur Cayley, and James Joseph Sylvester, Gregory questioned the foundations, examined the general laws, and pushed the boundaries of the then existing conceptions of algebra. Studies by Elaine Koppelman and Patricia Allaire have convincingly argued that the algebraic insights of Gregory, as well as those of other central figures in the development of the British approach to algebra, were influenced by the exposure these mathematicians had to the calculus of operations.[1]

In viewing the work of these algebraic pioneers, it is important to consider the wider context in which their accomplishments occurred. Gregory's work, as

---

[1] See [Koppelman, 1971] and [Allaire, 1997]. For more on the connections between the calculus of operations and the British approach to algebra, also see [Panteki, 1991]. For a discussion of Gregory's conception of the calculus of operations as a foundation for the calculus, see [Allaire and Bradley, 2002].

the above comments by De Morgan and Herschel suggest, are intimately linked to the *Cambridge Mathematical Journal*. In fact, all but one of Gregory's papers appeared in the *Journal*, of which he was a co-founder. As the *Journal*'s editor and most active contributor from November 1837 to November 1843, Gregory initiated a conversation about the calculus of operations. In this discussion, junior mathematicians, many of them first-time researchers, explored the method, responded to Gregory, built on his articles, and affected Gregory's own views of his work.

This paper explores the discussion between Gregory and other junior, British mathematicians on the calculus of operations as well as the publication venue in which this exchange took place. The examination of this work provides an intriguing window into the evolution of a research area fundamental to the development of the British approach to algebra. Considering it, moreover, within the context of the *Cambridge Mathematical Journal* reveals the role that a nascent form of mathematical communication played in promoting research by a new generation of British mathematicians.[2]

The Establishment of a "Proper Channel" for the Research of Junior British Mathematicians

Before the 1837 foundation of the *Cambridge Mathematical Journal*, the publication options for nineteenth-century British mathematicians, and especially British mathematicians new to research, were considerably limited. British mathematical outlets certainly included monographs and encyclopedia articles; however, these formats were better suited for the synthesis and promotion of well thought-out mathematical ideas than for the quick dissemination of new research by young, unproven researchers.

The *Transactions* of several British scientific societies were open to mathematical articles during this period; in fact, the *Philosophical Transactions* of the Royal Society of London had accepted such articles since the seventeenth century. These society journals presented significant obstacles to new researchers trying to publish mathematics rapidly, however. For example, a mathematician wanting to publish in these society venues needed either to belong to the society or to have a member formally "communicate" the paper. After successfully submitting a paper, an author still risked never seeing the manuscript again. In the case of the Royal Society, if the paper was not accepted by the referees for publication, the manuscript was permanently deposited in the society's archives. A mathematician overcoming these hurdles might wait months or even years to see a communication in print, sandwiched between articles from other scientific fields. With no major mathematical society of their own before 1865, British mathematicians competed with scholars from these other fields for publication room in society journals.

Publishing in independent scientific journals, such as the *Philosophical Magazine* (1798-present), allowed early nineteenth-century British mathematicians to avoid society regulations and to publish their work more rapidly. However, this venue was still a general scientific one; mathematicians had to advocate for the publication of their articles to editors wary both of mathematics and its appeal to their readers.

---

[2]For a wider discussion of the role of British scientific journals in the development of nineteenth-century mathematics, see [Despeaux, 2002].

The independent mathematical journals in existence in Britain before the foundation of the *Cambridge Mathematical Journal* were generally short-lived affairs devoted primarily to questions for answer, enigmas, and puzzles. One notable exception was Thomas Leybourne's *Mathematical and Philosophical Respository*, launched in 1795. During the first decade of the nineteenth century, Leybourne's fellow mathematical masters at the Royal Military College, William Wallace and James Ivory, made innovative contributions to the *Repository* using continental mathematical approaches to differential calculus. Despite these contributions, the *Repository* still relied on questions for answer for much of its contents, and it had dissolved by 1835, two years before the foundation of the *Cambridge Mathematical Journal*.

Without the *Repository*, British mathematicians wanting to publish research in journals either had to negotiate scientific society hurdles or appeal to commercial scientific editors suspicious of mathematics; these prospects proved even more daunting to junior mathematicians new to the British scientific milieu [Crilly, 2004, pp. 459-461]. Confronted with these discouraging options, an ambitious group of mathematically inclined Cambridge students decided to create a new kind of publication venue. In December of 1836, the twenty-four-year-old Trinity College, Cambridge fellow, Archibald Smith, introduced the idea of founding a mathematical journal to Duncan Gregory, also of Trinity, the month before the latter was to take his mathematical Tripos examination. This examination, for over half of the nineteenth century, was a mandatory hurdle for the B. A. degree at Cambridge, even for those students interested in non-mathematical areas. A high ranking among the wranglers, the highest honor bracket for the examination, provided access to promising positions in the university, church, and government. Smith, the senior wrangler for 1836, seemed to have left his Tripos experience with a bitter taste in his mouth, writing just after the examination that he had become "quite tired of, I might almost say disgusted with, mathematics" [Smith, 1836]. A year's time, however, evidently relieved Smith of his distaste for mathematics and replaced it with a desire to encourage original mathematical research. Gregory agreed to act on Smith's publication idea after the completion of his own Tripos examination. He placed as fifth wrangler, a very respectable ranking considering that the higher ranked wranglers for 1837 included James Joseph Sylvester (second) and George Green (fourth) [Allaire, 1997, p. 76]. Samuel S. Greatheed, the fourth wrangler for 1835, was another Trinity fellow and is credited, along with Smith and Gregory, as a co-founder of the journal. Greatheed soon left Trinity for a career in the Church of England, and, in particular, in composing music after his marriage forced him to relinquish his fellowship. Smith followed another popular route for high-ranking wranglers and began a law career. These defections soon left Gregory as the driving force of the new *Journal* [Crilly, 2004, p. 463].

In the preface to the first number of the *Journal*, brought out in November of 1837 while he was still an undergraduate, Gregory outlined his goals for the new publication.[3] In this preface, Gregory pointed out that many recognized the lack of a "proper channel ... for the publication of papers on mathematical subjects, which did not appear to be of sufficient importance to be inserted in the

---

[3]The quotes that follow in this paragraph are all from [[Gregory], 1837a, p. 1]. Gregory's authorship of this preface was not explicitly indicated. Here and below, authors' names given in brackets in the references have been identified subsequently.

Transactions of any of the Scientific Societies." The two commercial "Philosophical Journals" were, in general, "devoted mainly to physical subjects."[4] Gregory was confident that Cambridge mathematicians were already "able and willing to communicate" material to the new *Journal*, "while," he noted, "the very existence of such a work is likely to draw out others, and make them direct their attention in some degree to original research." Reprints of abstracts and memoirs from abroad would inform the readers and keep them "on a level with the progressive state of mathematical science, and so lead them to feel a greater interest in the study of it." First and foremost, however, Gregory's new *Journal* would "supply a means of publication for original papers." The majority of the papers in the first volume were understandably provided by Gregory and his co-founders. Gregory, in particular, used this venue as a means to "draw out" young mathematicians to conduct original research on the calculus of operations.

The Calculus of Operations before the *Cambridge Mathematical Journal*

The calculus of operations involved separating symbols of operation—such as differentiation—from symbols of quantity. Using general properties of algebra, the separated symbols of operation were then simplified in order to reach a solution to an analytical problem or theorem.

For example, the "exponent" of $\frac{d}{dx}$, which records how many times differentiation occurs, and the "exponent" on Δ, which records the order of a finite difference, obey the same laws of exponentiation governing a variable x or a function f. In Koppleman's notation,

$$\frac{d^a}{dx^a}\left(\frac{d^b y}{dx^b}\right) = \frac{d^{a+b} y}{dx^{a+b}},$$

$$\Delta^n \Delta^m f(x) = \Delta^{n+m} f(x),$$

where $\Delta f(x) = f(x+h) - f(x)$ and $\Delta^m f(x) = \sum_{j=0}^{m} \binom{m}{j}(-1)^{m-j} f(x+jh)$, $x^n \cdot x^m = x^{n+m}$, and $f^n(f^m(x)) = f^{n+m}(x)$ [Koppelman, 1971, p. 156]. The analogy between exponentiation and the iteration of the derivative had been noticed as early as 1695 by Gottfried Leibniz [Koppelman, 1971, p. 157]. Since the 1770s, French mathematicians including Joseph-Louis Lagrange, Pierre Simon Laplace, Louis François Arbogast, Barnabé Brisson, Joseph Fourier, Joseph Liouville, François Joseph Servois, and Augustin-Louis Cauchy developed this analogy, applying it to differential, finite difference, and functional equations.[5]

An important result using this analogy that was investigated by Lagrange, Servois, Arbogast, and later Gregory involved Taylor's Theorem, which states that

$$f(x+h) = \sum_{k=0}^{\infty} \frac{h^k}{k!}\left(\frac{d}{dx}\right)^k f(x).$$

Rewriting

$$\sum_{k=0}^{\infty} \left[\frac{h^k}{k!}\left(\frac{d}{dx}\right)^k\right] f(x) \quad \text{as} \quad \sum_{k=0}^{\infty} \left[\frac{(h \cdot \frac{d}{dx})^k}{k!}\right] f(x),$$

---

[4]One of these two "Philosophical Journals" was most probably the *Philosophical Magazine*; the other could have been *The Edinburgh New Philosophical Journal* (1826–1864).

[5]For more on French contributions to the calculus of operations, see [Lusternik and Petrova, 1972].

and noticing that
$$\sum_{k=0}^{\infty} \frac{(h \cdot \frac{d}{dx})^k}{k!} = e^{h \cdot \frac{d}{dx}},$$
they reached the interesting result that $f(x+h) = e^{h \cdot \frac{d}{dx}} f(x)$. Since, as noted above, $\Delta f(x) = f(x+h) - f(x)$, they concluded that
$$\Delta f(x) = \left(e^{h \cdot \frac{d}{dx}} - 1\right) f(x),$$
and thus
$$\Delta = e^{h \cdot \frac{d}{dx}} - 1,$$
a result known as both Lagrange's Theorem and as the symbolic form of Taylor's Theorem.[6]

While continental mathematicians admired the calculus of operations as a tool for discovery and verification, they distrusted it as a rigorous technique of proof [Koppelman, 1971, p. 172]. These misgivings were summarized by Cauchy, who, in 1827, had written a series of articles that Koppelman considered as "the highest degree of development of the calculus of operations on the continent during the first half of the nineteenth century."[7] In 1843, Cauchy reflected that, "[n]evertheless, these formulas, deduced in this way from a symbolic equation, should not yet be considered as rigorously established; the method by which they will be discovered is, in reality, only an inductive method."[8] This distrust of the method appears to have stifled further continental research on the calculus of operations soon after Cauchy's culminating work of 1827; excepting a short article in 1853 by Barnabà Tortolini,[9] Koppelman stated that "there seems to have been nothing published on the Continent in this vein until the latter part of the century" [Koppelman, 1971, p. 167].

In spite of the continental distrust of the calculus of operations, the method was embraced and developed by mathematicians in Britain. As they had proven before with their adoption of the Lagrangian, algebraic view of the calculus, nineteenth-century British mathematicians accepted and developed some products of French mathematics that the French eventually eschewed.

In fact, twenty-five years before the foundation of the *Cambridge Mathematical Journal*, a group of Cambridge reformers, who called themselves the Analytical Society, established their own journal for the diffusion of what they saw as a radical new approach to mathematics. This approach was an algebraically based conception of the calculus developed by Lagrange that emphasized series expansions instead of limits or infinitesimals. With this new conception, Analytical Society members saw themselves as leaving behind the fluxional, Newtonian notation that had been tenaciously adhered to throughout the eighteenth century in Britain. At the same

---

[6] I thank an anonymous referee for pointing out this example. For more on this symbolical development of Taylor's Theorem, see [Koppelman, 1971, pp. 158-162], [Lusternik and Petrova, 1972, pp. 202-203], and [Allaire and Bradley, 2002, pp. 411-412]. Changing directions, Gregory proved Taylor's Theorem by assuming the existence of an operation D such that $D^h f(x) = f(x+h)$ in [Gregory, 1838c].

[7] See [Koppelman, 1971, p. 167] and [Cauchy, 1827a, 1827b, 1827c].

[8] "Toutefois, ces formules, ainsi déduite d'une équation symbolique, ne pourrait encore être considérées comme rigoureusement établies, la methode qui les aura découvrir n'étant en réalité qu'une méthode d'une induction" [Cauchy, 1843, p. 377].

[9] Tortolini's article concerned what he called the symbolic form of Taylor's Theorem, that is discussed above [Koppleman, 1971, p. 167].

time, the Society felt that it could connect with a "century of foreign improvement" in the calculus [Babbage and Herschel, 1813, p. xv].

One product of the Society's efforts to promote this new direction in British mathematics was a journal; the *Memoirs of the Analytical Society* appeared in November of 1813. This work contained three papers written anonymously by Herschel and Charles Babbage, who were the main presenters of papers during the Society's meetings [Enros, 1983, pp. 34-36]. These articles featured both the calculus of operations and the solution of functional equations, topics that would later gain popularity in the *Cambridge Mathematical Journal*.[10]

In Cambridge and beyond, the *Memoirs* received little notice. The high cost of publishing them added to the difficulties of sustaining the Analytical Society, the activity of which was preempted much of the time by the demands of the Tripos. Facing the graduation of almost all of its members and with few new recruits, the Analytical Society disintegrated in 1814.[11] After the Society's demise, Babbage, Herschel, and Peacock continued to write reform-minded publications. For example, in 1816, the three published an English translation of Silvestre Lacroix's *Calcul différentiel et intégral* as a way to bring Lacroix's formulation of the differential calculus to Cambridge and also as a platform from which to promote their own allegiance to the Lagrangian development of the calculus. In this as well as in a series of articles by Herschel from 1814 to 1822, the calculus of operations was ever present.[12] For example, in his 1814 *Philosophical Transactions* article, "Consideration of Various Points of Analysis," Herschel professed his belief in the fruitfulness of the method of separating the symbols of operation from those of quantity. He explained that "[t]his method I have, perhaps, extended and carried somewhat farther than has hitherto been customary; but, I trust, without losing sight of its grand and ultimate object, the union of extreme generality with conciseness of expression" [Herschel, 1814, p. 441].[13]

The interests of Herschel, Babbage, and Peacock moved away from the calculus of operations during the 1820s. Herschel turned towards astronomy and chemistry; Babbage gravitated towards astronomical and mathematical instruments; and Peacock began to consider the foundations of algebra. However, these former Analytical Society members began a discussion that would be revived by a new generation of Cambridge students eager to establish a publication venue for mathematical research.

The Revival: The Introduction of the Calculus of Operations into the *Cambridge Mathematical Journal*

As the Analytical Society had done a quarter of a century earlier, Gregory, Smith, and Greatheed launched a journal for the diffusion of mathematical research. Furthermore, the founders of the *Cambridge Mathematical Journal*, like the members of the Analytical Society, used their periodical for the publication of ideas on the calculus of operations. Unlike the *Memoirs*, however, the *Journal* survived well beyond its first volume, and thus became a sustained vehicle for

---

[10]Compare [Koppelman, 1971, pp. 181-183] and [Guicciardini, 1989, p. 138].
[11]See [Enros, 1983, p. 40] and [Enros, 1981, p. 141].
[12]Compare [Koppelman, 1971, pp. 184-187] and [Guicciardini, 1989, p. 182].
[13]For more on this article and Herschel's work on the calculus of operations from 1812 to 1820, see [Grattan-Guinness, 1992].

the transmission of mathematical ideas and, in particular, ideas surrounding the calculus of operations.

The *Journal*'s founders wasted no time in presenting the calculus of operations in their new endeavor; in fact, such an article, Samuel Greatheed's "On General Differentiation," appears just eleven pages into the first number. It is no surprise that Greatheed began the journal's discussion on the calculus of operations; Koppelman cites an 1837 article in the *Philosophical Magazine* by Greatheed as the first British publication on the calculus of operations after that by Herschel [Koppelman, 1971, p. 188]. In his *Journal* article of the same year, Greatheed was concerned with the nonintegral exponents of differentiation. He cited Liouville's 1832 work on the subject in the *Journal de l'École polytechnique*, and he defended this work against an unfavorable analysis made by George Peacock in his "Report on the Recent Progress and Present State of Certain Branches of Analysis" made to the British Association for the Advancement of Science in 1833.[14] Greatheed explained that "[t]he transition from differential coefficients whose indices are positive integers, to those whose indices are any whatever, should be made in the same manner as the transition in algebra, from symbols of quantity with positive integral indices to those with general indices" [Greatheed, 1837, p. 11]. He then gave definitions for differential coefficients with nonintegral exponents that coincided with those of positive integral coefficients [Greatheed, 1837, p. 12]:

$$\frac{d^\alpha(u+v)}{dx^\alpha} = \frac{d^\alpha u}{dx^\alpha} + \frac{d^\alpha v}{dx^\alpha} \cdots,$$

$$\frac{d^\alpha}{dx^\alpha} \cdot \frac{d^\beta}{dx^\beta} u = \frac{d^{\alpha+\beta} u}{dx^{\alpha+\beta}} \cdots,$$

and

$$\frac{d^\alpha}{dx^\alpha} \cdot \frac{d^\beta u}{dx^\beta} = \frac{d^\beta}{dx^\beta} \cdot \frac{d^\alpha u}{dx^\alpha}.$$

From definitions such as these, he proceeded to find a formula for $\frac{d^\alpha x^n}{dx^\alpha}$, where α and n are unrelated.[15]

Greatheed's article certainly caught the interest of the *Journal*'s co-founder, Gregory. Displaying his knowledge of continental mathematical research, Gregory commented on this article in a "Mathematical Note" in the *Journal*. There, he presented what seemed to him a better proof of one of Greatheed's equations involving gamma functions and Eulerian integrals (now called beta integrals), that Gregory had found in the work of "Professor Jacobi of Koenigsberg" [Gregory, 1838a, p. 94].

Gregory showed familiarity with continental work on the calculus of operations in his first of what would become a steady stream of articles on the topic in the *Journal*. In this article, Gregory concerned himself with the solution of linear differential equations with constant coefficients. Taking such an equation,

$$\frac{d^n y}{dx^n} + A\frac{d^{n-1} y}{dx^{n-1}} + B\frac{d^{n-2} y}{dx^{n-2}} + \cdots + R\frac{dy}{dx} + Sy = X,$$

where X is an arbitrary function of x, Gregory separated "the signs of operation from those of quantity," obtaining,

$$\left(\frac{d^n}{dx^n} + A\frac{d^{n-1}}{dx^{n-1}} + B\frac{d^{n-2}}{dx^{n-2}} + \cdots + R\frac{d}{dx} + S\right)y = X.$$

---

[14] See [Liouville, 1832] and [Peacock, 1833].
[15] Compare [Greatheed, 1837, p. 20] and [Koppelman, 1971, p. 211].

Because the expression in parenthesis solely involves constants, and, as Gregory noted, "the signs of operation may be considered as one operation performed on y" [Gregory, 1837b, p. 23], he rewrote the equation as

$$f\left(\frac{d}{dx}\right)y = X$$

Finding y, he reasoned, then reduced to finding "the inverse operation of $f(\frac{d}{dx})$" [Gregory, 1837b, p. 23]. With this inverse, $\{f(\frac{d}{dx})\}^{-1}$ in hand,

$$y = \left\{f\left(\frac{d}{dx}\right)\right\}^{-1} X.$$

Consider, for example, the linear differential equation,[16]

$$\frac{d^2 y}{dx^2} + \frac{dy}{dx} - 6y = 0.$$

Separation of the symbols of operation from quantity gives

$$\left(\frac{d^2}{dx} + \frac{d}{dx} - 6\right) y = 0 \quad \text{or} \quad \left(\frac{d}{dx} - 2\right)\left(\frac{d}{dx} + 3\right) y = 0.$$

Considering the simpler equations

$$\left(\frac{d}{dx} - 2\right) y_1 = 0 \quad \text{and} \quad \left(\frac{d}{dx} + 3\right) y_2 = 0,$$

yields the solutions $y_1 = C_1 e^{2x}$ and $y_2 = C_2 e^{-3x}$, respectively. From this, we can conclude that the general solution to the original equation consists of linear combinations of these equations' particular solutions, that is, $y = C_1 e^{2x} + C_2 e^{-3x}$. Wishing to generalize his method of solution further, Gregory lamented that its application to linear differential equations with variable coefficients was "attended with considerable difficulty, and indeed neither Brisson nor Cauchy seem to have made any progress" [Gregory, 1837b, p. 30].

In this article, Gregory promoted the calculus of operations by virtually providing a "primer" on the subject, a tactic that he would use repeatedly in his articles in the *Journal*. He also spread the gospel of the calculus of operations by arguing for the method's legitimacy. It was in these discussions of legitimacy that Gregory was confronted with and generalized the algebraic properties involved in his method.

Gregory contended that with an argument armed not with mere analogy, but with reasoning that "is perfectly strict and logical[,] ... we might with propriety call ... [symbols of quantity] also symbols of operation" [Gregory, 1837b, p. 30]. For example, x^n could be viewed as an operation x iterated n times on unity. In this light, Gregory wrote, "we shall have little difficulty in seeing the correctness of the principle by which other operations, such as we represent by $(\frac{d}{dx})$, (Δ), &c., are treated in the same way as a, b, &c. For whatever is proved of the latter symbols, from the known laws of combination, must be equally true of all other symbols which are subject to the same laws of combination" [Gregory, 1837b, p. 30]. Gregory gave the example of the logarithm function as an operation that does not preserve addition; the binomial theorem, a result from algebra, thus does not hold for it.

---

[16]Again, I thank an anonymous referee for pointing out this demonstration. For another application of Gregory's techniques to a similar linear differential equation, see [Allaire and Bradley, 2002, p. 415].

The binomial theorem, does, he wrote, hold for differentiation, an operation that preserves addition and obeys other general laws. The fact that linear differential equations with variable coefficients disobeyed the laws of combination, Gregory continued, was precisely why they were so difficult. The operation x then applied to the derivative of z with respect to x is not the same as the derivative with respect to x of x applied to z; that is, "the second law of combination [commutativity] does not hold with regard to these symbols of operation" [Gregory, 1837b, pp. 31-32].

Gregory signed his first article on the calculus of operations under the pseudonym Γ. The practice of using pseudonyms to sign scientific journal articles had been common during the eighteenth century and was carried on into the early nineteenth century. Anonymity allowed a contributor to avoid criticism, and, in the case of the *Cambridge Mathematical Journal*, it hid the fact that the co-founders wrote the majority of the articles in first volume [Crilly, 2004, p. 466].

Throughout the *Journal*'s volumes, under his own name and under pseudonyms, Gregory continued his efforts to endorse the calculus of operations. His promotional tactics took several forms, one of which was an appeal for the method's applicability. In two articles in the first volume, Gregory applied the calculus of operations to questions about the motion of heat. In the first of these, he pointed out that Fourier had used the calculus of operations for partial differential equations in his 1822 *Traité de la chaleur*, but that "he appears to have had some unwillingness to give himself up to it entirely as a guide in his investigations, as if he were not familiar with the principles on which it is founded" [Gregory, 1838b, p. 123]. Furthermore, he noted that "[o]ther French writers seem to have avoided carefully entering at all on the track which Fourier opened" [Gregory, 1838b, pp. 123-124]. Mr. Greatheed had not avoided this path, nor, Gregory reported, had he. Gregory then went on to give examples of the use of the calculus of operations for the solution of partial differential equations, and in particular, those found in the study of the motion of heat in a ring. He again lamented the difficulty involved in solving differential equations with variable coefficients [Gregory, 1838b, pp. 128-129].

In a second article in the first volume, "On the Integration of Simultaneous Differential Equations," Gregory used the calculus of operations for solving systems of differential equations. He explained that "[w]e have ... only to separate the symbol of differentiation from its subject, and then proceed to eliminate one of the variables between the given equations, exactly as if the symbol of differentiation were an ordinary coefficient. Thus the difficulty of elimination becomes reduced to that between ordinary and algebraical equations" [Gregory, 1838d, pp. 173-174]. After giving his readers the rudiments of this application, he used it to solve a system of equations given by the Astronomer Royal, George Airy, for a problem in mathematical astronomy [Gregory, 1838d, p. 177]. Gregory's confidence in the method and his zealous desire to promote it shines through his closing remarks to the article: "[w]e have now given a sufficient number of examples to enable the student to understand thoroughly the method, and we think that they show clearly the advantages of a process, which, to some persons, might appear to carry out to a startling extent the principles on which it is founded" [Gregory, 1838d, p. 181].

A further application of the use of the calculus of operations in solving systems of differential equations appeared in two articles in the second volume of the *Journal*, "On the Sympathy of Pendulums" [Gregory and Smith, 1840] and "On the Motion of a Pendulum When Its Point of Suspension Is Disturbed" [Smith,

1840]. Interestingly, these papers are signed "D.G.S" and "G.S.," respectively. In the index to the second edition of the first volume, which identifies many of the pseudonyms, "D.G.S." is listed as Duncan Gregory and Archibald Smith, making this first article a very early British example of a jointly written, published mathematical article. "G.S.," the author of the second article, is identified as Smith only, although the "G." in the pseudonym suggests that Gregory may have influenced it at least somewhat.

While actively employing the calculus of operations in the solution of applied mathematical questions, Gregory was, at the same time, arguing that the method was a useful theoretical tool for pure mathematics. In an article "On the Impossible Logarithms of Quantities" that appeared in the February 1839 number of the *Journal*, Gregory outlined the paper, "On the Real Nature of Symbolical Algebra" [Gregory 1840a], that he had presented to the Royal Society of Edinburgh on 7 May, 1838, the only article in his collected works that did not first appear in the *Cambridge Mathematical Journal*.[17]

The true algebra, Gregory contended, "is the science of symbols, defined not by their nature, but by the laws of combination to which they are subject" [Gregory, 1839b, p. 226]. Specifically, he discussed how this viewpoint could clear up "several disputed points in Algebra" such as uncertainties about the logarithms of negative numbers [Gregory, 1839b, p. 226].[18] To this end, Gregory carefully laid down the laws of combination governing the operation of taking the logarithm as well as those governing the operation of addition. The operation denoted by the + symbol, Gregory wrote, must not be considered "merely as an affectation of other symbols, which we call symbols of quantity, but as a distinct operation possessing certain properties peculiar to itself, and subject, like the more ordinary symbols, to be acted on by any other operations." An understanding of laws of this operation, in turn, clarified "certain analytical anomalies, which do not at first sight appear to be connected" [Gregory, 1839b, p. 226]. For example, "[s]ince by the rule of signs $- \times -$ gives $+$, we ought to have $\sqrt{-a} \times \sqrt{-a} = \sqrt{+a^2} = \pm a$; whereas we know that it must be only $-a$." The flaw, Gregory maintained, was the "consequence of sometimes tacitly assuming the existence of $+$, and at another time neglecting it." Thus, if we do not neglect $+$ in the above equation, we obtain "$\sqrt{-a} \times \sqrt{-a} = \sqrt{+a^2} = \sqrt{+}\sqrt{a^2} = -a$; for $\ldots - \times - = +$ or $-^2 = +$, which of course gives us $+^{\frac{1}{2}} = -$" [Gregory, 1839b, p. 228].

Gregory wrote that while Peacock had used the symbol 1 to be "the recipient of the affectations of a^m," he had instead chosen to use the symbol $+$. Although the difference between the two symbols was little in arithmetic, "in the general Theory of Algebra there is a considerable difference; for 1 being an arithmetical symbol necessarily recals [*sic*] arithmetical notations; and as the circumstances in which its peculiar nature is evolved occur in general symbolical Algebra, and may be wholly independent of arithmetic, it is of importance to avoid the confusion which must

---

[17] For interesting discussions of both this *Journal* article and Gregory's Royal Society of Edinburgh article, see [Allaire, pp. 88-100 and 101-104]. For more about the early nineteenth-century debate in Britain on negative numbers and their relationship to algebra, see [Pycior, 1982], [Pycior, 1984], and [Richards, 1980].

[18] For more on the uncertainties surrounding negative numbers considered by early nineteenth-century British mathematicians, see [Rice, 2001].

be caused by the introduction into general symbolical Algebra of symbols limited in their signification" [Gregory, 1839b, p. 229].

Peacock's views on this issue appeared in his 1830 *Treatise on Algebra* and in his 1833 "Report on the Recent Progress and Present State of Certain Branches of Analysis," which represented two of the first steps in the British movement to solidify the foundations of algebra [Richards, 1980, p. 346]. Peacock's idea of algebra was related to arithmetic through a process that was not "an ascent from particulars to generals ... but one which is essentially arbitrary." It was, however, "restricted with a specific view to its operations and their results admitting of such interpretations as may make its applications most generally useful" [Peacock, 1833, p. 194]. Central to this process was the principle of the permanence of equivalent forms, which required an agreement at any intersection of arithmetic and symbolic algebra. Gregory's disagreement of Peacock's use of the symbol 1 is indicative of his more general divergence from this principle [Allaire and Bradley, p. 404]. Gregory, as well as other *Journal* contributors, sometimes differed with Peacock's foundations of algebra; however, Peacock's work helped turn their focus towards operations and provided a provocative foil for their findings.

In a later *Journal* article, "On a Difficulty in the Theory of Algebra," Gregory returned to the subject of $+$, writing that confusion surrounding the symbol results from "writing a *sum* ... in a manner different from that in which we express the performance of any other operation." Instead of prefixing the "symbol of operation to the subject," such as $f(a)$, we write the operation of the addition of x after its subject a, such as $a + x$. Gregory pointed out that this reversal of order, while confusing in this case, is totally acceptable, and "indeed Mr. Murphy prefixes the subject to the symbol of operation, apparently for the purpose of avoiding the prejudices which our ordinary mode of writing is apt to produce" [Gregory, 1842, pp. 154-155]. Robert Murphy, third wrangler in 1829, had employed this right-handed operational notation in his 1837 *Philosophical Transactions* publication, "First Memoir on the Theory of Analytic Operations" [Murphy, 1837]. This notation clarified the ordering when multiple operations were applied to a subject [Allaire and Bradley, 2002, p. 423].[19]

While his "First Memoir" was influential to Gregory, Murphy never published in the *Journal*.[20] Although Murphy clearly belonged to the network of Cambridge mathematicians, he was seven years Gregory's senior and perhaps felt more connected to the Cambridge Philosophical Society than to the young contributors to the *Journal*. In fact, of the twenty-four *Cambridge Mathematical Journal* authors for which birthdates were available, all but eight were younger than thirty when they made their first contributions.[21]

---

[19]For more about Murphy's "Memoir," noted by Koppelman for its for "extreme degree of generality and abstraction," see [Koppelman, 1971, pp. 195-197] and [Allaire and Bradley, 2002, pp. 422-423].

[20]His papers were limited to the *Transactions of the Cambridge Philosophical Society* (7), the *Philosophical Transactions* (2), and the *Philosophical Magazine* (10), plus two books [Cannell, 2001, pp. 273-274].

[21]Birthdate information was unavailable for two of the identified contributors to the *Cambridge Mathematical Journal*. Since each of the *Journal*'s volumes ran over a two-year period (volume one ran from 1837 to 1839, and volume two ran from 1839 to 1841, for example), the middle date of this period was used to date an author at his first contribution (James Cockle, for example, was born in 1819 and made his first contribution to volume two, so his age is given as twenty-one. The average age for a first contribution to the *Cambridge Mathematical Journal*

While an operational point of view could clear up difficulties in algebra, Gregory argued that its generality could provide a single foundation of proof for theorems in different mathematical fields. In an article in the first volume of the *Journal*, Gregory argued that "the more important of the theorems in the Differentiable Calculus and the Calculus of Finite Differences ... can be proved by the method of the separation of symbols," that is, the calculus of operations. Gregory noted as an example that the binomial theorem giving the series expansion of $(a+b)^n$ is valid for all operations a, b, and n that obey the commutative, distributive, and index laws [Gregory, 1839a, p. 212].

Gregory closed his article by acknowledging that he had earlier overlooked Herschel's work on the application of the calculus of operations to differential equations, and that he had erroneously given priority to Brisson. This correction afforded him yet another opportunity to promote the method. He wrote that

> [i]t is much to be regretted, that neither Sir John Herschel himself, nor any other person, followed up this method, which is calculated to be of so much use in the higher analysis. Perhaps this may have arisen from the theory of the method not having been properly laid down, so that a certain degree of doubt existed as to the correctness of the principle. I trust, however, that the various developments which I have given in several articles in this Journal, of the principles of the method as well as the proofs of its utility, are sufficient for removing all doubts on this head, and that it will now be regarded as a powerful instrument in the hands of mathematicians [Gregory, 1839a, p. 222].

The fact that Gregory publicized the calculus of operations by giving "proofs of its utility" in both pure and applied mathematics strongly reflected the Cambridge mathematical environment at that time. While Cambridge, during the first third of the nineteenth century, witnessed the infiltration of Lagrangian methods of analysis into its mathematics curriculum, it nonetheless remained a bastion of applied or "mixed" mathematics. Both the Mathematical Tripos and the Smith's Prize Examination encouraged successive generations of Cambridge students to study mixed mathematics. Gregory's pursuit and promotion of a mathematical program that emphasized generality and abstraction was done within an environment highly attuned to the utility of mathematics. Moreover, the appeal of the utility of the calculus of operations, perhaps more than any argument for its legitimacy, encouraged the readers of the *Cambridge Mathematical Journal* to implement the method in their own research.

The Conversation Begins: The Adoption of the Calculus of Operations by Contributors to the *Cambridge Mathematical Journal*

As the above discussion has shown, the founders of the *Journal*, and especially Gregory, vigorously promoted the calculus of operations by teaching the method to the *Journal*'s readers, attempting to legitimize it, and demonstrating its utility. What came of these efforts? As the next few examples demonstrate, junior mathematicians from Cambridge and beyond soon joined in the discussion in the *Journal* surrounding the calculus of operations.

---

was 29.4 (without its oldest contributor, Gregory's former teacher, William Wallace, this average drops to 27.7).

One such mathematician was Alexander Craufurd, the fifteenth wrangler in Gregory's Tripos year of 1837. Craufurd's article, "On a Method of Algebraic Elimination," appeared at the end of the *Journal*'s second volume and gave the rules for his "much more simple and satisfactory" method of elimination for systems of algebraic as well as differential equations "of all orders and degrees" [Craufurd, 1841, p. 276].

Craufurd believed that his method placed the problem of elimination "in a simple and luminous point of view" in which "different cases treated are referred to one common principle" [Craufurd, 1841, p. 281]. In his postscript, Craufurd wrote that the motivation for his method came from reading the article Gregory had published in the first volume on the solution of simultaneous differential equations. Appraising the method of elimination used there as more "natural and properly applicable" than the usual method of elimination for algebraic equations, Craufurd explained that "[t]he question was thus raised, whether a method similar to that used in treating differential equations might not be discovered for algebraic. This question being once asked, the answer to it was soon found; especially as the paper which had suggested these reflections pointed out an analogy between the processes of Differentiation and Multiplication" [Craufurd, 1841, p. 281]. One of the fundamental tenets in Gregory's use of the calculus of operations, the transferability of mathematical techniques to different sets of symbols that obey the same laws of combination, had struck a chord with a reader of the *Journal*.[22]

The similarities between algebraic and differential equations was also discussed in the *Journal* by Robert Leslie Ellis. The senior wrangler for 1840 and fellow of Trinity College, Ellis was second only to Gregory in the number of contributions made to the *Journal*.[23] To the *Journal*'s third volume, Ellis submitted some "Remarks on the Distinction Between Algebraical and Functional Equations," in which he examined the relationship between algebraic and functional equations in what he said was "a more general manner" than earlier such examinations. He wrote that algebra, or "[t]he science of symbols is conversant with operations, and not with quantities; and an equation, of whatever species, may be defined to be a congeries of operations, known and unknown, equated to the symbol zero." Echoing the argument in Gregory's first *Journal* article, Ellis argued, "[i]f one symbol is said to be a function of another, it is, in reality, the result of an operation performed upon it. Thus the idea of functional dependence pervades the whole science of symbols" [Ellis, 1842a, p. 92].

With these principles in mind, Ellis outlined a classification system for equations based on the nature of their operations and the order in which they appear. For example, given the following equations:

(1) $$x^2 + ax + b = 0,$$

---

[22]This postscript demonstrates that the *Journal* was reaching readers and inspiring them to enter into mathematical discussions in print. We can thank James Joseph Sylvester for this postscript; Craufurd authored it to acknowledge that Sylvester had already anticipated his results. In fact, the issue for February 1841, the same month that Craufurd had originally submitted the article, contained a paper by Sylvester in which he, in Craufurd's words, "not only anticipated me in the fundamental idea, but ... likewise devised some very ingenious rules for the more expeditious, and even merely mechanical, application of it" [Craufurd, 1841, p. 281].

[23]Ellis made twenty-two contributions, while Gregory made twenty-six.

$$(2) \qquad \frac{dy}{dx} - x = 0,$$

$$(3) \qquad \phi(mx) + x = 0,$$

$$(4) \qquad \phi\phi x - x = 0 \quad \text{and,}$$

$$(5) \qquad \phi \frac{dy}{dx} \phi x + x = 0,$$

Ellis explained that the algebraic equations (1) and (2) are "discriminated by the nature of the operations," but for both equations, the unknown operations are performed immediately on the bases [Ellis, 1842a, p. 93]. However, this is not the case in (3), where the unknown operation ϕ is performed on "the result of a previous operation" [Ellis, 1842a, p. 93]. This previous operation could also be unknown, as it is in (4). Moreover, the operation $\frac{d}{dx}$ could intervene between unknown operations as in (5). In (3), (4), and (5), the cohesive principle is the fact that the unknown operation is not performed directly on the base. Thus, Ellis called (3), (4), and (5) functional equations, ordinary of the first order, ordinary of the second order, and differential, respectively. Functional notation, Ellis noted, does not make an equation functional; the algebraic equations (1) and (2) rewritten in the following functional garb, respectively,

$$\{\phi(ab)\}^2 + a\phi(ab) + b = 0,$$

$$\frac{d}{dx}\phi(x) - x = 0$$

are still algebraic [Ellis, 1842a, pp. 92-93]. "The name of functional equation is not happy," wrote Ellis, for "it refers to the notation, and not to the essence of the thing" [Ellis, 1842a, p. 94].

Ellis evaluated his article as "the outline of a natural arrangement of the science of symbols" [Ellis, 1942a, p. 94]. As with Gregory, utility was a point that Ellis felt he needed to address. He wrote that "[i]t is not difficult to overrate the importance of a mere classification; but I hope to be able to show, that the considerations now suggested are not without some degree of utility" [Ellis, 1842a, p. 94]. One useful consequence of this classification, in his eyes, was the connection it made between the solutions of different types of equations: "[a]s the distinction between functional and common equations depends on the order of operations, it follows that, when part of the solution of an equation does not vary with the nature of the operation subjected to the resolving process, this part is applicable as much to functional equations as to any other" [Ellis, 1842a, p. 94]. He illustrated this point in a subsequent *Journal* article wherein he solved a class of differential functional equations by appealing to differential algebraic equations whose solutions are independent of any knowledge of the function in question [Ellis, 1842b].

Gregory applauded Ellis's work, considering it as "not only correct, but of very great importance for the proper understanding of the higher departments of analysis" [Gregory, 1843, p. 239]. Gregory's difficult linear differential equations with variable coefficients were, in Ellis's classification, types of functional equations, while the more manageable linear differential equations with constant coefficients were algebraic. With this new classification, Gregory was able to examine his difficulties in a new light. With Ellis's theory, Gregory wrote, "it is easy to see *why* the solution of functional equations must involve difficulties of a higher order than equations of the other class." The "whole difficulty" in solving those of the other class, for example, linear differential equations with constant coefficients, "lies in the performing of the inverse operation" [Gregory, 1843, p. 240 (his emphasis)].

Recall from above that in his first article to the *Journal* on the calculus of operations, [Gregory, 1837b], Gregory discussed solving equations of this form. For example,

$$\left(\frac{d^n y}{dx^n} + A\frac{d^{n-1}y}{dx^{n-1}} + B\frac{d^{n-2}y}{dx^{n-2}} + \cdots + R\frac{dy}{dx} + S\right) = X,$$

where X is any function of x, involved separating the symbols and factoring the equation as

$$\left(\frac{d}{dx} - a_1\right)\left(\frac{d}{dx} - a_2\right)\cdots\left(\frac{d}{dx} - a_n\right)y = X,$$

then solving n simpler equations of the form

$$\left(\frac{d}{dx} - a_i\right)y = X, \quad i = 1, 2, \ldots n,$$

and summing linear combinations of these solutions. Gregory justified his solution to the first-order equations above by considering $(\frac{d}{dx} - a)^n X$.[24] By the binomial theorem,

(6) $$\left(\frac{d}{dx} - a\right)^n X = \left\{\frac{d^n}{dx^n} - na\frac{d^{n-1}}{dx^{n-1}} + \frac{n(n-1)}{1\cdot 2}a^2\frac{d^{n-2}}{dx^{n-2}} - \cdots\right\}X.$$

Since $\left(\frac{d}{dx}\right)^p(e^{-ax}) = (-a)^p e^{-ax}$, and thus $(-a)^p = e^{ax}\left(\frac{d}{dx}\right)^p e^{-ax}$, we may rewrite the righthand side of (6) as

$$e^{ax}\left\{\frac{d^n}{dx^n} + \frac{d^{n-1}}{dx^{n-1}}\cdot\frac{d'}{dx} + \frac{n(n-1)}{1\cdot 2}\frac{d^{n-2}}{dx^{n-2}}\cdot\frac{d'^2}{dx} + \cdots\right\}e^{-ax}X,$$

where $\frac{d}{dx}$ operates on X and $\frac{d'}{dx}$ operates on e^{-ax}. By the generalized product rule (called by Gregory the Theorem of Leibniz), we can conclude that

$$\left(\frac{d}{dx} - a\right)^n X = e^{ax}\left(\frac{d}{dx}\right)^n (e^{-ax}X)$$

[Gregory, 1837b, pp. 25-26]. When $n = -1$, $\left(\frac{d}{dx}\right)^n = \int dx$, and

$$\left(\frac{d}{dx} - a\right)^{-1} X = e^{ax}\int e^{-ax}X\,dx.$$

Thus, Gregory's inverse operation yields the solution of the first-order equation $(\frac{d}{dx} - 2)y = 0$, namely,

$$y = \left(\frac{d}{dx} - 2\right)^{-1}(0) = e^{2x}\int e^{-2x}\cdot(0)dx = Ce^{2x}.$$

---

[24]Gregory actually considered $(\frac{d}{dx} \pm a)^n X$, but for clarity, I restrict simply to $(\frac{d}{dx} - a)^n X$. I thank an anonymous referee for suggesting this illustration.

While finding inverse operations for Ellis's class of algebraic equations presented a "difficulty ... [that] may be very great," explained Gregory, "it is a difficulty wholly different in *kind* from that which we meet in trying to solve an equation in which the unknown operation is performed on that which is known [i.e., a functional equation]" [Gregory, 1843, p. 240 (his emphasis)]. Gregory gave as an example the equation $\phi(ax) - \phi(x) = 0$, "where the object is to determine the form of ϕ." It is invalid, Gregory pointed out, to rewrite the equation as $\phi(ax - x) = 0$, "since the form of ϕ is unknown, and we therefore cannot assume it to be subject to the distributive law; neither can we we write $a\phi(x) - \phi(x) = 0$, since we cannot assume that ϕ and a are commutative operations" [Gregory, 1843, p. 240].

Gregory and Ellis disseminated their mathematical ideas not only through their articles in the *Journal*, but as Cambridge moderators. The design and management of the Mathematical Tripos came from within the colleges of Cambridge, with junior fellows being appointed as examiners and moderators for short, rotational periods [Becher, 1980a, p. 38]. Moderators set the problem papers, and examiners helped give and grade the examination. For example, this system allowed Peacock, Tripos moderator in 1817, 1819, and 1821, to introduce questions on Lagrange's algebraic formulation of the differential calculus that had been advocated by his fellow members of the Analytical Society. While young Cambridge innovators saw this system as a means to introduce important reforms, William Whewell, Cambridge professor and later Master of Trinity College, criticized it as enabling the moderators to "constantly write new books, juvenile, hasty, worthless, which take their places in the examinations and exclude all steady standard works."[25] Gregory served as moderator in 1841 and then again in 1845, while Ellis served in these roles in 1844 and 1845, respectively.

In a letter to Gregory after the 1841 Tripos, Ellis indicated the power the moderators had over the mathematical directions taken by Cambridge students. He wrote that "I can imagine the Johnians read the journal diligently; it must have done them good. But I was pleased to see how fairly in your problem paper you kept clear of the parts of analysis to which your own taste inclined you the most, especially as I had heard it surmised that you would not be able to avoid giving them an undue preponderance" [TCCM1].

Besides the Tripos examination, the Smith's Prize examination at Cambridge also encouraged interest in the calculus of operations and the *Journal*.[26] Arthur Cayley, First Smith's Prizeman for 1842, almost certainly encountered the calculus of operations in his studies leading up to the examination. In fact, one problem on his examination involved the symbolic formulation of Lagrange's Theorem that $\Delta u = e^{h\frac{du}{dx}} - 1$. This question, set by William Whewell, asked if "the consequences of the separation of symbols of operation from those of quantity, [are] universally true? or within what limits?"[27] One year later, in a *Journal* article, Cayley returned to the subject of this formula and presented Lagrange's Theorem "in rather a remarkable symbolical form" [Cayley, 1843, p. 283]. Cayley's initial exposure to the calculus of operations left a lasting impression; according to Cayley biographer,

---

[25]For the quote, see [Becher, 1980a, p. 38].

[26]Created by the bequest of Trinity College, Cambridge Master Robert Smith in 1798, these two prizes were to be awarded annually to bachelors of arts for ability in mathematics and natural philosophy. For more on these prizes, see [Barrow-Green, 1999].

[27]As quoted in [Crilly, 2006, p. 52].

Tony Crilly, "[t]hroughout Cayley's career, the phrase ['separation of symbols'] reoccurs" [Crilly, 2006, p. 69].

Cambridge students were not the only ones interested in the *Journal*. George Boole, for example, was altogether outside the university sphere, Cambridge or otherwise. The *Cambridge Mathematical Journal* was admittedly primarily a venue for Cambridge mathematicians; in fact, twenty-one of the twenty-six identified contributors to the *Journal* had a Cambridge affiliation.[28] However, it also represented a means for the outsider Boole to get his work noticed by other mathematicians. Though mathematically gifted, Boole had not attended university, and had taught from the age of sixteen at various institutions in or near Lincoln. After contacting Gregory in 1839, he began publishing in the *Journal* and thereby introduced himself to Cambridge mathematicians [Smith, 1982, p. 7]. Soon enough, Boole, too, was conducting research that involved the calculus of operations.

Specifically, Boole attacked the problem of the solution of linear differential equations with constant coefficients, the same problem that Gregory had discussed in 1837 in his first *Journal* article on the calculus of operations. In an 1840 letter to Boole, Gregory praised this research, calling it "exceedingly ingenious," and capable of reducing "the problem to the greatest degree of simplicity of which it admits" [Gregory, 1840b]. Boole had tried to get this work published in the *Philosophical Magazine* to no avail. Giving a telling opinion on the existing publication environment for junior mathematicians, Gregory responded that "I do not think that the non-insertion of your paper in the Phil. Mag. was due to any other cause than this: that the editor is ignorant of mathematics, and is very unwilling to risk the publication of any mathematical communication, unless a previous knowledge of the author gives him some security for the correctness of the paper" [Gregory, 1840b].

Although the *Journal* was not Boole's first choice for publication, and Boole's result was a great improvement on his own, Gregory was both magnanimous and helpful, writing that "I shall be very happy to get your article inserted in the journal, but I have some doubts whether the paper, as you have sent it to me, is in the best form .... If it be agreeable to you I will draw up the paper in the way which I think is best fitted for publication, and will transmit [it] to you for your inspection" [Gregory, 1840b]. Boole gladly accepted Gregory's offer to rewrite the paper, which appeared in the second volume of the *Journal* [Boole, 1840]. Boole later described this paper as being actually written by Gregory, "from notes furnished by the author of this work, whose name the memoir bears. The illustrations were supplied by Mr. Gregory. In mentioning these circumstances the author recalls to memory a brief but valued friendship."[29]

The problem of differential equations with variable coefficients, the difficulty of which Gregory had lamented in 1837, had not apparently been far from his mind three years later as he was writing to Boole. He explained that "I shall be glad to hear that you have made progress in the solution of equations, with variable co-efficients. The question is a very difficult one, and of the highest importance, as it is in that direction that we must look for some extension of our means of

---

[28]This group provided at least two-thirds of the contributions. Even after William Thomson published an index of the *Journal* which identified many pseudonyms, sixty-three of the 278 articles remained anonymous.

[29]See [Boole, 1859, p. 381], [Harley, 1866, p. 149], or [Boole, 1952, p. 435].

analysis" [Gregory, 1840b]. Boole soon successfully tackled this problem, and again more experienced mathematicians helped him bring his solution to print. Unsure of where to publish this memoir, Boole asked the advice of Gregory and Augustus De Morgan. Their responses give further contemporary appraisals of the publication sphere for research-level mathematics in the early 1840s.

Gregory responded that "I have been prevented from answering your letter by a severe attack of illness, from which I have not yet recovered. My advice certainly is, that you should endeavour to get your paper printed by the Royal Society, both because you will thereby avoid a considerable expense, and, because a paper in the 'Philosophical Transactions' is more likely to be known and read than one printed separately."[30] He then explained that Boole would have to get a Society member to communicate the paper and offered to get Airy to do so, if needed.

De Morgan similarly advised that

> [w]ith regard to the manner of printing: I see no channel in this country except the Phil. Trans. the Cambridge Phil. Trans. or the Cambr. Journal. It is probably too long for the third & I am afraid Gregory is in no state to attend to or decide upon it. Whether the R[oyal] S[ociety] would print it or not is a question. I think they ought to do so, but in sending it to them there is the nuisance of keeping a copy ... as they are very dog-in-the-mangerish about what they call their archives and will not return a paper even when they do not print it. The Cambr. Soc. labour under want of funds and would look suspiciously I suspect, upon anything long. I think if you do not mind copying it out you should try the R. S. in the first instance. The Phil[osophical] Mag[azine] I have no doubt would print a summary but it would be decidedly too long for that periodical [De Morgan, 1843].[31]

Boole took the advice of De Morgan and Gregory and submitted his paper, "On a General Method of Analysis," to the Royal Society [Boole, 1844]. Not only did the Society print the memoir, but on its merits they also awarded Boole a Royal Medal in 1844 [Smith, 1982 p. 2]. Koppelman has appraised the work as Boole's "most important work in the calculus of operations" [Koppelman, 1971, p. 197]. The breakthrough in Boole's memoir was similar to the key to Sir William Rowan Hamilton's invention of 16 October, 1843: a loosening of the requirement of commutativity. Instead of applying non-commutativity to quaternions, Boole applied it to operations [Koppelman, 1971, pp. 197-198]. Boole emphasized in the paper that "[t]he position which I am most anxious to establish is, that any great advances in the higher analysis must be sought for by an increased attention to the laws of the combination of symbols. The value of this principle can scarcely be overrated."[32]

George Boole serves as an example of a mathematician totally outside the university sphere who both made his publication début and gained some of his motivation in research-level mathematics from the *Cambridge Mathematical Journal*.

---

[30]See Gregory to Boole, 19 June 1843, as quoted in [Harley, 1866, p. 155] or [Boole, 1952, p 443].

[31]For an example of the Royal Society rejecting one of De Morgan's papers and their general printing policies, see [Rice, 1996], especially pp. 211-217.

[32]See [Boole, 1844, p. 282] or [Koppelman, 1971, p. 199].

The discussion in which he participated in the *Journal* exposed him to a pressing research problem among young British mathematicians and pushed him towards ideas concerning general laws of combination, ideas that he would later apply to his symbolic logic.

As De Morgan and Gregory mentioned in their letters to Boole, Gregory, by 1843, was in the midst of a recurring illness. Gregory was first attacked by this sickness late in 1842, and by the following spring he had left Cambridge, never to return. The November 1843 number of the *Journal* was the last issue he edited; Gregory finally succumbed to his illness in 1844 at age of thirty [Ellis, 1844, p. 151]. In Gregory's absence, the direction of the *Cambridge Mathematical Journal* was assumed by Ellis.[33] After editing the *Journal*'s fourth volume, Ellis sought someone to whom he could entrust his editorial duties. William Thomson, who would later become Lord Kelvin, agreed to Ellis's proposal and began a new series of the *Journal*, which he renamed the *Cambridge and Dublin Mathematical Journal*.[34]

Although Gregory had played such a vital role in its encouragement, British research on the calculus of operations did not die with him. Brice Bronwin, Arthur Curtis, Francis Newman, William Donkin, W. H. L. Russell, William Spottiswoode, Robert Carmichael, and William Center all published papers involving the calculus of operations in the new series of the *Journal*.[35] While these mathematicians are certainly not all well-known, or even known, names in the history of mathematics, their activity indicates that the discussion so dependent on Gregory in the first volume of the *Journal* had, by the time of his death, strengthened enough to continue without him.

Conclusion

In their discussion of the generation of Cambridge mathematicians coming of age during the tenure of the *Cambridge Mathematical Journal*, Crosbie Smith and Norton Wise describe this

> generation's programme for obtaining mathematical power and efficiency. Through their emphasis on generality they would release the power of symbolic representations from the cumbrous constrictions of particular interpretations. By demanding simplicity along with generality they would make the essential features of derivations stand out, revealing the forms common to problems normally considered different. Finally, simplicity made generality useful in solving particular problems, so that utility ever attended the other two emphases [Smith and Wise, 1989, p. 180].

Possibly because it was a method that scored high in generality, simplicity, and utility, the calculus of operations enjoyed a remarkably positive reception and proliferation by the junior mathematicians publishing in the *Cambridge Mathematical Journal*, even while it was viewed suspiciously on the continent. Clearly, Gregory's efforts to teach, legitimize, and show the usefulness of the method—from a platform

---

[33] William Walton, the eighth wrangler for 1836, initially helped Ellis with this task, and the two men jointly edited the number for February 1844 [Crilly, 2004, p. 468].

[34] For more on this and a third reincarnation of the *Journal*, see [Crilly, 2004].

[35] For more on these papers, see [Koppelman, 1971, pp. 201, 203, 205-206, and 212-214].

for the publication of mathematical research done by junior researchers—helped revive the discussion on the calculus of operations in Britain.

Did the discussion about the calculus of operations among the young contributors to the *Journal* drown out the call to study other fertile areas of study? Perhaps this question should be extended to all those British mathematicians during the middle third of the nineteenth century who were beguiled by the calculus of operations. Indeed, the discussion of the calculus of operations did not begin with the *Journal*, nor was it limited to it. During the late 1830s and the 1840s, articles on the calculus of operations appeared, for example, in the *Philosophical Magazine*, the *Philosophical Transactions* of the Royal Society of London, the *Transactions* of the Royal Society of Edinburgh and the *Proceedings* of the Royal Irish Academy. Niccolò Guicciardini has argued that the popularity of the calculus of operations and, in general, the Lagrangian, algebraically based formulation of the calculus, while sparking "important contributions to algebra and logic, ... had its own drawback: it did not allow many British mathematicians influenced by the Analytical Society to appreciate the importance of Cauchy's rigorization of the calculus, which was motivated by the desire to avoid the 'generalities of algebra.' The shift from the fluxional calculus to the Lagrangian calculus, which marks the definitive death of the Newtonian tradition, once again left the British isolated" [Guicciardini, 1989, p. 138]. Thus, while many of the contributors to the *Journal* were blinkered by their interest in the calculus of operations, they were in good company.

In the thick of a general infection of *mania analytica*[36] and surrounded by a publication environment that often proved daunting to inexperienced researchers, the *Journal* provided a constant and welcoming forum for work by young mathematicians on the calculus of operations. Sir William Rowan Hamilton noted the open-minded approach to symbols that the *Journal* contributors shared in an 1846 draft letter to Peacock. Hamilton admitted that "[m]y views respecting the nature, extent, and importance of symbolical science may have approximated gradually" to Peacock's, but he wondered if Peacock would "still refuse, perhaps, and with justice, to recognize me as belonging to your school." Hamilton, however, had no illusions about membership to another school, writing that "[a]t least I am sure that the school which produced, with such admitted ability, many articles on Symbolical Algebra, or on subjects connected therewith, in the Cambridge Mathematical Journal, would not concede to me the honour of their fellowship. For I still look more and more habitually beyond the symbols than they would choose to do. To use De Morgan's image, I turn up the fronts of the dissected map, and try to learn what countries they denote, as well as how they fit together" [Hamilton, 1846].[37]

While Hamilton had no desire to agree with the approach of these young mathematicians, his letter to Peacock shows a recognition of them as a cohesive school. Brought together through the pages of the *Journal* and immersed in the discussion fostered by its editor, junior contributors began to leave the map pieces face down and turn their thoughts to what would become known as the British approach to algebra.[38]

---

[36]Babbage coined the phrase *mania analytica* in 1817 [Becher, 1980b, p. 398].

[37]For De Morgan's complete quote, see [Richards, 1980, p. 356].

[38]See the third issue of volume 31 of *Historia Mathematica* for discussions of schools from a variety of perspectives.

References

Archival Sources

TCCM: Trinity College, Cambridge Additional Manuscripts.
TCCM1: Robert Leslie Ellis to Duncan Gregory, Add Ms. C. 67, no date.

Printed Sources

Allaire, Patricia R. 1997. "The Development of British Symbolical Algebra as a Response to 'The Problem of Negatives' with an Emphasis on the Contribution of Duncan Farquharson Gregory." Unpublished D. A. Dissertation: Adelphi University.

Allaire, Patricia R. and Bradley, Robert E. 2002. "Symbolic Algebra as a Foundation for Calulus: D. F. Gregory's Contribution." *Historia Mathematica* 29, 395-426.

[Babbage, Charles and Herschel, John F. W.].[39] 1813. "Preface." *Memoirs of the Analytical Society* 1, i-xxii.

Barrow-Green, June. "'A Corrective to the Spirit of Too Exclusively Pure Mathematics': Robert Smith (1689–1768) and his Prizes at Cambridge University." *Annals of Science* 56, 271-316.

Becher, Harvey W. 1980a. "William Whewell and Cambridge Mathematics." *Historical Studies in the Physical Sciences* 11, 1-48.

———. 1980b. "Woodhouse, Babbage, Peacock, and Modern Algebra." *Historia Mathematica* 7, 389-400.

Boole, George. 1840. "On the Integration of Linear Differential Equations." *Cambridge Mathematical Journal* 2, 114-119.

———. 1844. "On a General Method in Analysis." *Philosophical Transactions of the Royal Society of London* 134, 225-282.

———. 1859. *A Treatise on Differential Equations.* Cambridge: Macmillan and Co.

———. 1952. *Studies in Logic and Probability.* La Salle, Ill.: Open Court Publishing Co.

Cannell, Doris Mary. 2002. *George Green: Mathematician & Physicist, 1793-1841: The Background to His Life and Work.* 2nd Ed. Philadelphia: SIAM.

Cauchy, Augustin. 1827a. "Sur l'analogie des puissances et des différences." *Exercises de mathématiques, Seconde année.* Paris: Chez de Bure Frères, or [Cauchy, 1882–1974, 2.7:198-235].

———. 1827b. "Addition à l'article précédent." *Exercises de mathématiques, Seconde année.* Paris: Chez de Bure Frères, or [Cauchy, 1882–1974, 2.7:235-254].

———. 1827c. "Sur la transformation des fonctions qui représentent les intégrales générales des équations différentielles linéaires." *Exercises de mathématiques, Seconde année.* Paris: Chez de Bure Frères, or [Cauchy, 1882–1974, 2.7:255-266].

---

[39] Authors whose names are in square brackets have been identified subsequently; the articles originally appeared anonymously.

———. 1843. "Note sur des théorèmes nouveaux et de nouvelles formules qui se déduisent de quelques équations symoliques." *Comptes rendus de l'Académie des Sciences de Paris* 17, 377, or [Cauchy, 1882–1974, 1.8:26-28].

———. 1882–1974. *Œuvres complètes d'Augustin Cauchy*. 2 Ser. 27 Vols. Paris: Gauthier-Villars et fils.

[Craufurd, Alexander]. 1841. "On a Method of Algebraic Elimination." *Cambridge Mathematical Journal* 2, 276-282.

Crilly, Tony. 2004. "The *Cambridge Mathematical Journal* and Its Descendants: 1830–1870." *Historia Mathematica* 31, 455-497.

———. 2006. *Arthur Cayley: Mathematician Laureate of the Victorian Age*. Baltimore: Johns Hopkins University Press.

De Morgan, Augustus. 1843. Augustus De Morgan to George Boole, 24 November, 1843. Quoted in [Smith, 1982, p. 13].

———. 1845. Augustus De Morgan to John F. W. Herschel, 28 May, 1845. Quoted in [Smith and Wise, 1981, p. 173].

Despeaux, Sloan Evans. 2002. "The Development of a Publication Community: Nineteenth-Century Mathematics in British Scientific Journals." Unpublished Doctoral Dissertation: University of Virginia.

Ellis, Robert Leslie. 1842a. "Remarks on the Distinction Between Algebraical and Functional Equations." *Cambridge Mathematical Journal* 3, 92-94.

———. 1842b. "On the Solution of Functional Differential Equations." *Cambridge Mathematical Journal* 3, 131-143.

———. 1844. "Memoir of the Late D. F. Gregory, M.A., Fellow of Trinity College, Cambridge." *Cambridge Mathematical Journal* 4, 145-152.

Enros, Philip C. 1981. "Cambridge University and the Adoption of Analytics in Early Nineteenth-century England." In *Social History of Nineteenth Century Mathematics*. Ed. Herbert Mehrtens, Henk Bos, and Ivo Schneider. Boston: Birkhäuser Verlag, 135-164.

———. 1983. "The Analytical Society (1812–1813): Precursor of the Renewal of Cambridge Mathematics." *Historia Mathematica* 10, 26-37.

Grattan-Guinness, Ivor. 1992. "The Young Mathematician." In *John Herschel 1792–1971: A Bicentennial Commemoration*. Ed. D. G. King-Hele. Bristol: J.W. Arrowsmith, Ltd., 17-28.

Graves, Robert Perceval. 1885. *Life of Sir William Rowan Hamilton*. Vol. 2. Dublin: Hodges, Figgis.

[Greatheed, Samuel]. 1837. "On General Differentiation." *Cambridge Mathematical Journal* 1, 11-21.

[Gregory, Duncan]. 1837a. "Preface." *Cambridge Mathematical Journal* 1, 1-2.

[———]. 1837b. "On the Solution of Linear Differential Equations with Constant Coefficients." *Cambridge Mathematical Journal* 1, 22-32.

[———]. 1838a. "Mathematical Notes." *Cambridge Mathematical Journal* 1, 94.

[———]. 1838b. "On the Solution of Partial Differential Equations." *Cambridge Mathematical Journal* 1, 123-131.

———. 1838c. "Mathematical Notes: 1. Taylor's Theorem." *Cambridge Mathematical Journal* 1, 143-144.

REFERENCES

_____. 1838d. "On the Integration of Simultaneous Differential Equations." *Cambridge Mathematical Journal* 1, 173-181.

[_____]. 1839a. "Demonstrations of Theorems in the Differential Calculus and Calculus of Finite Differences." *Cambridge Mathematical Journal* 1, 212-222.

_____. 1839b. "On the Impossible Logarithms of Quantities." *Cambridge Mathematical Journal* 1, 226-234.

_____. 1840a. "On the Real Nature of Symbolical Algebra." *Transactions of the Royal Society of Edinburgh* 14, 208-216.

_____. 1840b. Duncan Gregory to George Boole, 16 February, 1840. Quoted in [Harley, 1866, pp. 148-149] or [Boole, 1952, pp. 434-435].

_____. 1842. "On a Difficulty in the Theory of Algebra." *Cambridge Mathematical Journal* 3, 153-159.

_____. 1843. "On the Solution of Certain Functional Equations." *Cambridge Mathematical Journal* 3, 239-246.

[Gregory, Duncan and Smith, Archibald]. 1840. "On the Sympathy of Pendulums." *Cambridge Mathematical Journal* 2, 120-128.

Guicciardini, Niccolò. 1989. *The Development of Newtonian Calculus in Britain 1700-1800* Cambridge: Cambridge University Press.

Hamilton, William Rowan. 1846. William Rowan Hamilton to George Peacock, draft letter, 13 October, 1846. Quoted in [Graves, 1885, p. 528].

Harley, Robert. 1866. "George Boole, F.R.S." *British Quarterly Review.* 44, 141-181. Reprinted in [Boole, 1952, pp. 425-472].

Herschel, John F. W. 1814. "Consideration of Various Points of Analysis." *Philosophical Transactions of the Royal Society of London* 104, 440-468.

_____. 1845. "Address." *Report of the Fifteenth Meeting of the British Association for the Advancement of Science at Cambridge.* London: J. Murray, pp. xxvii-xliv.

Koppelman, Elaine. 1971. "The Calculus of Operations and the Rise of Abstract Algebra." *Archive for History of Exact Sciences* 8, 155-242.

Liouville, Joseph. 1832. "Sur quelques questions de géométrie et de méchanique et sur un nouveau genre de calcul pour résoudre ces questions." *Journal de l'École polytechnique* 13, 1-69.

Lusternik, Lazar' Aronovich and Petrova, Svetlana S. 1972. "Les premières étapes du calcul symbolique." *Revue d'histoire des sciences* 25, 201-206.

Murphy, Robert. 1837. "First Memoir on the Theory of Analytic Operations." *Philosophical Transactions of the Royal Society of London* 127, 179-210.

Panteki, Maria. 1991. "Relationships Between Algebra, Differential Equations and Logic in England: 1800-1860." Unpublished Ph.D. Thesis: Middlesex University.

Peacock, George. 1833. "Report on the Recent Progress and Present State of Certain Branches of Analysis." *Report of the First and Second Meetings of the British Association for the Advancement of Science Report at York in 1831 and at Oxford in 1832.* London: J. Murray, pp. 185-352.

Pycior, Helena M. 1981. "George Peacock and the British Origins of Symbolical Algebra." *Historia Mathematica* 8, 23-45.

_____. 1982. "Early Criticism of the Symbolical Approach to Algebra." *Historia Mathematica* 9, 392-412.

_____. 1984. "Internalism, Externalism, and Beyond: 19th-Century British Algebra." *Historia Mathematica* 11, 424-441.

Rice, Adrian C. 1996. "Augustus De Morgan: Historian of Science." *History of Science* 34, 201-240.

_____. 2000. "A Gradual Innovation: The Introduction of Cauchian Calculus into Mid-Nineteenth-Century Britain." *Proceedings of the Canadian Society for the History and Philosophy of Mathematics* 13, 48-63.

_____. 2001. "Inexplicable? The Status of Complex Numbers in Britain, 1750–1850." *Matematisk-fysiske Meddelelser Det Kongelige Danske Videnskabernes Selskab* 46, 147-180.

Richards, Joan L. 1980. "The Art and Science of British Algebra: A Study in the Perception of Mathematical Truth." *Historia Mathematica* 7, 343-365.

Smith, Archibald. 1836. Archibald Smith to William Ramsay, 22 January, 1836, Strathclyde Regional Archives. Quoted in [Smith and Wise, 1989, p. 56].

[_____]. 1840. "On the Motion of a Pendulum When Its Point of Suspension Is Disturbed." *Cambridge Mathematical Journal* 2, 204-208.

[_____]. 1845. Archibald Smith to William Thomson, 16 July, 1845, Thomson Papers, University Library, Cambridge. Quoted in [Smith and Wise, 1989, p. 176].

Smith, Crosbie and Wise, M. Norton. 1989. *Energy and Empire: A Biographical Study of Lord Kelvin*. Cambridge: Cambridge University Press.

Smith, Gordon C. 1982. *The Boole–De Morgan Correspondence, 1842–1864*. Oxford: Clarendon Press.

Wilson, David B. Ed. 1990. *The Correspondence Between Sir George Gabriel Stokes and Sir William Thomson, Baron Kelvin of Largs*. 2 Vols. Cambridge: Cambridge University Press.

CHAPTER 4

Divisibility Theories in the Early History of Commutative Algebra and the Foundations of Algebraic Geometry

Olaf Neumann
Friedrich-Schiller-Universität Jena, Germany

This chapter explores the origins and role of divisibility theories in nineteenth-century commutative algebra. As is well known, the concept of divisibility makes sense only if some realm of objects admitting a multiplication is specified. In the late eighteenth century, divisibility was defined for rational integers, certain special types of quadratic irrationalities like $a+b\sqrt{-3}$ (where $a, b \in \mathbb{Z}$), polynomials or "entire rational functions" of one or several variables, and power series in one variable. For rational integers as well as for polynomials of one indeterminate with real or complex coefficients, the *unique* decomposition into indecomposable—that is, *prime* or *irreducible* factors—was taken for granted. In both cases, the Euclidean algorithm obtains. These prominent instances of unique factorization domains (UFDs in modern terminology) served as a *pattern* for later divisibility theories. Relative to algebra, the vague concepts of "quantity" and "magnitude" were still of central importance. In fact, a mathematician of no less influence and importance than Leonhard Euler argued that "all magnitudes may be expressed by numbers; and that the foundation of all Mathematical Sciences must be laid in a complete treatise on the science of numbers, and in an accurate examination of the different possible methods of calculation. This fundamental part of mathematics is called Analysis, or Algebra. In Algebra, then," he continued, "we consider only numbers, which represent quantities, without regarding the different kinds of quantity. These are the subjects of other branches of mathematics." As for arithmetic, it "treats of numbers in particular, and is the science of numbers properly so called; but this science extends only to certain methods of calculation, which occur in common practice; Algebra, on the contrary, comprehends in general all the cases that can exist in the doctrine and calculation of numbers" [Euler, 1770/1972, p. 2]. Moreover, *all* "magnitudes" under consideration admitted a *commutative* law of multiplication that the nineteenth-century, English mathematician James Joseph Sylvester would style the "yoke" of algebra [Sylvester, 1904–1912/1973, 4:209]. It was William Rowan Hamilton's discovery of quaternions in 1843, the development of the calculus of matrices later in the century, the publication of Hermann Grassmann's *Ausdehnungslehre* in 1844 and then in a second edition in 1862, and, last but not least, the emergence of group theory that ultimately brought the possibility of "sound" non-commutative multiplications to the fore.[1]

In what follows, I consider divisibility in realms with a commutative-associative law of multiplication. The term "commutative algebra" should thus be understood nearly in the modern sense, that is, as the theory of commutative rings, and the term "domain" will take on the technical sense of "commutative ring with unit element and without zero divisors."

Paraphrasing Euler, algebra may be thought of as having originated from the "formal" or operational aspects of other branches of mathematics. Therefore, the historian of algebra (like any other historian of mathematics) should first of all pay due attention to the basic mathematical problems of a given epoch. From this point of view, the history of mathematical knowledge is primarily a story of success and failure in solving problems; the introduction of new concepts should, therefore, be seen not as an end in itself but rather as a tool for making problems more precise, for organizing results, for building theories, and for supplying or simplifying arguments. Of import is the specific or even indispensable use of a concept and not its merely "implicit" use. The analysis that follows takes this as its point of departure.

On Some Developments Rooted in the Eighteenth Century

Around the turn of the nineteenth century, the solution of three main types of equations largely defined what may be considered algebraic research: (systems of) algebraic, that is, polynomial equations in one or several unknown "quantities," and, in particular, the solution of equations by "radicals" (otherwise known as "algebraic" solutions); differential equations; and Diophantine equations, especially, representations of numbers by quadratic forms with integer coefficients. A number of achievements connected with these problems, dating from the eighteenth century and to roughly 1850, are relevant to the present account of divisibility theories, but in order to frame the discussion, we need only consider several.

First, it almost goes without saying that the recognition of the role of complex numbers in issues of factorization as well as the realization of the connection between the zeroes and the linear factors of a polynomial were key.[2] In particular, mathematicians appreciated that for a polynomial $f(z)$, $f(a) = 0$ if and only if $f(z)$ is divisible by $z - a$. Moreover, the number of times that $z - a$ divides $f(z)$ determined the multiplicity of the zero.

The so-called fundamental theorem of algebra—as a theorem accepted and proven according to the standards of the day—also had several key consequences for a polynomial $f(z)$. Algebraically, it implied that $f(z)$ splits completely into linear factors with complex coefficients[3] and that $f(z)$ has exactly $\deg(f)$ zeroes in total, if all zeroes are counted with their multiplicities. This, of course, can also be interpreted geometrically, namely, the curves $w - f(z) = 0$ and $w = 0$ intersect in precisely $\deg(f)$ points, if they are counted with suitable multiplicities. The fundamental theorem of algebra also had an important application in the differential

---

[1]Compare [Alten, 2003, ch. 7], [Dieudonné, 1978], [Kolmogorov and Yushkevich, 1992, ch. 2], [Nový, 1973], and [Scholz, 1990].

[2]On complex numbers, see [Tropfke, 1980, §§1.3.8]. Despite the term "Elementarmathematik" in its title, the fourth edition of Tropfke's book gave a very well documented and extensive overview of many topics relevant to matters of divisibility.

[3]See Harold Edwards's contribution to the present volume.

and integral calculus: partial fraction expansion of rational functions $g(z)$ of z yields a *complete* integration of differentials $g(z)dz$. And, finally, relative to systems of polynomial equations in one unknown, the theorem says that $f_1(z) = 0, \ldots, f_m(z) = 0$ have a common solution in the complex numbers if and only if the greatest common divisor of $f_1, \ldots, f_m$ is not constant. This is also equivalent to the fact that there is no identity $g_1 \cdot f_1 + \cdots + g_m \cdot f_m = 1$ with polynomials $g_1(z), \ldots, g_m(z)$.[4]

Systems of linear equations also underwent intense mathematical scrutiny and were shown to be solvable by the successive elimination of the unknowns. This procedure contributed, during the first four decades of the nineteenth century, to the creation of a free-standing theory of determinants that was then pursued for its applications in solving systems of polynomial equations as well as for its inherent algebraic interest [Tropfke, 1980, §3.3.3].

More generally, the elimination of unknowns in arbitrary systems of polynomial equations had actually played an extremely important role as early as the seventeenth century. By the eighteenth century, this line of investigation had developed particularly at the hands of the French mathematician, Étienne Bezout.[5] In the nineteenth century, and specifically in 1841, Sylvester extended the theory through his introduction and subsequent analysis of the important determinant called the resultant (or "eliminant") [Sylvester, 1904–1912/1973, 1:61-65]. In particular, he showed that two homogeneous polynomials or "forms" $f(x_1, x_2)$, $g(x_1, x_2)$ have a common zero $(x_1, x_2) \neq (0, 0)$ if and only if their resultant $R(f, g)$ equals 0. This result was then successfully extended to systems of n forms

$$f_1(x_1, \ldots, x_n), \ldots, f_n(x_1, \ldots, x_n)$$

in n variables.[6] Throughout the nineteenth and into the twentieth century, in fact, resultants continued to play a prominent role in all manner of elimination problems.[7] In particular, Kronecker used them decisively in his "arithmetic theory of algebraic quantities" [Kronecker, 1882, §10], while David Hilbert and Emanuel Lasker used them in their seminal work in invariant theory and ring theory, respectively.[8]

In the context of multidimensional geometry, moreover, elimination means, in modern terms, the projection of algebraic sets into subspaces of the ambient space. Resultants allow for the explicit description in terms of defining equations of the images of such projections. For instance, if in the plane two algebraic curves $f(x, y) = 0$ and $g(x, y) = 0$ without common component are given, then Sylvester's resultant $R_y(f, g)$ of f and g with regard to y (elimination of y) describes how the intersection of the given curves is projected onto the x-axis since the zeroes of R are the x-coordinates of the intersection points. The special case of $f(x, y) = a_0 y^n +$

---

[4]If z is replaced by n complex unknowns $z_1, \ldots, z_n$, then *mutatis mutandis* the equivalence of the first and third assertions remains true. This, the so-called "Nullstellensatz," was proved by David Hilbert as late as 1893 [Hilbert, 1893, §3]. Hilbert's proof relies on Kronecker's elimination theory.

[5]See [Bezout, 1779], [Brill and Noether, 1894, §I], [Serret, 1854, 1:ch. 4 and 2:ch. 5], and [Tropfke, 1980, §3.3.7].

[6]See [Salmon, 1866, art. 78], [Netto, 1899], and [Netto, 1900, pp. 33-154].

[7]See, among many possible examples, [Salmon, 1866, Lessons VIII-X], [Landsberg, 1899], [Netto, 1896–1900], [Weber, 1895, §§53-57], [Macaulay, 1916, §II]. See also [Van der Waerden, 1939, §15] and [Jacobson, 1985, ch. 5.4.].

[8]See [Hilbert, 1890], [Hilbert, 1893], and [Lasker, 1905].

$\cdots + a_n = 0$ and $g(x, y) = x - (b_1 y^{n-1} + b_2 y^{n-2} + \cdots + b_n) = 0$ yields the equation $R_y(f, g) = h(x) = 0$ satisfied by x (the so-called Tschirnhausen transformation).[9]

A number of algebraic techniques and results were also developed as a result of trying to understand better how to solve equations in one unknown of degree n and how to interpret their roots. For example, the coefficients of such equations were recognized as elementary symmetric polynomials of the roots, and theorems such as the fundamental theorem on symmetric polynomials were brought to bear in the analysis of solvability. Auxiliary equations termed "resolvents" were developed in order to isolate and analyze roots. Polynomials with coefficients in some fixed realm of "rationally known quantities," that is, in some field in today's terminology, were studied as were their divisibility properties via the Euclidean algorithm, the concept of irreducibility, and the notion of unique factorization into irreducible factors. Finally, "algebraic" solutions of equations of degree less than or equal to 4 were also given.[10] It should be stressed that irreducibility arguments, in particular, were indispensable in Gauss's cyclotomy theory [Gauss, 1801, §VII] as well as in the general theories of equations formulated by Niels Henrik Abel and Évariste Galois.[11] This marked an essential difference from previous theories of equations [Neumann, 2006].

Efforts to deal effectively with power series expansions also contributed significantly to algebraic developments. The analysis of Newton's polygon and, in particular, of Newton-Puiseux series, for example, led to the local uniformization of irreducible plane algebraic curves which, in turn, led after 1850 to the distinction between points on an irreducible curve and branches of the curve at a given point. This distinction underlies the appropriate theory of singular points as well as intersection theory and the divisor theories for curves developed in the second half of the nineteenth century as special divisibility theories.[12]

Finally, Euler, Joseph-Louis Lagrange, and Adrien-Marie Legendre created a genuinely arithmetical theory of binary quadratic forms [Weil, 1983, chs. 3-4]. These mathematicians employed quadratic irrationalities provided by the decomposition

$$(6) \quad \begin{aligned} 4a(ax^2 + bxy + cy^2) &= (2ax + by)^2 - (b^2 - 4ac)y^2 \\ &= (2ax + (b + \sqrt{d})y)(2ax + (b - \sqrt{d})y), \end{aligned}$$

where $a, b, c \in \mathbb{Z}$, where $d := b^2 - 4ac$ is the discriminant, and where, without loss of generality, $\gcd(a, b, c) = 1$. For every non-square $d \in \mathbb{Z}$, the set

$$\mathbb{Z}[\sqrt{d}] = \{s + \sqrt{d} \cdot t \mid s, t \in \mathbb{Z}\}$$

is obviously a domain equipped with the additive and multiplicative map of conjugation

$$\alpha = s + \sqrt{d} \cdot t \mapsto \alpha' := s - \sqrt{d} \cdot t$$

---

[9] For further comments on this and related topics, see the section on "Kronecker's Modular Systems" below.

[10] On these various ideas, see [Dieudonné, 1978, ch. 2], [Neumann, 2006], [Nový, 1973, ch. 3], [Scholz, 1990], and [Tropfke, 1980, §3.3.6].

[11] Ludvig Sylow was apparently the first to emphasize the irreducibility arguments in the works of Abel and Galois. See his annotations in [Abel, 1881]. I owe this reference to Christian Skau (Trondheim). I cannot agree, however, with Sylow's claim that Abel and Galois (and not Gauss) would have been the first to use irreducibility as a guiding principle of argumentation. See Dedekind's judgment in [Dedekind, 1930–1932, 3:408-419].

[12] Compare [Brill and Noether, 1894, §§1 and 6], [Jung, 1923], and [Noether, 1919].

and the multiplicative norm map
$$\alpha = s + \sqrt{d} \cdot t \mapsto N(\alpha) := \alpha \cdot \alpha' = s^2 - d \cdot t^2.$$
In view of (1), the values of any given quadratic form with discriminant d as defined are norms of numbers in $\mathbb{Z}[\sqrt{d}]$ up to the factor $4a$, that is, they are special divisors of norms. Although the question naturally suggested itself to use the domains $\mathbb{Z}[\sqrt{d}]$ and to ask about their divisibility properties, Lagrange had observed that a divisor (in $\mathbb{Z}$) of a norm need not be a norm itself. He gave the explicit example
$$1313 = 13 \cdot 101 = 36^2 + 17 \cdot 1^2 = (36 + \sqrt{-17})(36 - \sqrt{-17}) = N(36 + \sqrt{-17}),$$
where the factors 13 and 101 are not of the form $u^2 + 17v^2$ [Lagrange, 1769, p. 78]. Using the norm, it is easy to see that the factors 13, 101, $36+\sqrt{-17}$, and $36-\sqrt{-17}$ are indecomposable in the domain $\mathbb{Z}[\sqrt{-17}]$. This thus represents an example in which the unique decomposition into indecomposable factors fails to hold. Lagrange and Legendre had already found a certain way out of this difficulty; they could prove that if $\gcd(x, y) = 1$, then any odd divisor (coprime to $2d$) of $ax^2 + bxy + cy^2$ can be represented (perhaps up to sign) by another form $a'x'^2 + b'x'y' + c'y'^2$ with the same discriminant $d = b'^2 - 4a'c' = b^2 - 4ac$. In Lagrange's numerical example, taking $a' = 2, b' = 2$, and $c' = 9$ yields $13 = 2 \cdot 1^2 + 2 \cdot 1 \cdot 1 + 9 \cdot 1^2$ and $101 = 2 \cdot 2^2 + 2 \cdot 2 \cdot 3 + 9 \cdot 3^2$. This fact hints at why *all* binary quadratic forms of a given discriminant must be considered simultaneously. Indeed, this task was carried out to a great extent by Lagrange and Legendre and later by Gauss in an unsurpassable way in his *Disquisitiones arithmeticæ*.[13]

Developments Inspired by Gauss

One key development inspired by Gauss was the notion and exploitation of the notion of congruence. In the opening article of his *Disquisitiones arithmeticæ*, Gauss defined congruence for rational integers this way: $a \equiv b \pmod{m}$ if and only if $m \mid (a - b)$ [Gauss, 1801, art. 1].[14] As is well known, $\equiv$ is an equivalence relation compatible with addition, subtraction, and multiplication. Despite its seeming simplicity, the concept of congruence may be fruitfully generalized in a wide variety of mathematical settings by varying both the realm of underlying objects and the modulus.

Gauss himself was one of the first to exploit the possibilities of this generalization when he considered $\alpha \equiv \beta \pmod{\gamma}$ if and only if $\gamma \mid (\alpha - \beta)$, where instead of $\alpha, \beta, \gamma \in \mathbb{Z}$, he took $\alpha, \beta, \gamma \in \mathbb{Z}[\sqrt{-1}]$.[15] This relation is a special case of the analogous relations studied by Augustin-Louis Cauchy, Carl Gustav Jacob Jacobi, Gotthold Eisenstein, and Eduard Kummer, all of whom chose cyclotomic integers for α, β, γ, that is, numbers in domains $\mathbb{Z}[\zeta_n]$, where ζ_n denotes a primitive nth root of unity [Neumann, 1979-1980, pt. 2-3]. In particular, Kummer based his consistent introduction of ideal complex numbers entirely on such congruences.[16]

Following Gauss, the mathematicians Eisenstein, Theodor Schönemann, and Dedekind initiated the study of the so-called "higher congruences with respect

---

[13] See the analyses in [Dieudonné, 1978, §5.4], [Neumann, 1979, pt. 1], [Neumann, 2005], and [Weil, 1983, ch. 4].

[14] The number m was called the "modulus," although this should be translated as "measure."

[15] Compare [Gauss, 1828] and [Gauss, 1832].

[16] See [Kummer, 1975, pp. 165-192], [Bourbaki, 1994], [Dieudonné, 1978, §5.5.2], [Edwards, 1980], and [Haubrich, 1992, ch. 3].

to a double module." They considered a rational prime p and polynomials with coefficients in $\mathbb{Z}$. The congruence with respect to the "double module" $(p, f(x))$ is defined by

$$g(x) \equiv h(x) \pmod{p, f(x)} :\Longleftrightarrow h(x) - g(x) = q_1(x) \cdot p + q_2(x) \cdot f(x),$$

with $q_1(x), q_2(x) \in \mathbb{Z}[x]$.[17] In this way, they successfully built the foundation of the theory of finite fields (or Galois fields) that had previously been published by Galois.

The theory of higher congruences is closely related to the theory of polynomial congruences, that is, congruences $f(x) \equiv 0 \pmod{m}$, where $m \in \mathbb{Z}$ and $f(x) \in \mathbb{Z}[x]$ [Serret, 1849, pt. 3]. After Kummer had constructed his theory of ideal numbers, Dedekind linked it with polynomial congruences.[18] To use Kummer's notation, let λ be an odd prime, α a primitive λth root of unity, and $\Phi_\lambda(x) = x^{\lambda-1} + \cdots + x + 1$ the (irreducible) λth cyclotomic polynomial such that $\Phi_\lambda(\alpha) = 0$. For p any prime unequal to λ, the decomposition of p into ideal prime factors in the domain $\mathbb{Z}[\alpha]$ is in one-to-one correspondence with the factorization

$$\Phi_\lambda(x) \equiv P_1(x) \cdot P_2(x) \cdots P_e(x) \pmod{p},$$

where $P_1(x), \ldots, P_e(x)$ denote different polynomials in $\mathbb{Z}[x]$ which are irreducible $\pmod{p}$. It proved very tempting to carry over this fact to arbitrary number fields $\mathbb{Q}(\alpha)$ with an algebraic integer α satisfying an irreducible monic equation $\Phi(x) = 0$ with rational integer coefficients and to define ideal prime factors of a rational prime p by factorizing $\Phi(x) \pmod{p}$.

Pursuing this approach in the late 1850s and early 1860s, Dedekind encountered serious difficulties in the form of those primes which divide the discriminant of the polynomial $\Phi(x)$. Although Kronecker apparently recognized the same problem at roughly the same time, little precise evidence remains as to how far either pursued this line of research [Haubrich, 1992]. Eventually, they abandoned it in favor of more general and intrinsic considerations.

Eduard Selling was actually the first to publish a theory of ideal numbers in arbitrary finite (Galois) number fields based on higher congruences [Selling, 1865]. Although, on the whole, his theory was abortive, he did have the novel idea of introducing algebraic numbers "integral with respect to a given rational prime."[19] This idea of semi-localization, to use modern terminology, reappeared independently in hidden form in the consistent theory of Egor Ivanovič Zolotarev (Zolotareff) [Zolotarev, 1880].[20]

Congruences also allowed for the construction of roots of polynomials. In 1847, for example, Cauchy introduced $\sqrt{-1}$ as the residue class of x modulo the polynomial $(x^2 + 1)$ in the domain $\mathbb{R}[x]$ [Tropfke, 1980, p. 156]. It would, however, be almost fifty more years before Heinrich Weber would publish a general construction of roots. Letting k be any field and $f(x)$ be an irreducible polynomial with coefficients in k, Weber showed that the residue class ring $k[x]/(f(x))$ is also a field, and the residue class $x \pmod{f(x)}$ is a root of $f(x)$ [Weber, 1893].[21] This construction has since passed into all recent textbooks of algebra, although its basic

---

[17] See [Serret, 1855, pp. 343-370] and [Dedekind, 1930-1932, 1:40-67].
[18] See [Dedekind, 1930–1932, 3:418-419] and [Haubrich, 1992, ch. 8].
[19] See [Selling, 1865, p. 31] and [Haubrich, 1992, ch. 4.2].
[20] For more on Zolotarev's theory, see below.
[21] See also Harold Edwards's contribution to the present volume.

idea is, apparently, due to Kronecker and is an excellent example of how Kronecker understood the arithmetical existence of algebraic quantities [Kronecker, 1882, §13].

In his "Grundzüge," Kronecker defined various congruences modulo "modular systems [Modulsysteme]." Let D be a domain of algebraic numbers or algebraic functions in several indeterminates. If $f_1, \ldots, f_m, g, h$ are in D, then the most refined congruence relation was of the form

$$g \equiv h \pmod{f_1, \ldots, f_m} :\Longleftrightarrow g - h = q_1 \cdot f_1 + \cdots + q_m \cdot f_m$$

for $q_1, \ldots, q_m$ in D.[22] The finite sequence $(f_1, \ldots, f_m)$ was termed a "modular system," in which the order of the terms does not matter.[23]

Dedekind approached these matters in a way that was both decisive for the whole of algebra and very much characteristic of his mathematical persona. Instead of elements or finite sequences of elements in a domain playing the role of modules, he considered sets of a special kind, which he called "modules [Moduln]" and which have been so termed ever since. A module in Dedekind's sense is an additive subgroup $\mathcal{M}$ of $\mathbb{C}$ with no further restrictions [Dirichlet, 1863, §161]. He defined congruence modulo $\mathcal{M}$ as

$$\alpha \equiv \beta \pmod{\mathcal{M}} :\Longleftrightarrow \alpha - \beta \in \mathcal{M}$$

and justified his use of the term module in a footnote by giving explicit reference to Gauss's concept of congruence. Dedekind's guiding notion was clearly the observation that, in all previous instances, the congruence relation $\equiv$ had always been determined by the set $\{\xi \mid \xi \equiv 0\}$ which is closed under subtraction. For Dedekind, the relation $\equiv$ (mod $\mathcal{M}$) was still an equivalence relation compatible with addition and subtraction but, in general, *not* with multiplication. Furthermore, he defined operations for modules—sums, intersections, products, and quotients—thus opening up an entirely new field of research. It is for this reason that Dedekind is considered one of the founding fathers of lattice theory [Dedekind, 1930–1932, 2:112-120 and 236-271]. In particular, the product of modules proved important. If $\mathcal{M}_1, \mathcal{M}_2$ denote arbitrary modules, then their product is defined by

$$(7) \qquad \mathcal{M}_1 \cdot \mathcal{M}_2 := \left\{ \sum_{i=1}^{k} m_{1,i} \cdot m_{2,i} \mid k \geq 0, m_{1,i} \in \mathcal{M}_1, m_{2,i} \in \mathcal{M}_2 \right\}.$$

Ideals are thus special instances of modules. To any module $\mathcal{M}$, Dedekind associated the set

$$(8) \qquad \mathcal{M}^0 := \{\xi \in \mathbb{C} \mid \xi \cdot \mathcal{M} \subseteq \mathcal{M}\},$$

which is evidently a subdomain of $\mathbb{C}$ that embraces $\mathbb{Z}$ and which Dedekind dubbed the order of $\mathcal{M}$ [Dedekind, 1930–1932, 3:72]. Thus, $\mathcal{M}$ is closed under multiplication by elements of $\mathcal{M}^0$, which marks the origin of the phrase "module (over a ring)." In addition to this idea, Dedekind and his coauthor Weber introduced, in their treatment of algebraic functions of one variable (over $\mathbb{C}$), the notion of "function modules [Funktionenmoduln]" as additive groups of algebraic functions closed under multiplication by polynomials in the chosen independent variable [Dedekind, 1930–1932, 1:251-252]. Indeed, this interpretation provides a nineteenth-century instance

---

[22]Compare [Kronecker, 1882, p. 77] and [Kronecker, 1901, lecture 12].

[23]Another Kroneckerian congruence relation will be described in the section on "Kronecker's Modular Systems" below.

of modules in the recent sense of modules over non-numerical rings [Jacobson, 1985, ch. 3.2].

If the additive group $\mathcal{M} \neq \{0\}$ is generated by *finitely* many elements, that is, if $\mathcal{M}$ is a finite module in Dedekind's sense, then its order $\mathcal{M}^0$ contains only *algebraic* integers and is a finite module as well [Dedekind, 1930–1932, 3:93]. In the first and second drafts of his theory of algebraic number fields, Dedekind restricted his concept of order to this special case. More precisely, he spoke of orders in a given finite number field Ω as domains consisting of (not necessarily all) algebraic integers in Ω and containing $\mathbb{Q}$-linear bases of Ω generating the given domain as an additive group.

If Gauss's work directly inspired this wide range of work on congruence, it also sparked interest in binary quadratic forms. Instead of forms $ax^2 + bxy + cy^2$, Gauss considered forms with *even* middle coefficients,

$$(a, b, c) := ax^2 + 2bxy + cy^2,$$

with determinant $D := b^2 - ac$ [Gauss, 1801, §V]. Without loss of generality, one may assume that $\gcd(a, b, c) = 1$, that is, that the forms are primitive, that D is not square, and that $a > 0$ for $D < 0$, namely, that the forms are definite. Under these assumptions, there are two possibilities: either $\gcd(a, 2b, c) = 1$, in which case the forms are called properly primitive or proper; or $\gcd(a, 2b, c) = 2$, in which case the forms are termed improperly primitive or improper. The latter condition entails $D \equiv 1 \pmod 4$. The set of all primitive (and definite, if $D < 0$) forms with fixed value of D and fixed value of $\gcd(a, 2b, c)$ is called an order [Gauss, 1801, art. 226].[24] Gauss's further elaboration of this theory far surpassed the earlier work of Lagrange and Legendre. For Gauss, every order is split up into (finitely many) genera, and each genus is split up in its turn into a finite number of classes with respect to proper equivalence [Gauss, 1801, arts. 157, 158, and 223-233]. One of Gauss's most significant innovations here was his definition of a commutative composition of the classes such that the classes in a given proper order form a finite commutative *group*, whereas the genera form a factor group of explicitly known structure [Gauss, 1801, arts. 234-261]. He represented the inverse of the class of a form (a, b, c) by the form $(a, -b, c)$, and the unit or principal class by the form $(1, 0, -D) = x^2 - Dy^2$, if the order is proper.[25]

In contrast to Euler, Lagrange, and Legendre, Gauss, in his *Disquisitiones arithmeticæ*, restricted the use of quadratic irrationalities to his discussion of the reduction of forms, where some estimates were unavoidable [Gauss, 1801, arts. 171-205]. In his subsequent investigations into classes, orders, and genera, he *bypassed* totally quadratic irrationalities and their divisibility problems. Instead, he worked

---

[24]To my regret, I omitted the definition of orders in [Neumann, 1979, pp. 32-33], leaving the discussion of Gauss's results incomplete at this point.

[25]Moreover, Gauss proved that every improperly primitive form composes the "simplest" improperly primitive form $(2, 1, \frac{1-D}{2}) = 2x^2 + 2xy + \frac{1-D}{2}y^2$ and a suitable primitive form of determinant D [Gauss, 1801, pp. 383-386]. Hence, the set of improper classes is equipped with a (non-canonical) group structure which is isomorphic to "the" class group of the quadratic field $\mathbb{Q}(\sqrt{D})$, in the modern understanding of square-free non-squares D with $D \equiv 1 \bmod 4$. It seems that this fact has been neglected by most of Gauss's commentators.

with explicit relations between forms regarded as polynomials with integer coefficients. In this sense, one may say that Gauss's theory of binary quadratic forms served as a substitute for a divisibility theory of quadratic irrationalities.[26]

Gauss's theory was deemed notoriously difficult and sophisticated by his contemporaries and immediate successors. Roughly speaking, the conceptual structure of the theories of binary quadratic forms of both Gustav Peter Lejeune Dirichlet and Henry J. S. Smith still followed very closely the lines of Gauss's work,[27] although Smith succeeded in generalizing and simplifying Gauss's theory of composition considerably in the spirit of invariant theory [Smith, 1859–1865, art. 105-119].

With the second edition of Dirichlet's *Vorlesungen* in 1871, however, Dedekind began to lay out in his various supplements to Dirichlet's work his highly original theory of algebraic integers and ideals (in the tenth supplement of 1871) as well as his applications to binary quadratic forms (in the twelfth supplement of 1879 and 1894). In essence, Dedekind returned to quadratic irrationalities and their divisibility properties and brought the earlier approach of Euler, Lagrange, and Legendre to mathematical completion. Like these earlier authors, Dedekind considered quadratic forms $ax^2 + bxy + cy^2$ with $a, b, c \in \mathbb{Z}$, $\gcd(a, b, c) = 1$ (without loss of generality), and d, a non-square, as defined above. To understand his conceptual innovation, consider equation (1) above. Dividing each term of (1) by 4 yields

$$(9) \quad \begin{aligned} a(ax^2 + bxy + cy^2) &= \left(ax + \frac{b+\sqrt{d}}{2}y\right)\left(ax + \frac{b-\sqrt{d}}{2}y\right) \\ &= N\left(ax + \frac{b-\sqrt{d}}{2}y\right), \end{aligned}$$

where the quantities

$$\omega := \frac{b+\sqrt{d}}{2}, \quad \omega' := \frac{b-\sqrt{d}}{2}$$

are the roots of the monic polynomial $x^2 - bx + \frac{b^2-d}{4} = x^2 - bx + ac$ with integer coefficients. They are thus algebraic integers of the field $\mathbb{Q}(\sqrt{d})$, according to the definitions Dedekind gave in his tenth supplement to Dirichlet's work [Dirichlet-Dedekind, 1871, §160].] It is easy to see that the set $\mathbb{Z}[\omega] := \{s + t \cdot \omega \mid s, t \in \mathbb{Z}\}$ is a domain which is "twice as large" as the domain $\mathbb{Z}[\sqrt{d}]$ and which consists of algebraic integers of the field $\mathbb{Q}(\sqrt{d})$. On the other hand, the set $\mathcal{M} := \{a \cdot x + \omega \cdot y \mid x, y \in \mathbb{Z}\}$ is a Dedekindian module in the sense explained above. It is not difficult to visualize $\mathcal{M}$. If $d < 0$, then the points $\alpha \in \mathcal{M}$ form a lattice in the complex plane, while if $d > 0$, then the points (α, α') with $\alpha \in \mathcal{M}$ form a lattice in the real plane. Henri Poincaré [Poincaré, 1880] and Felix Klein [Klein, 1893] modified this visualization in such a way that they were able to obtain a geometric representation of Dedekind's theory.

Moreover, in the notation established in (3) above, $\mathcal{M} \subseteq \mathbb{Z}[\omega]$ and $\mathcal{M}^0 = \mathbb{Z}[\omega]$. Indeed, because $\mathbb{Q}(\sqrt{d}) = \mathbb{Q}(\omega)$ from $\xi = s + t \cdot \omega$, $s, t \in \mathbb{Q}$, $\xi \cdot \mathcal{M} \subseteq \mathcal{M}$, it follows

---

[26]There is a story that Dirichlet once told Kronecker to the effect that Gauss already used something like ideal factors of quadratic irrationalities around 1799 [Kronecker, 1975, 1:98]. This story, however, does not stand up to a critical examination. See [Waterhouse, 1984].

[27]See [Dirichlet, 1863, §4, supp. IV and X] and [Smith, 1859–1865, arts. 84-138]. The fourth and tenth supplements on genera and composition, respectively, to [Dirichlet, 1863] were written by Dedekind.

that $\xi \cdot a \in \mathcal{M}$, $\xi \cdot \omega \in \mathcal{M}$, $a|sa$, $s \in \mathbb{Z}$, and $at, bt, ct \in \mathbb{Z}$. Hence, $t \in \mathbb{Z}$ with regard to $\gcd(a,b,c) = 1$. In Dedekind's terminology, "$\mathcal{M}$ belongs to $\mathbb{Z}[\omega]$" or "$\mathcal{M}$ is an ideal of $\mathbb{Z}[\omega]$" [Dedekind, 1930–1932, 1:127-128]. Conversely, if η is an arbitrary irrational algebraic integer in $\mathbb{Q}(\sqrt{d})$, then $\mathbb{Z}[\eta]$ is the order of a suitable module; putting $\mathcal{M} = \mathbb{Z}[\eta]$, then $\mathcal{M}^0 = \mathbb{Z}[\eta]$. From the explicit description of the algebraic integers in the quadratic field $\mathbb{Q}(\sqrt{d})$, it can be deduced that *all* orders of $\mathbb{Q}(\sqrt{d})$ (recall the definition above) have this peculiar form $\mathbb{Z}[\eta]$ (for η as above).[28]

The domain $\mathbb{Z}[\omega]$ and its norm map into $\mathbb{Z}$ turn out to be the appropriate objects needed to handle the quadratic forms $ax^2 + bxy + cy^2$ with $d = b^2 - 4ac$. All Dedekindian "orders [Ordnungen]" $\mathbb{Z}[\omega]$ are in one-to-one correspondence with Gaussian orders.[29] For his orders, Dedekind developed a complete divisibility theory based on the multiplication (2) of the modules belonging to a given order or, in other words, of the ideals of the order. Under mild and quite natural restrictions, Dedekind got a unique prime factorization of modules [Dedekind, 1930–1932, 1:105-158 and 3:303-306].[30] With this, the whole theory of binary quadratic forms could be treated completely within the framework of modules, reflecting the full strength of Dedekind's new concepts.[31]

It is open for speculation why men like Dirichlet and Gauss did not consider the quantities $\frac{b+\sqrt{d}}{2}$ as integral irrationalities. Remarkably, Dirichlet discussed the divisibility properties of domains $\mathbb{Z}[\sqrt{-a}], a > 0$, *en passant* in his *Vorlesungen* [Dirichlet, 1863, §16]. He noted that such a domain for $a = -1$ is completely analogous to $\mathbb{Z}$, whereas for $a = -11$, the unique prime factorization fails to hold. He explicitly exhibited the decompositions $15 = 3 \cdot 5 = (2 + \sqrt{-11})(2 - \sqrt{-11})$, in which all factors are indecomposable in the domain $\mathbb{Z}[\sqrt{-11}]$. However, thanks to Dedekind, we now understand $\frac{1+\sqrt{-11}}{2}$ as an algebraic integer with unique prime factorization restored as in the larger domain $\mathbb{Z}[\frac{1+\sqrt{-11}}{2}]$:

$$3 = \frac{1+\sqrt{-11}}{2} \cdot \frac{1-\sqrt{-11}}{2}, \quad 5 = \frac{3+\sqrt{-11}}{2} \cdot \frac{3-\sqrt{-11}}{2},$$

$$2 + \sqrt{-11} = -\frac{1-\sqrt{-11}}{2} \cdot \frac{3-\sqrt{-11}}{2}.$$

Gauss's work also informed subsequent research on polynomials with integer coefficients. Consider the product of two polynomials

(10) $$(a_0 + a_1 \cdot X + \ldots + a_k \cdot X^k) \cdot (b_0 + b_1 \cdot X + \ldots + b_l \cdot X^l)$$
$$= c_0 + c_1 \cdot X + \ldots + c_{k+l} \cdot X^{k+l},$$

where X denotes the indeterminate. The question naturally arose, what can be said about the mutual relationship between the coefficients of the polynomials involved in this equation?

---

[28]Compare [Borevich and Shafarevich, 1986, pp. 136-145] and [Dedekind, 1930–1932, 3:208].

[29]It seems that this was the reason why Dedekind chose the term "Ordnung."

[30]Kummer's casual statement that "[t]he whole theory of binary forms in two variables can be interpreted via the theory of complex numbers of the form $x + y\sqrt{D}$ [Die ganze Theorie der Formen vom zweiten Grade, mit zwei Variablen, kann nämlich als Theorie der komplexen Zahlen von der Form $x + y\sqrt{D}$ aufgefaßt werden]" did not embrace all relevant cases [Kummer, 1975, pp. 208-209].

[31]See also [Borevich and Shafarevich, 1986, pp. 136-145].

Gauss had already obtained a crucial result in this direction. He assumed that all of the coefficients occurring in (5) are rational numbers with $a_k = b_l = c_{k+l} = 1$. It is thus true that if at least one of the a_is ($0 \leq i \leq k$) or b_js ($0 \leq j \leq l$) is not an integer, then at least one of the c_ms ($0 \leq m \leq k + l$) is also not an integer [Gauss, 1801, pp. 36-38].

Since the concept of the greatest common divisor clearly carries over from the integers to rational numbers, Gauss's proposition is equivalent to the assertion that "[i]f $\gcd(a_0, a_1, \ldots, a_k) = \gcd(b_0, b_1, \ldots, b_l) = 1$, then $\gcd(c_0, c_1, \ldots, c_{k+l}) = 1$" [Jacobson, 1985, p. 152]. In modern terms, this simply says that the product of two primitive polynomials is also primitive. In this version, Gauss's proposition is usually called "Gauss's lemma." Gauss's proof hinged on the unique prime decomposition of the rational integers. The same arguments led to the following theorem: "[i]f R is a factorial domain and X an indeterminate over R then the domain of polynomials $R[X]$ is factorial too" [Jacobson, 1985, p. 153].[32]

The theorem raised the question of how effectively to factorize a polynomial in $R[X_1, \ldots, X_n]$, when the factorization in R is supposed to be known. Kronecker settled this question very ingeniously when $R = \mathbb{Z}$ and $R = \mathbb{Q}$ by employing the transformation $X_i \mapsto X^{d^{i-1}}$ ($1 \leq i \leq n$), where d depends on the given polynomial and must be sufficiently large. This transformation reduced the problem to the case of one indeterminate, where Kronecker determined the factors by an interpolation procedure [Kronecker, 1882, §4]. A more subtle problem occurs if R is an algebraic function field over $\mathbb{Q}$, that is, if R is a finite extension of one of the rational function fields $\mathbb{Q}(Y_1, \ldots, Y_k)$ of k variables. (Note that if $k = 0$ here, then R is a finite number field.) In these cases, Kronecker sketched an algorithm which was later expounded in greater detail by various authors.[33]

By 1883, Kronecker had succeeded in proving a very far-reaching and important generalization of Gauss's lemma, namely, "[i]f all coefficients are algebraic integers or algebraic functions of several indeterminates, then all products $a_i \cdot b_j$ are integrally dependent on the c_ms" [Kronecker, 1895-1931, 2:419-424]. In other words, Kronecker saw how to solve the obvious equations $\sum_{i+j=m} a_i b_j = c_m$ (where $0 \leq m \leq k + l$) that represent the products $a_i b_j$, in the form of integral algebraic functions of the c_ms. The following conclusion is then immediate, namely, "[i]f all of the c_ms are divisible by some element d, then all products $a_i \cdot b_j$ are divisible by d as well." It was in this formulation that Kronecker's result was rediscovered independently by Dedekind in 1892 [Dedekind, 1930–1932, 2:28-39]. Kronecker, Dedekind, Adolf Hurwitz, and Julius König were all well aware of how to simplify the divisibility theories—divisor theory and ideal theory—by means of these generalizations of Gauss's lemma.[34] The theorem's most advanced generalized version read: if in (5), all coefficients belong to the quotient field of some integrally closed domain D and if some element d of the quotient field of D divides all of the c_ms, then d also divides each of the products $a_i b_j$. A proof of the theorem in this form results

---

[32]Special cases of this may be found in [Kronecker, 1882, §4], [Molk, 1885, ch. 2], and [Weber, 1895, 1:§20] for the domains $\mathbb{Z}[X_1, \ldots, X_n]$, and in [Netto, 1900, pp. 10-25] for the domains $K[X_1, \ldots, X_n]$, where K denoted a field.

[33]See [Kronecker, 1882, pp. 12-13], [Molk, 1885, pp. 41-49], [Weber, 1899, pp. 563-567] (and only in the 1899 2d ed.), [Netto, 1896, pp. 51-64], [Netto, 1900, pp. 10-25], and [König, 1903, pp. 164-170].

[34]This has been discussed in great detail in [Dedekind, 1930–1932, 2:50-58], [König, 1903, pp. 78-83], [Edwards, 1990, pt. 0], [Gray, 1997, pp. 28-29], and [Neumann, 2002, §2.7].

more or less obviously from [Dedekind, 1930–1932, 2:28-39].[35] The generalization of Gauss's lemma must, in fact, be regarded as one of the most basic theorems in commutative algebra discovered in the nineteenth century.

One final influence of Gauss's work merits mention before turning to later nineteenth-century developments relevant to the present discussion. In 1828 and 1831, Gauss published a divisibility theory which had every feature one could possibly wish for [Gauss, 1828; 1831]. He proved that the domain $\mathbb{Z}[\sqrt{-1}]$, that is, the set of complex numbers of the form $a + b\sqrt{-1}$, for $a, b \in \mathbb{Z}$) has a Euclidean algorithm and is factorial for this reason. Gauss's contemporaries like Jacobi were strongly impressed by the amazing simplicity of Gauss's final results and, in particular, by his theory of congruences in $\mathbb{Z}[\sqrt{-1}]$ and its number-theoretic implications such as the law of reciprocity for biquadratic residues [Neumann, 1980, pt. 3, p. 39]. Furthermore, since $\sqrt{-1}$ is just a primitive fourth root of unity, Gauss suggested in a footnote that for the search for the reciprocity laws of nth power residues, it would be necessary to use the nth roots of unity and, in this way, "amplify the field of higher arithmetic" [Gauss, 1832, art. 30]. Indeed, after the publication of Gauss's papers, a great many mathematicians tried to generalize his results to other rings $\mathbb{Z}[\zeta_n]$, where ζ_n denotes a primitive nth root of unity [Neumann, 1980, pt. 3]. Only Ernst Eduard Kummer succeeded in making a real breakthrough via his invention of the ideal complex numbers. This marked a turning point in algebraic number theory [Kummer, 1975, pp. 165ff].

From Kummer to Zolotarev

While the general history of algebraic integers and the genesis of their various divisibility theories—ideal theory, divisor theory, valuation theory, and p-adic completions—is well documented,[36] little is known about some of the intermediate stages in the development of the theories of Dedekind and Kronecker. Judgments diverge, for example, on the reception and real influence of the theories listed above. In [Neumann, 2002, pp. 151-154], I concluded that until 1910 Kronecker's theory had greater influence than Dedekind's. Edmund Landau had already pointed out that during Dedekind's lifetime, the influence of his algebraic number theory and ideal theory remained restricted to a small circle of scholars almost exclusively in Germany [Landau, 1917, p. 64]. Furthermore, as late as 1912 in the one-volume version of his textbook on algebra, Heinrich Weber expounded his own "functional" version of Kronecker's theory first and only then passed to Dedekind's theory with the remark that Weber's "functionals," in contrast to Dedekind's ideal theory, "do not require completely new concepts from the very beginning [nicht von vornherein ganz neue Begriffsbildungen nötig machen]" [Weber, 1912, p. 489 (note 1)]. Kurt Hensel's p-adic numbers only became influential after 1920 thanks to his student, Helmut Hasse.

Concurrent with the work of Kronecker and Dedekind in Germany, Egor Zolotarev pursued similar lines of research in Russia. His ideas, however, gained little attention and recognition outside of Russia, even though they were published in 1880 in the *Journal de mathématiques pures et appliquées*, a widely disseminated Western journal. Kronecker held that Zolotarev's theory was based entirely on

---

[35]See also [Neumann, 2002, p. 183].

[36]See, for example, [Bourbaki, 1994], [Dieudonné, 1978], [Edwards, 1980], [Haubrich, 1992], [Neumann, 1979–1980; 1981; 2002], and [Ullrich, 1998; 1999].

higher congruences and, therefore, should be considered "unsuccessful [verfehlt]" [Kronecker, 1882, p. 118]. Dedekind mentioned Zolotarev's paper in the preface (dated 11 November, 1880) to the third edition of Dirichlet's *Vorlesungen über Zahlentheorie* but refrained from expressing a definite opinion.[37] Apparently, neither Kronecker nor Dedekind could appreciate the new notion of the arithmetic of algebraic integers with regard to a rational prime.[38] From 1880 to 1930, Zolotarev's ideas were diffused almost exclusively in Slavonic countries. In 1930, Oystein Øre called Zolotarev's theory one of the consistent foundations of algebraic number theory without exceptional cases [Dedekind, 1930–1932, 1:231]. In the same year, three papers on Zolotarev's theory appeared in key American journals.[39]

All of the theories just mentioned, except Hensel's, emerged from an examination of Kummer's theory of ideal numbers, a new type of divisibility theory. More specifically, for domains D of cyclotomic integers, Kummer had constructed a semi-group $\mathcal{D}$ of ideal numbers admitting unique prime factorization and a multiplicative map $\mathcal{J} : D \to \mathcal{D}$ such that $\alpha \mid \beta \iff \mathcal{J}(\alpha) \mid \mathcal{J}(\beta)$ together with some further well-known properties. Although Dedekind, Kronecker, and Zolotarev all approached their work from the point of view of arbitrary finite number fields rather than relying solely on cyclotomic fields, they differed from each other considerably in the actual approaches they chose. Both Dedekind and Kronecker based their theories on the domain $\mathcal{O}$ of *all* algebraic integers in some given finite number field K. They both recognized that, in general, $\mathcal{O}$ was not a domain of the special form $\mathbb{Z}[\omega]$, where ω denotes an algebraic integer. Dedekind, however, worked within $\mathcal{O}$, while Kronecker adjoined indeterminates to K and constructed a semi-group of integral divisors from there. Each integral divisor is represented by some quotient $\frac{F}{P}$, where F denotes a polynomial (in arbitrarily many indeterminates) with algebraic integer coefficients and where P is a polynomial with rational integer coefficients the greatest common divisor of which is 1. In other words, P is a primitive polynomial [Kronecker, 1882, pp. 45-55]. The quotient $\frac{F}{P}$ represents an ideal greatest common divisor of the coefficients of F in the sense of Kronecker's theory. Weber called those quotients $\frac{F}{P}$ "integral functionals," and he realized that they form a domain $\mathcal{F}(\mathcal{O})$ (my notation) by virtue of the generalized version of Gauss's lemma.[40] This domain contains $\mathcal{O}$, and any finite set of its elements as a greatest common divisor which is a linear combination of the given elements. Hence $\mathcal{F}(\mathcal{O})$ is a factorial domain, and its divisibility theory is, in general, much simpler than that of $\mathcal{O}$.[41] In this setting, integral divisors are just in one-to-one correspondence with the classes of associated elements of $\mathcal{F}(\mathcal{O})$. Moreover, the adjunction of indeterminates

---

[37]See [Dirichlet, 1879, p. viii] and [Dedekind, 1930–1932, 3:425].

[38]See [Zolotarev, 1880, pp. 141-166], [Tchebotarev, 1930, pp. 118-126], and [Bashmakova, 1949, pp. 319-332].

[39]See [Engström, 1930a,b] and [Tchebotarev, 1930]. On the reception of Zolotarev's work, see [Piazza, 1998].

[40]See [Weber, 1896, §77] and [Weber, 1912, pp. 435-451].

[41]The domain $\mathcal{F}(\mathcal{O})$ is termed a Bezout domain. Actually, this domain turns out to be, in modern terminology, a principal ideal domain. See [Weber, 1896, §§148-160; 1912] and [Neumann, 2002]. König further generalized Kronecker's and Weber's constructions in an axiomatic way [König, 1903, pp. 461-552]. We are thus justified in calling König's book the first textbook on commutative algebra.

enabled Kronecker to restore the method of higher congruences for the explicit determination of the prime divisors dividing a given rational prime.[42] Relative to this approach, some difficult questions connected with the discriminant—that were left open by Kronecker—were eventually settled in [Hensel, 1894] and [Mertens, 1894]. As Alexander Ostrowski wrote, it was Mertens who succeeded in founding the arithmetic of number fields entirely on the notion of "higher congruences" [Ostrowski, 1919, p. 282 (note 4)].[43]

Zolotarev also worked in the domain $\mathcal{O}$, but for a rational prime p, he introduced some specific concepts with regard to p [Zolotarev, 1880, pp. 141-166]. He proved the existence of ideal prime factors of p represented by algebraic integers $\nu, \nu_1, \ldots, \nu_s$ such that for every algebraic integer α, the equality $H\alpha = \beta \nu^m \nu_1^{m_1} \cdots \nu_s^{m_s}$ obtains, for some rational integer H coprime with p, some algebraic integer β, the norm of which is coprime with p, and some uniquely determined exponents $m, m_1, \ldots, m_s$ [Zolotarev, 1880, pp. 165-166]. Zolotarev's ideal numbers are formal power products of those new prime elements, and so his entirely constructive theory generalized that of Kummer. This theory was later simplified by Nikolaĭ G. Tchebotarev and Izabella G. Bashmakova. In particular, Bashmakova introduced the domain

$$\mathcal{O}_p := \left\{ \frac{\alpha}{q} \mid \alpha \in \mathcal{O}, \, q \in \mathbf{Z}, \, p \nmid q \right\},$$

or the integral closure of the discrete valuation ring $\mathbf{Z}_{(p)} := \{a \cdot b^{-1} \mid a, b \in \mathbf{Z}, \, p \nmid q\}$ [Bashmakova, 1949, pp. 319-321]. In modern terms, then, Zolotarev's theory says that $\mathcal{O}_p$ is a principal ideal domain with only finitely many prime ideals, and his main argument can be reformulated in this way. For any $\alpha, \beta \in \mathcal{O}_p$, let $\gamma = \xi \cdot \alpha + \eta \cdot \beta \neq 0$, for $\xi, \eta \in \mathcal{O}_p$, be chosen such that the highest p-power dividing $\mathrm{Norm}(\gamma)$ is as small as possible. Each such γ can thus serve as a greatest common denominator (in $\mathcal{O}_p$) of α, β. Today, the domain $\mathcal{O}_p$ is sometimes called a semilocal ring, since it has only finitely many maximal ideals. Zolotarev's approach, modified in this way, has survived in at least one advanced modern textbook on number theory [Borevich and Shafarevich, 1986, pp. 188-192].

Up to this point, we have seen that to algebraic integers (of a given field K) we can attach, in various ways, new objects (Dedekindian ideals, Kroneckerian divisors, Zolotarev's ideal numbers), the divisibility properties of which are rather transparent. Dedekind and Kronecker addressed the special (and difficult) question of whether those new objects could be represented immediately by algebraic integers in some larger field L containing K. In other words, they asked, does every ideal or divisor in K become a principal ideal or principal divisor, respectively, in L?

---

[42]Compare [Kronecker, 1882, §25] and [Edwards, 1990, §2.4].

[43]It should be noted that Kronecker's theory fits into a broader picture where, for much more general domains, "divisors of first level" are defined. See, for example, [Kronecker, 1882, pt. 2], [Edwards, 1990], and [Neumann, 2002]. Here again, certain questions connected with discriminants turned out to be rather subtle and led Kronecker as early as 1886 to correct a claim he had made in his "Grundzüge." Compare [Kronecker, 1895–1931, 4:389-470] and [Ostrowski, 1919]. In modern terms, Kronecker had discovered that a factor of the discriminant could correspond to an inseparable residue class extension. Moreover, relative to the method of "higher congruences" applied to algebraic functions of more than one indeterminate, König detected and corrected a subtle fallacy in Kronecker's work. See [Kronecker, 1882, pp. 115-116], [König, 1903, pp. vi and 504-547], and [Ostrowski, 1919, p. 287].

Kronecker called this question the problem of "the species to be associated with a given species [die Frage der zu assoziierenden Gattungen]" [Kronecker, 1882, §19]. In modern number-theoretical terms, a "species [Gattung]" is a set of primitive elements of a given finite number field, thus Kronecker's problem can be formulated as follows: construct a finite extension field L such that all divisors of K become principal divisors in L. Relative to the finiteness of the class number of divisors, it is not difficult to construct at least one such field L. If h denotes the class number of K and $\mathcal{D}$ is a divisor of K, then $\mathcal{D}^h$ is a principal divisor, call it (α), and in the larger field $K(\sqrt[h]{\alpha})$, the divisor $\mathcal{D}$ is principal, too. A finite number of steps yields a field L, where *all* divisors of K turn out to be principal. Dedekind had already published this construction in his supplements to Dirichlet's *Vorlesungen über Zahlentheorie*.[44] From it, he concluded that the domain of *all* algebraic integers in the field $\mathbb{C}$ is a Bezout domain, that is, every finitely generated ideal is a principal ideal. This beautiful and smooth-looking result was obtained, however, only after much conceptual effort and argumentation. Some twenty years later, Hurwitz found a slightly more direct proof [Hurwitz, 1895].

Kronecker, however, had apparently envisaged a more refined or somehow more natural construction of such fields L. He claimed that as early as 1856 he had observed the existence of special species to be associated with imaginary quadratic number fields by means of the complex multiplication of elliptic functions [Kronecker, 1882, §19].[45] Much later, this remarkable discovery of Kronecker's was absorbed into class field theory in the form of the principal ideal theorem.

Complex Analytic and Algebraic Functions

Alexander von Brill and Max Noether, the authors of an extensive report on the history of algebraic functions (of one variable), called Abel the founding father of that field [Brill and Noether, 1894, p. 212]. In view of his results on integrals of algebraic functions, algebraic functions were initially treated in the theory of complex analytic functions, whereas the *independent* development of the theory was the work of the last four decades of the nineteenth century [Brill and Noether, 1894, pp. 285-288].

The greatest challenge came from Bernhard Riemann's grand results [Riemann, 1857], which were founded on transcendental arguments and which called for alternative and, in some regards, more rigorous proofs [Brill and Noether, 1894, pp. 282-286]. In addition to the visionary Riemann, the most important figure was Karl Weierstrass, who, beginning in the 1860s, taught his subtle and rigorous theory of "analytische Gebilde" and Abelian functions in his lectures at the University of Berlin. In particular, he formulated a local divisibility theory for convergent power

---

[44]See [Dirichlet, 1871, §164; 1879, §175; 1894, §181].

[45]In more detail, he had realized that in the case of complex multiplication, the transformations of elliptic functions could be interpreted as multiplications of the arguments by some algebraic integers. See [Kronecker, 1895–1931, 4:177-183] as well as [Landsberg, 1899, p. 296]. Those algebraic integers are obtained by adjunction of the so-called singular moduli. Thus far, I have been unable to find a complete algebraic interpretation of this claim in Kronecker's *Werke*. For the field $\mathbb{Q}(\sqrt{-31})$ with class number 3, he showed that the adjunction of the singular moduli gives an unramified cyclic cubic extension. See [Kronecker, 1895–1931, 4:123-129].

series of several complex variables and proved his famous "preparation theorem [Vorbereitungssatz]."[46]

For the present purposes, a consideration of a work by Kronecker on discriminants of algebraic functions of one variable published will prove instructive [Kronecker, 1895–1931, 2:193-236]. In the opening pages of the paper, which was written in the late 1850s but not published until 1881, Kronecker gave an account of the arithmetic investigations into algebraic numbers he had conducted in 1857. At that time, Kronecker had communicated his results to Weierstrass, who had suggested that he carry the results over to algebraic functions. It was in this way that Kronecker was led to the arithmetic theory of algebraic functions, which he elaborated essentially in 1858 in the paper under consideration. Prior to Kronecker, and particularly in the work of Riemann, algebraic curves were studied locally by means of the Newton-Puiseux series in the neighborhood of any point (as mentioned in the first section above) with some specific difficulties arising at the singular points. Kronecker overcame these difficulties and succeeded in replacing some of Riemann's analytical arguments by suitable algebraic transformations of $F(w, z) = 0$ that left z fixed.[47] In particular, he studied the discriminants of all functions w which are integral over the domain $\mathbb{C}[z]$. Such a discriminant $\Delta_w(z)$ is just the resultant (up to sign) of $F(w, z)$ and its partial derivative $\frac{\partial F}{\partial w}$; the zeroes of $\Delta_w(z)$ give all (finite) regular points with "vertical" tangents (that is, ramified over the complex z-plane) by virtue of the fact that $\frac{\partial F}{\partial z} \neq 0$, and they give all (finite) singular points on the curve $F(w, z) = 0$, which are described by the equations $F(w, z) = \frac{\partial F}{\partial w} = \frac{\partial F}{\partial z} = 0$. After suitably changing w, Kronecker was able to reduce the singular points in such a way that only multiple points with separated tangents remained. This then allowed him to distinguish the ramification over $\mathbb{C}[z]$ and the influence of the singular points. Soon after Kronecker's initial discoveries, Weierstrass discovered his prime functions representing ideal prime factors of algebraic functions, which Kronecker saw as rendering his own paper obsolete. It was for this reason that Kronecker only published his work decades later in 1881.[48]

Another response to Riemann's work came from Dedekind and Weber in 1880. The points of the Riemann surface corresponding to an algebraic function field $K = \mathbb{C}(z, w)$ defined by an irreducible equation $F(w, z) = 0$ were interpreted by them as additive-multiplicative maps from K into $\mathbb{C} \cup \infty$ (under the rules $1/0 = \infty$ and $1/\infty = 0$) or, in modern terms, as discrete valuations of K. They then transplanted the ideal theory they needed from number fields to the domain of all functions integral over the domain $\mathbb{C}[z]$, where z is a fixed independent variable [Dedekind, 1930–1932, 1:238-350]. It turned out that this ideal theory obeyed the same laws as relative to number fields but was simpler than in that context. In his *Vorlesungen über Riemannsche Flächen*, Felix Klein compared Dedekind and Weber's ideal-theoretic and Kronecker's divisor-theoretic approaches and, once again, stressed the analogy between algebraic functions and algebraic numbers [Klein, 1892, pp. 43-46;

---

[46]See [Weierstrass, 1894–1927, 2:135-142]. The domain of the convergent power series centered at a given point would later be interpreted as a noetherian factorial domain. Compare [Lasker, 1905, p. 89], [Dieudonné, 1978, §4.7.2], and [Ullrich 1994]. The domain of formal power series also has the same properties. On Weierstrass in historical context, see [Dieudonné, 1978, §§4.6-4.7] and [Ullrich, 2003].

[47]Compare [Kronecker, 1895–1931, 2:193-236] and [Brill and Noether, 1894, pp 367-402].

[48]See [Weierstrass, 1894–1927, 4:387-393], [Landsberg, 1899, pp. 296-298], and [Ullrich, 1999, pp. 120-121].

Klein, 1986, pp. 89-94].[49] Dedekind and Weber had emphasized, moreover, that their theory would also work over the field of all algebraic numbers [Dedekind, 1930–1932, 1:240], while Kronecker had initiated the systematic study of algebraic functions (of one or several indeterminates) from the opposite point of view. Thus, Kronecker defined algebraic functions only over finitely generated (algebraic or transcendental) extensions of $\mathbb{Q}$, an approach that stemmed from his ambitious goal to build up a "general arithmetic" in which there was no place for a field like $\mathbb{R}$ or $\mathbb{C}$ [Kronecker, 1882].[50]

Kronecker's Modular Systems

Kronecker introduced his central concept of "modular system" (which he sometimes termed "divisor system") in the second part of his extensive treatise, "Grundzüge einer arithmetischen Theorie der algebraischen Grössen" [Kronecker, 1882]; he later added several important observations and complementary results to this treatise. Reading this work, however, is notoriously difficult, since, in many places, it fails to follow a clear-cut, logical structure of the form "definition, premises, assertion, proof" [Edwards/Neumann/Purkert, 1982]. Moreover, since precise references both to Kronecker's own papers and to the work of other authors are often lacking, it is often difficult to judge how original Kronecker's contributions actually were. Luckily, though, Kronecker's lectures on number theory, edited posthumously by Hensel, do contain more detailed explanations of some concepts [Kronecker, 1901, part II], and Jules Molk, a French mathematician who had studied with Kronecker between 1882 and 1884, treated many topics in the "Grundzüge" in a comprehensive and detailed manner.[51]

The "Grundzüge" reflect fully Kronecker's program of a general arithmetic, which is to be understood as the arithmetical theory of the entire functions of indeterminates with integer coefficients or polynomials with coefficients in $\mathbb{Z}$ [Kronecker, 1895–1931, 3:249-274]. The basic objects are domains of integrity—like $\mathbb{Z}[x_1,\ldots,x_n]$ or $\mathbb{Q}[x_1,\ldots,x_n]$—and domains of rationality—like $\mathbb{Q}(x_1,\ldots,x_n)$, in which $x_1,\ldots,x_n$ denote independent indeterminates. Kronecker also considered finitely generated integral or algebraic extensions of these domains, respectively.[52] His restriction to *finitely generated* domains and their quotient fields must be stressed. In addition to this, he rejected the naïve set theory of both Georg Cantor and Dedekind. The most drastic consequence of this rejection was that he did not accept the Dedekindian concepts of module and ideal as legitimate constituents of his own theory. Given these facts, Nicolas Bourbaki's contention that Kronecker's "central theme [in the "Grundzüge"] is (in modern language) the study of the ideals of a finite integral algebra over one of the rings of polynomials $\mathbb{C}[X_1,\ldots,X_n]$

---

[49]By contrast, Emmy Noether gave a detailed overview of the arithmetic theory of algebraic functions of one variable in [Noether, 1919]. In particular, she interpreted the singular points on a curve in terms of ideal theory. Heinrich Jung wrote a textbook based on Newton-Puiseux series, in the opening pages of which he gave a nice sketch of the historical development [Jung, 1923, pp. 1-4]. Jung also established a divisor theory of algebraic functions of two variables over $\mathbb{C}$ and applied it with success to algebraic surfaces [Gray, 1994, pp. 176-180].

[50]Harold Edwards wrote his book, *Divisor Theory*, in much the same spirit as Kronecker and proved the Riemann-Roch theorem there [Edwards, 1990].

[51]Compare [Molk, 1885] and [Gray, 1997, pp. 22-25].

[52]In what follows, I shall speak of "domains" instead of "domains of integrity." Edwards's notion of "natural domains" could be of use in this context [Edwards, 1990].

or $\mathbb{Z}[X_1,\ldots,X_n]$" and that "Kronecker limits himself *a priori* to those of these ideals which are of finite type" [Bourbaki, 1994, p. 105] contradicts Kronecker's way of thinking in a twofold way: first, there was no place for the field $\mathbb{C}$ in the "Grundzüge," and, second, Kronecker did not consider ideals there. This is not merely a matter of language but, above all, an issue of ontology. What Kronecker *really* did was to consider finite sequences of elements (or "Modulsysteme" in his terminology) and *several* equivalence relations between them plus congruence relations resulting from these. It is *not* permissible in all cases to pass from a module system to the ideal generated by it.[53] Even conceding the necessity of a translation of Kroneckerian concepts into modern-day concepts and vice versa, the dictionary will still have gaps. In particular, the theory of modular systems has no counterpart to the intersection of given ideals, for, in ideal theory, it is not a priori clear how to calculate a finite system of generators (if one exists at all) in the intersection of two finitely generated ideals.[54]

Letting D be some fixed domain, a modular system is a finite sequence of elements of D, considered nearly always without regard to the order of the elements and under the tacit assumption that at least one element is nonzero. The most refined equivalence relation $\sim$ between such sequences is defined as follows: $(f_1,\ldots,f_k) \sim (g_1,\ldots,g_l)$ if and only if there are relations $f_i = \sum_{\lambda=1}^{l} x_{i,\lambda} \cdot g_\lambda$ and $g_j = \sum_{\kappa=1}^{k} y_{j,\kappa} \cdot f_\kappa$, where $1 \le i \le k, 1 \le j \le l$ and where $x_{i,\lambda}, y_{j,\kappa} \in D$. This relation, in fact, corresponds to the equality of ideals generated in D by the elements of the sequences, respectively. The corresponding congruence relation $\equiv$ (mod $f_1,\ldots,f_k$) is then defined this way for arbitrary α,β in D: $\alpha \equiv \beta$ (mod $f_1,\ldots,f_k$) if and only if $(\alpha - \beta, f_1,\ldots,f_k) \sim (f_1,\ldots,f_k)$.

A different situation obtains using an equivalence relation $\sim_d$ that is, in general, coarser than $\sim$. For example and for the sake of simplicity, let D be a factorial domain which is not a Bezout domain; either of the domains $\mathbb{Z}[x]$ or $\mathbb{Q}[x,y]$ furnishes a case in point. It is then possible to define $(f_1,\ldots,f_k) \sim_d (g_1,\ldots,g_l)$ if and only if both sequences have the same greatest common divisor in D and introduce a congruence relation $\equiv_d$ belonging to $\sim_d$ defined by $\alpha \equiv_d \beta$ (mod $f_1,\ldots,f_k$) if and only if $\alpha - \beta$ is a multiple of the greatest common divisor of $f_1,\ldots,f_k$. It is easy to see that the two relations $\sim$ and $\sim_d$ are really different: $(x,y) \not\sim 1$ but $(x,y) \sim_d 1$ in $\mathbb{Q}[x,y]$. One of Kronecker's great achievements was to generalize $\sim_d$ and $\equiv_d$ to all domains he considered. This is precisely the theory of divisors of the first level and, obviously, the definition of greatest common divisors. As noted in the previous section, this was done by the adjunction of indeterminates.[55] Using the notion of the functional domain $\mathcal{F}(D)$ alluded to above, we then have the following equivalences: $(f_1,\ldots,f_k) \sim_d (g_1,\ldots,g_l)$ if and only if in $\mathcal{F}(D)$, both sequences

---

[53]For example, in the case of divisors of the first level represented by modular systems, it is, in general, not permissible to pass immediately from the modular system to the ideals generated by them. Instead, those ideals must be further subjected to certain closure operations [Neumann, 2002, pp. 169-176].

[54]The corresponding inverse problem is relevant to the ideal theory of Lasker, Macaulay, and Emmy Noether, and it reads as follows: given a finitely generated ideal $\mathcal{I}$, construct two larger finitely generated ideals $\mathcal{I}_1, \mathcal{I}_2$ such that $\mathcal{I} = \mathcal{I}_1 \cap \mathcal{I}_2$. This problem has its counterpart in Kroneckerian terms insofar as, for modular systems $\mathcal{M}, \mathcal{M}_1, \mathcal{M}_2$, a symbolic equivalence $\mathcal{M} \sim \mathcal{M}_1 \cap \mathcal{M}_2$ can be defined by means of congruences $f \equiv 0 \bmod \mathcal{M} \Leftrightarrow f \equiv 0 \bmod \mathcal{M}_1$ and $f \equiv 0 \bmod \mathcal{M}_2$. It seems, however, that Kronecker never addressed this problem in full generality.

[55]See also [Edwards, 1990] and [Neumann, 2002].

have the same greatest common divisor up to units. This is also equivalent to $(f_1, \ldots, f_k) \sim (g_1, \ldots, g_l)$ in $\mathcal{F}(D)$, since $\mathcal{F}(D)$ is a Bezout domain.

For modular systems $\underline{\mathcal{M}}_1 = (f_1, \ldots, f_k)$ and $\underline{\mathcal{M}}_2 = (g_1, \ldots, g_l)$, Kronecker introduced a sum $\underline{\mathcal{M}}_1 + \underline{\mathcal{M}}_2 := (f_1, \ldots, f_k, g_1, \ldots, g_l)$ and a product $\underline{\mathcal{M}}_1 \cdot \underline{\mathcal{M}}_2$ defined for all $f_i \cdot g_j$ in any order [Kronecker, 1882, §21]. Heuristically, these operations correspond to the intersection and the union of the algebraic sets determined by the systems of equations $f_i = 0$, where $1 \le i \le k$, and $g_j = 0$, for $1 \le j \le l$, respectively.[56] Kronecker also defined a relation of inclusion, namely, $\underline{\mathcal{M}}_1$ is contained in $\underline{\mathcal{M}}_2$, which he understood to mean that from $f \equiv 0 \pmod{\underline{\mathcal{M}}_2}$, it follows that $f \equiv 0 \pmod{\underline{\mathcal{M}}_1}$. Notice that this relation induces the set-theoretic inclusion of the algebraic sets in the *converse* order. Although algebraic sets occurred for heuristic reasons in several sections of Kronecker's treatise as well as in his lectures on number theory, those algebraic sets had no place in his systematic deductive exposition of the theory. In other words, Kronecker did not present "algebraic geometry" in the sense of a "theory of algebraic sets." Instead, his theory served as a machinery for obtaining defining equations for the components of algebraic sets. It thus has important consequences for algebraic geometry.[57]

On the whole, Kronecker's achievements are encapsulated in these three, key concepts: first, the algebraic definition of the "level [Stufe]" or "rank [Rang]" of a modular system, that is, of the codimension of the algebraic set defined by that system; second, the definition and algorithmic determination of the indecomposable divisors contained in a modular system; and, third, the fact that any modular system in n indeterminates can be replaced by a system of $n + 1$ elements without changing the corresponding algebraic set. Kronecker's theory does not, however, give suitable multiplicities of the factors in general. Furthermore, starting from a given modular system, equations could result that would fail to add anything new to the algebraic sets already defined by the remaining equations. In other words, the algebraic picture could possibly be more complicated than the geometric one. With these caveats, consider in greater detail Kronecker's set-up.

Kronecker's entire theory rests on two pillars: divisibility theory and elimination theory. As just noted, the level or rank of a modular system constituted a key idea [Kronecker, 1882, §20]. *Geometrically* speaking, the level is nothing more than the codimension of the corresponding algebraic set relative to the ambient space, that is, the difference between the number of variables and the dimension of the algebraic set [Landsberg, 1899, p. 302].[58] The *algebraic* definition of the level requires the tools of elimination theory, including the introduction of auxiliary indeterminates.[59] Basically, Kronecker's method, with minor alterations, survived in Bartel

---

[56] For more on these connections, see the next section below.

[57] Compare [Bourbaki, 1994], [Dieudonné, 1974], and [Gray, 1997].

[58] It does not, however, denote the dimensions of the components as erroneously claimed, for instance, in [Dieudonné, 1974, p. 60] and [Edwards *et al.*, 1982, p. 76].

[59] For more details on this aspect of Kronecker's work, see [Landsberg, 1899, pp. 302-305]. It is interesting to note that Kronecker was not the first to use new indeterminates to handle systems of equations [Landsberg, 1899, p. 303 (note 55)]. As early as 1847, Joseph Liouville had introduced the quantity $x = u_1 \cdot x_1 + \cdots + u_n \cdot x_n$ with new indeterminates $u_1, \ldots, u_n$. Kronecker also used this with no mention of Liouville. Compare [Liouville, 1847] and [Kronecker, 1882, §10].

In connection with Bezout's famous theorem on the intersection of plane curves, Molk quoted Siméon Denis Poisson, who, apparently, was the first to use auxiliary indeterminates to prove this theorem [Molk, 1885, p. 86]. But Molk did not give a specific reference to Poisson's work, whereas

Leendert van der Waerden's textbook on algebraic geometry [van der Waerden, 1939, §31].[60]

Kronecker aimed at the separation of modular systems or divisors of different levels contained in some given modular system. Geometrically, this means splitting up a given algebraic set into a finite number of smaller ones [Kronecker, 1882, §§20-21]. For any given modular system, the associated divisor of the first level simply describes the components of codimension 1 in the algebraic set. A peculiarity and strong point of Kronecker's theory is the modified concept of level in the case of arithmetic domains like $\mathbb{Z}[x_1, \ldots, x_n]$. In this case, level 1 is always ascribed to the modular systems (f). In particular, this is done for the modular system (a) with $a \in \mathbb{Z}, a \neq 0$, even though the algebraic set $a = 0$ will be empty. Ascribing level 1 to (a) contradicts the naïve geometric intuition that would reasonably call for codimension $n + 1$ of the empty set. However, with respect to divisibility, there is, in fact, no reason to differentiate, for instance, between a rational prime and a non-constant prime polynomimal in $\mathbb{Z}[x]$. Thus, the divisors of the first level will correspond exactly to the factors in the usual sense of divisibility in $\mathbb{Z}[x_1, \ldots, x_n]$, as expected in a factorial domain.

For convenience, identify the positive natural numbers a with the modular systems (a), and let $(f_1, \ldots, f_m)$ be any modular system over $\mathbb{Z}[x_1, \ldots, x_n]$. Then there are two possibilities: either $(f_1, \ldots, f_m)$ is not contained in any positive natural number, and so, by definition, the level of $(f_1, \ldots, f_m)$ just equals the same level as indicated above in the geometric sense; or $(f_1, \ldots, f_m)$ is contained in some positive natural number a_0 chosen as small as possible. In the latter case, it can happen that $(f_1, \ldots, f_m)$ is equivalent to $(a_0, g_1, \ldots, g_k)$, where $(g_1, \ldots, g_k)$ is a modular system not contained in any natural number. This means that there is an identity $a_0 = h_1 f_1 + \cdots + h_m f_m$, for $h_1, \ldots, h_m \in \mathbb{Z}[x_1, \ldots, x_n]$, whereas there is no such identity for the $g_1, \ldots, g_k$. The level of $(g_1, \ldots, g_k)$ thus has a geometric meaning, and, by definition, the level of $(f_1, \ldots, f_m)$ is one more than the level of $(g_1, \ldots, g_k)$.[61] For instance, $(a, x_1, \ldots, x_k)$ has level $k + 1$, if $a \in \mathbb{Z}$ for $a \neq 0$ ($1 \leq k \leq n$). Hence, the highest possible level is $n + 1$ (instead of n). It thus makes sense to assign dimension $n + 1$, in Kronecker's sense, to the whole ring $\mathbb{Z}[x_1, \ldots, x_n]$. In modern terms, this Kronecker dimension is nothing other than the Krull dimension of noetherian rings [Jacobson, 1989, pp. 450-455].

In the case of divisors of higher level, a particular difficulty arises insofar as a modular system of higher level could be contained in some other modular system without being a factor of it. Kronecker himself exhibited an example of this phenomenon, specifically the modular systems $(x^2 + p, p^2)$ and (x, p) of level 2 in $\mathbb{Z}[x]$, where p denotes a rational prime. The former contains the latter, but the latter is not a factor of the former [Kronecker, 1882, p. 83]. Therefore, in general, one cannot always expect to separate the divisors by means of a purely multiplicative decomposition. The questions raised by this situation were eventually answered

---

Landsberg cited a paper in the *Journal de l'École polytechnique* (11 (1811):199) [Landsberg, 1899, p. 303 (note 55)]. I have unfortunately been unable to check this reference against the original.

[60]Section 33 is entitled precisely "Die effektive Zerlegung einer Mannigfaltigkeit in irreduzible mittels der Eliminationstheorie [The Effective Decomposition of a Manifold into Irreducible Submanifolds by Means of Elimination]."

[61]Compare [Kronecker, 1882, p. 82] and [Landsberg, 1889, p. 304].

by Emanuel Lasker, Francis Macaulay, and Emmy Noether.[62] The domain $\mathbb{Z}[x]$ of dimension 2 had already been studied extensively by Kronecker in his lectures on number theory, where, in a certain sense, he arrived at a definitive result (factorization into simple divisor systems) [Kronecker, 1901, pp. 225-228].

To illustrate the role of divisibility theory and elimination, consider several examples of the decomposition of modular systems, for simplicity, over the domain $\mathbb{Q}[x_1, \ldots, x_n]$.[63] If the modular system consists of one non-constant member $f(x_1, \ldots, x_n)$, only then will the factorization of f into irreducible factors $p_1, \ldots, p_m$ (where $f = p_1^{e_1} \cdots p_m^{e_m}$) correspond to the decomposition of the hypersurface $f = 0$ into irreducible components $p_i = 0$ for $1 \leq i \leq m$ of codimension 1. The factorization of f is a matter that belongs solely to the topic of divisibility theory.

To demonstrate the role of elimination explicitly, consider the proof of the famous theorem named after Bezout: any two projective algebraic curves without common component have exactly as many points of intersection (in the projective plane) as the product of their degrees indicates, if the points are counted with suitable multiplicities. Let $F(x:y:z) = 0$ and $G(x:y:z) = 0$ denote the homogeneous equations of the curves with $F(x:y:z)$ and $G(x:y:z) \in \mathbb{Q}[x,y,z]$. There is a suitable line with equation $ax + by + cz = 0$ which does not contain any point of intersection of $F = 0$ and $G = 0$ since, by assumption, F and G have no common factor.[64] Choosing this line as the "infinite line" and applying a suitable linear transformation to the coordinates x, y, z, we can, without loss of generality, suppose that all points of intersection of $F = 0$ and $G = 0$ are of the form $(x:y:z)$ with $z \neq 0$. Specializing $z \mapsto 1$, we can restrict the context to the affine plane with coordinates x and y and to polynomials $f(x,y) = F(x:y:1)$ and $g(x,y) = G(x:y:1)$. By assumption, f and g have no common factor. The resultants $R_x(f,g) \in \mathbb{Q}[y]$ and $R_y(f,g) \in \mathbb{Q}[x]$ show that there will be only finitely many intersection points and that their coordinates will be algebraic numbers. Let (ξ_i, η_i) for $1 \leq i \leq N$ be those intersection points, and let K be a finite Galois extension of $\mathbb{Q}$ containing all coordinates ξ_i, η_i. Then the automorphisms of $K/\mathbb{Q}$ permute the N intersection points. Now choose new indeterminates u, v and set $w := u \cdot x + v \cdot y$. Via w, the point coordinates ξ_i and η_i can be "encoded" as coefficients of a linear polynomial in u, v, namely, $\zeta_i = u \cdot \xi_i + v \cdot \eta_i$. Moreover, the polynomial

$$R(w) := \prod_{i=1}^{N}(w - \zeta_i) = \prod_{i=1}^{N}(w - u \cdot \xi_i - v \cdot \eta_i) = \prod_{i=1}^{N}[u \cdot (x - \xi_i) + v \cdot (y - \eta_i)]$$

can be defined such that $R(w)$ has *rational* coefficients, since, under the automorphisms of $K/\mathbb{Q}$, its linear factors are interchanged. This fact suggests a search for *rational procedures* applied to the initially given polynomials $f(x,y)$ and $g(x,y)$ in order to obtain $R(w)$ or some other polynomial (also with rational coefficients) with

---

[62]See [Lasker, 1905], [Macaulay, 1916, pp. 40-42], and [Noether, 1921] as well as the sections that follow.

[63]It would be natural actually to take into account all domains $K[x_1, \ldots, x_n]$, where K runs through all finite number fields. The factorization over those domains would lead to the concept of absolute irreducibility. I have not, however, found the idea of this kind of irreducibility in Kronecker's *Werke*.

[64]The resultant of $F(x:y:z)$, $G(x:y:z)$ and the linear form $ax + by + cz$ is different from zero for suitably chosen rational coefficients a, b, c. Hence, on $ax + by + cz = 0$, there will be no intersection point of $F = 0$ and $G = 0$ at all.

the same linear factors as $R(w)$ but with altered exponents. The following recipe shows how this can be done.[65]

First, apply a linear transformation of x, y such that in each of the polynomials f, g there occur powers of x and y with the highest possible exponents. Next, specialize by mapping v to 1 and by setting $w = u \cdot x + y$ to get

$$\phi(x, w, u) := f(x, w - u \cdot x), \text{ and } \psi(x, w, u) := g(x, w - u \cdot x).$$

Finally, form the resultant $R(w, u)$ of ϕ and ψ with respect to x. It is not hard to see that this resultant has the desired properties [Molk, 1885, pp. 79-86]. According to Jules Molk, Poisson was the first to arrive at this result, having introduced new indeterminates. This procedure can be further refined to obtain the full version of Bezout's theorem in the projective plane.[66]

The Work of David Hilbert and Emanuel Lasker

Hilbert and Lasker both made more precise the relationship between modular systems and algebraic sets. Although both did *projective* algebraic geometry, for simplicity, we restrict the discussion here to the affine case, working in the polynomial ring $\overline{\mathbb{Q}}[x_1, \ldots, x_n]$ and the *affine* space $\overline{\mathbb{Q}}^n$, where $\overline{\mathbb{Q}}$ denotes the field of all algebraic numbers.[67] By the Hilbert Basis Theorem, every ideal of $\overline{\mathbb{Q}}[x_1, \ldots, x_n]$ has a finite basis, that is, is generated by some modular system.[68] To every ideal $\mathcal{I}$, there is an associated algebraic set

$$V(\mathcal{I}) := \{(\alpha_1, \ldots, \alpha_n) \in \overline{\mathbb{Q}}^n \mid f(\alpha_1, \ldots, \alpha_n) = 0, \text{ for all } f \in \mathcal{I}\}.$$

Conversely, to every subset $\mathcal{A}$ of $\overline{\mathbb{Q}}$, there is an associated ideal

$$I(\mathcal{A}) := \{f \in \overline{\mathbb{Q}}[x_1, \ldots, x_n] \mid f(\alpha_1, \ldots, \alpha_n) = 0, \text{ for all } (\alpha_1, \ldots, \alpha_n) \in \mathcal{A}\}.$$

It thus comes as no surprise that the operations on ideals are reflected in relations between the algebraic sets. From the chains of inclusions

$$\mathcal{I}_1 \supseteq \mathcal{I}_1 \cap \mathcal{I}_2 \supseteq \mathcal{I}_1 \cdot \mathcal{I}_2, \quad \mathcal{I}_2 \supseteq \mathcal{I}_1 \cap \mathcal{I}_2 \supseteq \mathcal{I}_1 \cdot \mathcal{I}_2$$

for two ideals $\mathcal{I}_1$, $\mathcal{I}_2$, it follows that

$$V(\mathcal{I}_1) \cup V(\mathcal{I}_2) \subseteq V(\mathcal{I}_1 \cap \mathcal{I}_2) \subseteq V(\mathcal{I}_1 \cdot \mathcal{I}_2) = V(\mathcal{I}_1) \cup V(\mathcal{I}_2),$$

hence

$$V(\mathcal{I}_1 \cap \mathcal{I}_2) = V(\mathcal{I}_1 \cdot \mathcal{I}_2) = V(\mathcal{I}_1) \cup V(\mathcal{I}_2).$$

These equalities suggest that an algebraic set $\mathcal{A} = V(\mathcal{I})$ may be decomposed into smaller pieces either by factoring $\mathcal{I}$ in the form $\mathcal{I} = \mathcal{I}_1 \cdot \mathcal{I}_2$ or by representing $\mathcal{I}$ as an intersection $\mathcal{I} = \mathcal{I}_1 \cap \mathcal{I}_2$. The first technique, combined with elimination processes, was Kronecker's approach, whereas the second was pursued by Lasker.

---

[65] Compare [Molk, 1885, pp. 79-86].

[66] See [van der Waerden, 1939, §17]. Molk also interpreted the factorization of $R(w, u)$ in terms of Kronecker's divisor theory as $(f(x,y), g(x,y)) \sim \prod_{(k)} [u \cdot (x - \xi_k) + (y - \eta_k)]$, where the altered indices k were chosen such that the multiplicities of the linear factors were taken into account [Molk, 1885, pp. 85-86]. On either side, one has a divisor of level 2.

[67] Note that moving to the projective case would not, in principle, be difficult, but this would require the introduction of technicalities that would deflect us from the main ideas. Note, too, that Hilbert and Lasker worked over the field $\mathbb{C}$ of complex numbers instead of $\overline{\mathbb{Q}}$. The choice of $\overline{\mathbb{Q}}$ does not change anything substantial in their argumentation and follows more closely the approach taken earlier by Kronecker.

[68] Compare [Hilbert, 1890, pp. 474-479] and [Jacobson, 1989, pp. 420-421].

To round off these considerations, we remark that the sum $\mathcal{I}_1 + \mathcal{I}_2 = (\mathcal{I}_1, \mathcal{I}_2)$ corresponds to the intersection of algebraic sets: $V(\mathcal{I}_1 + \mathcal{I}_2) = V(\mathcal{I}_1) \cap V(\mathcal{I}_2)$. The Hilbert Nullstellensatz asserts that

$$I(V(\mathcal{I})) = \sqrt{\mathcal{I}} := \{ f \in \overline{\mathbb{Q}}[x_1, \ldots, x_n] \mid f^k \in \mathcal{I} \text{ for some, } k \geq 1 \},$$

for each ideal $\mathcal{I}$.[69] It is not hard to see that $\sqrt{\sqrt{\mathcal{I}}} = \sqrt{\mathcal{I}} \supseteq \mathcal{I}$, and $V(\sqrt{\mathcal{I}}) = V(\mathcal{I})$. This means that, with respect to algebraic sets, the ideal $\mathcal{I}$ may be replaced by $\sqrt{\mathcal{I}}$, or what is today known as the radical of the ideal $\mathcal{I}$. From Lasker's results, it can be shown that $\sqrt{\mathcal{I}}$ is just the intersection of all prime ideals containing $\mathcal{I}$. Moreover, in the presence of the axiom of choice, the same assertion can be proved for ideals in arbitrary commutative rings, as Wolfgang Krull observed no later than 1929.[70]

In some respects, Lasker's terminology and his approaches were close to those used by Kronecker and Hilbert. Up to the modifications due to projective geometry, Lasker's modules are our ideals, and for algebraic sets, he used the more imaginative expression "configurations of algebraic structures [Konfigurationen von algebraischen Gebilden]." He took his concept of "prime ideal [Primmodul]" from Dedekind; it is the same as the modern concept, if we tacitly assume that a prime ideal be *proper*, that is, that it be different from the whole ring [Lasker, 1905, p. 50]. To a great extent, Lasker, like Kronecker, still worked explicitly with defining equations of algebraic sets, whereas, in contrast to Kronecker, he accepted naïve set theory. By means of resultants, he introduced the concept of an (absolutely) irreducible algebraic set by beginning with hypersurfaces $f = 0$ defined by irreducible fs and then by constructing chains of irreducible sets of descending dimension. Thus, Lasker could simultaneously define irreducibility and dimension based entirely on the unique prime factorization of suitable polynomials (resultants as a rule). (The Kroneckerian concept of level occurred once again here, but on a new foundation.) Finally, Lasker was able to prove that every algebraic set is a union of a finite number of uniquely determined irreducible algebraic sets [Lasker, 1905, pp. 35-38]. His definition of irreducibility allowed him to show that if $\mathcal{C}$ is irreducible, then $I(\mathcal{C})$ is a prime ideal. According to Lasker, the converse is also true, namely, if $\mathcal{P}$ is a prime ideal, then $V(\mathcal{P})$ is irreducible and $I(V(\mathcal{P})) = \mathcal{P}$, the latter following from Hilbert's Nullstellensatz [Lasker, 1905, pp. 51 and 607-608].

Now letting $\mathcal{I}$ be an arbitrary ideal, it is natural to ask how to split up the algebraic set $\mathcal{A} = V(\mathcal{I})$ into its irreducible components starting from the ideal $\mathcal{I}$. Lasker answered this question through his innovative concept of primary ideal. It would be reasonable to try to describe the components $\mathcal{C}$ of $\mathcal{A}$ by suitable ideals $\mathcal{Q}$ with the properties $\mathcal{C} = V(\mathcal{Q})$ and $\mathcal{Q} \supseteq \mathcal{I}$. According to Lasker, the ideal $\mathcal{P} := I(\mathcal{C}) \supseteq \mathcal{Q}$ is a prime ideal, and $\mathcal{Q}$ has a special relationship to $\mathcal{P}$, namely, $V(\mathcal{Q}) = V(\mathcal{P})$ and $\mathcal{P} = \sqrt{\mathcal{Q}}$. From the fact that $fg \in \mathcal{Q}$, for $f \notin \mathcal{P}$, it follows that $g \in \mathcal{Q}$. Now, Lasker's definition of primary ideals was suggested by these consequences of the Nullstellensatz. It reads: for a prime ideal $\mathcal{P}$, an ideal $\mathcal{Q}$ is called a primary ideal *belonging to* $\mathcal{P}$ if and only if $\dim V(\mathcal{Q}) \leq \dim V(\mathcal{P})$ and from $fg \in \mathcal{Q}$, for $f \notin \mathcal{P}$, it follows that $g \in \mathcal{Q}$ [Lasker, 1905, p. 51]. It is not difficult to see that either $\mathcal{Q} \subseteq \mathcal{P}$ or $\mathcal{Q}$ is the whole ring. Relying on his definition, Lasker was able to prove his so-called "main" theorem: every ideal $\mathcal{I}$ has the form

---

[69]See [Hilbert, 1893, pp. 320-325] and [Jacobson, 1989, p. 429].

[70]See [Krull, 1939, pp. 9-10] as well as the account of it in [Jacobson, 1989, p. 392].

$\mathcal{I} = \mathcal{Q}_1 \cap \cdots \cap \mathcal{Q}_k$, where $\mathcal{Q}_1, \ldots, \mathcal{Q}_k$ denote primary ideals (belonging, in general, to different prime ideals) [Lasker, 1905, p. 51-54 (theorem VII)].[71]

Lasker went on to carry over his results to the domains $\mathbb{Z}[x_1, \ldots, x_n]$ [Lasker, 1905, pp. 59-84]. To every rational prime p, he attached the algebra modulo p, that is, the factor ring $\mathbb{Z}[x_1, \ldots, x_n]/(p) \simeq \mathbb{F}_p[x_1, \ldots, x_n]$ (evidently, a factorial domain), where $\mathbb{F}_p$ denotes the finite field with p elements. He then introduced—all modulo p—algebraic sets, irreducible algebraic sets, and the unique decomposition of algebraic sets into irreducible components. He again defined primary ideals in $\mathbb{Z}[x_1, \ldots, x_n]$ with respect to some given prime ideal and formulated a key result analogous to the main theorem above: every ideal is the intersection of primary ideals [Lasker, 1905, p. 62 (theorem XIII)]. A remarkable feature of Lasker's theory is that, for every ideal $\mathcal{I} \subseteq \mathbb{Z}[x_1, \ldots, x_n]$, it takes into account the algebraic set $V(\mathcal{I})$ as well as the collection of the algebraic sets modulo p, that is, the algebraic sets $V(\mathcal{I} \bmod p)$, where p runs through all rational primes. This allowed him to attach a "dimension [Maximalmannigfaltigkeit]" to the ideal $\mathcal{I}$, which equals the maximum of the dimensions of all components occurring in the algebraic sets under consideration [Lasker, 1905, p. 61].[72]

Dedekind and Kronecker had already begun to study objects similar to the primary ideals Lasker would later develop. In §172 of the third edition of Dirichlet's *Vorlesungen über Zahlentheorie* of 1879, Dedekind introduced the concept of an "einartiges Ideal" in the domain $\mathcal{O}$ of all algebraic integers in some given finite number field, namely, an ideal $\neq \mathcal{O}$ contained in a single prime ideal. He went on to state without proof that "every ideal $\mathcal{A}$ different from $\mathcal{O}$ is either 'einartig' or it can be written in a unique way as a product of coprime 'einartig' ideals."[73] Ideals $\mathcal{I}_1, \mathcal{I}_2$ are coprime if and only if $\mathcal{I}_1 + \mathcal{I}_2 = (\mathcal{I}_1, \mathcal{I}_2) = (1)$. It is well-known that, under this assumption, $\mathcal{I}_1 \cap \mathcal{I}_2 = \mathcal{I}_1 \cdot \mathcal{I}_2$. Therefore, Dedekind proved that any ideal different from (0) and $\mathcal{O}$ is the intersection of so-called "einartige Ideale." Furthermore, among other things, he remarked that this fact is not restricted to the domain $\mathcal{O}$—the maximal Dedekindian order of the given number field—but can be proved instead in any order of the given number field [Dedekind, 1930–1932, 3:305]. (Recall the discussion of Dedekind's ideas in the section "Developments Inspired by Gauss" above.) Specializing this theorem to the domain $\mathbb{Z}$, it is clear that this is nothing more than the unique factorization of rational integers into powers of primes, where the "einartige Ideale" are just the ideals generated by powers

---

[71] Due to the framework of projective geometry, Lasker's original formulation of his main theorem explicitly emphasized "one module the forms of which have no common zero [ein Modul, dessen Formen keinen gemeinsamen Nullpunkt besitzen]" [Lasker, 1905, p. 51]. This (homogeneous) ideal is, however, also primary in the modern sense and belongs to the prime ideal of all polynomials in the homogeneous indeterminates with the constant term 0.

[72] In some respects, Alexander Grothendieck's complete redefinition of algebraic geometry in the 1950s resembles Lasker's and Kronecker's views on systems of equations. It is fair to say, moreover, that, especially with Lasker, geometry modulo p was introduced in a systematic way into the theory of special arithmetic rings like $\mathbb{Z}[x_1, \ldots, x_n]$.

This discussion far from exhausts the wealth of ideas in Lasker's paper. At several points, he aptly commented on the work of numerous authors, emphasizing, in particular, the importance of Max Noether's so-called "Fundamental" or "$A\phi + B\psi$" Theorem for the "method of modular systems" [Lasker, 1905, p. 44-46].

[73] "jedes von $\mathcal{O}$ verschiedene Ideal $\mathcal{A}$ ist entweder einartig, oder es läßt sich, und zwar nur auf eine einzige Weise, als ein Produkt von lauter einartigen Idealen darstellen, die zugleich relative Primideale sind." See [Dirichlet, 1863, 3rd ed. 1879, §172] or [Dedekind, 1930–1932, 3:303].

of primes. A literal translation of Dedekind's definition of "einartige Ideale" to more general rings would fail to convey the sense of what he had in mind, since rings like $\mathbb{Z}[x_1,\ldots,x_n]$ or $\mathbb{Q}[x_1,\ldots,x_n]$ have ascending chains of two or more prime ideals different from (0). For instance, $(x_1) \subset (x_1,x_2) \subset \cdots \subset (x_1,\ldots,x_n)$. It was Lasker who succeeded in generalizing Dedekind's "einartige Ideale" via his concept of primary ideals, the main concept of his theory [Lasker, 1905, pp. 51-65].

In his lectures on number theory, Kronecker had tried to generalize the unique prime factorization of rational integers to modular systems $(f_1,\ldots,f_k)$ over the domains $\mathbb{Z}[x_1,\ldots,x_n]$, where $n \geq 1$ [Kronecker, 1901, pp. 170-241]. He was very much concerned with the reduction (via equivalence) of these systems to simpler ones. As an obvious first step, he divided all members of the given modular system by their greatest common divisor. He next considered the natural numbers m with the property $m \equiv 0 \pmod{(f_1,\ldots,f_k)}$. There are two cases: if 0 is the only number of this kind, and if m_0 is the least positive number with $m_0 \equiv 0 \pmod{(f_1,\ldots,f_k)}$. Together with the canonical factorization $m_0 = \prod_p p^{e(p)}$ (p runs through the rational primes), Kronecker derived the decomposition

$$(f_1,\ldots,f_k) \sim (m_0,f_1,\ldots,f_k) \sim \prod_p (p^{e(p)}, f_1,\ldots,f_k).$$

Assuming $n = 1$, the Euclidean algorithm applies in the domain $\mathbb{Q}[x_1]$.[74] From $\gcd(f_1,\ldots,f_k) = 1$, it follows that the second case with $m_0 \geq 1$ always obtains. The modular systems $(p^{e(p)}, f_1,\ldots,f_k)$ give rise to the prime modular systems $(p, P(x_1))$, where $P(x_1)$ is irreducible modulo p, and to certain simple modular systems $(p^{e(p)}, g_1,\ldots,g_h)$ which contain only one prime modular systems by definition. By means of these concepts, Kronecker proved that every modular system $(f_1,\ldots,f_k)$ consisting of non-constant members with $\gcd(f_1,\ldots,f_k) = 1$ is a product of what he termed coprime "simple systems" [Kronecker, 1901, pp. 227-228]. Moving from Kronecker's modular systems to the ideals generated by their members allows for the reformulation of Kronecker's result according to the following dictionary: prime modular systems are prime ideals, simple modular systems are primary ideals, and products of coprime modular systems are products or intersections of coprime ideals. Stated in these new terms, Kronecker's result is: in the domain $\mathbb{Z}[x_1]$, every ideal is the intersection of primary ideals.

The Work of Emmy Noether and Her Successors

In her ground-breaking paper, "Idealtheorie in Ringbereichen," Emmy Noether proved this central result: "Let D be any domain with the ascending chain condition for ideals [that is, D is noetherian]. Then every ideal is the 'least common multiple' [or intersection] of 'primary" ideals'" [Noether, 1921, pp. 33-41, or Noether, 1983, pp. 363-371]. She defined primary ideals this way: "An ideal $\mathcal{Q}$ is called 'primary' if, from $a \cdot b \equiv 0\ (\mathcal{Q});\ a \not\equiv 0\ (\mathcal{Q})$, it necessarily follows that $b^\kappa \equiv 0\ (\mathcal{Q})$, where κ is finite."[75] Why did she use the terminology least common multiple instead of intersection here? And where did the notion of primary ideal come from? Both questions trace back first to Lasker and then to Dedekind and Kronecker. In fact, Noether adopted all of Dedekind's terminology for the divisibility relations between

---

[74]The special case $m_0 = 1$ can be left aside.

[75]"Ein Ideal $\mathcal{Q}$ heißt primär, wenn aus $a \cdot b \equiv 0\ (\mathcal{Q});\ a \not\equiv 0\ (\mathcal{Q})$ notwendig folgt: $b^\kappa \equiv 0\ (\mathcal{Q})$, wo der Exponent κ eine endliche Zahl ist." See [Noether, 1921, p. 37] or [Noether, 1983, p. 367].

rational integers in her theory of modules and ideals [Dedekind, 1930–1932, 3:62-75]. Indeed, for rational integers a, b, m, the equality $m = \text{lcm}(a, b)$ can be replaced by the set-theoretic equality of ideals $(m) = (a) \cap (b)$. This answers the first question.

Relative to the second question, as noted above, Lasker's concept of primary ideals involved explicit reference to some given prime ideal. This was not the case with Noether. She succeeded in freeing that concept from any such reference. It is not difficult to see, however, that, for example, for the rings $\overline{\mathbb{Q}}[x_1, \ldots, x_n]$, Lasker's and Noether's definitions of primary ideals are equivalent. Let $\mathcal{Q}$ be a primary ideal according to Noether's definition. Then it follows easily that $\mathcal{P} := \sqrt{\mathcal{Q}}$ is a prime ideal. From Hilbert's Nullstellensatz, moreover, $V(\mathcal{Q}) = V(\mathcal{P})$. Now suppose that $fg \in \mathcal{Q}$, for $f \notin \mathcal{P}$, and assume for the moment that $g \notin \mathcal{Q}$. Then, by Noether's definition, $f^m \in \mathcal{Q}$, for some $m \geq 1$. From $f \notin \mathcal{P}$, however, it follows that $f^k \notin \mathcal{Q}$, for all $k \geq 1$. This contradiction shows that $g \in \mathcal{Q}$. Hence, $\mathcal{Q}$ is primary in Lasker's sense. Conversely, let $\mathcal{Q}$ be a primary ideal in Lasker's sense of "belonging to the prime ideal $\mathcal{P}$." Lasker proved that $\mathcal{P}^k \subseteq \mathcal{Q}$, for some exponent $k \geq 1$ [Lasker, 1905, p. 56 (theorem X)]. Suppose $fg \in \mathcal{Q}$, for $f \notin \mathcal{Q}$, and assume that $g^m \notin \mathcal{Q}$, for all $m \geq 1$. Then $g \notin \mathcal{P}$ because $\mathcal{P}^k \subseteq \mathcal{Q}$. Therefore, by Lasker's definition, $f \in \mathcal{Q}$, which contradicts the assumption that $f \notin \mathcal{Q}$. Hence $g^m \in \mathcal{Q}$, for some $m \geq 1$, and $\mathcal{Q}$ is primary in Noether's sense as well. In conclusion, Noether's definition can be reformulated as follows: an ideal $\mathcal{I}$ is primary if and only if its radical $\sqrt{\mathcal{I}}$ is a prime ideal or, equivalently, if and only if there is a smallest prime ideal containing $\mathcal{I}$, in other words, if and only if the intersection of all prime ideals containing $\mathcal{I}$ is a prime ideal, too. This formulation comes very close to Dedekind's definition of "einartige Ideale."

Elementary number theory shows that the following are equivalent and that this chain of equivalences characterizes the rational prime powers:

(1) $\pm q \neq 0, \pm 1$ is the power of a rational prime;
(2) there is a single rational prime dividing q;
(3) from $ab \equiv 0$ (q); $a \not\equiv 0$ (q) it follows that $b^\kappa \equiv 0$ (q), where κ is finite;
(4) q is lcm-irreducible, that is, from $q = \text{lcm}(a, b)$, it follows $\pm a = q$ or $\pm b = q$;
(5) (q) is $\cap$-irreducible;
(6) there exists a single prime ideal (p) with $(q) \subseteq (p)$; and
(7) there is a rational prime p such that $ab \equiv 0(q)$ and $a \not\equiv 0(p)$ entails $b \equiv 0(q)$.

From this, it is clear that Noether's primary ideals generalize the rational prime powers. Thus, for noetherian domains, Emmy Noether's theorem can be regarded as the farthest-reaching generalization of the main theorem of elementary multiplicative number theory. Noether herself stressed this fact in the introductory pages of her paper.

Some nice properties do get lost, however. First, primary ideals are no longer necessarily powers of prime ideals. Second, primary ideals are, in general, not uniquely determined, although some uniqueness assertions will hold, if one cancels all primary components not containing the intersection of the remaining ones. A further reduction can be achieved if all primary components associated with the same prime ideal $\mathcal{P}$ are replaced by their intersection, which turns out to be primary and which also belongs to $\mathcal{P}$. This allows for a non-redundant decomposition $\mathcal{I} = \mathcal{Q}_1 \cap \cdots \cap \mathcal{Q}_k$ into primary components $\mathcal{Q}_1, \ldots, \mathcal{Q}_k$ with distinct associated prime ideals $\mathcal{P}_1, \ldots, \mathcal{P}_k$, respectively. Noether proved that the set $\{\mathcal{P}_1, \ldots, \mathcal{P}_k\}$ is uniquely determined by the ideal $\mathcal{I}$ and, moreover, that among the $\mathcal{Q}_i$s, those components are uniquely determined whose associated prime ideals $\mathcal{P}_i$ are minimal in the set $\{\mathcal{P}_1, \ldots, \mathcal{P}_k\}$.[76] These unique primary components are termed non-embedded or isolated in allusion to the geometry of the situation (consider, for example, the rings $\overline{\mathbb{Q}}[x_1, \ldots, x_n]$ and the algebraic sets in $\overline{\mathbb{Q}}^n$ discussed in the previous section). The decomposition $\mathcal{I} = \mathcal{Q}_1 \cap \cdots \cap \mathcal{Q}_k$ provides the representation $V(\mathcal{I}) = V(\mathcal{Q}_1) \cup \cdots \cup V(\mathcal{Q}_k)$, where the sets $V(\mathcal{Q}_i) = V(\mathcal{P}_i)$, for $1 \leq i \leq k$, are irreducible. Among the $V(\mathcal{P}_i)$s, there can be proper inclusions. Hence some of the $V(\mathcal{P}_i)$s can be redundant, but the maximal ones among them are not redundant and are uniquely determined. Similarly, the corresponding prime ideals are uniquely determined. The maximal components occurring in the representation above are simply those which are not embedded in any other components. The ideal-theoretic uniqueness properties in question are thus highly plausible in light of the geometric picture. It is no exaggeration to say that Noether's 1921 paper was a work of genius which showed the amazing consequences of the ascending chain condition for ideals.

Another definitive result is due to Krull, who, in the early 1930s, characterized those domains—now called Krull domains—which admit a sound divisor theory like that of Kronecker [Krull, 1939, pp. 26-45].[77] Also in the 1930s, the American mathematician Joseph Ritt inaugurated a new branch of algebra now called differential algebra [Ritt, 1950].[78] This area focuses on differential equations from an algebraic point of view in close analogy to algebraic equations. It is remarkable how Ritt and his pupils, especially Ellis R. Kolchin, used the methods due to Kronecker, Hilbert, König, Lasker, Emmy Noether, and other algebraists to connect the differential calculus and the ideas of Galois theory and algebraic geometry.

Concluding Remarks

Looking back at the results and questions discussed in the first section above, it becomes clear that many problems in mathematics amount to problems of divisibility in certain commutative-associative domains of "quantities" like algebraic numbers or algebraic functions. In the nineteenth century, the efforts of numerous outstanding mathematicians were directed towards the supreme goal of building divisibility theories for the domains in question in direct analogy to elementary number theory. Kummer's invention of ideal complex numbers played a crucial role in this connection. After Kummer, the development of divisibility theories can be described as proceeding along two lines. In today's terminology, the first line involved embedding domains in principal ideal domains or direct products of such domains (Weierstrass, Kronecker, Zolotarev, Hensel, Weber, König, Krull); the second was inaugurated by Dedekind's theory of ideals and culminated in the contributions of Hilbert, Hurwitz, Mertens, Lasker, Macaulay, and Emmy Noether. It was Emmy Noether's theory of rings with ascending chain condition that gave the farthest-reaching generalization of elementary multiplicative number theory and, in a certain sense, unified all preceding divisibility theories.

---

[76]Compare [Jacobson, 1989, p. 439].

[77]We omit the technical details here and refer the reader to [Neumann, 2002, pp. 173-174 (theorem 2.5.11)].

[78]Compare [Alten et al., 2003, pp. 604-606].

Acknowledgments

I am very grateful to Jeremy Gray and Karen Parshall for their linguistic help during the preparation of this chapter. During the workshop on the history of algebra at the Mathematical Sciences Research Institute, Berkeley at which I delivered a talk on this topic, I was generously supported by a grant from the National Science Foundation. Special thanks go to the referees for their constructive and helpful criticism and to Lincoln C. Durst for his reference to and quotations from the English translation of Euler's *Algebra*.

References

Abel, Niels Henrik. 1881. *Oeuvres complètes*. Ed. Sophus Lie and Ludwig Sylow. 2 Vols. Christiana: Imprimerie de Grondahl & Son.

Alten, Heinz-Wilhelm *et al.* 2003. *4000 Jahre Algebra: Geschichte, Kulturen, Menschen*. Berlin: Springer-Verlag.

Bashmakova, Izabella G. 1949. "Obosnovanie delimosti v trudakh E. I. Zolotareva." *Istoriko-matematičeskie issledovaniya* 2, 233-351.

Bezout, Étienne. 1779. *Théorie générale des équations algébriques*. Paris: Imprimerie de Philippe-Denis Pierres; English Trans. 2006. *General Theory of Algebraic Equations*. Trans. Eric Feron. Princeton: Princeton University Press.

Borevich, Zenon I. and Shafarevich, Igor R. 1986. *Number Theory*. New York: Academic Press, Inc.

Bourbaki, Nicolas. 1994. *Elements of the History of Mathematics*. Trans. John Meldrum. Berlin: Springer-Verlag.

Brill, Alexander and Noether, Max. 1894. "Die Entwicklung der Theorie der algebraischen Functionen in älterer und neuerer Zeit." *Jahresbericht der Deutscher Mathematiker-Vereinigung* 3, 109-566.

Dedekind, Richard. 1930–1932. *Gesammelte mathematische Werke*. 3 Vols. Ed. Robert Fricke, Emmy Noether, and Oystein Øre. Braunschweig: Vieweg Verlag.

Dieudonné, Jean. 1974. *Cours de géométrie algébrique*. Vol. 1. *Aperçu historique sur la développement de la géométrie algébrique*. Paris: Presses universitaires de France.

———. 1978. *Abrégé d'histoire des mathématiques*. 2 Vols. Paris: Hermann.

———. 1985. *Geschichte der Mathematik 1700–1900*. Braunschweig/Wiesbaden: Vieweg Verlag.[79]

Dirichlet, Peter Gustav Lejeune. 1863. *Vorlesungen über Zahlentheorie*. Braunschweig: Vieweg Verlag; Ed. and Supp. Richard Dedekind. 2nd Ed. 1871. 3rd Ed. 1879. 4th Ed. 1894.

Edwards, Harold M. 1980. "The Genesis of Ideal Theory." *Archive for History of Exact Sciences* 23, 321-378.

———. 1990. *Divisor Theory*. Boston/Basel/Berlin: Birkhäuser Verlag.

---

[79]The German translation contains numerous annotations by the translators and additional references, especially, relative to chapter 5 on number theory by the author of the present chapter.

Edwards, Harold M.; Neumann, Olaf; and Purkert, Walter. 1982. "Dedekinds 'Bunte Bemerkungen' zu Kroneckers 'Grundzüge.'" *Archive for History of Exact Sciences* 27, 49-85.

Engström, H. T. 1930a. "The Theorem of Dedekind in the Ideal Theory of Zolotarev." *American Mathematical Monthly* 37, 879-887.

──────────. 1930b. "An Example of the Ideal Theory of Zolotarev." *Transactions of the American Mathematical Society* 32, 128-129.

Euler, Leonhard. 1770. *Vollständige Anleitung zur Algebra*. St. Petersburg: Kayserliche Academie der Wissenschaften; Rev. German Ed. 1959. Ed. Joseph E. Hofmann. Stuttgart: Philipp Reclam; English Trans. 1840. *Elements of Algebra*. Trans. Rev. John Hewlett. London: Longman and Orme; Reprint Ed. 1972. New York: Springer-Verlag.

Gauss, Carl Friedrich. 1801. *Disquisitiones arithmeticæ*. Leipzig: Gerhardt Fleischer. In *Werke*. Vol. 1. German Trans. in [Maser, 1889, pp. 1-453].

──────────. 1828. "Theoria residuorum biquadraticorum: Commentatio prima." In *Werke*. 2:65-92. German Trans. in [Maser, 1889, pp. 511-533].

──────────. 1832. "Theoria residuorum biquadraticorum: Commentatio secunda." In *Werke*. 2:93-148. German Trans. in [Maser, 1889, pp. 534-586].

──────────. 1863–1929. *Werke*. Ed. Gesellschaft der Wissenschaften zu Göttingen. 12 Vols. Göttingen: Akademie der Wissenschaften.

Gray, Jeremy. 1989. "Algebraic Geometry in the Late Nineteenth Century." In *The History of Modern Mathematics*. Ed. David E. Rowe and John McCleary. 2 Vols. Boston: Academic Press, Inc., 1:361-385.

──────────. 1994. "German and Italian Algebraic Geometry." *Rendiconti del Circolo matematico di Palermo*. 2d Ser. Supp. 36, 151-184.

──────────. 1997. "Algebraic Geometry between Noether and Noether: A Forgotten Chapter in the History of Algebraic Geometry." *Revue d'histoire des mathématiques* 3, 1-48.

Haubrich, Ralf. 1992. *Zur Entstehung der algebraischen Zahlentheorie Richard Dedekinds*. Unpublished Doctoral Dissertation: Georg-August-Universität Göttingen.

Hensel, Kurt. 1894. "Untersuchung der Fundamentalgleichung einer Gattung für eine reelle Primzahl als Modul und Bestimmung der Theiler ihrer Discriminante." *Journal für die reine und angewandte Mathematik* 113, 61-83.

Hilbert, David. 1890. "Über die Theorie der algebraischen Formen." *Mathematische Annalen* 36, 473-534, or [Hilbert, 1932–1935, 2:199-257].

──────────. 1893. "Über die vollen Invariantensysteme." *Mathematische Annalen* 42, 313-373, or [Hilbert, 1932–1935, 2:287-344].

──────────. 1932–1935. *Gesammelte Abhandlungen*. 2 Vols. Berlin: Springer-Verlag.

Hurwitz, Adolf. 1895. "Zur Theorie der algebraischen Zahlen." *Nachrichten der Gesellschaft der Wissenschaften zu Göttingen: Mathematisch-physikalische Klasse*, 324-331, or Hurwitz, Adolf. 1932–1933. *Mathematische Werke*. 2 Vols. Basel: Birkhäuser Verlag, 2:236-243; Reprint Ed. 1963. Basel: Birkhäuser Verlag.

Jacobson, Nathan. 1985–1989. *Basic Algebra*. 2nd Ed. New York: W. H. Freeman.

Jung, Heinrich W. E. 1923. *Einführung in die Theorie der algebraischen Funktionen einer Veränderlichen*. Berlin/Leipzig: Walter de Gruyter.

Klein, Felix. 1892. *Riemannsche Flächen: Vorlesungen, gehalten in Göttingen 1891/92*. Reprint Ed. 1986. Ed. Günther Eisenreich and Walter Purkert. Leipzig: B. G. Teubner Verlag.

_____. 1893. "Über die Komposition der binären quadratischen Formen." *Nachrichten der Königlichen Gesellschaft der Wissenschaften zu Göttingen: Mathematisch-physikalische Klasse*. In Klein, Felix. 1973. *Gesammelte mathematische Abhandlungen*. 3 Vols. Berlin: Springer-Verlag, 3:283-286.

König, Julius. 1903. *Einleitung in die allgemeine Theorie der algebraischen Gröszen*. Leipzig: B. G. Teubner Verlag.

Kolmogorov, Andrei and Yushkevich, Adolf. 1992–1998. *Mathematics of the 19th Century*. Trans. Roger Cooke. 3 Vols. 1992. *Mathematical Logic, Algebra, Number Theory, Probability Theory*. Vol. 1. Basel: Birkhäuser Verlag.

Kronecker, Leopold. 1882. "Grundzüge einer arithmetischen Theorie der algebraischen Grössen." *Journal für die reine und angewandte Mathematik* 92, 1-122. In Kronecker, Leopold. 1895–1931. *Werke*. Ed. Kurt Hensel. 5 Vols. Leipzig/Berlin: B. G. Teubner Verlag, 2:237-387.

_____. 1901. *Vorlesungen über Zahlentheorie*. Ed. Kurt Hensel. Leipzig: B. G. Teubner Verlag.

Krull, Wolfgang. 1939. "Allgemeine Modul-, Ring- und Idealtheorie." In *Enzyklopädie der mathematischen Wissenschaften*. 2d Ed. Leipzig/Berlin: B. G. Teubner Verlag, No. 5, Art. 11.

Kummer, Ernst Eduard. 1975. *Collected Papers*. Ed. André Weil. 2 Vols. *Contributions to Number Theory*. Vol. I. Berlin: Springer-Verlag.

Lagrange, Joseph Louis. 1769. "Sur la solution des problèmes indéterminés du second degré." *Mémoires de l'Académie royale des sciences et des belles-lettres de Berlin* 22, 1165-1319, or Lagrange, Joseph-Louis. 1867–1893. *Oeuvres*. 14 Vols. Paris: Gauthier-Villars, 2:377-535.

Landau, Edmund. 1917. "Richard Dedekind: Gedächtnisrede." *Nachrichten der Königlichen Gesellschaft der Wissenschaften zu Göttingen: Geschäftliche Mitteilungen*, 50-70.

Landsberg, Georg. 1899. "Algebraische Gebilde: Arithmetische Theorie algebraischer Grössen." *Encyklopädie der mathematischen Wissenschaften*. 6 Vols. Leipzig: B. G. Teubner Verlag, 1.1:283-319.

Lasker, Emanuel. 1905. "Zur Theorie der Moduln und Ideale." *Mathematische Annalen* 60, 20-116 and 607-608.

Liouville, Joseph. 1847. "Sur les équations algébriques à plusieurs inconnues." *Journal de mathématiques pures et appliquées* 12, 68-72.

Macaulay, Francis S. 1916. *The Algebraic Theory of Modular Systems*. Cambridge: Cambridge University Press; 2d Ed. 1996.

Maser, Hermann, Ed. 1889. *Untersuchungen über höhere Arithmetik*. Berlin: Julius Springer Verlag: Berlin; Reprint Ed. 1981. New York: Chelsea Publishing Company.

Mertens, Franz. 1894. "Über die Fundamentalgleichung eines Gattungsbereiches algebraischer Zahlen." *Sitzungsberichte der Akademie der Wissenschaften zu Wien: Mathematisch-Naturwissenschaftiche Klasse* 103, 5-40.

Molk, Jules. 1885. "Sur une notion qui comprend celle de la divisibilité et sur la théorie générale de l'élimination." *Acta Mathematica* 6, 1-166.

Netto, Eugen. 1896–1900. *Vorlesungen über Algebra*. 2 Vols. Leipzig: B. G. Teubner Verlag.

—————. 1899. "Rationale Funktionen mehrerer Veränderlichen." *Encyklopädie der mathematischen Wissenschaften*. 6 Vols. Leipzig: B. G. Teubner Verlag, 1.1:255-282.

Neumann, Olaf. 1979–1980. "Bemerkungen aus heutiger Sicht über Gauß' Beiträge zu Zahlentheorie, Algebra und Funktionentheorie." 3 Parts. *NTM. Schriftenreihe für Geschichte der Naturwissenschaften, Technik und Medezin* 16(2), 22-39; 17(1), 32-48; 17(2), 38-58. (Parts 2 and 3 are also under the title "Zur Genesis der algebraischen Zahlentheorie.")

—————. 1981. "Über die Anstöße zu Kummers Schöpfung der 'Idealen Complexen Zahlen.'" In *Mathematical Perspectives: Essays on Mathematics and Its Historical Development*. Ed. Joseph W. Dauben. New York: Academic Press, Inc., 179-199.

—————. 2002. "Was sollen und was sind Divisoren?" *Mathematische Semesterberichte* 48, 139-192.

—————. 2005. "Carl Friedrich Gauß: *Disqusitiones Arithmeticæ* (1801)." In *Landmark Writings in Western Mathematics, 1640–1940*. Ed. Ivor Grattan-Guinness. Amsterdam: Elsevier, pp. 303-315.

—————. 2006. "C. F. Gauss's *Disquisitiones Arithmeticæ* and the Theory of Equations." In *The Shaping of Arithmetic: Number Theory After Carl Friedrich Gauss's Disquisitiones Arithmeticæ*. Ed. Catherine Goldstein, Norbert Schappacher, and Joachim Schwermer, Berlin: Springer-Verlag, pp. .

Noether, Emmy. 1919. "Die arithmetische Theorie der algebraischen Funktionen einer Veränderlichen in ihrer Beziehung zu den übrigen Theorien und zu der Zahlkörpertheorie." *Jahresbericht der Deutschen Mathematiker-Vereinigung* 38, 182-203, or [Noether, 1983, pp. 271-292].

—————. 1921. "Idealtheorie in Ringbereichen." *Mathematische Annalen* 83, 24-66, or [Noether, 1983, pp. 354-396].

—————. 1983. *Gesammelte Abhandlungen/Collected Papers*. Ed. Nathan Jacobson. Berlin: Springer-Verlag.

Nový, Luboš. 1973. *Origins of Modern Algebra*. Prague: Academia.

Ostrowski, Alexander. 1919. "Zur arithmetischen Theorie der algebraischen Grössen." *Nachrichten der Gesellschaft der Wissenschaften zu Göttingen: Mathematisch-physikalische Klasse*, 279-298.

Piazza, Paola. 1998. "Zolotarev's Foundation of Algebraic Number Theory." Unpublished Doctoral Dissertation: Università degli studi di Messina.

Poincaré, Henri. 1880. "Sur un mode nouveau de représentation géométrique des formes quadratiques définies ou indéfinies." *Journal de l'École polytechnique* 47, 177-245, or Poincaré, Henri. *Oeuvres de Henri Poincaré*. Ed. Paul Appell *et al*. 11 Vols. Paris: Gauthier-Villars, 5:117-180.

Riemann, Bernhard. 1857. "Theorie der Abel'schen Functionen." *Journal für die reine und angewandte Mathematik* 54, 101-155, or Riemann, Bernhard. 1990. *Gesammelte mathematische Werke, wissenschaftlicher Nachlaß und Nachträge/Collected Papers.* Ed. Heinrich Weber, Richard Dedekind, and Raghavan Narasimhan. 3d Ed. Berlin: Springer-Verlag, pp. 120-174.

Ritt, Joseph Fels. 1950. *Differential Algebra.* Colloquium Publications. Vol. 33. Providence: American Mathematical Society; Reprint Ed. 1966. New York: Dover Publications, Inc.

Salmon, George. 1866. *Lessons Introductory to the Modern Higher Algebra.* Dublin: Hodges. German Trans. 1873. *Vorlesungen über die Algebra der linearen Transformationen.* Ed. and Supp. Wilhelm Fiedler. 2d Ed. Leipzig: B. G. Teubner Verlag.

Scholz, Erhard. 1990. *Geschichte der Algebra.* Mannheim/Zürich/Vienna: B. I., Wissenschaftsverlag.

Serret, Joseph Alfred. 1849. *Cours d'algèbre supérieure.* Paris: Gauthier-Villars. 2d Ed. 1854. 3d Ed. 1866. 5th Ed. 1885.

Selling, Eduard. 1865. "Ueber die idealen Primfactoren der complexen Zahlen, welche aus den Wurzeln einer beliebigen irreductiblen Gleichung rational gebildet werden." *Zeitschrift für Mathematik und Physik* 10, 17-47.

Smith, Henry John Stephen. 1859–1865. "Report on the Theory of Numbers." Parts 1-6. *Report of the British Association for the Advancement of Science* 28, 228-267; 29, 120-172; 30, 292-340; 31, 503-526; 32, 768-786; and 34, 322-374, or Smith, Henry John Stephen. 1894. *The Collected Mathematical Papers of Henry John Stephen Smith.* Ed. James W. L. Glaisher. 2 Vols. Oxford: Clarendon Press, 1, 38-364; Reprint Ed. 1965. New York: Chelsea Publishing Co.

Sylvester, James Joseph. 1904–1912. *The Collected Mathematical Papers.* 4 Vols. Cambridge: Cambridge University Press. Reprint Ed. 1973. New York: Chelsea Publishing Company.

Tchebotarev, Nicolai G. 1930. "The Foundations of the Ideal Theory of Zolotarev." *American Mathematical Monthly* 37, 117-128.

Tropfke, Johannes. 1980. *Geschichte der Elementarmathematik.* 4th Ed. *Arithmetik und Algebra.* Ed. Kurt Vogel, Karin Reich, and Helmuth Gericke. Berlin/New York: Walter de Gruyter.

Ullrich, Peter. 1994. "Zur Algebraisierung der komplexen Analysis seit Weierstraß." Unpublished Seminar Talk. 20 October, 1994. Universität Leipzig.

—————. 1998. "The Genesis of Hensel's p-adic Numbers." In *Charlemagne and His Heritage: 1200 Years of Civilization and Science in Europe.* Ed. Paul L. Butzer, Hubertus Jongen, and Walter Oberschelp. 2 Vols. *Mathematical Arts.* Vol. 2. Turnhout: Brepols Publishers, 163-178.

—————. 1999. "Die Entdeckung der Analogie zwischen Zahl- und Funktionenkörpern: Der Ursprung der 'Dedekind-Ringe.'" *Jahresbericht der Deutschen Mathematiker-Vereinigung* 101, 116-134.

—————. 2003. "Die Weierstraßschen 'analytischen Gebilde': Alternativen zu Riemanns 'Flächen' und Vorboten der komplexen Räume." *Jahresbericht der Deutschen Mathematiker-Vereinigung* 105, 30-59.

Van der Waerden, Bartel L. 1939. *Einführung in die algebraische Geometrie.* Berlin: Springer-Verlag. 3d Ed. 1973.

Waterhouse, William. 1984. "A Note by Gauss Referring to Ideals?" *Historia Mathematica* 11, 142-146.

Weber, Heinrich. 1893. "Die allgemeinen Grundlagen der Galois'schen Gleichungstheorie." *Mathematische Annalen* 43, 521-549.

———. 1895–1908. *Lehrbuch der Algebra.* 3 Vols. Braunschweig: Vieweg Verlag. Reprint Ed. 1961. Braunschweig: Vieweg Verlag. Reprint Ed. 1979. New York: Chelsea Publishing Company.

Weber, Heinrich. 1912. *Lehrbuch der Algebra: Kleine Ausgabe in einem Bande.* Braunschweig: Vieweg Verlag.

Weierstrass, Karl. 1894–1927. *Mathematische Werke.* 7 Vols. Berlin: Mayer & Müller. Reprint Ed. 1967. Hildesheim/New York: Georg Olms Verlag.

Weil, André. 1983. *Number Theory: An Approach through History from Hammurapi to Legendre.* Boston: Birkhäuser Verlag.

Zolotarev (Zolotareff), Egor (G.) Ivanovič. 1880. "Sur la théorie des nombres complexes." *Journal de mathématiques pures et appliquées* 6, 51-84, 115-128, and 129-166.

CHAPTER 5

Kronecker's Fundamental Theorem of General Arithmetic

Harold M. Edwards
Courant Institute, United States

Introduction

The theorem of my title is the theorem Leopold Kronecker published in 1887 in the 100th volume of Crelle's *Journal* under the title *"Ein Fundamentalsatz der allgemeinen Arithmetik"* [Kronecker, 1887]. At the time, Kronecker was an editor[1] of Crelle, and he assigned particular importance to the 100th volume, soliciting papers from the most distinguished mathematicians of the day. The publication of his "Fundamentalsatz" in this volume shows that it—the 107th paper in the chronological list of his publications—was a paper to which he assigned special significance.

Further proof of the importance of the paper to him is provided by a letter he wrote to Gösta Mittag-Leffler on 4 April, 1886. Kronecker referred in the letter both to a major grievance he felt against Mittag-Leffler in connection with a prize competition Mittag-Leffler had organized the previous year and to an earlier exchange of opinions that the two had had about the foundations of mathematics; in the earlier discussion, Kronecker had promised a paper for Mittag-Leffler's newly founded *Acta Mathematica* about foundations that would elaborate on his objections to new views that were being advanced at the time.

> I promised Hermite I would not bring up again the circumstance that so angered me last year. ... I've been working extremely hard. ... In my two courses, on higher arithmetic and on the algebra of elliptic functions, I have newly worked out a great deal of the material, and beyond this I have put the foundations of algebra in an entirely new form. ... You will recall that I originally intended to give you my works on this subject for publication. ... The so-called fundamental theorem of algebra is replaced by my new "fundamental theorem of general arithmetic." ... I owe this beautiful and sure foundation of algebra to my sharp critique of the method of defining quantities that

---

[1] He and Karl Weierstrass were joint editors, but, according to Biermann's history of mathematics in Berlin, relations between the two were extremely strained by this time [Biermann, 1973, pp. 100-102]. It is my impression that Kronecker was the more energetic editor and that he was the moving force behind making the 100th volume a special event. Compare Kronecker's letter to Rudolph Lipschitz of 13 November, 1885 in [Lipschitz, 1986, pp. 185-186].

began with Heine and to the precious *Galois principle*. It will appear in the 100th volume of our journal [Archives Mittag-Leffler].

Today, Kronecker's reference to Eduard Heine in this connection seems surprising. Instead of Heine, we are more likely today to associate the ideas Kronecker referred to here with Karl Weierstrass, Richard Dedekind, and Georg Cantor. But one of Kronecker's very few published remarks on the foundations of mathematics, a footnote [Kronecker, 1895–1930, 3a:156] to a paper on module systems, called Heine the first to have attempted to grasp the concept of the "irrational" in complete generality, so the meaning of Kronecker's reference to Heine in the letter is unmistakable. The footnote deals specifically with the treatment of infinites. In it, he took a position that seems extreme in today's climate of opinion: infinite series should be avoided unless a specific rule for generating their terms is provided.

In the letter, he expressed another opinion that is radical by today's standards: the "modern" method of defining quantities (his quotation marks) *cannot be used* for the definition of the coefficients of two polynomials if one intends to use the algorithm for finding their greatest common divisor.[2] He did not elaborate, but he surely meant that if, for example, one wants to find the reciprocal of a nonzero element of the field $\mathbb{Q}[\sqrt{2}]$, which involves finding the greatest common divisor of $x^2 - 2$ and a nonzero polynomial $f(x)$ of degree less than 2, it is not enough to know the coefficients of $f(x)$ as *real numbers;* they must be known *exactly* as rational numbers, because arbitrarily small variations in the coefficients can cause major changes in the greatest common divisor of the polynomials. As long as one is considering the problem of "finding the reciprocal of $f(\sqrt{2})$," it is natural to assume that the coefficients are rational numbers, but Kronecker's point no doubt is that if one considers instead the problem of "finding the greatest common divisor of $x^2 - 2$ and $f(x)$" the answer *cannot be found* if $f(x) = x + a$ for a number a that is very close to $\sqrt{2}$ but may not be exactly $\sqrt{2}$.

In other words, in arithmetic and algebra, which were Kronecker's greatest interests, real numbers as we now conceive of them are not only inappropriate but

---

[2]Here is the quote in full: "I now know quite specifically that the 'modern' way of defining quantities is not applicable if one wants to define coefficients of polynomials to which the algorithm for finding the greatest common denominator will be applicable. *Only* polynomials with rational coefficients are permissible in this case, and everything that has been published about this is at the least inexact. My treatment of algebra, as I set it forth in my *Festschrift*, is in all subjects in which one makes use of common divisors the only one possible. The place of the so-called fundamental theorem of algebra, which is not applicable in such subjects, is taken by my new 'fundamental theorem of general arithmetic,' that every polynomial in x, whose coefficients are polynomials in variables, can *in the sense of congruence for a prime module system* actually be decomposed as a product of linear factors, so that it has a complete set of *rational* congruence roots [Ich weiss jetzt ganz genau, dass die 'moderne' Grossenerklärung nicht anwendbar ist, wenn man Coefficienten ganzer Functionen definirien will, auf die das Verfahren der Auffindung des grossten gemeinsamen Theilers angewendet werden soll. *Nur* ganze Functionen mit rationalen Coefficienten sind dabei zulassig und Alles, was darüber publicirt worden, ist mindestens ungenau. Meine Behandlung der Algebra, wie ich sie in der Festschrift angebahnt habe, ist für alle diejenigen Gebiete, in denen man den Begriff des gemeinsamen Theilers braucht, die einzig mögliche. An Stelle des sogenanten Fundamentalsatzes der Algebra, der für die erwähnten Gebiete unbrauchbar ist, tritt mein neuer 'Fundamentalsatz der allgemeinen Arithmetik,' dass jede ganze Function von x, deren Coefficienten ganzzahlige Functionen von Variabeln sind, sich *im Sinne der Congruenz für ein Primmodulsystem* in ein Product von Linearfactoren wirklich zerlegen lässt, also vollständig *rationale* Congruenzwurzeln besitzt]."

unacceptable. Nonetheless, I believe that David Hilbert's condemnation of Kronecker as a "Verbotsdiktator" is unwarranted because, contrary to the caricature of Kronecker so often repeated in popular mathematical writing, he does not seem to have *forbidden* the use of infinite series but simply to have recommended *avoiding* them [Hilbert, 1922, pp. 159-161]. His criticism of the "modern" ways of dealing with infinites was profoundly *constructive* criticism, in the sense that he did not denounce[3] them. The footnote just mentioned [Kronecker, 1895–1930, 3a:156], in which Kronecker says that arbitrary infinite series are not "permissible [zulässig]," is almost the only prohibition to be found in his writings, and even that, since it occurs in conjunction with the word "avoid [vermeiden]," could be regarded as advice rather than a stern injunction. He simply said the new ways were *unnecessary* and demonstrated in his own work the profound results he was able to obtain without them.

Although Kronecker's views in this matter had scant influence on the development of algebra in the following decades, I believe they are worth our attention both because they bring to a successful conclusion, as I hope to show in this chapter, well over a century of thought about "the fundamental theorem of algebra" and because they are central to Kronecker's approach to algebra, an approach that, valuable as it is, is in danger of remaining forever, in André Weil's words, "buried in the impressive but seldom opened volumes of his *Complete Works*" [Weil, 1950, p. 90].

Statement and Proof of Kronecker's Theorem

Gauss's doctoral dissertation of 1799 was devoted to what we call the fundamental theorem of algebra and what Kronecker in his letter to Mittag-Leffler termed the so-called fundamental theorem of algebra: A polynomial of degree n has n complex roots, or, as Gauss stated it, "a polynomial in one variable with integer coefficients can be written as a product of polynomials of degree one or two with real coefficients" [Gauss, 1799/1876, 3:1]. In the introduction to his thesis, Gauss criticized earlier proofs of this theorem by Leonhard Euler and Joseph-Louis Lagrange, among others, saying that the proofs were circular because they used *computations with the roots* of the polynomial in the course of proving that the roots were complex numbers. How can computations with the roots be justified without first showing that they are complex numbers, the very theorem to be proved?

This key question—how can you compute with the roots of a polynomial?—is answered by Kronecker's "Fundamentalsatz." Gauss himself seems to have reconsidered his criticism because his *second* proof of the theorem, published in 1815, takes a quite different approach [Gauss, 1815]. In this proof, complex numbers

---

[3]Kurt Hensel's foreword to Kronecker's *Vorlesungen über Zahlentheorie* says that "Kronecker was far from wanting to reject altogether a definition or proof that did not meet these highest standards, but he believed that such a proof did lack something, and that supplying this lack was an important task, through which our understanding would be deepened in an essential way [Kronecker war weit davon entfernt, eine Definition oder einen Beweis vollständig zu verwerfen, der jenen höchsten Anforderungen nicht entsprach, aber er glaubte, daß dann eben noch etwas fehle, und er hielt eine Ergänzung nach dieser Richtung hin für eine wichtige Aufgabe, durch die unsere Erkenntnis in einem wesentlichen Punkte erweitert würde]." More than once, I have heard colleagues express surprise at this characterization of Kronecker's views, which is so different from the usual depiction of Kronecker's views and personality. Having failed over many years to find evidence in primary sources for the usual depiction, I believe Hensel's.

enter only at the end. The first part of the proof is devoted to developing a method of computing with the roots of a given polynomial in a rather limited way, and the second uses computations with the roots to show that the roots can be described concretely as complex numbers. In other words, the essence of the theorem has nothing to do with complex numbers but everything to do with computations with the roots. Some of the eighteenth-century proofs Gauss criticized had already arrived at this crucial question without resolving it: *how can you compute with the roots of a given polynomial?* In a peculiar way, Évariste Galois gave the pragmatic answer. His answer was, on the one hand, unfounded, in a sense I will explain, but it was nonetheless the very foundation of Galois theory.

The essential idea was what Kronecker was referring to when he wrote of "the precious Galois principle [das köstliche Galoissche Princip]." It is to make use of the fact that symmetric polynomials of the roots can be evaluated to find a quantity with the properties that (1) it is a root of a known polynomial, and (2) each root of the given polynomial can be expressed rationally in terms of it. I will show how this method makes it possible to compute with the roots of $x^3 - 2$.

By the theory of symmetric polynomials, which was certainly well known to the mathematicians of the eighteenth century, the three roots a, b, and c of $x^3 - 2$ satisfy $a + b + c = 0$, $ab + bc + ca = 0$, and $abc = 2$. We may not know what these roots *are* in the sense that we do not know how to compute with them, but we do know that if we *did* know how to compute with them we would surely find that these three equations held. The "quantity" in the Galois principle as it is stated above can be taken—for a simple and general reason (see below)—to be $a - b$, call it t. We may not know what t *is*, but we can find a polynomial of which it is a root in the following way.

The polynomial $(X-a+b)(X+a-b)(X-b+c)(X+b-c)(X-c+a)(X+c-a)$ is a polynomial in X of degree 6 which, on the one hand, has t as a root, and, on the other hand, has coefficients that are symmetric in a, b, and c and can therefore be evaluated by expressing them in terms of the elementary symmetric polynomials $a + b + c$, $ab + bc + ca$, and abc and using the known values 0, 0, and 2 of these polynomials. The result $X^6 + 108$ can be found this way.

First, combine the three pairs of factors to find $(X^2-(a-b)^2)(X^2-(b-c)^2)(X^2-(c-a)^2)$. Thus, other than the coefficient 1 of X^6, we need only find the coefficients of X^4, X^2, and the constant term. The first of these is $-(a-b)^2-(b-c)^2-(c-a)^2 = -2(a^2+b^2+c^2)+2(ab+bc+ca) = -2(a+b+c)^2+6(ab+bc+ca) = 0$. A similarly simple computation shows that the coefficient of X^2 is also 0. The proof that the constant term $-(a-b)^2(b-c)^2(c-a)^2$ is 108 is easy using a method that is useful in many other contexts as well. The identity

$$(a-b)(b-c)(c-a) = \begin{vmatrix} 1 & 1 & 1 \\ a & b & c \\ a^2 & b^2 & c^2 \end{vmatrix}$$

follows from the observation that both sides are homogeneous polynomials of degree 3 that contain bc^2 and that change sign whenever two of three values a, b, and c are interchanged. Therefore, the square of this determinant is the determinant of

the product
$$\begin{bmatrix} 1 & 1 & 1 \\ a & b & c \\ a^2 & b^2 & c^2 \end{bmatrix} \begin{bmatrix} 1 & a & a^2 \\ 1 & b & b^2 \\ 1 & c & c^2 \end{bmatrix} = \begin{bmatrix} 1+1+1 & a+b+c & a^2+b^2+c^2 \\ a+b+c & a^2+b^2+c^2 & a^3+b^3+c^3 \\ a^2+b^2+c^2 & a^3+b^3+c^3 & a^4+b^4+c^4 \end{bmatrix}.$$

The entries of this matrix are easy to evaluate using the fact that a, b, and c are cube roots of 2 and the identities $a+b+c=0$ and $a^2+b^2+c^2 = (a+b+c)^2 - 4(ab+bc+ca) = 0$, to find

$$\begin{bmatrix} 3 & 0 & 0 \\ 0 & 0 & 2+2+2 \\ 0 & 2+2+2 & 2a+2b+2c \end{bmatrix} = \begin{bmatrix} 3 & 0 & 0 \\ 0 & 0 & 6 \\ 0 & 6 & 0 \end{bmatrix},$$

a matrix whose determinant is -108, and to conclude that the difference $t = a - b$ of any two roots of $x^3 - 2$ is a root of $X^6 + 108$. Knowing that t is a root of this irreducible polynomial tells you how to compute with it in a very concrete sense: compute with polynomials in t with rational coefficients making use of the one relation $t^6 = -108$ to reduce such polynomials until their degree is less than 6. In other words, adjoin one root t of $X^6 + 108$ to the field of rational numbers.

But this is the easy part of Galois's program. The hard part is to express a, b, and c rationally in terms of t. Although the reconstruction I propose has been called "far-fetched" [Neumann, 1986, p. 411], the following remark of Niels Henrik Abel, published a few years before Galois's work, provides new evidence for it: "When a quantity satisfies, at the same time, two given algebraic equations, these equations have a common factor of the first degree. When one supposes that they have no other common factor than this one, one can always, *as one knows*, express the unknown as a rational function of the coefficients of the two equations" [Abel, 1827/1881, p. 212 (my emphasis)].[4] These are the first two sentences of a paper Abel published in Gergonne's *Annales de mathématiques pures et appliquées*. It seems entirely plausible that Galois might have seen and been inspired by this paper of Abel's, but the point is that Abel regarded this principle as well-known and that my ascription of it to Galois is not in the least "far-fetched." The argument is little more than the Euclidean algorithm for polynomials. Since a satisfies two polynomial equations—$a^3 - 2 = 0$ and the polynomial equation with coefficients that involve t (derived below)—the principle can be used to express a rationally in terms of t.

Galois's basic construction uses the formula
$$(V-b)(V-c) = \frac{(V-a)(V-b)(V-c)}{V-a} = \frac{V^3 - 2}{V-a} = \frac{V^3 - a^3}{V-a} = V^2 + aV + a^2$$
or, more generally,
$$(V-b)(V-c)\cdots(V-e) = \frac{f(V) - f(a)}{V-a},$$
where f is the given polynomial and V is a new unknown, to express the elementary symmetric polynomials (and therefore all symmetric polynomials) in b, c, ..., e as polynomials in a. With $V = -t + a$, we have $(-t+a-b)(-t+a-c) = (-t+a)^2 + a(-t+a) + a^2 = t^2 - 2at + a^2 - at + a^2 + a^2 = t^2 - 3at + 3a^2$. Since the

---

[4]"Lorsqu'une quantité satisfait, à la fois, à deux équations algébriques données, ces deux équations ont un facteur commun du premier degré. En supposant quelles n'ont pas d'autre facteur commun que celui-là, on peut toujours, comme l'on sait, exprimer rationellement l'inconnue en fonction des coefficiens des deux équations."

first factor on the left is zero by the definition of t, we have $t^2 - 3at + 3a^2 = 0$, which can be combined with $a^3 = 2$ to determine a rationally in terms of t. Explicitly, $0 = (t+a)(t^2 - 3at + 3a^2) - 3(a^3 - 2) = -2at^2 + t^3 + 6$, from which we find $a = \frac{t^3+6}{2t^2}$. Then $b = a - t = \frac{t^3+6-2t^3}{2t^2} = \frac{-t^3+6}{2t^2}$ and $c = -a - b = -\frac{6}{t^2}$, and we have expressed *all three* roots of $x^3 - 2$ rationally in terms of one root t of $t^6 + 108$.

Kronecker once wrote that "I recognize a true scientific value—in the field of mathematics—only in concrete mathematical truths, or, to put it more pointedly, only in mathematical formulas."[5] The mathematical truth just derived is given by the formula

$$x^3 - 2 \equiv \left(x - \frac{t^3+6}{2t^2}\right)\left(x - \frac{-t^3+6}{2t^2}\right)\left(x + \frac{6}{t^2}\right) \pmod{t^6 + 108},$$

or, expressed without denominators,

$$432(x^3 - 2) \equiv -4t^6(x^3-2) \equiv -(2t^2 x - t^3 - 6)(2t^2 x + t^3 - 6)(t^2 x + 6) \pmod{t^6+108}.$$

It simply states that expansion of the product on the right followed by use of the relation $t^6 + 108 = 0$ to reduce the degree of the resulting expression in t gives the expression on the left.[6] The important feature of the formula is, of course, the fact that the factors on the right are all linear in x. In modern parlance, the adjunction of a root t of $t^6 + 108 = 0$ to the field of rational numbers gives a field over which $x^3 - 2$ becomes a product of linear factors.

The same method exactly can be used to derive, for any given polynomial $f(x)$, say with integer coefficients, a formula of the same form

$$f(x) \equiv (x - \rho_1(t))(x - \rho_2(t)) \cdots (x - \rho_n(t)) \pmod{G(t)},$$

in which $G(t)$ is an irreducible, monic polynomial with integer coefficients and $\rho_1(t)$, $\rho_2(t)$, ..., $\rho_n(t)$ are rational functions of t with integer coefficients.

The construction proceeds in this way. First, set $t = Aa + Bb + \cdots + Ee$, where $a, b, \ldots, e$ are the roots of the given $f(x)$ and where $A, B, \ldots, E$ are strategically chosen integers. (In the example above, $A = 1$, $B = -1$, and $C = 0$.) Next, multiply together the $(n-1)!$ polynomials obtained from $t - (Aa + Bb + \cdots + Cc)$ by permuting $b, c, \ldots, e$ in all possible ways. On the one hand, the result is zero because it is a product in which one factor is zero by the definition of t. On the other hand, it is a polynomial in t and a because it is symmetric in $b, c, \ldots, e$. In this way, one finds a polynomial equation $F(t, a) = 0$. Finally, use the two relations $f(a) = 0$ and $F(t, a) = 0$, of which a is the only common root, to express a rationally in terms of the coefficients of the two relations and therefore as a rational function of t, call it $\rho_a(t)$. The same procedure can be followed for all roots, not just for a, to find $f(x) \equiv (x - \rho_a(t)) \cdots (x - \rho_e(t)) \pmod{G(t)}$, where G is the irreducible polynomial of which t is a root.

The evasive requirement that the integer multipliers $A, B, \ldots, E$ be "chosen strategically" is easily explained. Galois required that the $n!$ quantities $Aa + Bb + \cdots + Ee$ obtained by permuting the roots $a, b, \ldots, e$ be distinct, and this is easy (in theory, not computationally) to guarantee. The product of the $n!(n!-1)$ differences

---

[5]Kronecker to Georg Cantor, 21 August, 1884 in [Meschkowski, 1967, 237-239]. For the English translation, see [Edwards, 1995, p. 45].

[6]Kronecker actually gave the more general formula

$$4 \cdot (9c)^3 \cdot (x^3 - c) \equiv (9cx - t^4)(18cx + 9ct + t^4)(18cx - 9ct + t^4) \pmod{t^6 + 27c^2},$$

except he used c_3 and z instead of c and t [Kronecker, 1887, p. 494].

of these quantities is symmetric in the roots and is therefore a known polynomial in A, B, ..., E with integer coefficients. To fulfill Galois's condition that the $n!$ quantities $Aa + Bb + \cdots + Ee$ be distinct, one only needs to give the integers A, B, ..., E values that make this product of their differences nonzero, something that is obviously possible provided this polynomial in A, B, ..., E with integer coefficients is not identically zero. This, in turn, is true if and only if the roots a, b, ..., e of the given polynomial are distinct because a product of nonzero polynomials is not zero. Since the roots of a polynomial are distinct if and only if it is relatively prime to its derivative, one can, as Kronecker put it, "free a polynomial of multiple roots" by dividing it by the greatest common divisor of it and its derivative and, in this way, reduce the problem of computing with the roots of a given polynomial to the case in which the polynomial has distinct roots.[7]

What one needs to be uneasy about is the very concept of *computing with the roots of the given polynomial*. Once again, in my opinion, the truly fundamental theorem of algebra is the statement that there is a valid way to do this.

The argument I have just outlined is Galois's proof (presented, admittedly, in more detail than Galois gave, but still, I maintain, what Galois clearly intended) that *if there is any valid way to compute* with the roots of a given polynomial $f(x)$, then they can be described as *explicit* elements of the field $\mathbb{Q}[t]$ modulo $G(t)$ obtained by adjoining to the rationals $\mathbb{Q}$ a single root t of an explicitly computable (in theory) irreducible and monic polynomial $G(t)$ with integer coefficients.[8]

Of course, Galois did not put it this way. On the contrary, he seems tacitly to have assumed—as did his eighteenth-century predecessors—that one *can* compute with the roots and arrived at this simple scheme for doing the computations without ever proving that the computations are valid. As far as I know, nothing he wrote shows any wish to justify this rather amazing assumption.

Again, Galois's method constructs for a given $f(x)$ with integer coefficients a formula

$$f(x) \equiv (x - \rho_a(t)) \cdots (x - \rho_e(t)) \pmod{G(t)},$$

where G is irreducible and monic with integer coefficients. This not only constructs a splitting field $\mathbb{Q}[t]$ modulo $G(t)$ of $f(x)$, it also constructs the n roots $\rho_a(t)$, ..., $\rho_e(t)$ of $f(x)$ within it.

Kronecker's "Fundamental Theorem of General Arithmetic" is essentially this same statement freed from reliance on any tacit assumptions, but it, and Galois's theorem too, is more general than the one I have stated. I have been assuming $f(x)$ has *integer* coefficients, but once the theorem has been formulated in this way, it is quite natural to allow the coefficients of $f(x)$ to contain *letters* as well as integers. Galois had such polynomials very much in mind because, among other things, he wanted to prove that the roots of $x^5 + Ax^4 + Bx^3 + Cx^2 + Dx + E$, a polynomial whose coefficients are "letters," cannot be expressed by radicals. In fact, the term "general arithmetic" in Kronecker's title refers to the algebra of polynomials with integer coefficients. The fundamental theorem says that, given a polynomial with integer coefficients in several indeterminates, and given one of the indeterminates

---

[7]If a polynomial $F(x)$ had multiple roots, "one could free it of repeated factors by dividing it by the greatest common divisor of the function $F(x)$ and its derivative [würde man dieselbe von gleichen Factoren dadurch befreien können, dass man sie durch den grössten Theiler, den die Function $F(x)$ mit ihrer Ableitung gemein hat, dividirt]" [Kronecker, 1882, p. 12].

[8]See [Galois, 1831, Lemma III] as translated in [Edwards, 1984, p. 103].

x, one can find a monic, irreducible polynomial $G(t)$ in a new indeterminate t and the indeterminates of the given polynomial other than x, which has the property that adjunction of *one* root of $G(t)$ splits the given polynomial into factors linear in x.

This presentation deviates from Kronecker's in two important ways. First, I have taken some liberties with Kronecker's statement of the theorem in the hopes of making it simpler. I have ignored Kronecker's "module systems" in favor of Galois's notion of *adjoining a root of an irreducible polynomial*. It is not that there is anything wrong with module systems, but Kronecker's manipulations with them are difficult to follow, while, on the other hand, the adjunction notion is simple and clear and achieves the needed construction easily. An adjunction can be regarded as a module system of a particularly simple type, and there is no need for more general module systems in the statement of the theorem. Second, the proof I have sketched is entirely different from his, which, to be honest, I cannot follow. My proof is nonetheless inspired by his in the following way. There is a simple, naïve construction of a splitting field of a polynomial. Factor the polynomial. If all factors are linear, the construction terminates. Otherwise, adjoin one root of one irreducible factor of degree greater than 1 and begin again, regarding the polynomial as now having its coefficients in the extended field. With each step, the field is extended and the number of linear factors of the polynomial increases, so the construction must terminate after a finite number of adjunctions with a field over which the polynomial splits into linear factors. The difficulty with this proof from a Kroneckerian point of view—there is no difficulty from the "modern" point of view, and this is the proof van der Waerden gives—is that it requires that one be able to factor a given polynomial with rational coefficients when it is regarded as a polynomial with coefficients in a given algebraic number field.

This problem, factoring polynomials with coefficients in an algebraic number field, was an obstacle for me in my book, *Galois Theory*. Kronecker gives a factorization method in his *Grundzüge* [Kronecker 1882, pp. 12-13] but, at the time I wrote my book, I was not satisfied with my understanding of Kronecker's factorization method, and my exposition of it was labored. In recent years, I have worked out a factorization method that does satisfy me.[9] In retrospect, I can see that this method is the one Kronecker is describing in the *Grundzüge*, but even now I find his explanation too sketchy. Nonetheless, with my new understanding of his factorization method I am able to prove the "Fundamentalsatz" with full Kroneckerian constructivist rigor, so, despite my discomfort with his quite different proof in his paper about the theorem, I am entirely comfortable with the theorem itself.

Conclusions

To conclude, I would like to justify my statement at the outset that Kronecker's theorem can be regarded as bringing to a successful conclusion the study of the so-called fundamental theorem of algebra.

Briefly put, the idea is that the "Fundamentalsatz" tells us that a polynomial $F(x)$ of degree n with rational coefficients has n roots, all of which can be expressed rationally in terms of a single root t of a polynomial $G(x)$, which can even be assumed to be monic and irreducible, with integer coefficients. This amounts to an

---

[9] See [Edwards, 2004, essay 1] for a full discussion of the method, Kronecker's statement of it, and an English translation.

algebraic solution of the problem of constructing the roots of $F(x)$—constructing a field over which $F(x)$ splits into linear factors—because the construction of a field in which $G(t)$ has one root is elementary.

As was mentioned, Gauss in 1815 had reduced the proof of the (so-called) fundamental theorem of algebra to the justification of certain computations with the roots; since the "Fundamentalsatz" fully justifies computation with the roots, it can be used to complete Gauss's proof.

As far as I am aware, Kronecker never made use of his "Fundamentalsatz" to prove that a polynomial of degree n with rational coefficients has n complex roots, but this can be done in a much more direct way than using Gauss's rather difficult 1815 construction. The direct construction consists of two steps. First, the integral of $d \log |G(z)|$ around the boundary of any square in the complex z-plane throughout which $|G(z)|$ is bounded away from zero is zero because $\log |G(z)|$ is single-valued on such a square. On the other hand, the integral of $d \log |G(z)|$ around the boundary of a very large square centered at the origin is nearly the same as the integral of $d \log |z^n|$ around the same boundary, where n is the degree of $G(z)$. Since this second integral is $2\pi n i$, it follows that one can, by partitioning a big square into small squares, find small squares on which $|G(z)|$ is nearly zero. Second, given a value of z_0 for which $|G(z)|$ is nearly zero, the usual Newton's method iteration

$$z_{n+1} = z_n - \frac{G(z_n)}{G'(z_n)}$$

can be used to construct a sequence that converges to a root of G.[10]

The complex root of G found in this way, and all of the complex roots of F it implies, conform to Kronecker's requirement that, although they are by their nature limits of infinite sequences, these sequences are explicitly generated by specific rules. Thus, for the (so-called) fundamental theorem of algebra, the "modern" approach to infinites that Kronecker claimed was unnecessary is indeed unnecessary.

References

Archival Sources[11]

[Archives Mittag-Leffler] Archives of the Mittag-Leffler Institute, Djursholm, Sweden.

Printed Sources

Abel, Niels Henrik. 1827. "Recherche de la quantité qui satisfait à la fois à deux équations algébriques données." *Annales de mathématiques pures et appliquées* 12, or Abel, Niels Henrik. *Oeuvres complètes de Niels Henrik Abel*. 2 Vols. Christiania: Imprimerie de Grondahl & Son, 1:212-218.

Biermann, Kurt-R. 1973. *Die Mathematik und ihre Dozenten an der Berliner Universität*. Berlin: Akademie-Verlag.

Edwards, Harold M. 1984. *Galois Theory*. Graduate Texts in Mathematics. Vol. 101. New York/Berlin: Springer-Verlag.

---

[10]However, in constructing the estimates needed for a rigorous proof, it is easier to use the less rapidly converging iteration $z_{n+1} = z_n - \frac{G(z_n)}{G'(z_0)}$.

[11]I thank the Mittag-Leffler Institute for permission to quote from its archives.

———. 1995. "Kronecker on the Foundations of Mathematics." In *From Dedekind to Gödel*. Ed. Jean Hintikka. Dordrecht: Kluwer Academic Publishers, 45-52.

———. 2004. *Essays in Constructive Mathematics*. New York-Berlin: Springer-Verlag.

Galois, Évariste. [ca. 1831]. "Mémoire sur les conditions de résolubilité des équations par radicaux." *Journal de mathématiques pures et appliquées* 11 (1846), 417-433. English Trans. [Edwards, 1984, Appendix 1].

Gauss, Carl Friedrich. 1799. "Demonstratio nova theorematis omnem functionem rationalem integram unius variabilis in factores reales primi vel secundi gradus resolvi posse." Helmstadt, or [Gauss, 1876, 3:1-30].

———. 1815. "Demonstratio nova altera theorematis omnem functionem rationalem integram unius variabilis in factores reales primi vel secundi gradus resolvi posse." *Commentationes societatis regiae scientiarum Gottingensis recentiores* 3, or [Gauss, 1876, 3:31-56].

———. 1876. *Werke*. 12 Vols. Göttingen: Königliche Gesellschaft der Wissenschaften zu Göttingen.

Hilbert, David. 1922. "Neubegrundung der Mathematik: Erste Mitteilung," *Abhandlungen der Mathematischen Seminar Hamburg Universität* 1, 157-177, or Hilbert, David. 1932–1935. *Gesammelte Abhandlungen*. 3 Vols. Berlin: Springer-Verlag, 3:157-177.

Kronecker, Leopold. 1882. *Grundzüge einer arithmetischen Theorie der algebraischen Grössen*. Berlin: Reimer. See also *Journal für die reine und angewandte Mathematik* 92, 1-122, or [Kronecker, 1895–1930, 2:237-388].

———. 1886. "Über einige Anwendungen der Modulsysteme auf elementare algebraische Fragen." *Journal für die reine und angewandte Mathematik* 99, 329-371, or [Kronecker, 1895–1930, 3a:145-208].

———. 1887. "Ein Fundamentalsatz der allgemeinen Arithmetik," *Journal für die reine und angewandte Mathematik* 100, 490-510, or [Kronecker, 1895–1930, 3a:209-240].

———. 1895–1930. *Mathematische Werke*. Leipzig: Preussische Akademie der Wissenschaften. Reprint Ed. 1968. New York: Chelsea Publishing Co.

———. 1901. *Vorlesungen über Zahlentheorie*. Ed. Kurt Hensel. Leipzig: B. G. Teubner Verlag. Reprint Ed. 1978. Berlin: Springer-Verlag.

Lipschitz, Rudolph. 1986. *Briefwechsel mit Cantor, Dedekind, Helmholtz, Kronecker, Weierstrass und anderen*. Ed. Winfried Scharlau. Dokumente zur Geschichte der Mathematik. Vol. 2. Braunschweig/Wiesbaden: Vieweg Verlag.

Meschkowski, Herbert. 1967. *Probleme des Unendlichen*. Braunschweig: Vieweg Verlag.

Neumann, Peter. 1986. "Review of *Galois Theory* by Harold M. Edwards." *American Mathematical Monthly* 93, 407-411.

Weil, André. 1950. "Number Theory and Algebraic Geometry." *Proceedings of the International Congress of Mathematicians (Cambridge, MA)*. 2 Vols. New York: American Mathematical Society, 2:90-100, or Weil, André. 1979. *Œuvres scientifiques/Collected Papers*. 3 Vols. New York/Berlin: Springer-Verlag, 2:442-452.

CHAPTER 6

Developments in the Theory of Algebras over Number Fields: A New Foundation for the Hasse Norm Residue Symbol and New Approaches to Both the Artin Reciprocity Law and Class Field Theory

Günther Frei
Hombrechtikon, Switzerland

Dedicated to Johann Jakob Burckhardt on his 100th birthday, 13 July, 2003

Introduction

In 1927, Leonard Eugene Dickson's book *Algebren und ihre Zahlentheorie* was published by Orell Füssli in Zürich [Dickson, 1927]. It was a completely revised and considerably enlarged version of the first English edition, entitled *Algebras and Their Arithmetics* [Dickson, 1923], and translated into German essentially by Johann Jakob Burckhardt at the suggestion of Andreas Speiser, who also added a chapter entitled "Idealtheorie in rationalen Algebren." The book strongly influenced the development of the abstract theory of noncommutative algebra—mostly fostered by Emmy Noether and Emil Artin—as well as on Helmut Hasse's work in number theory. The publications of the American school of the theory of algebras had been largely ignored by mathematicians in the German-speaking part of Europe, and in Europe in general.[1] One of the few exceptions was Emmy Noether. She had already cited Dickson's first edition in her course on hypercomplex numbers in Göttingen in 1924. She also reviewed Joseph H. Maclagan Wedderburn's 1923 article, "Algebraic Fields," for the *Jahrbuch über die Fortschritte der Mathematik*.[2] She might initially have heard of the work on algebras being done in America from the young American mathematicians who visited Göttingen after having obtained their American degree.

After having been informed by Artin of the appearance of Dickson's book, Hasse immediately worked out a summary for himself—something he did whenever he studied a field new to him—and published an extensive review in the *Jahresbericht der Deutschen Mathematiker-Vereinigung* [Hasse, 1927b]. Artin, Hasse, and Noether cherished the hope that the theory of noncommutative algebras would give

---

[1]On the American school of algebra, see [Parshall, 2004].
[2]See [Wedderburn, 1923] and [Noether, 1923].

a clue as to how to extend class field theory—which, by the theory of Teiji Takagi, amounts to the theory of abelian extensions of algebraic number fields [Takagi, 1920]—to the noncommutative case of algebraic extensions with non-abelian Galois group. Although this idea did not lead to the desired goal, it inspired a new approach to the Reciprocity Law and to class field theory. It was responsible for the creation of cohomological methods in class field theory by Emmy Noether and Hasse and paved the way for the cohomology of groups.

In this chapter, which is only a sketch for a more extensive study of the subject, I will first review some of the main events pertinent for the sequel that led from Hamilton's discovery of the quaternions in 1843 to the two books by Dickson. I will then give a general indication of the influence these had, mainly on the work of Hasse, but also on that of Artin, Emmy Noether, A. Adrian Albert, and Richard Brauer. I will also show how this led to Hasse's new approach to the norm residue symbol, to the Reciprocity Law, and to class field theory.[3]

The Beginnings of Structure Theory

The path to a theory of algebras was prepared by Leonhard Euler in his *Introductio in analysin infinitorum*. There, he viewed a complex number z as a point in the complex plane with right-angular real coordinates (x, y) as $z = x + yi$ or, in polar coordinates, (r, ϕ) as $z = r(\cos \phi + i \sin \phi)$ and treated i as a number linearly independent from 1 [Euler, 1748, §132]. Euler established, by means of power series [Euler, 1748, §§133-134] and starting from the factorization of $1 = (\cos \phi)^2 + (\sin \phi)^2 = (\cos \phi + i \sin \phi)(\cos \phi - i \sin \phi)$ [Euler, 1748, §132], the fundamental relation $e^{i\phi} = (\cos \phi + i \sin \phi)$ from which the remarkable formula $e^{i\pi} = -1$ is an immediate consequence [Euler, 1748, §138]. His great innovation was the discovery of the connection between the trigonometric functions and the exponential function. His discovery of this relation was also a prerequisite for the creation of another field closely related to the theory of algebras, namely, the theory of algebraic number fields, which built on Carl Friedrich Gauss's theory of quadratic forms, on the one hand, and Gauss's theory of cyclotomic fields, on the other. In the theory of cyclotomy he presented in Section 7 of the *Disquisitiones arithmeticæ* [Gauss, 1801], Gauss did not mention explicitly the exponential function, since he wanted his theory to be elementary and independent of analysis [Frei, 2001b/2007]. It is clear, however, that he was guided by Euler's formulas [Gauss, 1801, arts. 337 and 339]. Incidentally, the terminology "complex number" seems to have been first introduced by Gauss in his second paper on biquadratic residues [Gauss, 1831, art. 31].

---

[3]It is a great pleasure to thank Peter Roquette for his hospitality, his support, many interesting discussions, and his comments on a first draft of this sketch. My thanks go also to the University of Heidelberg for their practical support and to the Deutsche Forschungsgemeinschaft for their financial support. I am also indebted to Jeremy Gray for his offer to polish my English of this article, to Karen Parshall for her great patience and for encouraging me to write this sketch for the proceedings of the conference, and to Charles Curtis for his kind and helpful review.

The article is part of a project, joint with Peter Roquette, on the Artin-Hasse correspondence and on the work of Helmut Hasse. It is also a continuation of [Frei, 2001a]. A shorter version appeared in German in October 2003 for the Festschrift for J. J. Burckhardt on the occasion of his 100th birthday [Frei, 2003a]. A more detailed exposition was given in [Frei, 2003b]. The more extensive study will appear elsewhere.

Addition and multiplication of complex numbers in the complex plane were first introduced by the Norwegian-Danish mathematician Caspar Wessel [Wessel, 1897] and by the Swiss Robert Argand [Argand, 1806]. In modern terminology, they considered them as operations in a two-dimensional vector space over the real numbers, where the complex numbers were thought of as directed line segments (vectors) attached to the origin or as points in the complex plane. Addition was defined component-wise, and multiplication was defined so that it obeys the relation $i^2 = -1$. From all of this, it became evident that Euler's formulas imply that multiplication with i can be interpreted as a rotation of 90° around the origin in the complex x-y-plane.

These properties motivated the Irish mathematician William Rowan Hamilton to look for a three-dimensional generalization of the complex numbers, namely, for another "complex number" j such that a point $\mathbf{p}$ in the three-dimensional space can be represented as $\mathbf{p} = x + yi + zj$ with real coordinates (x, y, z), and such that multiplication with j induces a rotation of 90° in a direction orthogonal to the x-y-plane. After thinking about the problem for many years, Hamilton finally realized on 16 October, 1843 that, while this cannot be achieved, it is possible to construct in a four-dimensional real vector space $\mathbb{H} = \{\mathbf{q} = x + yi + zj + tk \mid x, y, z, t \in \mathbb{R}\}$ a three-dimensional subspace, namely, the imaginary part $\mathbb{J} = \{yi + zj + tk\}$ of $\mathbb{H}$, in which multiplication with i, j and k induces rotations of 90° about the origin.[4] However, this was only possible by dropping the condition that multiplication be commutative. For the "basis elements" $1, i, j, k$, Hamilton found the fundamental properties:

$$i^2 = j^2 = k^2 = -1,$$

$$ij = k,\ jk = i,\ ki = j;\ ij = -ji,\ jk = -kj,\ ki = -ik.$$

He called the elements $\mathbf{q}$ of the algebra $\mathbb{H} := \{\mathbf{q} = x + yi + zj + tk \mid x, y, z, t \in \mathbb{R}\}$ thus defined *quaternions*, and he proved that the multiplication of quaternions satisfies the associative law.[5]

As a result of this, other algebras were soon discovered, mainly by British mathematicians. Already by December 1843, John Thomas Graves had found the octonions [Graves, 1848, p. 338],[6] a real vector space of dimension 8 with a multiplication satisfying neither the commutative nor the associative law, but satisfying the so-called alternative law, a weak form of associativity.[7] However, like every

---

[4]Hamilton already used the terms "vector," "radius vector," and "scalar" in this connection. See [Hamilton, 1967, p. 211].

[5]See [Hamilton, 1967, p. 114]. Henceforth the term "algebra" will denote a vector space on which a bilinear multiplication (of vectors) is defined. The multiplication law of quaternions had already been discovered by Gauss in 1819 in connection with transformations of a three-dimensional space imbedded in a four-dimensional space, but Gauss did not publish it [Gauss, 1819].

[6]See also [Hamilton, 1967, p. 648] and [Dickson, 1914b, p. 14]. Graves's results were only published in 1848 by Hamilton. In the meantime, Arthur Cayley had discovered the octonions independently in 1845 [Cayley, 1889–1898, 1:127 and 11:368-371]. For this reason they are sometimes also called the Cayley numbers.

[7]The dimension of an algebra was termed its "rank" by Hasse [Hasse, 1931a, p. 499] and its "order" by Wedderburn [Wedderburn, 1907, p. 79]. We will use the terminology "dimension" not only for a vector space but also for an algebra defined on it.

nonzero quaternion, every nonzero octonion **q** admits a uniquely determined multiplicative inverse $\mathbf{q}^{-1}$. Therefore, neither the quaternions nor the octonions have zero divisors.[8]

In 1853, Hamilton introduced the biquaternions $\mathbb{B} := \{\mathbf{q} = x + yi + zj + tk \mid x, y, z, t \in \mathbb{C}\}$, the quaternions over the complex numbers $\mathbb{C}$ [Hamilton, 1853, p. 650]. These also form a real vector space of dimension 8, but with a multiplication completely different from that of the octonions, since the biquaternions contain zero divisors. For that reason, the *complex* quaternions $\mathbb{H}(\mathbb{C}) = \mathbb{B}$ have quite a different structure from the *real* quaternions $\mathbb{H}(\mathbb{R}) = \mathbb{H}$. This is a result of the different behavior of the quadratic form $\Phi = x^2 + y^2 + z^2 + t^2$ with respect to $\mathbb{R}$ and $\mathbb{C}$. Over $\mathbb{R}$, Φ is anisotropic, that is, $x^2 + y^2 + z^2 + t^2 = 0$ holds only in the trivial case $x = y = z = t = 0$, whereas over $\mathbb{C}$, Φ is isotropic, that is, $x^2 + y^2 + z^2 + t^2 = 0$ admits non-trivial solutions $(x, y, z, t) \in \mathbb{C}$. Since Φ is anisotropic over $\mathbb{R}$, every nonzero real quaternion $\mathbf{q} = x + yi + zj + tk$ in $\mathbb{H}(\mathbb{R})$ admits a uniquely determined inverse $\mathbf{q}^{-1}$, namely, $\mathbf{q}^{-1} = \frac{\mathbf{q}'}{N(\mathbf{q})}$, where $\mathbf{q}' = x - yi - zj - tk$ denotes the *conjugate*, and $N(\mathbf{q}) = \mathbf{q}'\mathbf{q} = \mathbf{q}\mathbf{q}' = x^2 + y^2 + z^2 + t^2$ denotes the *norm* of $\mathbf{q}$.[9] Moreover, $\mathbf{q} = \mathbf{0}$, that is, $x = y = z = t = 0$ if and only if $N(\mathbf{q}) = 0$. However, $\mathbb{H}(\mathbb{C})$ has non-trivial zero divisors, namely, the quaternions $\mathbf{q} = x + yi + zj + tk$, for which $x^2 + y^2 + z^2 + t^2 = 0$ admits a non-trivial solution $(x, y, z, t) \in \mathbb{C}$. Those satisfy $\mathbf{q}\mathbf{q}' = N(\mathbf{q}) = 0$ with $\mathbf{q}, \mathbf{q}' \neq 0$. That the algebra of biquaternions has non-trivial zero divisors can also be deduced from the fact discovered by Cayley in 1858, that the biquaternions are isomorphic to the (full) matrix algebra $\mathcal{M}(2, \mathbb{C})$ over $\mathbb{C}$ [Cayley, 1889–1898, 2:475-496 on p. 491]. Cayley showed that to the quaternion $\mathbf{q} = x + yi + zj + tk$ in $\mathbb{B}$ there corresponds the matrix

$$\mathbf{M} = \begin{pmatrix} x + yi & -z + ti \\ z + ti & x - yi \end{pmatrix}$$

in $\mathcal{M}(2, \mathbb{C})$ and vice versa. Moreover, the norm of $\mathbf{q}$ is equal to the determinant of $\mathbf{M}$. This isomorphism implies that the algebra of biquaternions is associative, unlike the octonions.

The theory of matrices was initiated around 1855 by Cayley in order to render the theory of determinants and the theory of invariants more transparent [Cayley, 1889–1898, 2:185-188, 475, and 604]. However, the term "matrix" had already been coined by James Joseph Sylvester in 1850 [Sylvester, 1904–1912/1973, 1:150]. In a fundamental paper of 1858 [Cayley, 1889–1898, 2:475-496], Cayley showed that the (square) matrices (with elements from a field K) can be viewed as a vector space (over K) equipped with a multiplication determined by the properties of the composition of linear transformations.[10] Thus, they constitute an algebra (over K), which is associative but not commutative and which has zero divisors. A set of basis elements is given by the matrices E_{ij} with zeros everywhere, except at the place corresponding to the i-th row and j-th column where they have the entry 1. Therefore, the $n \times n$ matrices $\mathcal{M}(n, K)$ (over K) form an algebra of dimension n^2 (over K). Cayley was also the first to treat hypercomplex numbers as matrices.

---

[8]An element $q \neq 0$ is called a zero divisor, if there exists an element $r \neq 0$ such that $qr = 0$.

[9]The left inverse coincides with the right inverse.

[10]In connection with his theory of equivalence of ternary quadratic forms, Gauss, in 1801, had already obtained the composition matrix of two square matrices of rank 3 attached to a linear transformation of ternary quadratic forms [Gauss, 1801, art. 270].

In an important paper of 1854, he had even already considered the group algebra over a finite group after having introduced the abstract notion of a group [Cayley, 1889–1898, 2:123-130 on p. 129].

Wedderburn's General Structure Theorems

New important algebras or hypercomplex systems, as they were often called particularly in Germany, were constructed by Hermann Günter Grassmann in [Grassmann, 1844; 1862] and by William Kingdon Clifford in 1873 and 1878.[11] The former, named after Grassmann, play an important role in geometry, while the latter, named for Clifford, arise in number theory [O'Meara, 1963, pp. 131-141], in the theory of Lie groups [Chevalley, 1946, pp. 16-67], and in physics [Lounesto, 2001]. The multitude of algebras found in the thirty-five years after Hamilton's discovery of the quaternions naturally led to the problem of classifying all possible algebras (with unit), for which the associative law and the two distributive laws hold. A first important investigation was undertaken by Benjamin Peirce in his work "Linear Associative Algebras," privately printed in 1870 but published only posthumously in 1881 by his son Charles S. Peirce [Peirce, 1881]. Peirce introduced the notions of nilpotent and idempotent elements and deduced the first structure theorems for algebras. An element a of an algebra $\mathcal{A}$ is called nilpotent, if $a^n = 0$ holds for some $n \in \mathbb{N}$, and idempotent, if $a^2 = a$. These concepts allowed Peirce to classify many algebras up to dimension 6. Important structure theorems for associative algebras with unit over $\mathbb{C}$ were obtained in 1891 by Georg Scheffers [Scheffers, 1891] and Theodor Molien. Molien showed that every simple[12] algebra over $\mathbb{C}$ is a full matrix algebra, that is, it is isomorphic to an algebra of square matrices with coefficients in $\mathbb{C}$ [Molien, 1893]. This theorem was also proved independently and more completely seven years later by Élie Cartan [Cartan, 1898].[13]

A general theory of associative algebras with unit, henceforth simply called algebras, over an arbitrary field was developed in 1907 by Wedderburn in his paper "On Hypercomplex Numbers" [Wedderburn, 1907]. This paper was fundamental for the structure of algebras and for the relation between algebras and number fields. There, Wedderburn first defined a number of notions associated with a given algebra $\mathcal{A}$, among them, the notions of the sum and product of linear subspaces—called complexes by Wedderburn [Wedderburn, 1907, p. 79]. He called a subalgebra $\mathcal{B}$ of $\mathcal{A}$ invariant, if $\mathcal{AB} \subseteq \mathcal{B}$ and $\mathcal{BA} \subseteq \mathcal{B}$ holds, that is, if $\mathcal{B}$ is a two-sided ideal in $\mathcal{A}$ [Wedderburn, 1907, p. 81]. An algebra $\mathcal{A}$ was termed simple, if it does not contain a non-trivial invariant subalgebra—that is, if it contains no other two-sided ideals besides the null ideal (0) and the unit ideal (1) = $\mathcal{A}$ [Wedderburn, 1907, p. 81; Dickson, 1923, p. 42]—and semisimple, if it does not contain an invariant nilpotent subalgebra—that is, if its radical is equal to the null ideal (0).[14] The radical $\mathcal{N}$ of $\mathcal{A}$ appears in Wedderburn's paper as the maximal nilpotent invariant subalgebra of

---

[11] See [Klein, 1926, pp. 173-187], [Zaddach, 1994], [Clifford, 1882, pp. 181-200 and 266-176], and [Lounesto, 2001].

[12] See below.

[13] For the work of Scheffers, Molien, and Cartan, see [van der Waerden, 1985, pp. 205-209], [Hawkins, 1972], [Curtis, 1999], and [Parshall, 1985]. For the work of Élie Cartan on Lie algebras and for further developments, see [Borel, 2001] and [Hawkins, 2000].

[14] See [Wedderburn, 1907, pp. 89 and 94] and [Dickson, 1923, p. 51]. An algebra is nilpotent, if all of its elements are nilpotent [Wedderburn, 1907, p. 91]. Originally, Wedderburn called an algebra $\mathcal{A}$ nilpotent, if $\mathcal{A}^n = 0$ for some natural number n [Wedderburn, 1907, p. 87].

$\mathcal{A}$, as Wedderburn showed that $\mathcal{N}$ is the union of all nilpotent invariant subalgebras in $\mathcal{A}$ [Wedderburn, 1907, p. 89].[15]

After these preliminaries, Wedderburn next proved what we will call the "Main Theorem" [Wedderburn, 1907, p. 109]:

(1) Every algebra $\mathcal{A}$ over a field K is the sum of its uniquely determined radical and a semisimple algebra over K uniquely determined up to an isomorphism;

(2) Every semisimple algebra $\mathcal{A}$ over a field K is the direct sum of uniquely determined simple algebras over K;

(3) Every simple algebra $\mathcal{A}$ over a field K is isomorphic to a full matrix algebra $\mathcal{M}(n, \mathcal{S})$ over a skew field $\mathcal{S}$ over K;

(4) $\mathcal{M}$ and $\mathcal{S}$ are uniquely determined up to an inner automorphism of $\mathcal{A}$; and

(5) The center Z of a simple algebra $\mathcal{A}$ over a field K is a field over K.[16]

The term "skew field [Schiefkörper]" was coined by B. L. van der Waerden [Artin, 1927a, p. 245; Hasse, 1927b]. For Wedderburn in 1907, a skew field was called a "primitive" or "division" algebra [Wedderburn, 1907, p. 91].[17] This is an algebra for which all field axioms are satisfied, except that the multiplication does not have to be commutative. Hence a skew field does not contain zero divisors. In particular, every nonzero element in a skew field admits a uniquely determined left and right inverse, and these coincide. Georg Frobenius showed in 1878 that the only skew fields over $\mathbb{R}$ are $\mathbb{R}$ itself, $\mathbb{C}$, and the quaternions $\mathbb{H}$. He also showed that over $\mathbb{C}$, there are no other skew fields besides $\mathbb{C}$ itself [Frobenius, 1878, p. 163]. This theorem was also proved independently three years later by C. S. Peirce [Peirce, 1881, Appendix].[18]

Hurwitz and the Arithmetic of Quaternions

In the tenth supplement to the second edition of Dirichlet's *Vorlesungen über Zahlentheorie* (1871), Richard Dedekind introduced the formal concept of a "field [Körper]" and showed that an algebraic number field K of degree n over the rational numbers $\mathbb{Q}$ can be viewed as a commutative algebra of dimension n over $\mathbb{Q}$, generated by the powers of a root of an irreducible rational polynomial of degree n over $\mathbb{Q}$. Dedekind called n (linearly over $\mathbb{Q}$) independent elements of the field K a "basis" of the field and, later, "principal units" of the field [Dirichlet, 1871, §159; Dedekind, 1885].

In the same supplement, Dedekind also developed the arithmetic of an algebraic number field K over $\mathbb{Q}$, that is, the ideal theory in the ring R of integers in K [Dirichlet, 1871, §163]. Fundamental was his notion of an order in K, a concept that went back to Gauss [Gauss, 1801]. Orders are subrings of K which are free

---

[15]See [Wedderburn, 1907, p. 89], [Dickson, 1923, p. 44], and [Dickson, 1927, p. 95]. The term radical was coined by Gottfried Köthe [van der Waerden, 1985, p. 211].

[16]For the latter two results, (4) and (5), respectively, see [Dickson, 1923, pp. 78 and 80, resp.] as well as [Hasse, 1932, pp. 177 and 186-187] and [Hasse, 1931a, p. 497], respectively.

[17]This terminology was also used by American authors such as Dickson.

[18]On the origins of Wedderburn's work, see [Parshall, 1985]. [van der Waerden, 1959, §§141-152] and [O'Meara, 1963, pp. 118-131] provide modern presentations of Wedderburn's theory. On Wedderburn's influence on later developments in algebra, see [Artin, 1950], and on Frobenius's contributions, consult [Curtis, 1999].

$\mathbb{Z}$-modules of rank n. The ring of integers R in K is characterized by the property that it is the unique maximal order in K. Dedekind's theory motivated Adolf Hurwitz to develop, following Dedekind, an analogous theory of rings and ideals for a noncommutative algebra over an algebraic number field, in particular, for the rational quaternions $\mathbb{H}(\mathbb{Q}) := \{\mathbf{a} = a_0 + a_1 i + a_2 j + a_3 k \mid a_\nu \in \mathbb{Q},\ \nu = 0, 1, 2, 3\}$ over $\mathbb{Q}$ [Hurwitz, 1896; Hurwitz, 1919]. Already in 1885, Rudolf Lipschitz had begun to study this problem [Lipschitz, 1886], but it was Hurwitz who succeeded in finding the appropriate definition of an *integer* rational quaternion, namely, one in which the integer quaternions constitute a maximal order in the algebra of rational quaternions $\mathbb{H}(\mathbb{Q})$ [Hurwitz, 1896, §3].

Hurwitz called a rational quaternion $\mathbf{a} = a_0 + a_1 i + a_2 j + a_3 k$, with $a_\nu \in \mathbb{Q}$, $\nu = 0, 1, 2, 3$, an *integer* quaternion, if either all the a_ν are integer numbers or all the a_ν are half-integer numbers, that is, of the form $a_\nu = n + \frac{1}{2}$ with $n \in \mathbb{Z}$. The integer rational quaternions form a ring $\mathcal{R}$ within the skew field $\mathbb{H}(\mathbb{Q}) = \mathcal{Q}$ of rational quaternions. More precisely, $\mathcal{R}$ is a maximal order over $\mathbb{Z}$ in $\mathcal{Q}$ containing the basis $\{1, i, j, k\}$ of the skew field $\mathcal{Q}$, and $\mathcal{R}$ is generated over $\mathbb{Z}$ by i, j, k and $l = \frac{1+i+j+k}{2}$.

In order to derive the arithmetic of $\mathcal{R}$, Hurwitz first proved the fundamental theorem that every automorphism ϕ of $\mathcal{Q}$—which he called a permutation—is an inner automorphism, that is, $\phi(\mathbf{a}) = \mathbf{q}\mathbf{a}\mathbf{q}^{-1}$ for some quaternion $\mathbf{q} = q_0 + q_1 i + q_2 j + q_3 k \in \mathcal{Q}$ [Hurwitz, 1896, §2].[19] The uniquely determined inverse $\mathbf{q}^{-1}$ of $\mathbf{q}$ has the form $\mathbf{q}^{-1} = \frac{\mathbf{q}'}{N(\mathbf{q})}$, where $\mathbf{q}' = q_0 - q_1 i - q_2 j - q_3 k$ denotes the conjugate, and $N(\mathbf{q}) = \mathbf{q}'\mathbf{q} = \mathbf{q}\mathbf{q}' = q_0^2 + q_1^2 + q_2^2 + q_3^2$ denotes the norm of $\mathbf{q}$.

Hurwitz next determined the units in $\mathcal{R}$, that is, the elements $\mathbf{e}$ in $\mathcal{R}$, for which $N(\mathbf{e}) = 1$. There are precisely 24 such units, namely, $\pm 1, \pm i, \pm j, \pm k$ and $\frac{\pm 1 \pm i \pm j \pm k}{2}$.[20] This implies that $\mathcal{R}$ admits precisely 24 automorphisms, and these are given by conjugation with the units $\mathbf{q} = \mathbf{e}$ of $\mathcal{R}$ and their products with $\mathbf{g} = 1 + i$.

Hurwitz was also able to show that $\mathcal{R}$ is Euclidean, that is, $\mathcal{R}$ admits a right and a left division algorithm [Hurwitz, 1896, §5]. Hence, every right (respectively, left) ideal in $\mathcal{R}$ is a principal right (respectively, left) ideal. Finally, using the norm properties, Hurwitz obtained the Fundamental Theorem of Arithmetic for the ring $\mathcal{R}$ of integer quaternions, namely, every integer quaternion can be uniquely decomposed into prime quaternions, up to the order of the primes and up to units [Hurwitz, 1896, §9]. He obtained this from the property that a quaternion is prime in $\mathcal{R}$ if and only if its norm is a prime number [Hurwitz, 1896, §8], and from the fact that the norm is multiplicative, that is, $N(\mathbf{q_1 q_2}) = N(\mathbf{q_1})N(\mathbf{q_2})$ for $\mathbf{q_1}, \mathbf{q_2} \in \mathcal{Q}$.

In addition, Hurwitz used the number of so-called "primary [primäre]" quaternions—that is, those quaternions congruent to 1 or $1+2r$ modulo $2(1+i)$ [Hurwitz, 1896, §6]—whose norm is equal to a given (positive) integer number m to deduce Jacobi's Theorem of 1828 that the number of representations of m as a sum of four squares of integer numbers is either 8 or 24 times the sum of the odd divisors of m, depending on whether m is odd or even.[21] As Hurwitz noted, this is because

---

[19]This important theorem was later extended by Albert Skolem and Emmy Noether to simple algebras [Noether, 1933a].

[20]The group of units is isomorphic to the group of homogeneous substitutions of a tetrahedron. This group plays an important role in the theory of modular functions [Hurwitz, 1896, §7].

[21]See [Hurwitz, 1896, §9] and [Jacobi, 1881–1891, 1:247 and 6:245].

among the 24 associates **eq** or **qe** of **q** with odd norm, there is precisely one which is primary [Hurwitz, 1896, § 6].[22]

Following Hurwitz's idea, Louis-Gustave Du Pasquier, one of Hurwitz's doctoral students, extended Hurwitz's results on quaternions to an arithmetic theory of linear substitutions [Du Pasquier, 1906; 1909; 1916; 1920],[23] building on Frobenius's fundamental paper on linear substitutions and bilinear forms [Frobenius, 1878]. Thus, Du Pasquier essentially initiated the study of the arithmetic of semisimple algebras. Dickson extended this theory to an arithmetic of associative algebras, which amounts to an arithmetic theory of maximal orders in an associative algebra [Dickson, 1923, Ch. 10; 1927, Ch. 10]. This last fact had already been put forward by Du Pasquier in the case of semisimple algebras [Du Pasquier, 1916] and taken over by Dickson in the general case.[24] The theory was developed further and in different directions by Speiser [Speiser, 1926; 1927; 1935], Artin [Artin, 1927b; 1927c], Hasse [Hasse, 1931a], and Brandt.[25]

The Structure of Skew Fields: Connections with Algebraic Number Theory

According to part (3) of Wedderburn's Main Theorem, the structure of a simple algebra $\mathcal{A}$ over a field K depends on the structure of the corresponding skew field $\mathcal{S}$, but before 1906 no other skew fields were known besides the quaternions. In that year, however, a completely new kind of skew field was described by Dickson in a short abstract [Dickson, 1906; 1914a, pp. 31-33 and 38; 1923, § 47]. These skew fields (over algebraic number fields) later played an important role in class field theory and in the cohomology of groups.

The starting point for Dickson's discovery was the fact that the real quaternions $\mathbb{H} = \mathbb{H}(\mathbb{R})$ can be viewed as an algebra of dimension 2 over $\mathbb{C}$ in the following way:

$$\mathbb{H}(\mathbb{R}) = \{x + yi + zj + tk = x + yi + zj + tij = (x + yi) + (z + ti)j = r + sj\},$$

with $x, y, z, t \in \mathbb{R}$ and $r, s \in \mathbb{C}$. Here, i and j are two independent quadratic elements over $\mathbb{R}$, satisfying the algebraic equation $x^2 + 1 = 0$. This means that $\mathbb{H}(\mathbb{R})$ is a skew field over $\mathbb{R}$, containing $\mathbb{C}$ as a quadratic subfield $\mathbb{C} = \mathbb{R}(i)$ over $\mathbb{R}$.

Generalizing this idea, Dickson constructed in 1912 what he called a generalized quaternion algebra [Dickson, 1923, § 47; 1914a, p. 38; 1912, p. 66]. To do this, he replaced $\mathbb{R}$ by an arbitrary field K of characteristic different from 2 and i and j by two arbitrary elements α and β linearly independent over K and both quadratic over K. More precisely, $\alpha^2 = a$ and $\beta^2 = b$ are in K, but α and β are not. The vector space generated over K by $1, \alpha, \beta, \alpha\beta$ becomes an (associative) algebra $\mathcal{A}$ (with unit element) over the field K by introducing a multiplication defined by $\alpha^2 = a$, $\beta^2 = b$, $\alpha\beta = -\beta\alpha$, and thus $(\alpha\beta)^2 = -ab$. We will denote this algebra $\mathcal{A} = \{\mathbf{q} = x + y\alpha + z\beta + t\alpha\beta \mid x, y, z, t \in K\}$ by $\mathcal{A} = K[a, b]$. As before, let $\mathbf{q}' = x - y\alpha - z\beta - t\alpha\beta$ stand for the conjugate of $\mathbf{q}$. The norm of $\mathbf{q}$, that is,

---

[22]On the life and work of Hurwitz, see [Hilbert, 1921] and [Frei, 1995].

[23]See also [Dickson, 1927, p. 201; 1923, pp. 145 and 157].

[24]The ring of integers, in the sense of Dickson, of an associative algebra with unit coincides with the ring of integers as defined by Du Pasquier in the case where the associative algebra is semisimple [Dickson, 1923, §§ 92 and 96]. For Dickson's work on the arithmetic of algebras, see [Fenster, 1998], and for some early work of Dickson on algebras, see [Fenster, 1999].

[25]For the literature to Brandt, see [Deuring, 1935, p. 139]. I will consider the development of the arithmetic of algebras and the arithmetic of orders from Gauss to Eichler in a separate article.

$N(\mathbf{q}) = \mathbf{q}'\mathbf{q} = \mathbf{q}\mathbf{q}' = x^2 - ay^2 - bz^2 + abt^2$, represents a quaternary quadratic form over K. This algebra $\mathcal{A}$ is of dimension 4 over K and contains the subfield $L = K(\sqrt{a}) = K(\alpha)$ which is quadratic over K. Dickson also studied—following the ideas of Hurwitz—the arithmetic of $\mathcal{A}$ in the special case where $\mathcal{A} = \mathbb{Q}[-1, -b]$ with $b \in \mathbb{Z}$, and he used the structure of $\mathcal{A}$ in order to investigate the diophantine equation $x^2 + y^2 + z^2 + t^2 = uv$ [Dickson, 1923, §§ 105-106].

Dickson gave a decisive new turn to the theory by also taking into account the conjugate $-\sqrt{a}$ of $\sqrt{a} = \alpha$ over K for the construction of $\mathcal{A} = K[a, b]$ over $L = K(\sqrt{a}) = K(\alpha)$ by means of $\beta = \sqrt{b}$. He denoted this conjugate $\sigma(\sqrt{a})$, where σ stands for a polynomial with coefficients in K. However, in what follows it will be more appropriate to adopt the newer and more abstract point of view of Emmy Noether; for her, σ should be seen as an element of the Galois group G of L/K.[26] G is cyclic of order 2 and is generated by the automorphism σ determined by $\sigma(\alpha) = -\alpha$, or more explicitly by $\sigma(x + y\alpha) = x - y\alpha$, for $x + y\alpha \in L$ and for $x, y \in K$. In this way, the multiplication in $\mathcal{A}$ is determined by the rules

$$\alpha^2 = a \in K, \quad \beta^2 = b \in K, \quad \beta\alpha = -\alpha\beta = \sigma(\alpha)\beta.$$

This means that the action of σ in L is given by conjugation with β: $\sigma(\alpha) = \beta\alpha\beta^{-1}$. The arithmetic of L is thus directly connected with the multiplicative structure of $\mathcal{A}$.

This idea of Emmy Noether's follows Hilbert's program, which had already been propagated by Kronecker. It postulates that the arithmetic in a number field L should be expressed solely by the Galois group G and the discriminant D of L/K. This was in direct analogy to Riemann's concept that the Riemann surface of a function field is determined by its covering transformations—that is, by its fundamental group—and by the ramification points.

Dickson also examined the important problem of finding conditions for his algebra $\mathcal{A} = K[a, b] = L(\beta)$ to be a skew field [Dickson, 1923, § 47]. For this to be true, every nonzero element $\mathbf{q} = r + s\beta \in \mathcal{A}$, with $r, s \in L$, has to admit an inverse. If $s = 0$ and $r \neq 0$, then $\mathbf{q}^{-1} = r^{-1} \in L$ is the inverse of $\mathbf{q}$. If $s \neq 0$, then $\mathbf{q} = r + s\beta = s(rs^{-1} + \beta) = s\gamma$ admits an inverse if and only if $\gamma = rs^{-1} + \beta = t + \beta$ has an inverse, where $t = rs^{-1}$ is in L; hence $t = x + y\alpha$ with $x, y \in K$. Because $\alpha\beta = \beta\sigma(\alpha)$, for every $t = x + y\alpha \in L$, $t\beta = (x+y\alpha)\beta = x\beta + y\alpha\beta = x\beta + y\beta\sigma(\alpha) = \beta(x + y\sigma(\alpha)) = \beta\sigma(x+y\alpha) = \beta\sigma(t)$. The existence of an inverse of $\gamma = t + \beta = \beta + t$ is now determined by $(\beta + t)(\beta - \sigma(t)) = \beta^2 + t\beta - \beta\sigma(t) - t\sigma(t) = \beta^2 - t\sigma(t) = b - \mathfrak{N}(t) \in K$, where $\mathfrak{N}(t) = t\sigma(t)$ denotes the norm of t with respect to L/K. Since $b - \mathfrak{N}(t)$ is an element of K, it follows that $\gamma = t + \beta$ and thus $\mathbf{q} = s\gamma$ has an inverse $\mathbf{q}^{-1}$ in $\mathcal{A} = L(\beta) = K[a, b]$, namely,

$$\mathbf{q}^{-1} = \frac{1}{s} \cdot \frac{\beta - \sigma(t)}{b - \mathfrak{N}(t)},$$

if $b \neq \mathfrak{N}(t)$ for every $t \in L = K(\alpha)$, that is, if b is not the norm of an element $t \in L = K(\sqrt{a})$. From this, Dickson deduced the important result that the algebra of generalized quaternions $\mathcal{A} = K[a, b]$ is a skew field, if $b \in K$ is not the norm of an element in $K(\sqrt{a})$. Therefore, the multiplicative structure of $\mathcal{A} = K[a, b]$ depends

---

[26]This ultimately led to Noether's theory of crossed products, first expounded in her lectures "Nichtkommutative Arithmetik" in the summer term of 1929 and in "Algebra hyperkomplexer Grössen" in the winter term of 1929–1930 at the University of Göttingen [Noether, 1930; Hasse, 1932, p. 180].

on the group of norms of $K(\sqrt{a})/K$. But since the norm group of relatively abelian number fields is determined by the theory of class fields, it follows that the structure of the algebra $\mathcal{A} = K[a,b]$ is also determined by class field theory. This was the decisive point of departure for Hasse which later led to his general norm residue symbol and then to the cohomological formulation of class field theory.

Dickson had noticed as early as 1906 that his construction can be generalized to a finite cyclic algebraic extension L/K [Dickson, 1906], but it was only later, in 1914, 1923, and 1927, that he furnished a more precise description of this construction [Dickson, 1914a; 1923, §47 and Appendix I; 1927, §§36-43]. Note that Dickson's symbol $\sigma(\alpha)$ for a conjugate of α stands for a polynomial in α with coefficients in K, and this was still the case in his 1927 book.[27] However, if we adopt the more abstract point of view of Emmy Noether, in which σ denotes an automorphism,[28] Dickson's results can be summarized as follows.

Let K be any field, let $L = K(\alpha)$ be a finite cyclic eextension over K of degree n, and let σ be a generator of the cyclic Galois group G of L over K. Let, furthermore, β be an element independent of α over K, such that $\beta^n = b \in K$, but $\beta^r \notin K$ for $1 \leq r < n$. Then

(1) the n-dimensional vector space $\mathcal{A} = L(\beta) = K(\alpha, \beta)$ over L, generated by β, becomes an associative algebra with unit element of dimension n over L and of dimension n^2 over K with a K-basis $\{\alpha^\nu \beta^\mu | \nu, \mu = 0, 1, \ldots, n-1\}$, if the multiplication satisfies the property: $\beta\alpha = \sigma(\alpha)\beta$, that is, if the action of σ in L is given by conjugation with β, namely, $\sigma(\alpha) = \beta\alpha\beta^{-1}$, and

(2) $\mathcal{A} = K(\alpha, \beta)$ is a skew field over K, if b^r for $1 \leq r < n$ is not the norm of an element t in $L = K(\alpha)$.

Property (1) as well as property (2) in the cases $n = 2, 3$ was proved by Dickson in 1914 [Dickson, 1914a, pp. 31-33]. Property (2) in the general case for arbitrary n was established by Wedderburn in the same year and in the same journal [Wedderburn, 1914, pp. 164-166].[29] In the case where K is a number field, Hasse showed in 1931 that (2) is not only sufficient for $\mathcal{A}$ to be a skew field but also necessary [Hasse, 1932a, p. 180]. He also showed that the element $b \in K$ is the norm of an element t in $L = K(\alpha)$ if and only if the algebra $\mathcal{A} = K(\alpha, \beta)$ is isomorphic to a full matrix algebra $\mathcal{M}(n, K)$ of $n \times n$-matrices over the ground field K [Hasse, 1932, pp. 175, 179, and 199].[30] This connection of the norm with the splitting of Dickson's algebra was crucial for the later development leading to the theorem of Brauer-Hasse-Noether and to a new definition of the norm residue symbol by Hasse (see below).

Dickson's theorem established for the first time that there exist skew fields that are not fields, quaternions, or generalized quaternions and whose dimension n^2 over the ground field K is greater than 4. Dickson constructed the following example for $n = 3$ [Dickson, 1923, §48]. He took for the ground field K the rational number

---

[27]See [Dickson, 1923, p. 65; 1927, p. 52]. Dickson uses $\theta(\xi)$, respectively $\theta(i)$, instead of our $\sigma(\alpha)$.

[28]This later led to Emmy Noether's general point of view that the Galois group G of L/K should be imbedded into an algebra $\mathcal{A}$, the crossed product of G and $L^\times$, $\mathcal{A} = G \times L^\times$, such that an automorphism of L/K becomes an inner automorphism of $\mathcal{A}$ [Noether, 1932, p. 191]. See below.

[29]Compare also [Dickson, 1923, Appendix I].

[30]This theorem will be referred to as the "Norm Splitting Theorem" below.

field $\mathbb{Q}$ (thus $K = \mathbb{Q}$) and for the cyclic field L of degree 3 over $\mathbb{Q}$ the maximal real subfield of the cyclotomic field of the 7-th roots of unity $\mathbb{Q}(\zeta)$, $\zeta = e^{\frac{2\pi i}{7}}$, generated by $\zeta + \zeta^{-1}$. Hence $L = K(\alpha) = \mathbb{Q}(\alpha)$ with $\alpha = \zeta + \frac{1}{\zeta} = 2\cos\frac{2\pi}{7}$. The number α is a root of the (irreducible) cyclic polynomial $x^3 + x^2 - 2x - 1 = 0$. If β is an element linearly independent of α over $\mathbb{Q}$, such that $\beta^3 = b$ is an even integer not divisible by 8, then $\mathcal{A} = L(\beta) = \mathbb{Q}(\alpha, \beta)$ is a skew field of dimension 9 over $\mathbb{Q}$. For, if $t = p\alpha^2 + q\alpha + r$ is an arbitrary element in $\mathbb{Q}(\alpha)$ with $p, q, r \in \mathbb{Z}$, then the norm $\mathfrak{N}$ of t with respect to $\mathbb{Q}(\alpha)/\mathbb{Q}$ is $\mathfrak{N}(t) \equiv 1 + (p+1)(q+1)(r+1)$ modulo 2. Therefore, if p or q or r are odd, then $\mathfrak{N}(t)$ is also odd. But if p, q, r all are even, then $\mathfrak{N}(t)$ is divisible by 8.

The Theory of Semisimple Algebras

Andreas Speiser was aware of Dickson's book, *Algebras and Their Arithmetics*, soon after its publication in 1923, owing to his work on group theory and on algebraic number theory and owing, in particular, to work in connection with his paper "Allgemeine Zahlentheorie" [Speiser, 1926]. Dickson's book gave the first textbook treatment of Wedderburn's theory as well as of the theory of associative algebras over arbitrary fields together with the cyclic algebras constructed by Dickson.[31] Since Dickson's book was not widely known in Germany, Speiser wrote to Dickson and proposed to have the book translated into German. Dickson agreed and sent a completely revised and considerably enlarged version, which was translated into German essentially by Speiser's young student, Johann Jakob Burckhardt,[32] and published by Orell Füssli in Zurich in 1927 under the title *Algebren und ihre Zahlentheorie*. The book won the Cole Prize of the American Mathematical Society in 1928.

In his 1927 review of Dickson's book in the *Jahresbericht der Deutschen Mathematiker-Vereinigung*, Hasse wrote that "[t]his German translation of the textbook on 'Algebras and Their Arithmetics' or the 'Theory of hypercomplex number systems'—as we call it in Germany—and which appeared in Chicago in English in 1923, is most welcome. After all, it is the first presentation in German in a textbook of a recently created, highly important theory, which increasingly attracts the interest of algebraists and number theorists" [Hasse, 1927b].[33] Indeed, the German translation had a very strong influence on the work of a number of German-speaking number theorists. It led Emil Artin to a general arithmetical theory of maximal orders [Artin, 1927b; 1927c], and it inspired Helmut Hasse's work on the local theory of algebras [Hasse, 1931a] as well as Emmy Noether's general theory of noncommutative rings [Noether, 1929; 1933a]. Artin and Hasse also thought that the new

---

[31]The book by Gaetano Scorza, *Corpi numerici e algebre*, published in Messina in 1921, also presents an introduction to the theory of algebras, but it does not contain Wedderburn's Main Theorem in full generality.

[32]Burckhardt was born on 13 July, 1903, and he is still in good health at this writing. His one-hundredth birthday was celebrated by the University of Zürich on 31 October, 2003 [Frei, 2003c].

[33]"Diese deutsche Übersetzung des 1923 in Chikago in englischer Sprache erschienenen Lehrbuchs der 'Algebren und ihrer Zahlentheorie', oder—wie wir in Deutschland zu sagen gewohnt sind—der 'Theorie der hyperkomplexen Zahlensysteme', ist lebhaft zu begrüßen. Handelt es sich doch um die erste deutschsprachige Darstellung in Buchform einer in neuerer Zeit entstandenen, hochbedeutenden Theorie, die in wachsendem Maße das Interesse der Algebraiker und Zahlentheoretiker auf sich zieht."

theory of noncommutative algebras would give a clue as to how to extend class field theory to non-abelian extensions of algebraic number fields.

In 1927, Artin had just proved his general Reciprocity Law for relatively abelian number fields [Artin, 1927d], thereby bringing class field theory to a certain conclusion. That theory had turned out to be the theory of relatively abelian number fields, due to Takagi's discovery that every (finite) abelian extension over an algebraic number field K is a class field over K, and vice versa [Takagi, 1920]. As early as 1923, Artin had made an important move into the non-abelian territory of number theory by introducing L-functions for relatively Galois extensions of algebraic number fields [Artin, 1923]. It was the investigation of the properties of these L-functions that led Artin to his general Reciprocity Law, a law he could initially prove only in special cases [Frei, 2002].

As for Hasse, he was occupied in 1927 with the second part of his report on class field theory, dealing with the reciprocity laws [Hasse, 1930a]. After having been informed by Artin about Dickson's book, however, he immediately read it and wrote up his review for the *Jahresbericht*. He was particularly interested in the study of the corresponding local theory and in the connection with the theory of norms on which he had done pioneering work. He had been led to the theory of norms by his theory of quadratic forms over algebraic number fields and by his Local-Global Principle.[34] He then got naturally to the reciprocity laws and to class field theory [Frei, 2001a], since class field theory can also be viewed as a theory of norm forms in algebraic number fields. According to the Artin Reciprocity Law, to every ideal class of an algebraic number field K modulo the norms with respect to an abelian extension L/K, there corresponds an element of the abelian Galois group G of L/K, whereby the norm group in K corresponds to the unit element of G. Since the norm group determines the structure of the Dickson algebra $\mathcal{A} = L(\beta)$, generated over L by means of G for a cyclic group G, Hasse pursued the possibility that the class field theory of abelian extensions of algebraic number fields might admit a generalization to non-abelian Galois extensions by means of the theory of (noncommutative) algebras over K. Although this goal was finally achieved only with respect to some special properties, the theory of noncommutative algebras, on the one hand, and class field theory, on the other, profited greatly from these investigations; they led to the cohomological formulation of class field theory and to the cohomology of groups.

Inspired by Dickson's book, great progress in the theory of algebras was made in Germany in the short period from 1927 to 1931, mostly due to Artin, Brauer, Hasse, and Emmy Noether. Some of these important discoveries were presented by Hasse in a long paper, sent to the *Transactions of the American Mathematical Society* on 29 May, 1931 [Hasse, 1932].[35] It was written in English, since Hasse intended to reach, in particular, the American algebraists and to inform them about the advances made in Germany. A preliminary announcement of Hasse's results presented on the 24 April, 1931 had already appeared in the *Göttinger Nachrichten* under the title "Theorie der zyklischen Algebren über einem algebraischen Zahlkörper." Even

---

[34]Compare [Hasse, 1932, p. 211], where Hasse writes: "Let me note once more the analogy between the foregoing theory of cyclic representable algebras and my theory of general quadratic forms which I have developed in some previous papers, ... as one of the starting points of my present work."

[35]See also [Deuring, 1935] for a report on this progress.

earlier, however, at the beginning of April and one month after the so-called "Skew Field Congress [Schiefkörperkongress]" had taken place in Marburg 1931, Hasse had communicated his discoveries to Emmy Noether.[36] As she put it tersely in a letter to Hasse, "I have read your theorems with great enthusiasm, like an exciting novel; you went really very far, indeed! Now (Deuring long ago, it crosses my mind) I also wish for the converse: the direct hypercomplex rationale of the invariants, that is, of the correlation between the decomposition group and the group of the noncommutative fields at each particular place; so that it is induced by a single global correlation; and thus the hypercomplex rationale of the reciprocity law!"[37]

Hasse presented his results in [Hasse, 1932], based partially on Emmy Noether's lecture notes [Noether, 1930], as follows. According to Wedderburn, the center of a skew field $\mathcal{S}$ over a field K is a finite algebraic extension Z over K [Hasse, 1931a, p. 497]. If $\mathcal{A}$ is a full matrix algebra over $\mathcal{S}$, then Z is also the center of $\mathcal{A}$. If $Z = K$, then $\mathcal{S}$ is said to be central over K, following a suggestion made by van der Waerden [van der Waerden, 1959, p. 193]. (Wedderburn, like Dickson [Dickson, 1927, p. 138], used the term "normal" instead of "central.") If $\mathcal{A}$ is a full matrix algebra over $\mathcal{S}$ and $\mathcal{S}$ is central over K, then $\mathcal{A}$ is also called central over K. Quite generally, a simple algebra $\mathcal{A}$ over a field K is said to be central over K, if K is the center of $\mathcal{A}$.

We have used the notation $\mathcal{A} = K(\alpha, \beta)$ for the algebra $\mathcal{A}$ constructed by Dickson over an algebraic number field K by means of a cyclic extension L/K of degree n, $L = K(\alpha)$. This algebra has a generating automorphism σ satisfying $\beta\alpha = \sigma(\alpha)\beta$, where $\beta^n = b \neq 0$ is in K, but β^r is not in K, for $1 \leq r < n$. Hasse called such an algebra "cyclic"[38] of degree n over K, and he denoted it by $\mathcal{A} = (b, K(\alpha), \sigma) = (b, L, \sigma)$. If $\alpha_1, \ldots, \alpha_n$ is a basis of L/K, then $\{\beta^\nu \alpha_\mu \mid \nu = 0, 1, \ldots, n-1, \mu = 1, 2, \ldots, n\}$ is a basis of $\mathcal{A}$ over K. $\mathcal{A}$ is of dimension—or as Hasse put it of rank—n^2 over K.

Based on Emmy Noether's lecture notes, Hasse proved (Theorem A):[39]

(1) Every cyclic algebra $\mathcal{A} = (b, K(\alpha), \sigma)$ over an algebraic number field K is a central simple algebra over K.

(2) $L = K(\alpha)$ is a maximal sub-field of $\mathcal{A}$.

Conversely, Dickson had proved for $n = 2$, that every central skew field over an algebraic number field K is cyclic, that is, is a cyclic algebra over K.[40] As for $n = 3$, the theorem for central skew fields was proved by Wedderburn in 1921.[41] Albert studied the case $n = 4$ by means of Hasse's p-adic theory of isotropic quadratic

---

[36]The congress took place from 26 to 28 February, 1931. See [Roquette, 2005, §7.3].

[37]"Ihre Sätze habe ich mit großer Begeisterung, wie einen spannenden Roman gelesen; Sie sind wirklich weit gekommen! Jetzt (Deuring schon lange, wie mir einfällt) wünsche ich mir noch die Umkehrung: direkte hyperkomplexe Begründung der Invarianten, d. h. der Zuordnung von Zerlegungsgruppe und Gruppe der nichtkommutativen Körper zu den einzelnen Stellen; sodaß dies aus einer einzigen Zuordnung im Großen induziert wird; und damit hyperkomplexe Begründung des Reziprozitätsgesetzes!" See [Lemmermeyer-Roquette, 2003, Letter 27 dated 12 April, 1931, p. 109].

[38]Wedderburn, and sometimes also Hasse, called it a "Dickson algebra."

[39]See [Hasse, 1932, p. 172]. For convenience and for later reference we call it "Theorem A," similarly in the sequel "Theorem B," etc.

[40]See [Dickson, 1927, pp. 45 and 47]. Dickson proved the theorem for $n = 2$ even for arbitrary fields of characteristic $\neq 2$.

[41]See [Dickson, 1923, pp. 232-233]. Wedderburn proved it even for arbitrary fields of characteristic 0.

forms and Hasse's Local-Global Principle [Albert, 1932; Albert and Hasse, 1932, p. 722].[42]

Whether the converse of this theorem remains true for arbitrary n and for any algebraic number field—that is, whether every central simple algebra over a number field K is a cyclic algebra—was one of the main open problems of the theory of algebras. It was settled in 1932 in [Brauer-Hasse-Noether, 1932, p. 399] after Hasse succeeded in proving the analogous theorem for p-adic central skew fields over an algebraic p-adic number field [Hasse, 1931a, p. 514; Albert and Hasse, 1932, p. 722].[43] It was probably in a letter written to Artin prior to 27 November, 1930 that Hasse reported the results he ultimately published in [Hasse, 1931a] and conjectured that every central simple algebra over a number field is a cyclic algebra. We will refer to this as the "Main Theorem of Algebras." In a letter dated 27 November, 1930, Artin wrote to Hasse, referring to Hasse's paper [Hasse, 1931a]:

> Thank you for sending the proof sheets of your beautiful and interesting article on the hypercomplex arithmetic. Through this everything has really become very simple. As far as the theorem on skew fields is concerned, I believe that every skew field (finite) over a number field is cyclic. Of course generalized accordingly: K/k cyclic of degree n, generating substitution σ, skew field relations for all $\alpha : \sigma \cdot \alpha = \sigma(\alpha) \cdot \sigma = \alpha^\sigma \cdot \sigma$ and $\sigma^n = \beta$ with appropriate β. Can you not prove this with your method, or do you not believe it, or do you have a counterexample?[44]

In the *Transactions* paper, Hasse also proved, again based on Emmy Noether's lecture notes (Theorem B) [Hasse, 1932, p. 173]: If $\mathcal{A} = (b, K(\alpha), \sigma)$ and $\mathcal{A}' = (b', K(\alpha), \sigma^r)$ are two cyclic algebras over K of degree n with $(r, n) = 1$, then $\mathcal{A}$ and $\mathcal{A}'$ are isomorphic if and only if b'/b^r is the norm of an element $t \neq 0$ in $L = K(\alpha)$. If this is the case, then $\beta' = \beta^r t$, where $\sigma(\alpha) = \beta\alpha\beta^{-1}$, and similarly for β'. For $r = 1$, this had been discovered 1921 by Wedderburn [Wedderburn, 1921]. Hasse's principal goal in his paper in the *Transactions* was the construction of a full system of invariants for a cyclic algebra $\mathcal{A}$, which characterizes $\mathcal{A}$ (that is,

---

[42]Charles Curtis informed me that Albert's paper was based on Albert's Ph.D. thesis, but the latter does not refer explicitly to Hasse's theory of p-adic quadratic forms. I would like to thank Charles Curtis for this information. For Hasse's critique of Albert's proof, see [Frei, 2003b].

[43]For Albert's contribution to this theorem, see [Roquette, 2005, pp. 68-71] and compare Fenster's chapter below. As for the rectification of the oft-repeated but erroneous statements that Albert's contribution was not mentioned in [Brauer-Hasse-Noether, 1932], that Albert was nosed out in a photo finish (insinuating that Albert had obtained the same result without the decisive help he had received from Hasse regarding the local theory and Hasse's theory of quadratic forms), and that Brauer, Hasse, and Noether had deprived him of a fundamental result, see [Roquette, 2005, pp. 61-75] and [Frei, 2003b].

[44]"Ich danke Ihnen auch für die Uebersendung der Korrekturen Ihrer schönen und interessanten Arbeit über die Hyperkomplexe Arithmetik. Dadurch ist wirklich alles sehr einfach geworden. Was den Satz über die Schiefkörper betrifft so glaube ich, dass jeder Schiefkörper (endlicher) über einem Zahlkörper zyklisch sein wird. Natürlich entsprechend verallgemeinert: K/k zyklischer Körper n-ten Grades, erzeugende Substitution σ, Schiefkörperrelationen für alle $\alpha : \sigma \cdot \alpha = \sigma(\alpha) \cdot \sigma = \alpha^\sigma \cdot \sigma$ und $\sigma^n = \beta$ mit passendem β. Könnten Sie das nicht etwa mit Ihrer Methode beweisen oder glauben Sie das nicht oder haben Sie gar ein Gegenbeispiel?" [Frei, 1981, Letter 35]. For the history of the conjecture of the Main Theorem of Algebras and Curtis's erroneous statement in [Curtis, 1999, p. 203] that it was first made by Dickson, see [Frei, 2003b] and [Roquette, 2005, pp. 54-57].

its Brauer class) completely, by means of his new norm residue symbol,[45] analogous to what he had done in the local case in his paper on p-adic skew fields.[46]

The theory of cyclic algebras $\mathcal{A} = (b, K(\alpha), \sigma)$ was generalized by Emmy Noether from a cyclic extension L/K to an arbitrary Galois extension L/K, where the cyclic Galois group $G = \langle \sigma \rangle$ is replaced by an arbitrary (Galois) group G. In addition, she introduced the idea of crossing the multiplicative group $L^\times$ of L with the group G. Emmy Noether first presented this new object, which she called a "crossed product [verschränktes Produkt]," in her Göttingen lectures in the winter semester of 1929–1930 [Noether, 1930, Ch. 6]. However, it was Hasse who first published it—with her approval—in his long article in the *Transactions* [Hasse, 1932]. It is also in this article that we find for the first time the corresponding English term "crossed product."[47] In the *Transactions* paper, Hasse had supposed that the base field K is perfect, that is, finite or of characteristic 0. However, Noether remarked in a letter to Hasse on 2 June, 1931 that many parts of his investigations hold without this condition [Lemmermeyer-Roquette, 2003, Letter 28].

If L/K is a Galois extension of degree n with Galois group G, then Noether's construction of a crossed product $\mathcal{A} = L \times G$ over L of the multiplicative group $L^\times$ of L with the group G runs as follows. She associates to each element σ in G a basis element u_σ in $\mathcal{A}$ over L, so that $\mathcal{A}$ becomes a vector space of dimension n over L. $\mathcal{A}$ is then turned into an algebra by requiring the following multiplication rules:

(1) $au_\sigma = u_\sigma a^\sigma$, for every a in L.[48]
(2) $u_\sigma u_\tau = u_{\sigma\tau} a_{\sigma,\tau}$ with $a_{\sigma,\tau} \neq 0$ in L.

Following Issai Schur, the set (a) of coefficients $a_{\sigma,\tau}$ in L is called a factor system of $\mathcal{A}$.[49] $\mathcal{A}$ is an algebra of dimension n^2 over K with basis $\{u_\sigma \alpha_k\}$, where σ runs through the group G, and $\{\alpha_1, \ldots, \alpha_n\}$ forms a basis of L/K. For this crossed product, Hasse writes $\mathcal{A} = (a, L)$. If the associative law holds in $\mathcal{A}$, then the factor system has further the property:[50]

(3) $a_{\sigma,\tau}^\rho = \frac{a_{\tau,\rho} a_{\sigma,\tau\rho}}{a_{\sigma\tau,\rho}}$.

Conversely, every factor system $(a) = \{a_{\sigma,\tau} \neq 0\}$ in L, satisfying (3), together with the multiplication rules (1) and (2), determines an associative algebra $\mathcal{A}$ of dimension n^2 over K, which can be represented as a crossed product $\mathcal{A} = (a, L)$. Two factor systems (a) and (a') over L determine the same algebra $\mathcal{A} = (a, L) = (a', L)$ if and only if the two factor systems satisfy:[51]

(4) $a'_{\sigma,\tau} = a_{\sigma,\tau} \frac{c_\tau c_\sigma^\tau}{c_{\sigma\tau}}$, for a $c_\sigma \neq 0$ in L.

---

[45] See below and [Hasse, 1932, pp. 173-180].

[46] See below and [Hasse, 1931a].

[47] See [Hasse, 1932, pp. 181-183]. See also the letter of Emmy Noether to Hasse, dated 12 April, 1931 and quoted above [Lemmermeyer-Roquette, 2003, Letter 27].

[48] Instead of $\sigma(a)$ we now write a^σ, following Kronecker.

[49] See [Hasse, 1932, pp. 190 and 214] and [Noether, 1930]. The notion of a factor system was introduced by Schur (1919), but the crossed representation (3) below was first considered by Speiser (1919). Brauer had used a different kind of factor system in his early work on the structure of central simple algebras and the Brauer group. See [Brauer, 1926; 1928] and [Curtis, 1999, pp. 228-230].

[50] See [Hasse, 1932, p. 181] and [Noether, 1930].

[51] See [Hasse, 1932, p. 183] and [Noether, 1930].

Thus, $u'_\sigma = u_\sigma c_\sigma$, if $\{u'_\sigma\}$ is a basis of $\mathcal{A}$ over L associated with (a'). Hasse termed two such factor systems "associated," and he wrote $(a) \sim (a')$. For the set of classes of associated factor systems in L with Galois group $G = \text{Gal}(L/K)$ we will write, following the modern notation, $H^2(G, L^\times)$ or $H^2(G, L/K)$. These form an abelian group, now called the second cohomology group of G with values in $L^\times$.

If $\mathcal{A}$ is a central simple algebra over the field K, and if L is a field extension of K, then the extended algebra $\mathcal{A} \otimes L$, obtained by extending the scalars (coefficients) from K to L, is also a central simple algebra over L.[52] If $\mathcal{A}$ is a full matrix algebra over the field L, then L is called a splitting field of $\mathcal{A}$. This notion was introduced in 1927 in [Brauer and Noether, 1927] and further developed in [Hasse, 1932, p. 183]. Brauer and Noether showed that a central skew field $\mathcal{S}$ over K is of dimension m^2 for some natural number m, called the Schur index of $\mathcal{S}$, and has a splitting field L of degree m over K. Furthermore, the splitting fields of degree m over K are splitting fields of minimal degree over K and are maximal subfields of $\mathcal{S}$. If L is a splitting field of a skew field $\mathcal{S}$, then L is also a splitting field for every central simple algebra which is isomorphic to a full matrix algebra over $\mathcal{S}$.

In 1931, Albert showed that every maximal subfield L of a skew field $\mathcal{S}$ finitely algebraic over a field K is a splitting field of $\mathcal{S}$ [Albert, 1931]. However, Brauer had already been aware of this property [Brauer and Noether, 1927], and Noether had proved that a field L finitely algebraic over a field K is a splitting field of a skew field $\mathcal{S}$ over K if and only if its irreducible representation over $\mathcal{S}$ yields a maximal subfield of the corresponding matrix ring over $\mathcal{S}$ [Noether, 1933a, p. 533].

At first, Noether had believed that each minimal splitting field of a skew field $\mathcal{S}$ over K is isomorphic to a maximal subfield of $\mathcal{S}$ [Roquette, 2005, pp. 51-54; Curtis, 1999, p. 227], but Brauer informed her that this is not so. This gave rise to the joint paper [Brauer and Noether, 1927], in which they proved that the dimensions of minimal splitting fields of a given central skew field are in general unbounded. A proof that such minimal splitting fields of unbounded dimension over a field K do exist, even for the rational quaternions, was furnished by Hasse based on his Local-Global Principle for the quaternary quadratic form $x_1^2 + x_2^2 + x_3^2 + x_4^2$. Using this, he constructed number fields L, where -1 is the sum of three squares and hence the sum of two squares, yielding cyclic splitting fields of degree 2^n over $\mathbb{Q}$ for every natural number n [Hasse, 1927c; Brauer and Noether, 1927]. In her lecture notes, Noether cautioned the reader that the guess that every splitting field of a skew field $\mathcal{S}$ contains a maximal subfield of $\mathcal{S}$ is not correct [Noether, 1930, p. 25].

The field of complex numbers $\mathbb{C}$ is a splitting field of the real quaternions $\mathbb{H} = \mathbb{H}(\mathbb{R})$ over the base field of real numbers $\mathbb{R}$, but the base field $\mathbb{R}$ itself is not a splitting field for $\mathbb{H}$. However, the base field $\mathbb{C}$ of the complex quaternions $\mathbb{B} = \mathbb{H}(\mathbb{C})$ is already a splitting field of $\mathbb{B}$, since $\mathbb{B}$ is isomorphic to a full matrix algebra (of dimension 4) over $\mathbb{C}$. Since every skew field $\mathcal{S}$ over a field K contains a splitting field L, the dimension $[\mathcal{S}:K]$ is always a square. This was already known to Wedderburn in 1907 [Wedderburn, 1907]. For crossed products, Hasse proved if L/K is a finite Galois extension of a perfect base field K with Galois group G and $\mathcal{A} = (a, L) = L \times G$ is a crossed product of $L^\times$ with G, then

(1) L is a maximal sub-field of $\mathcal{A}$,
(2) $\mathcal{A}$ is a central simple algebra over K, and

---

[52]See [Brauer and Noether, 1927], [Albert, 1931], and [Noether, 1933a].

(3) L is a splitting field of $\mathcal{A}$.[53]

This theorem admits a kind of converse, namely, that every central skew field $\mathcal{S}$ (and thus every central simple algebra $\mathcal{A}$) over a perfect field K is similar to a crossed product $\mathcal{A} = (a, L) = L \times G$. [Hasse, 1932, p. 185; Noether, 1930].

Hasse called two central simple algebras $\mathcal{A}$ and $\mathcal{A}'$ similar, in symbols $\mathcal{A} \sim \mathcal{A}'$, if the uniquely determined skew fields $\mathcal{S}$ and $\mathcal{S}'$, belonging to them according to Wedderburn's Main Theorem, are isomorphic [Hasse, 1932, p. 177]. This similarity relation was introduced by Brauer [Brauer, 1929a; 1929b]. By means of his theory of factor sets, developed for representations of groups of linear transformations [Brauer, 1926; 1928], Brauer had shown that the similarity classes of central simple algebras over a field K form an abelian group $\mathrm{Br}(K)$ with respect to the "direct product," and those having the same Galois splitting field L as a maximal subfield form a subgroup $\mathrm{Br}(L/K)$ [Brauer, 1929a; 1929b]. For this reason, Hasse called the first group the "Brauer group" of K, and the second the "Brauer group" of L/K [Brauer-Hasse-Noether, 1932, p. 403]. What was at the time called the "direct product" and denoted by "$\times$" is now usually called the tensor product and is denoted by "$\otimes$." For the second group, Brauer established in 1929 the important result that the Brauer group $\mathrm{Br}(L/K)$ is isomorphic to the group of classes of associated factor systems $H^2(G, L/K)$ in L with Galois group $G = \mathrm{Gal}(L/K)$.[54] For cyclic crossed products, Hasse proved, in addition, the following generalization of the Norm Splitting Theorem, connecting the norm theory of number fields with the splitting of algebras: If $\mathcal{A} = (b, K(\alpha), \sigma)$ is a cyclic algebra over the number field K as base field and (a), the corresponding factor system in $L = K(\alpha)$, that is, $\mathcal{A} = (a, K(\alpha))$, where $K(\alpha)/K$ is cyclic with Galois group $G = \langle \sigma \rangle$, then the class of the factor system (a) in $H^2(G, L/K)$ is trivial if and only if $\mathcal{A}$ splits over K, that is, if and only if $\mathcal{A}$ is isomorphic to a full matrix algebra over K. This happens if and only if $b \in K$ is the norm of an element in $L = K(\alpha)$ [Hasse, 1932, p. 199; Noether, 1930].

The Local Theory and the Theorem of Brauer-Hasse-Noether

Hasse gained a deeper understanding of the properties of a given algebra $\mathcal{A}$ over a number field K and of the proof of the preceding theorem by passing to the $\mathfrak{p}$-adic completion, also called $\mathfrak{p}$-adic localization, $K_\mathfrak{p}$ of K, that is, by passing to the $\mathfrak{p}$-adic extension $K_\mathfrak{p}$ of K and the resulting local algebra $\mathcal{A}_\mathfrak{p} = \mathcal{A} \otimes K_\mathfrak{p}$ of $\mathcal{A}$, where $\mathfrak{p}$ is a prime spot in K [Hasse, 1931a]. In particular, Hasse obtained—among other things—the proofs for Theorems A and B above [Hasse, 1932, pp. 172-173 (Theorems 1.3, 1.4, and 2.1)]. The key for this was the property that the Local-Global Principle, first discovered by Hasse for quadratic forms over fields, also holds for algebras. A central simple algebra $\mathcal{A}$ is called cyclically representable if it is similar to a cyclic algebra $(b, K(\alpha), \sigma)$ [Hasse, 1932, p. 178]. For these, Hasse proved the Local-Global Principle which states

(1) Two (central simple) cyclically representable algebras $\mathcal{A}$ and $\mathcal{A}'$ over an algebraic number field K belong to the same Brauer class in $\mathrm{Br}(L/K)$ if and only if their corresponding localizations $\mathcal{A}_\mathfrak{p}$ and $\mathcal{A}'_\mathfrak{p}$ belong to the same (local) Brauer class in $\mathrm{Br}(L_\mathfrak{p}/K_\mathfrak{p})$ for every prime spot $\mathfrak{p}$ of K.

---

[53]See [Hasse 1932, pp. 182 and 184] and [Noether, 1930].
[54]See [Hasse, 1932, p. 194] and [Brauer, 1929a; 1929b].

(2) A (central simple) cyclically representable algebra $\mathcal{A}$ over an algebraic number field K splits over K, that is, is isomorphic to a full matrix algebra over K, if and only if the corresponding local algebras $\mathcal{A}_\mathfrak{p}$ split over $K_\mathfrak{p}$ for every prime spot $\mathfrak{p}$ of K [Hasse, 1932, p. 211].

For the proof, Hasse had to refer to class field theory and to the Reciprocity Law for his norm residue symbol. Indeed, the second part of this theorem is a direct consequence of Hasse's Norm Theorem [Hasse, 1933, pp. 733 and 747], namely, if L is a cyclic extension of an algebraic number field K, then a number α in K is the norm of an element β in L if and only if for every prime spot $\mathfrak{p}$ in K, α (imbedded in $K_\mathfrak{p}$) is the norm of an element $\beta_\mathfrak{p}$ of the $\mathfrak{p}$-adic field $L^{(\mathfrak{p})}$ corresponding to L with respect to the $\mathfrak{p}$-adic completion $K_\mathfrak{p}$ of K.[55]

Hasse's next goal was to generalize the Local-Global Principle from cyclically representable algebras to any central simple algebra over an algebraic number field K, that is, to prove that a central simple algebra $\mathcal{A}$ over an algebraic number field K splits over K if and only if all the localizations $\mathcal{A}_\mathfrak{p}$ split over $K_\mathfrak{p}$ for all prime spots $\mathfrak{p}$ in K [Brauer-Hasse-Noether, 1932, p. 399]. Hasse realized that this generalization of his Local-Global Principle from cyclically representable (central simple) algebras to any central simple algebra over an algebraic number field implied the Main Theorem for central simple algebras over algebraic number fields, namely, every central skew field $\mathcal{S}$ over an algebraic number field K is a cyclic algebra $\mathcal{A} = (a, L)$ over K for some cyclic algebraic extension L/K and a factor system (a) in L.[56]

Here, we will only sketch the exposition as Hasse presented it in [Brauer-Hasse-Noether, 1932]. Brauer was able to reduce the proof of the Local-Global Principle for central simple algebras to the statement: every central simple algebra $\mathcal{A}$ over an algebraic number field K with solvable splitting field L/K, for which all the localizations $\mathcal{A}_\mathfrak{p}$ over $K_\mathfrak{p}$ split for every prime spot $\mathfrak{p}$ in K, splits over K [Brauer-Hasse-Noether, 1932, p. 400]. And, finally, it was Emmy Noether who completed the proof by further reducing it to the theorem: every central simple algebra $\mathcal{A}$ over an algebraic number field K with cyclic splitting field L/K of prime degree $q = [L : K]$, for which all the localizations $\mathcal{A}_\mathfrak{p}$ over $K_\mathfrak{p}$ split for every prime spot $\mathfrak{p}$ in K, splits over K [Brauer-Hasse-Noether, 1932, p. 401]. Since Hasse had already established Noether's result by means of his Local-Global Principle for cyclically representable algebras in his *Transactions* paper [Hasse, 1932, p. 211], the proof of the Main Theorem was now complete, except for an existence theorem of algebraic number fields that was mentioned in [Albert and Hasse, 1932] but furnished only later by Hasse [Hasse, 1950].[57] In a letter to Hasse, dated 1931, Artin wrote concerning this Main Theorem that "[y]ou cannot imagine how delighted I was about the proof finally happily achieved about the cyclic systems. This is the biggest advance made in number theory in recent years."[58]

---

[55]Compare [Hasse, 1930a, pp. 38-40] and [Hasse, 1931a].

[56]See [Brauer-Hasse-Noether, 1932, p. 399]. For the history of the so-called Brauer-Hasse-Noether theorem, the related existence theorem of Grunwald-Wang, and the contribution by Albert, we refer to the reader to the excellent booklet [Roquette, 2005] as well as [Frei, 2003b]. See also Curtis's chapter below.

[57]For the story of this existence theorem, see [Roquette, 2005, pp. 25-35].

[58]"Sie können sich gar nicht vorstellen, wie ich mich über den endlich geglückten Beweis für die zyklischen Systeme gefreut habe. Das ist der grösste Fortschritt in der Zahlentheorie der letzten Jahre." [Frei, 1981, p. 125].

The New Norm Residue Symbol and New Approaches to Both the Reciprocity Law and Class Field Theory

Hasse had made explicit use of class field theory and of the Artin Reciprocity Law for his norm residue symbol in order to prove his theorems on cyclic algebras over number fields [Hasse, 1932]. These led, in particular, to the Main Theorem of semisimple algebras over number fields [Brauer-Hasse-Noether, 1932]. Once this Main Theorem was established, it was possible to go in the other direction and give a direct definition of the norm residue symbol without making use of the Artin Reciprocity Law, as Hasse had done previously [Hasse, 1927a; 1930b]. It was also possible to give a proof of Artin's Reciprocity Law based on the theory of algebras and thereby to obtain a new direct approach to class field theory. It was Emmy Noether who remarked that the "[n]orm residue symbol is nothing else than cyclic algebra" [Weyl, 1935, p. 209] and that the theory of commutative algebras should be built on the theory of noncommutative algebras, since the latter are governed by simpler properties, in particular, relative to ramification [Hasse, 1933, pp. 731-732]. She suggested to Hasse that he should follow up this idea, saying "[n]ow the tables have to be turned."[59]

The norm residue symbol and its connection with the Reciprocity Law has a long history and appears first in the work of Gauss on quadratic forms as a character modulo p [Gauss, 1801, art. 230-231]. In Section Five of the *Disquisitiones arithmeticæ*, Gauss determined the conditions for an integral binary quadratic form $f = (a, b, c) = ax^2 + 2bxy + cy^2$ to represent an integer n by means of his genus theory.[60] For that purpose, Gauss attached to each odd prime number p dividing the determinant $d = b^2 - ac$ of f what he called a "character" of the form f and what we will denote $\varepsilon_p(f)$. The forms are always supposed to be non-degenerate—that is, d is not a square—and primitive—that is, the greatest common divisor of a, b, c is 1. If we introduce, following Dirichlet, the Legendre symbol $(\frac{\cdot}{p})$ [Dirichlet, 1839/1889, 1:421], then Gauss's character can be expressed as $\varepsilon_p(f) = (\frac{n}{p})$, where n is any integer prime to p represented by the form f, with integers x and y. Gauss showed that this definition is independent of the chosen number n represented by f. There is also a character corresponding to the prime $p = 2$, but this character requires special treatment, since it depends on the behavior of n (mod 8) [Frei, 1979; 1994]. Gauss said two quadratic forms f and f' with the same determinant d are of the same genus, if they have the same characters for all primes p dividing d, that is, if $\varepsilon_p(f) = \varepsilon_p(f')$ for all p dividing d. He also said that a form f belongs to the principal genus, if $\varepsilon_p(f) = 1$ for all p dividing d. Gauss determined that the number g of different genera for a given determinant d is 2^{t-1}, where t is the number of distinct prime divisors of d. Furthermore, he showed that for a given d, there is exactly one linear relation among the t characters $\varepsilon_p(f)$, p dividing d, which we will express as

$$\prod_{p|d} \varepsilon_p(f) = \prod_{p} \varepsilon_p(f) = 1,$$

---

[59]"Jetzt muss der Spiess umgedreht werden." Hasse in a conversation with the author. See also Emmy Noether's postcard to Hasse of 12 April, 1931 quoted above from [Lemmermeyer-Roquette, 2003, Brief 27].

[60]For the relevant details of the work of Gauss on quadratic forms, see [Frei, 1979]. For Gauss's second proof of the Reciprocity Law by means of the genera of quadratic forms, see [Frei, 1994].

where p ranges over all primes in the second product.[61] Gauss also took care to include what Hasse later called the infinite character—namely, $\varepsilon_\infty(f)$—by considering positive definite forms only in the case where the determinant d is negative. Gauss's fundamental theorem that the number of genera is exactly equal to 2^{t-1} [Gauss, 1801, arts. 261 and 287] must be viewed as a fundamental theorem of class field theory for the quadratic number field $\mathbb{Q}(\sqrt{d})$. In a first step, Gauss proved that $g \leq 2^{t-1}$ based on his theory of ambiguous forms and ambiguous classes. It is equivalent to the Law of Quadratic Reciprocity. Indeed, in article 262 of the *Disquisitiones arithmeticæ*, Gauss derived the Reciprocity Law from this inequality; this is essentially the first inequality of class field theory for quadratic number fields.

David Hilbert in his *Zahlbericht* made Gauss's theory more explicit by introducing what he called the norm residue symbol $\left(\frac{a,b}{p}\right)$ for two integers a and b with respect to a prime number p, where b is not a square [Hilbert, 1897, p. 162]. This symbol takes the value 1, if a is congruent modulo p and modulo every power of p to the norm of an integer β from the quadratic number field $K = \mathbb{Q}(\sqrt{b})$. Otherwise, it takes the value -1. The integers a for which $\left(\frac{a,b}{p}\right) = +1$ are called norm residues of the field K. Kurt Hensel realized that the definition of the (Hilbert) norm residue symbol should be given in terms of p-adic numbers [Hensel, 1913, p. 315], that is,

$$\left(\frac{a,b}{p}\right) = \begin{cases} +1 & \text{if } a = x^2 - by^2 \text{ is solvable with } x, y \in \mathbb{Q}_p, \\ -1 & \text{otherwise.} \end{cases}$$

Here $\mathbb{Q}_p$ denotes the field of p-adic numbers. Note tthat $x^2 - by^2$ is the norm form belonging to the field $K = \mathbb{Q}(\sqrt{b})$. Among the properties of the norm residue symbol proved by Hilbert are [Hilbert, 1897, §64]:

(1) Product Law: $\left(\frac{a_1 a_2, b}{p}\right) = \left(\frac{a_1, b}{p}\right)\left(\frac{a_2, b}{p}\right)$

(2) Symmetry Law or Permutation Law: $\left(\frac{b,a}{p}\right) = \left(\frac{a,b}{p}\right)$

(3) If $p \neq 2$ does not divide ab, then $\left(\frac{a,b}{p}\right) = +1$

(4) If $p \neq 2$ divides b, but does not divide a, then $\left(\frac{a,b}{p}\right) = \left(\frac{a}{p}\right)$,

where $\left(\frac{a}{p}\right)$ denotes the Legendre symbol. There are separate formulas for the prime $p = 2$. Hilbert then stated Gauss's Law of Quadratic Reciprocity in these terms [Hilbert, 1897, §69]: if a and b are two rational integers, not both negative, then $\prod_p \left(\frac{a,b}{p}\right) = 1$, the product taken over all prime numbers p. In the *Zahlbericht*, Hilbert formulated this result first as a lemma in order to prove Gauss's fundamental theorem on the number of genera in the quadratic number field $K = \mathbb{Q}(\sqrt{b})$.

Hasse discovered that Hilbert's condition "a and b not both negative" has to be interpreted as a condition at the infinite prime $p = p_\infty$ for which Hasse defined the following symbol [Hasse, 1924a, p. 120]:

$$\left(\frac{a,b}{p_\infty}\right) = \begin{cases} +1 & \text{if } a \text{ and } b \text{ are not both negative,} \\ -1 & \text{if } a \text{ and } b \text{ are both negative,} \end{cases}$$

or

$$\left(\frac{a,b}{p_\infty}\right) = \begin{cases} +1 & \text{if } a = x^2 - by^2 \text{ is solvable with } x, y \in \mathbb{Q}_{p_\infty} = \mathbb{R}, \\ -1 & \text{otherwise.} \end{cases}$$

---

[61] If p does not divide d, then $\varepsilon_p(f) = 1$. See [Frei, 1979; 1994].

With Hasse's definition, Hilbert's Reciprocity Law took on the smoother form [Hasse, 1924a, p. 121]:[62] if a and b are any two rational integers, then

$$\prod_p \left(\frac{a,b}{p}\right) = 1,$$

the product taken over all primes p, finite and infinite.

From this theorem, Hilbert first deduced Gauss's fundamental theorems on the genera of binary quadratic forms—or, what amounts to the same thing, on the genera of quadratic number fields [Frei, 1979]—and then Gauss's Reciprocity Law by following the path traced by Gauss in his second proof.[63] But Hilbert simplified it considerably by building on Theorem 90 in the *Zahlbericht* [Hilbert, 1897, §54], namely, if K/k is a relatively cyclic extension of algebraic number fields and α is a number in K with relative norm $N_{K/k}\alpha = 1$, then there exists an integer $\beta \in K$, such that $\alpha = \beta^{1-\sigma} = \frac{\beta}{\beta^\sigma}$, where σ is a generator of the Galois group G of K over k. Following the new ideas of Emmy Noether on the theory of algebras and crossed products, Hilbert's theorem can be formulated cohomologically in these terms: the first cohomology group of the Galois group G of K/k with coefficients in the multiplicative group $K^\times$ of K is trivial, that is, $H^1(G, K^\times) = 1$. Noether later generalized this theorem to any Galois extension K/k of algebraic number fields [Noether, 1933b]. She called it the "principal genus theorem [Hauptgeschlechtssatz]." It is a far-reaching generalization of Gauss's principal genus theorem, which asserts that every binary quadratic form f in the principal genus is the square (with respect to composition) of another form g.[64] Hilbert's Theorem 90 remains a key property for the proof of the existence of the cohomological Tate Reciprocity Isomorphism.[65]

In the fundamental paper "Über die Theorie des relativ-quadratischen Zahlkörpers," Hilbert extended the theory of quadratic number fields $\mathbb{Q}(\sqrt{b})$, as treated in Part Three of his *Zahlbericht*, to relative quadratic number fields K/k, where k is an arbitrary algebraic number field and K a relative quadratic extension of k, $K = k(\sqrt{\beta})$ with β in k [Hilbert, 1899b; 1899a]. Hilbert particularly aimed at a very general formulation of the Law of Quadratic Reciprocity in k. To this end, he started to outline a general theory of class fields K/k by means of the special case where K is quadratic over k. In this connection, he introduced the general quadratic (Hilbert) norm residue symbol $(\frac{\alpha,\beta}{\mathfrak{p}})$ with algebraic integers α, β in k and $\mathfrak{p}$ a prime ideal in k, as a generalization of the Hilbert symbol $(\frac{a,b}{p})$ in $\mathbb{Q}$. In [Hilbert, 1899a], the various reasons are explained for why this symbol is needed to formulate the Law of Quadratic Reciprocity in a more complete and appropriate way in k, one of the reasons being to get rid of the special role played by the prime 2. Hilbert's definition of $(\frac{\alpha,\beta}{\mathfrak{p}})$ is completely analogous to that of the symbol $(\frac{a,b}{p})$, and Hensel redefined it in terms of $\mathfrak{p}$-adic numbers in the local field $k_\mathfrak{p}$. The properties we have mentioned for $(\frac{a,b}{p})$ also hold for $(\frac{\alpha,\beta}{\mathfrak{p}})$. If the symbols for the infinite primes are defined appropriately [Hasse, 1924a, p. 120], then Hilbert's Law

---

[62]In [Hasse, 1924a], the results are derived more generally for a quadratic field $K = k(\sqrt{b})$ over an arbitrary algebraic number field k.

[63]See [Hilbert, 1897, §69], [Gauss, 1801], and [Frei, 1979; 1994].

[64]See [Gauss, 1801, arts. 247 and 286] and [Frei, 1979].

[65]See [Tate, 1952], [Artin-Tate, 1968, pp. 39-69], [Neukirch, 1969, p. 95], [Cassels-Fröhlich, 1967, p. 115], and [Gras, 2003, p. 108].

of Quadratic Reciprocity in a number field k again takes on the following closed form [Hilbert, 1899a, p. 90; Hasse, 1924a, p. 121]: if α and β are any two integers in a number field k, then $\prod_{\mathfrak{p}} \left(\frac{\alpha,\beta}{\mathfrak{p}}\right) = 1$, the product taken over all prime spots $\mathfrak{p}$ in k, finite and infinite. Hilbert pointed out that this formula should be seen as an analog of Cauchy's Theorem to the effect that the integral on a closed curve around all the singularities of a regular function in the complex plane (or on a Riemann Surface) is always equal to zero [Hilbert, 1899a, pp. 91-92].

Hilbert had also developed his theory of norm residues for what he called a Kummer extension K/k, where $k = \mathbb{Q}(\zeta)$. Here, ζ is a primitive lth root of unity for a prime number l, and $K = k(\sqrt[l]{\beta})$, for β an algebraic integer in k which is not the lth power of another integer in k [Hilbert, 1897, pp. 257-275]. Hilbert's theory of the lth power norm residues was extended by Hensel to general relatively abelian number fields K/k, where k is any algebraic number field containing the lth roots of unity for l a prime. He did this via his theory of $\mathfrak{p}$-adic numbers, analogous to what he had done in the relatively quadratic case [Hensel, 1922]. Thanks to his results on the multiplicative structure of the localizations $k_{\mathfrak{p}}$ of k, Hensel was in a position to define *explicitly* an lth power norm residue symbol in k, for a prime $\mathfrak{p}$ not dividing l, by means of the field $K = k(\sqrt[l]{\beta})$. For $\alpha \in k \subseteq k_{\mathfrak{p}}$, he put $\left(\frac{\alpha,\beta}{\mathfrak{p}}\right) = \zeta^{\lambda}$, where ζ is a (fixed) primitive lth root of unity, and λ is determined by means of the representation of α and β with respect to a properly chosen so-called Hensel basis in $k_{\mathfrak{p}}$ [Hensel, 1922, p. 9]. For this explicit lth power norm residue symbol, Hensel obtained immediately the so-called permutation law, namely, $\left(\frac{\beta,\alpha}{\mathfrak{p}}\right) \cdot \left(\frac{\alpha,\beta}{\mathfrak{p}}\right) = 1$. It follows directly from Hensel's definition of the symbol $\left(\frac{\alpha,\beta}{\mathfrak{p}}\right)$ that the *norm property* holds, that is, $\left(\frac{\alpha,\beta}{\mathfrak{p}}\right) = 1$ if and only if α is a norm residue of $K = k(\sqrt[l]{\beta})$ with respect to $\mathfrak{p}$ [Hensel, 1922, p. 6].

In a joint paper, Hensel and Hasse extended Hensel's result to the more difficult but also more important and critical case where $\mathfrak{p}$ is a divisor of l [Hasse-Hensel, 1923]. In a first step, they obtained the characterization of the norm residues with respect to $\mathfrak{p}$, when $\mathfrak{p}$ divides l, by means of an appropriate Hensel basis. The *explicit* definition of an lth power norm residue symbol in k, when $\mathfrak{p}$ divides l, as a well-determined power of a primitive lth root of unity ζ, was achieved by Hasse, first for $l = 2$ [Hasse, 1924b] and then for odd prime numbers l [Hasse, 1924c]. As for the case $l \neq 2$, Hasse could only get the existence and uniqueness of such an lth power norm residue symbol $\left(\frac{\alpha,\beta}{l}\right)$ in k in a rather roundabout way via the lth power Hilbert Reciprocity Law in k [Hasse, 1924c; 1925; Frei, 2002]. Proofs of this law had already been established by Furtwängler in 1912 and by Takagi in 1922 by transcendental (that is, analytic) means.[66] Later, Hasse came back to this fundamental problem of giving an explicit and canonical description of the norm residue symbol, first in [Hasse, 1930b] and then in [Hasse, 1933].

In the first paper, Hasse built on a symbol introduced by Artin and extended it to a new norm residue symbol analogous to the extension of the Legendre symbol to the Hilbert norm residue symbol. In order to define so-called L-functions for any Galois extension of number fields K/k, Artin in [Artin, 1923] had made use of what Hasse called the "Frobenius automorphism" [Hasse, 1930a, p. 6].[67] For each prime

---

[66]See [Furtwängler, 1912, p. 385] and [Takagi, 1922/1973, p. 209].

[67]On Artin's work on L-functions, see [Frei, 2002]. For Gauss's introduction of the Frobenius automorphism for finite fields, see [Frei, 2001b/2007].

ideal $\mathfrak{p}$ in k not dividing the relative discriminant $\mathfrak{d}(K/k)$ of K/k, the Frobenius automorphism $\sigma_\mathfrak{p}$ of $\mathfrak{p}$ is defined by the property $\sigma_\mathfrak{p}(\alpha) \equiv \alpha^{N(\mathfrak{p})} \pmod{\mathfrak{P}}$ for all $\alpha \in K$, where $\mathfrak{P}$ is a prime ideal in K dividing $\mathfrak{p}$, and $N(\mathfrak{p})$ is the (absolute) norm of $\mathfrak{p}$. The map $\sigma_\mathfrak{p}$ so defined is an element of the Galois group G of K/k. It depends on the prime ideal $\mathfrak{P}$ chosen in K, but the conjugacy class $[\sigma_\mathfrak{p}]$ of $\sigma_\mathfrak{p}$ in G does not. If K/k is abelian, then $[\sigma_\mathfrak{p}]$ contains only one element. In this case, Hasse called $\sigma_\mathfrak{p}$ the Artin automorphism or the Artin substitution [Hasse, 1930a; 1930b]. He wrote $\sigma_\mathfrak{p} = \left(\frac{K}{\mathfrak{p}}\right)$ and called $\left(\frac{K}{\mathfrak{p}}\right)$ the Artin symbol. Artin had conjectured in [Artin, 1923] and proved in [Artin, 1927d] that there is an *explicit* isomorphism between the Galois group G of K/k and the so-called Weber class group $\mathcal{C}_\mathfrak{f}$ in k with conductor $\mathfrak{f}$. It maps $\sigma_\mathfrak{p}$ onto the Weber congruence class $C_\mathfrak{p}$ attached to $\mathfrak{p}$.[68] This is a far-reaching generalization of Gauss's Reciprocity Law. The automorhism $\sigma_\mathfrak{p}$ is only defined for primes $\mathfrak{p}$ not dividing the discriminant of K/k. In [Hasse, 1930b], Hasse succeeded in extending $\sigma_\mathfrak{p}$ to the case where $\mathfrak{p}$ divides the discriminant in such a way that this new symbol extends the Artin symbol just as the Hilbert symbol extends the Legendre symbol. Hasse's new norm residue symbol $\left(\frac{\alpha, K}{\mathfrak{p}}\right)$ is an element of the Galois group G of K/k.[69] It satisfies the decomposition law with respect to α,[70] the norm property [Hasse, 1930b, p. 38 (Theorem 5)], and it is equal to the Artin symbol $\left(\frac{K}{\mathfrak{p}}\right)$, if $\mathfrak{p}$ neither divides the discriminant of K/k nor α. By means of Artin's Reciprocity Law, Hasse was able to derive the product formula (Hilbert's Reciprocity Law) for his new symbol.

In the second paper [Hasse, 1933], dedicated to Emmy Noether on the occasion of her fiftieth birthday, Hasse introduced the norm residue symbol, via the theory of algebras, as a fundamental 2-cycle. This prepared the way for the cohomological formulation of the Artin-Tate Reciprocity Law and of the fundamental theorems of class field theory. Hasse was following up on Emmy Noether's program and postulate [Noether, 1932, p. 189] that abelian theorems, that is, theorems for abelian extensions K/k with abelian Galois group G—and for that reason theorems of class field theory—have to be extended to normal extensions K/k. This is accomplished by crossing the Galois group G with the multiplicative group of the field K and by studying this crossed product, for instance, by imbedding the multiplicative group $K^\times$ and the Galois group G into a central simple (noncommutative) algebra $\mathcal{A}$ over k, so that the relative automorphisms of K/k become inner automorphisms of $\mathcal{A}$, or by viewing the Galois extension K/k as a Galois module. Because of this, Emmy Noether has to be considered the true founder of cohomological methods in algebra.[71]

Hasse carried out Noether's program by combining results from his paper on the local structure of $\mathfrak{p}$-adic skew fields [Hasse, 1931a] and his *Transactions* paper on cyclic algebras over an algebraic number field [Hasse, 1932]. If $\mathcal{A}$ is a central simple algebra of dimension n^2 over an algebraic number field K, then Hasse termed n the "degree" of $\mathcal{A}$ over K. The degree m of its corresponding skew field $\mathcal{S}$ (according

---

[68]See [Frei, 2002; 1989] for the details and the definition of a Weber congruence class, the Weber class group, and the conductor.

[69]More precisely, it is an element of the decomposition group of $\mathfrak{p}$.

[70]There is also a decomposition law with respect to K, stated appropriately [Hasse, 1930b, 140].

[71]Emmy Noether is also at the origin of homological methods in algebraic topology. See [Frei-Stammbach, 1999, p. 996].

to Wedderburn's Main Theorem) was called the (Schur) index of $\mathcal{A}$ [Hasse, 1931a, p. 517 (Theorem 41); 1933, p. 735 (Theorem 1.2)]. Brauer had proved that it is an invariant of the Brauer class of $\mathcal{A}$ [Brauer, 1929a; 1929b]. For a prime spot $\mathfrak{p}$ in K, Hasse denoted by $\mathcal{A}_\mathfrak{p} = \mathcal{A} \otimes K_\mathfrak{p}$ the localization of $\mathcal{A}$ at $\mathfrak{p}$ and by $K_\mathfrak{p}$ the localization at $\mathfrak{p}$ of K. In [Hasse, 1931a], he had shown that $\mathcal{A}_\mathfrak{p}$ is a central simple algebra over $K_\mathfrak{p}$ of degree n, and that the (local $\mathfrak{p}$-)index $m_\mathfrak{p}$ of $\mathcal{A}_\mathfrak{p}$ over $K_\mathfrak{p}$ is a divisor of m [Hasse, 1931a, pp. 498-499; 1933, p. 740 (Theorem 3.1)]. Furthermore, $\mathcal{A}_\mathfrak{p}$ is cyclic, and among the various cyclic generations $\mathcal{A}_\mathfrak{p} = (b_\mathfrak{p}, L_\mathfrak{p}, \sigma_\mathfrak{p}) = \sum_{\nu=0}^{n-1} \beta^\nu L_\mathfrak{p}$ of $\mathcal{A}_\mathfrak{p}$ over $K_\mathfrak{p}$, there is a uniquely determined canonical one, given by $b_\mathfrak{p} = \beta^n = \pi^{\nu_\mathfrak{p}}$, where π is an arbitrarily chosen prime element of $K_\mathfrak{p}$ and $\nu_\mathfrak{p}$ is a uniquely determined residue class modulo n with $(\nu_\mathfrak{p}, n) = r_\mathfrak{p}$ [Hasse, 1931a, pp. 507-517; 1933, pp. 742-743 (Theorems 4.1 and 4.2)]. Thus, $\frac{\nu_\mathfrak{p}}{n} \equiv \frac{\mu_\mathfrak{p}}{m_\mathfrak{p}}$ modulo $\mathbb{Z}$ with $(\mu_\mathfrak{p}, m_\mathfrak{p}) = 1$. $L_\mathfrak{p}$ is the uniquely determined unramified (cyclic) field extension of degree n of $K_\mathfrak{p}$, generated by a primitive $[\mathfrak{N}(\mathfrak{p})^n - 1]$th root of unity, where $\mathfrak{N}(\mathfrak{p})$ denotes the norm of $\mathfrak{p}$, and $\sigma_\mathfrak{p}$ is the Frobenius automorphism of $\mathfrak{p}$, which acts on $L_\mathfrak{p}$ by conjugation with β: $\sigma_\mathfrak{p}(\alpha) = \beta \alpha \beta^{-1}$ for α in $L_\mathfrak{p}$.

The rational number $\rho_\mathfrak{p} = \frac{\nu_\mathfrak{p}}{n}$ modulo $\mathbb{Z}$ is an invariant of the (local) Brauer class of $\mathcal{A}_\mathfrak{p}$ over $K_\mathfrak{p}$ in the Brauer group $\mathrm{Br}(K_\mathfrak{p})$. Hasse called it the "(local) $\mathfrak{p}$-invariant" of $\mathcal{A}$ over K. André Weil later dubbed it the Hasse invariant of $\mathcal{A}_\mathfrak{p}$ [Weil, 1967, pp. 221, 224, and 252]. For this invariant, Hasse proved the Local Isomorphism Theorem:[72] The mapping $h : \mathcal{A}_\mathfrak{p}/K_\mathfrak{p} \mapsto \frac{\mu_\mathfrak{p}}{m_\mathfrak{p}} = \rho_\mathfrak{p}$ modulo $\mathbb{Z}$ defines for any finite $\mathfrak{p}$ an isomorphism from the Brauer group $\mathrm{Br}(K_\mathfrak{p})$ over $K_\mathfrak{p}$ onto the (additive) rational residue class group $\mathbb{Q}/\mathbb{Z}$. For infinite real $\mathfrak{p}$, $\mathbb{Q}/\mathbb{Z}$ has to be replaced by $\frac{1}{2}\mathbb{Z}/\mathbb{Z}$. From the Norm Splitting Theorem and from his norm theorem of class field theory, that is, from the Local-Global Principle, Hasse now inferred the Global Isomorphism Theorem for a central simple algebra $\mathcal{A}$ of degree n over an algebraic number field K [Hasse, 1933, pp. 750-753 (Theorem 6.5)]: the mapping $H : \mathcal{A}/K \mapsto \{\frac{\mu_\mathfrak{p}}{m_\mathfrak{p}} = \rho_\mathfrak{p}$ modulo $\mathbb{Z}\}$ defines an isomorphism from the Brauer group $\mathrm{Br}(K)$ over K into the group of all sets $\{\rho_\mathfrak{p}$ modulo $\mathbb{Z}\}$, where $\rho_\mathfrak{p}$ is an arbitrary rational number if $\mathfrak{p}$ is finite, and $\rho_\mathfrak{p} = 0, \frac{1}{2}$ if $\mathfrak{p}$ is infinite. That is, the similarity class of every algebra in $\mathrm{Br}(K)$ is uniquely determined by its Hasse invariants. The image of H consists of the subgroup of all sets $\{\rho_\mathfrak{p}$ modulo $\mathbb{Z}\}$ satisfying:

(1) $\rho_\mathfrak{p} \not\equiv 0$ modulo $\mathbb{Z}$ only for a finite number of prime spots $\mathfrak{p}$, and
(2) $\sum_\mathfrak{p} \rho_\mathfrak{p} \equiv 0$ modulo $\mathbb{Z}$.

In light of this theorem, called the Sum Theorem by Hasse, the structure of the Brauer group $\mathrm{Br}(K)$ over an algebraic number field K is completely characterized. Weil called Hasse's Sum Theorem (written additively) Hasse's Law of Reciprocity [Weil, 1967, p. 255 (Theorem 2)]. Because Hasse did not make use of Artin's Reciprocity Law in his proof of the Global Isomorphism Theorem, he was now in a position to prove Artin's Reciprocity Law solely on the basis of the theory of semisimple algebras over algebraic number fields, as had been suggested by Emmy Noether.

From these studies of the structure of local algebras, Hasse obtained a direct definition of his norm residue symbol for an algebraic number field K, without making use of Artin's Reciprocity Law. He took a cyclic extension L of degree n over an algebraic number field K and σ a generating automorphism of the Galois

---

[72]see [Hasse, 1933, (5.1), in particular (5.12)].

group G of L/K. For an arbitrary number $\alpha \neq 0$ in K, he then considered the central simple algebra $\mathcal{A} = (\alpha, L, \sigma)$. If $\frac{\nu_{\mathfrak{p}}}{n} \equiv \frac{\mu_{\mathfrak{p}}}{m_{\mathfrak{p}}}$ modulo $\mathbb{Z}$ is the $\mathfrak{p}$-invariant of $\mathcal{A}$, he defined his norm residue symbol by $\left(\frac{\alpha, L/K}{\mathfrak{p}}\right) := \sigma^{-\nu_p}$ [Hasse, 1933, pp. 745-747 and 754-755 (Theorems 5.3 and 6.6)]. He showed that this symbol has all the desired properties, namely,

(1) Local Norm Property: $\left(\frac{\alpha, L/K}{\mathfrak{p}}\right) = 1$ if and only $\alpha \in K$ is the (local) norm of an element $\beta \neq 0$ in $L_{\mathfrak{P}}$ with $\mathfrak{P}|\mathfrak{p}$;

(2) Global Norm Property: $\left(\frac{\alpha, L/K}{\mathfrak{p}}\right) = 1$, for all $\mathfrak{p}$, if and only $\alpha \in K$ is the (global) norm of an element $\beta \neq 0$ in L;

(3) Multiplicative Property: $\left(\frac{\alpha_1 \alpha_2, L/K}{\mathfrak{p}}\right) = \left(\frac{\alpha_1, L/K}{\mathfrak{p}}\right)\left(\frac{\alpha_2, L/K}{\mathfrak{p}}\right)$;

(4) Product Formula (Reciprocity Law): $\prod_{\mathfrak{p}} \left(\frac{\alpha, L/K}{\mathfrak{p}}\right) = 1$;

(5) Relation to the Frobenius-Artin symbol: If L/K is unramified at a (finite) prime $\mathfrak{p}$ and $\sigma_{\mathfrak{p}}$ is the Frobenius-Artin substitution of $\mathfrak{p}$ in the Galois group G of L/K and $\nu_{\mathfrak{p}}$ is the order of α at $\mathfrak{p}$, then $\left(\frac{\alpha, L/K}{\mathfrak{p}}\right) = \sigma_{\mathfrak{p}}^{-\nu_{\mathfrak{p}}}$; and

(6) Relation to the Hilbert norm residue symbol: If $L = K(\sqrt[n]{\beta})$ is a Kummer extension of K, where K contains the nth roots of unity, then $\left(\frac{\alpha, L/K}{\mathfrak{p}}\right)\sqrt[n]{\beta} = \left(\frac{\alpha, \beta}{\mathfrak{p}}\right)\sqrt[n]{\beta}$, where $\left(\frac{\alpha, \beta}{\mathfrak{p}}\right)$ is the Hilbert norm residue symbol, which is a nth root of unity [Hasse, 1933, pp. 745-747 and 754-755 (Theorems 5.3 and 6.6)].

Properties (1) and (3), together with the surjectivity of the map $\alpha \to \left(\frac{\alpha, L/K}{\mathfrak{p}}\right)$, give the isomorphism theorem of local class field theory. The local norm property is equivalent to the property that the local algebra $\mathcal{A}_{\mathfrak{p}}$ is isomorphic to a full matrix algebra over the ground field $K_{\mathfrak{p}}$. The global norm property comes from the Hasse Norm Theorem of class field theory. The multiplicativity is a consequence of Hasse's Local Isomorphism Theorem, and the Product Formula (Reciprocity Law) is a consequence of Hasse's Sum Theorem. In fact, the Hasse Sum Theorem is equivalent to the Artin Reciprocity Law for cyclic extensions of algebraic number fields. Thus, Hasse obtained a new proof of Artin's Reciprocity Law for cyclic extensions by means of the theory of semisimple algebras.[73]

Summary and Conclusions

The Reciprocity Law is the central theorem of number theory. It has engaged mathematicians from Euler and Legendre in the eighteenth century to Gauss, who gave it eight different proofs in the first decade of the nineteenth century. Gauss's second proof—based on genus characters—turned out to be of special importance, given the fact that there he also introduced and developed the theory of orders, characters, and genera [Gauss, 1801]. By the end of the nineteenth century, Hilbert had given the law a new form, interpreting it as an infinite product over all primes of norm residue symbols, characters he introduced based on those of Gauss. Hilbert

---

[73]How class field theory can be based in the theory of simple algebras following Hasse's approach is treated in detail in [Weil, 1967, Part II]. For a more detailed exposition of the whole development, see [Frei, 2003b] and to a forthcoming publication which will also include the later contributions by Chevalley and Herbrand.

called this product the most complete form of the Reciprocity Law and saw it as an analog of Cauchy's Theorem to the effect that the integral on a closed curve around all the singularities of a regular function in the complex plane (or on a Riemann Surface) is equal to zero [Hilbert, 1899a]. This product form was essentially already contained in Gauss's second proof. The disadvantage of this closed form is that it is not explicit. In order to determine the norm residue symbol explicitly for each prime, a diophantine equation has to be solved modulo all its powers. It was Hasse who tackled the problem of finding an explicit expression for the norm residue symbols by means of Hensel's investigations on the structure of p-adic fields [Hensel, 1922]. Over the years from 1923 to 1930, Hasse succeeded in the quadratic case thanks to his results on rational quadratic forms [Hasse, 1924a-c], but in the higher cases his results were at first incomplete [Hasse 1927a; 1930b].

The clue for the solution came from quite an unexpected corner, namely, from the theory of associative algebras (with unity). This theory had been initiated and developed by the British in the mid-nineteenth century and later continued by the Americans, especially in the opening decades of the twentieth. Dickson, in particular, was responsible for characterizing a linear associative algebra abstractly by independent postulates and for giving a decisive new direction to the theory of associative algebras by introducing generalized quaternions over quadratic number fields [Dickson, 1912] and cyclic algebras over cyclic number fields [Dickson, 1906; 1914; 1923; 1927], thus bringing algebraic number theory into play. Whether these new algebras are skew fields or not depends on the norm theory and so on class field theory, since class field theory can be understood as a general theory of norm forms. It was this connection which attracted Hasse's interest; his work on quadratic forms and the norm residue symbol had led him both to Hilbert's norm theory and to class field theory. It had also attracted Artin's interest, however. The latter's discovery of a Reciprocity Law brought class field theory to a certain conclusion [Artin, 1927d].

Takagi had shown that class field theory is identical with the theory of abelian extensions of algebraic number fields, that is, algebraic extensions with commutative Galois group [Takagi, 1920]. For that reason, Hasse, Artin, and Emmy Noether believed that the theory of noncommutative algebras might give a clue as to how to extend class field theory to non-abelian class field theory, that is, number fields with noncommutative Galois group. Although this goal was finally not achieved, and the establishment of a non-abelian class field theory is still an open problem, Hasse's study of local associative algebras and the resulting proof of the fundamental theorem for semisimple algebras over algebraic number fields—that he proved together with Brauer and Noether—to the effect that every such algebra has a cyclic generation, led him to the long-sought, explicit determination of the norm residue symbol and to a new and direct approach to the Artin Reciprocity Law and to class field theory. With this, class field theory took on a new and (up to this day) final shape. On the other hand, the new shape was a prerequisite for subsequent developments that led to the Langlands conjectures, a project still ongoing.

Closely related to the Reciprocity Law is the arithmetic of algebraic number fields, which had been introduced by Gauss and subsequently elaborated by Kummer and Dedekind in the nineteenth century. Dedekind also took the important step of viewing an algebraic number field as a vector space over the rational number field [Dirichlet, 1871]. Dedekind's work, in particular, motivated Lipschitz to study the corresponding arithmetic theory of the rational quaternions. He took as integers the rational quaternions with integer coefficients thus generalizing the Gauss and Kummer rings of integers of cyclotomic fields in an obvious way [Lipschitz, 1886]. It was Hurwitz who realized that Lipschitz's definition does not give the full analog of Dedekind's theory; in particular, it does not give the fundamental theorem of arithmetic, since the Lipschitz ring is not a maximal order within the rational quaternions. To obtain a *maximal* order, Hurwitz had to add quaternions with half-integer coefficients [Hurwitz, 1896; 1919], similar to what Dedekind had done in the case of quadratic number fields [Dirichlet, 1871]. Hurwitz's arithmetic theory of maximal orders in a rational quaternion algebra was then extended by Du Pasquier to semisimple algebras [Du Pasquier, 1906; 1909; 1916; 1920] and by Dickson to associative algebras [Dickson, 1923] with Speiser treating the corresponding ideal theory of rational semisimple algebras [Speiser, 1926; 1927]. This led Artin to prove the finiteness of the class number in each maximal order of a semisimple algebra and to study more generally rings satisfying the descending chain condition [Artin, 1927b-c]. It also led to Emmy Noether's fundamental theory of ideals in rings with ascending chain condition. Artin's and Noether's general theory of rings and algebras received an authoritative presentation in 1930 in van der Waerden's fundamental treatise, *Moderne Algebra* [van der Waerden, 1959], which set the standard for all textbooks on algebra to come.

Both avenues, the one leading from associative algebras over algebraic number fields to the new norm residue symbol and the new class field theory, and the one leading from the abstract theory of associative algebras to the modern abstract theory of Artinian and Noetherian rings were essentially initiated by Dickson in the opening decade of the twentieth century. In addition, Dickson's *Algebras and Their Arithmetics* [Dickson, 1923]—with its summation of the results obtained by the American school of abstract algebra—as well as its German translation [Dickson, 1927] gave rise to an impressive and rapid development in the arithmetic-algebraic school in Germany around Artin, Hasse, and Noether. The latter introduced new contours to the theory of both algebraic number fields and rings. Thus, Dickson's book was crucial not only for bringing together two different schools but also for sparking the fruitful, mutual interaction that ultimately gave rise to new and powerful methods and to abstract theories, the strong influence of which continues up to the present.

References

Albert, A. Adrian. 1931. "On Direct Products." *Transactions of the American Mathematical Society* 33, 690-711.

——————. 1932. "Normal Division Algebras of Degree Four over an Algebraic Field." *Transactions of the American Mathematical Society* 34, 363-372.

——————. 1993. *Collected Mathematical Papers*. Ed. Richard E. Block, Nathan Jacobson, J. Marshall Osborn, David J. Saltman, and Daniel Zelinsky. 2 Vols. New York: American Mathematical Society.

Albert, A. Adrian and Hasse, Helmut. 1932. "A Determination of all Normal Division Algebras over an Algebraic Number Field." *Transactions of the American Mathematical Society* 34, 722-726.

Argand, Robert. 1806. *Essai sur une manière de représenter des quantités imaginaires dans les constructions géométriques.* Privately printed, or 1874. Paris: Gauthier-Villars.

Artin, Emil. 1923. "Über eine neue Art von *L*-Reihen." *Abhandlungen aus dem Mathematischen Seminar der Universität Hamburg* 3, 89-108.

———. 1927a. "Über einen Satz von Herrn J. H. Maclagan Wedderburn." *Abhandlungen aus dem Mathematischen Seminar der Universität Hamburg* 5, 245-250.

———. 1927b. "Zur Theorie der hyperkomplexen Zahlen." *Abhandlungen aus dem Mathematischen Seminar der Universität Hamburg* 5, 251-260.

———. 1927c. "Zur Arithmetik der hyperkomplexen Zahlen." *Abhandlungen aus dem Mathematischen Seminar der Universität Hamburg* 5, 261-289.

———. 1927d. "Beweis des allgemeinen Reziprozitätsgesetzes." *Abhandlungen aus dem Mathematischen Seminar der Universität Hamburg* 5, 353-363.

———. 1950. "The Influence of J. H. M. Wedderburn on the Development of Modern Algebra." *Bulletin of the American Mathematical Society* 56, 65-72.

Artin, Emil and Tate, John. 1968. *Class Field Theory.* New York-Amsterdam: W. A. Benjamin, Inc.

Borel, Armand. 2001. *Essays in the History of Lie Groups and Algebraic Groups.* HMATH. Vol. 21. Providence: American Mathematical Society and London: London Mathematical Society.

Brauer, Richard. 1926. "Über Zusammenhänge zwischen arithmetischen und invariantentheoretischen Eigenschaften von Gruppen linearer Substitutionen." *Sitzungsberichte der Preußischen Akademie der Wissenschaften zu Berlin,* 410-416.

———. 1928. "Untersuchungen über die arithmetischen Eigenschaften von Gruppen linearer Substitutionen: Erste Mitteilung." *Mathematische Zeitschrift* 28, 677-696.

———. 1929a. "Über Systeme hyperkomplexer Zahlen." *Mathematische Zeitschrift* 30, 79-107.

———. 1929b. "Über Systeme hyperkomplexer Grössen." *Jahresbericht der Deutschen Mathematiker-Vereinigung* 38, 47-48.

Brauer, Richard; Hasse, Helmut; and Noether, Emmy. 1932. "Beweis eines Hauptsatzes in der Theorie der Algebren." *Journal für die reine und angewandte Mathematik* 167, 399-404.

Brauer, Richard and Noether, Emmy. 1927. "Über minimale Zerfällungskörper irreduzibler Darstellungen." *Sitzungsberichte der Preußischen Akademie der Wissenschaften zu Berlin,* 1927, 221-228.

Cartan, Élie. 1898. "Les groupes linéaires et les systèmes de nombres complexes." *Annales de la Faculté des Sciences de Toulouse* 12B, B1-B99, or Cartan, Élie. 1952–1955. *Oeuvres complètes.* 3 Vols. in 6 Pts. Paris: Gauthier-Villars, 1(2):7-105.

Cassels, John W. S. and Fröhlich, Albrecht, Ed. 1967. *Algebraic Number Theory: Proceedings of an Instructional Conference Organized by the London*

Mathematical Society (a NATO Advanced Study Institute) with the Support of the International Mathematical Union. London and New York: Academic Press, Inc.

Cayley, Arthur. 1889–1898. *The Collected Mathematical Papers of Arthur Cayley.* Ed. Arthur Cayley and Andrew R. Forsyth. 14 Vols. Cambridge: Cambridge University Press.

Chevalley, Claude. 1946. *The Theory of Lie Groups.* Princeton: Princeton University Press.

Clifford, William Kingdon. 1882. *Mathematical Papers by William Kingdon Clifford.* Ed. Robert Tucker. Intro. by Henry J. S. Smith. London: Macmillan and Co.

Curtis, Charles W. 1999. *Pioneers of Representation Theory: Frobenius, Burnside, Schur, and Brauer.* HMATH. Vol. 15. Providence: American Mathematical Society and London: London Mathematical Society.

Dedekind, Richard. 1885. "Zur Theorie der aus n Haupteinheiten gebildeten komplexen Größen." *Nachrichten der Gesellschaft der Wissenschften zu Göttingen Jahrgang 1885*, 141-159, or Dedekind, Richard. 1930–1932. *Gesammelte mathematische Werke.* Ed. Robert Fricke, Emmy Noether, and Oystein Øre. 3 Vols. Braunschweig: F. Vieweg und Sohn, 2:1-22.

Deuring, Max. 1935. *Algebren.* Ergebnisse der Mathematik und ihrer Grenzgebiete. Vol. 4. Berlin: Julius Springer.

Dirichlet, Peter Lejeune. 1839. "Recherches sur diverses applications de l'analyse infinitésimale à la théorie des nombres." *Journal für die reine und angewandte Mathematik* 19, 324-369, or [Dirichlet, 1889, 1:411-496].

———. 1871. *Vorlesungen über Zahlentheorie.* Ed. and Supp. Richard Dedekind. 2nd Ed. Braunschweig: Vieweg Verlag.

———. 1889. *G. Lejeune Dirichlet's Werke.* Ed. Leopold Kronecker. 2 Vols. Berlin: Georg Reimer.

Dickson, Leonard Eugene. 1906. "Abstract 16, 14 April, 1906." *Bulletin of the American Mathematical Society* 12 (1905–1906), 441-442.

———. 1912. "Linear Algebras." *Transactions of the American Mathematical Society* 13, 59-73.

———. 1914a. "Linear Associative Algebras and Abelian Equations." *Transactions of the American Mathematical Society* 15, 31-46.

———. 1914b. *Linear Algebras.* Cambridge: Cambridge University Press. Reprint Ed. 1930.

———. 1923. *Algebras and Their Arithmetics.* The University of Chicago Science Series. Chicago: University of Chicago Press.

———. 1927. *Algebren und ihre Zahlentheorie.* Mit einem Kapitel über Idealtheorie von Andreas Speiser. Zürich: Orell Füssli Verlag.

Du Pasquier, Louis-Gustave. 1906. *Zahlentheorie der Tettarionen.* Dissertation: Zürich. 1906, or *Vierteljahrsschrift der Naturforschenden Gesellschaft in Zürich* 51, 55-129.

———. 1909. "Über holoide Systeme von Düotettarionen." *Vierteljahrsschrift der Naturforschenden Gesellschaft in Zürich* 54, 116-148.

———. 1916. "Sur l'arithmétique des nombres hypercomplexes." *L'Enseignement Mathématique* 18, 201-259 and 265-293.

———. 1920. "Sur la théorie des nombres hypercomplexes à coordonnées rationnelles." *Bulletin de la Société mathématique de France* 48, 1-24.

Euler, Leonhard. 1748/1922. *Introductio in analysin infinitorum. Opera omnia.* 1st Ser. Vol. 8. Basel: Birkhäuser Verlag.

Fenster, Della D. 1998. "Leonard Eugene Dickson and His Work in the Arithmetics of Algebras." *Archive for History of Exact Sciences* 52, 119–159.

———. 1999. "The Development of the Concept of an Algebra: Leonard Eugene Dickson's Role." *Studies in the History of Modern Mathematics.* Vol. 4. Supplementi dei Rendiconti del Circolo matematico di Palermo. 2d Ser. Ed. Umberto Bottazzini. Palermo: Circolo matematico di Palermo, 59-122.

Frei, Günther. 1979. "On the Development of the Genus of Quadratic Forms." *Annales des sciences mathématiques du Québec* 3, 5-62.

———. 1981. *Die Briefe von Emil Artin an Helmut Hasse, 1923–1953.* Collection Mathématique. Québec: Université Laval.

———. 1989. "Heinrich Weber and the Emergence of Class Field Theory." In *The History of Modern Mathematics.* Ed. David E. Rowe and John McCleary. 2 Vols. Boston: Academic Press, Inc., 1:425-450.

———. 1994. "The Reciprocity Law from Euler to Eisenstein." In *The Intersection of History and Mathematics.* Ed. Chikara Sasaki, Mitsuo Sugiura, and Joseph W. Dauben. Basel/Boston/ Berlin: Birkhäuser Verlag, 67-88.

———. 1995. "Adolf Hurwitz (1859–1919)." In *Die Albertus-Universität zu Königsberg und ihre Professoren: Jahrbuch der Albertus-Universität.* Vol. 29. Berlin: Duncker & Humblot, 527-542.

———. 2001a. "How Hasse Was Led to the Theory of Quadratic Forms, the Local-Global-Principle, the Theory of the Norm Residue Symbol, the Reciprocity Laws, and to Class Field Theory." In *Class Field Theory: Its Centenary and Prospect.* Ed. Katsuya Miyake. Advanced Studies in Pure Mathematics. Vol. 30. Tokyo: Mathematical Society of Japan, 31-62.

———. 2001b/2007. "Gauss's Unpublished Section Eight of the *Disquisitiones Arithmeticæ*: The Beginnings of a Theory of Function Fields over a Finite Field." Published in an abbreviated version as "The Unpublished Section Eight: On the Way to Function Fields over a Finite Field." In *The Shaping of Arithmetic after C. F. Gauss's Disquisitiones Arithmeticæ.* Ed. Catherine Goldstein, Norbert Schappacher, and Joachim Schwermer. Heidelberg: Springer Verlag, pp. 159-198.

———. 2002. "On the History of the Artin Reciprocity Law in Abelian Extensions of Algebraic Number Fields: How Artin Was Led to his Reciprocity Law." In *The Legacy of Niels Henrik Abel: The Abel Bicentennial, Oslo, 2002.* Ed. Olav Arnfinn Laudal and Ragni Piene. Heidelberg: Springer-Verlag, 267-294.

———. 2003a. "Zur Geschichte der Arithmetik der Algebren (1843–1932)." *Elemente der Mathematik* 58, 156-168.

———. [2003b]. *Zur Vorgeschichte und Entwicklung der Theorie der Algebren über algebraischen Zahlkörpern, die zum neuen Normenrestsymbol von Hasse führte. Mit einem Anhang zum Satz von Brauer-Hasse-Noether und einigen Bemerkungen zum Briefwechsel zwischen Adrian Albert und Helmut Hasse.* December 2003. To appear.

―――. 2003c. "Johann Jakob Burckhardt zum 100. Geburtstag am 13. Juli 2003." *Elemente der Mathematik* 58, 134-140.

Frei, Günther and Stammbach, Urs. 1999. "Heinz Hopf." In *History of Topology.* Ed. Ioan M. James. Amsterdam: Elsevier Science, 991-1008.

Frobenius, F. Georg. 1878/1968. "Über lineare Substitutionen und bilineare Formen." *Journal für die reine angewandte Mathematik* 84, 1-63, or Frobenius, F. Georg. 1968. *Gesammelte Abhandlungen.* Ed. Jean-Pierre Serre. 3 Vols. Heidelberg: Springer-Verlag, 1:343-405.

Furtwängler, Philipp. 1912. "Die Reziprozitätsgesetze für Potenzreste mit Primzahlexponenten in algebraischen Zahlkörpern, II." *Mathematische Annalen* 72, 346-386.

Gauss, Carl Friedrich. 1801. *Disquisitiones arithmeticæ.* In [Gauss, 1863–1903, 1:1-478].

―――. 1819. "Mutationen des Raumes." In [Gauss, 1863–1903, 8:357-362].

―――. 1831. "Theoria Residuorum Biquadraticorum." In [Gauss, 1863–1903, 2:95-148].

―――. 1863–1903. *Werke.* Ed. Königliche Gesellschaft der Wissenschaften zu Göttingen. 12 Vols. Göttingen: Akademie der Wissenschaften.

Gras, Georges. 2003. *Class Field Theory: From Theory to Practice.* Springer Monographs in Mathematics. Berlin: Springer-Verlag.

Grassmann, Hermann Günter. 1844. *Die lineale Ausdehnungslehre.* Leipzig: Verlag Wigand.

―――. 1862. *Die Ausdehnungslehre.* Berlin: Verlag Enslin.

Graves, John Thomas. 1848. In Young, John Radford. 1848. "On an Extension of a Theorem of Euler, with a Determination of the Limit beyond Which It Fails." *Transactions of the Royal Irish Academy* 21, 311-341. (See also Hamilton, William Rowan. 1848. "Additional Note Referred to in Page 336, Respecting the Researches of John T. Graves, Esq." *Transactions of the Royal Irish Academy* 21, 338-341.)

Hamilton, William Rowan. 1853. *Lectures on Quaternions.* Dublin: Hodges and Smith.

―――. 1967. *The Mathematical Papers of Sir William Rowan Hamilton.* Vol. 3. *Algebra.* Ed. Heini Halberstam and Richard E. Ingram. Cambridge: Cambridge University Press.

Hasse, Helmut. 1924a. "Darstellbarkeit von Zahlen durch quadratische Formen in einem beliebigen algebraischen Zahlkörper." *Journal für die reine und angewandte Mathematik* 153, 113-130.

―――. 1924b. "Zur Theorie des quadratischen *Hilbert*schen Normenrestsymbols in algebraischen Körpern." *Journal für die reine und angewandte Mathematik* 153, 76-93.

―――. 1924c. "Zur Theorie des *Hilbert*schen Normenrestsymbols in algebraischen Körpern." *Journal für die reine und angewandte Mathematik* 153, 184-191.

―――. 1925. "Direkter Beweis des Zerlegungs- und Vertauschungssatzes für das *Hilbert*sche Normenrestsymbol in einem algebraischen Zahlkörper im Falle eines Primteilers 𝔩 des Relativgrades l." *Journal für die reine und angewandte Mathematik* 154, 20-35.

———. 1927a. "Über das Reziprozitätsgesetz der m-ten Potenzreste." *Journal für die reine und angewandte Mathematik* 158, 228-259.

———. 1927b. Review of "L. E. Dickson, Algebren und ihre Zahlentheorie. Mit einem Kapitel über Idealtheorie von A. Speiser." *Jahresbericht der Deutschen Mathematiker-Vereinigung* 37, *Literarisches*, 90-97.

———. 1927c. "Existenz gewisser algebraischer Zahlkörper." *Sitzungsberichte der Preußischen Akademie der Wissenschaften zu Berlin*, 229-234.

———. 1930a. *Bericht über neuere Untersuchungen und Probleme aus der Theorie der algebraischen Zahlkörper, Teil II: Reziprozitätsgesetz. Jahresbericht der Deutschen Mathematiker-Vereinigung. Der Ergänzungsbände VI. Band.* Leipzig and Berlin: B. G. Teubner Verlag, 1-204.

———. 1930b. "Neue Begründung und Verallgemeinerung der Theorie des Normenrestsymbols." *Journal für die reine und angewandte Mathematik* 162, 134-144.

———. 1931a. "Über ℘-adische Schiefkörper und ihre Bedeutung für die Arithmetik hyperkomplexer Zahlsysteme." *Mathematische Annalen* 104, 495-534.

———. 1931b. "Beweis eines Satzes und Widerlegung einer Vermutung über das allgemeine Normenrestsymbol." *Nachrichten der Gesellschaft der Wissenschaften zu Göttingen, Mathematisch-physikalische Klasse* 1, 64-69.

———. 1932. "Theory of Cyclic Algebras over an Algebraic Number Field." *Transactions of the American Mathematical Society* 34, 171-214.

———. 1933. "Die Struktur der R. Brauerschen Algebrenklassengruppe über einem algebraischen Zahlkörper. Insbesondere Begründung der Theorie des Normenrestsymbols und die Herleitung des Reziprozitätsgesetzes mit nichtkommutativen Hilfsmitteln." *Mathematische Annalen* 107, 731-760.

———. 1950. "Zum Existenzsatz von Grunwald in der Klassenkörpertheorie." *Journal für die reine und angewandte Mathematik* 188, 40-64.

———. 1973. *Class Field Theory.* Ed. Günther Frei. Collection mathématique. Québec: Université Laval.

Hasse, Helmut and Hensel, Kurt. 1923. "Über die Normenreste eines relativzyklischen Körpers vom Primzahlgrad l nach einem Primteiler 𝔩 von l." *Mathematische Annalen* 90, 262-278.

Hawkins, Thomas. 1972. "Hypercomplex Numbers, Lie Groups, and the Creation of Group Representation Theory." *Archive for History of Exact Sciences* 8, 243-287.

———. 2000. *Emergence of the Theory of Lie Groups. An Essay in the History of Mathematics 1869–1926.* Sources and Studies in the History of Mathematics and Physical Sciences. New York: Springer-Verlag.

Hensel, Kurt. 1913. *Zahlentheorie.* Leipzig and Berlin: Göschen.

———. 1922. "Über die Normenreste in den allgemeinsten relativ-abelschen Zahlkörpern." *Mathematische Annalen* 85, 1-10.

Hilbert, David. 1897. "Die Theorie der algebraischen Zahlkörper." *Jahresbericht der Deutschen Mathematiker-Vereinigung* 4, 175-546, or Hilbert, David. 1970. *Gesammelte Abhandlungen.* 3 Vols. Heidelberg: Springer-Verlag, 1:63-363.

_____. 1899a. "Ueber die Theorie der relativquadratischen Zahlkörper." *Jahresbericht der Deutschen Mathematiker-Vereinigung* 6, 88-94.

_____. 1899b. "Ueber die Theorie des relativquadratischen Zahlkörpers." *Mathematische Annalen* 51, 1-127.

_____. 1921. "Adolf Hurwitz." *Mathematische Annalen* 83, 161-172.

Hurwitz, Adolf. 1896. "Über die Zahlentheorie der Quaternionen." *Nachrichten der Gesellschaft der Wissenschaften zu Göttingen, Mathematisch-physikalische Klasse,* 314-340.

_____. 1919. *Vorlesungen über die Zahlentheorie der Quaternionen.* Berlin: Springer-Verlag.

Jacobi, Carl G. J. 1881–1891. *Gesammelte Werke.* Ed. Königlich preussischen Akademie der Wissenschaften. 7 Vols. Berlin: G. Reimer.

Klein, Felix. 1926. *Vorlesungen über die Entwicklung der Mathematik im 19. Jahrhundert.* Vol. I. Berlin: Springer.

Lemmermeyer, Franz and Roquette, Peter. 2003. *Correspondence Helmut Hasse–Emmy Noether 1925–1935: Edited and Commented.* Manuscript of 9 November 2003, or 2006. *Helmut Hasse und Emmy Noether: Die Korrespondenz 1925–1935.* Göttingen: Universitätsverlag Göttingen.

Lipschitz, Rudolf. 1886. *Untersuchungen über die Summe von Quadraten.* Bonn: Cohen & Sohn.

Lounesto, Pertti. 1997. *Clifford Algebas and Spinors.* Cambridge: Cambridge University Press. 2d Ed. 2001.

Molien, Theodor. 1893. "Über Systeme höherer komplexer Zahlen." *Mathematische Annalen* 41, 83-156.

Neukirch, Jürgen. 1969. *Klassenkörpertheorie.* Mannheim: Bibliographisches Institut.

Noether, Emmy. 1923. Review of [Wedderburn, 1923]. *Jahrbuch über die Fortschritte der Mathematik* 49 (2), 82.

_____. 1929. "Hyperkomplexe Größen und Darstellungstheorie." *Mathematische Zeitschrift* 30, 641-692.

_____. 1930. *Algebra der hyperkomplexen Größen: Vorlesung von Prof. E. Noether. W.S. 1929/30.* Ed. Max Deuring, or [Noether, 1983, 711-763].

_____. 1932. "Hyperkomplexe Systeme in ihren Beziehungen zur kommutativen Algebra und Zahlentheorie." *Verhandlungen des Internationalen Mathematiker-Kongresses Zürich 1932.* Ed. Walter Saxer. 2 Vols. Zürich/Leipzig: Orell Füssli, 1:189-194, or [Noether, 1983, 636-641].

_____. 1933a. "Nichtkommutative Algebra." *Mathematische Zeitschrift* 37, 514-541, or [Noether, 1983, 642-669].

_____. 1933b. "Der Hauptgeschlechtssatz für relativ-galoissche Zahlkörper." *Mathematische Annalen* 108, 411-419, or [Noether, 1983, 670-678].

_____. 1983. *Emmy Noether: Gesammelte Abhandlungen/Collected Papers.* Ed. Nathan Jacobson. Berlin/New York: Springer-Verlag.

O'Meara, Timothy O. 1963. *Introduction to Quadratic Forms.* Heidelberg: Springer-Verlag.

Parshall, Karen. 1985. "Joseph H. M. Wedderburn and the Structure Theory of Algebras." *Archive for History of Exact Sciences* 32, 223-349.

——————. 2004. "Defining a Mathematical Research School: The Case of Algebra at the University of Chicago, 1892–1945." *Historia Mathematica* 31, 263-278.

Peirce, Benjamin. 1881. "Linear Associative Algebras." *American Journal of Mathematics* 4, 97-215.

Roquette, Peter. 2005. *The Brauer-Hasse-Noether Theorem in Historical Perspective.* Schriften der Mathematisch-naturwissenschaftlichen Klasse der Heidelberger Akademie der Wissenschaften. Heidelberg: Springer-Verlag.

Scheffers, Georg. 1891. "Zurückführung complexer Zahlensysteme auf typische Formen." *Mathematische Annalen* 39, 292-390.

Speiser, Andreas. 1926. "Allgemeine Zahlentheorie." *Vierteljahrsschrift der Naturforschenden Gesellschaft Zürich* 71, 8-48.

——————. 1927. *Idealtheorie in rationalen Algebren.* In [Dickson, 1927, pp. 269-303].

——————. 1935. "Zahlentheorie in rationalen Algebren." *Commentarii Mathematici Helvetici* 8, 391-406.

Sylvester, James Joseph. 1904–1912. *The Collected Mathematical Papers.* Cambridge: Cambridge University Press. Reprint Ed. 1973. New York: Chelsea Publishing Company.

Takagi, Teiji. 1920. "Über eine Theorie des relativ Abel'schen Zahlkörpers." *Journal of the College of Science, Imperial University of Tôkyo* 41, 1-133, or [Takagi, 1973, pp. 73-167].

——————. 1922. "Über das Reciprocitätsgesetz in einem beliebigen algebraischen Zahlkörper." *Journal of the College of Science, Imperial University of Tôkyo* 44, Art. 5, 1-50, or [Takagi, 1973, pp. 179-225].

——————. 1973. *Collected Papers of Teiji Takagi.* Ed. Sigekatu Kureda. Tokyo: Iwanami Shoten.

Tate, John. 1952. "The Higher Dimensional Cohomology Groups of Class Field Theory." *Annals of Mathematics* 56, 294-297.

van der Waerden, Bartel Leendert. 1959. *Algebra.* 2d Pt. 4th Ed. Berlin/Göttingen/Heidelberg: Springer-Verlag.

——————. 1985. *A History of Algebra.* Berlin/Heidelberg: Springer-Verlag.

Wedderburn, Joseph H. Maclagan. 1907. "On Hypercomplex Numbers." *Proceedings of the London Mathematical Society.* 2d Ser. 6, 77-118.

——————. 1914. "A Type of Primitive Algebra." *Transactions of the American Mathematical Society* 15, 162-166.

——————. 1921. "On Division Algebras." *Transactions of the American Mathematical Society* 22, 129-135.

——————. 1923. "Algebraic fields." *Annals of Mathematics* 24, 237-264.

Weil, André. 1967. *Basic Number Theory.* Berlin/Heidelberg: Springer-Verlag.

Wessel, Caspar. 1799/1897. *Essai sur la représentation analytique de la direction.* (Trans. of *Om Directionens analytiske Betegning* [Kjøbenhavn, 1799]). Copenhagen: Andr.-Fred, Höst & Sön.

Weyl, Hermann. 1935. "Emmy Noether." *Scripta Mathematica* 3 (1935), 201-220, or Weyl, Hermann. 1968. *Gesammelte Abhandlungen*. Ed. K Chandrasekharan. 4 Vols. Berlin/New York: Springer-Verlag, 3, 425-444.

Zaddach, Arno. 1994. *Grassmanns Algebra in der Geometrie*. Mannheim: BI-Wissenschaftsverlag.

CHAPTER 7

Minkowski, Hensel, and Hasse: On The Beginnings of the Local-Global Principle

Joachim Schwermer
University of Vienna and The Ernst Schrödinger International Institute for
Mathematical Physics, Austria

Introduction

The ring $\mathbb{Z}_p$ of p-adic integers, or more generally its field of fractions, the field $\mathbb{Q}_p$ of p-adic numbers is the completion of the field of rational numbers with respect to the p-adic valuation. Two rational numbers are very close in the p-adic metric if their difference, expressed as a fraction in lowest terms, has the property that its numerator is divisible by a high power of p. The normalized p-adic metric $|\ |_p$ can be characterized by the fact that it is multiplicative, that $|p|_p = 1/p$, and that if x is an integer not congruent to $0 \pmod{p}$, then $|x|_p = 1$.

However, this modern valuation-theoretic point of view was not the one taken by Kurt Hensel when he introduced p-adic numbers into mathematics at the end of the nineteenth century. He worked out his theory in detail in his book *Zahlentheorie*, published in 1913 [Hensel, 1913].

Hensel's idea was to transfer the strength of the calculus of series expansions from the theory of functions to number theory. Power or Laurent series expansions of a given function around any point c, that is, a series of the form

$$\sum_{\mu=N}^{\infty} a_\mu (z-c)^\mu, \quad c \in \mathbb{C}, N \in \mathbb{Z}, \text{ and } a_\mu \in \mathbb{C} \text{ for } \mu \in \mathbb{Z},$$

(where one takes $1/z$ instead of $z-c$ as the local parameter of the expansion around the point at infinity), encode virtually complete information on the analytic nature of the function in question. Hensel suggested an analogous program in his investigations of the arithmetic properties of algebraic or transcendental numbers. He aimed to find, instead of the simple representation by a decimal fraction, infinitely many others (parametrized by the primes p), each of which throws new light on the relation of that number to a specific integer. In the background of this approach is the close analogy between the case of function fields, that is, algebraic extensions of the field of rational functions, and the case of algebraic number fields, that is, finite algebraic extensions of the field of rational numbers. This analogy served to guide the work of Richard Dedekind, David Hilbert, and Hermann Minkowski.

It was Helmut Hasse who illustrated the importance of Hensel's theory of p-adic numbers for classical questions in number theory, in particular, the arithmetic theory of quadratic forms. Born in 1898, the young Hasse, attracted by Hensel's theory as presented in his book *Zahlentheorie*, matriculated in 1920 at Marburg University where Hensel held a professorship. In May of 1921, Hasse completed his thesis in which he achieved a quite general result pertaining to quadratic forms over the rationals.[1] The main objective was to give necessary and sufficient conditions under which a given number m is represented by a given quadratic form in n variables over $\mathbb{Q}$. The answer relates the representability over $\mathbb{Q}$ to that over all p-adic number fields $\mathbb{Q}_p$ and, in addition, the completion $\mathbb{R} = \mathbb{Q}_\infty$ of $\mathbb{Q}$ with respect to the normal absolute value.

This result and its methods of proof laid the foundations for another major theorem, contained in Hasse's *Habilitationsschrift* of December 1921. The problem he dealt with was to give necessary and sufficient conditions for two quadratic forms with rational coefficients to be related by an invertible linear substitution with rational coefficients, that is, to be equivalent over $\mathbb{Q}$. In working on this question, Hasse was well aware of Minkowski's result, obtained in 1890, in which he had described a complete set of invariants characterizing the equivalence class of a quadratic form over $\mathbb{Q}$. Yet Minkowski's approach hinged on his theory of integral quadratic forms worked out in 1882 [Minkowski, 1884]. It was Hasse's objective, following his earlier lines of thought in relating the behavior of f over $\mathbb{Q}$ with that over the fields $\mathbb{Q}_p$, for p a prime or $p = \infty$, to substantiate the theory of equivalence over $\mathbb{Q}$ independent of the quite complicated arithmetic theory of quadratic forms over $\mathbb{Z}$. Hasse's result was that quadratic forms are equivalent in $\mathbb{Q}$ if and only if they are equivalent in all $\mathbb{Q}_p$, p a prime or $p = \infty$. One might view this as one of the first manifestations of the so-called "local-global" principle, currently a leading principle in arithmetic and algebra.

In 1924, Hasse proved the obvious analogs of the two aforementioned results for any algebraic number field in addition to the field of rational numbers [Hasse, 1924a; 1924b]. The decisive point of these two papers lies in the distinction between the notions of *local* and *global* and their mutual influence. This idea of the p-adic transfer from the "small" to the "large," as Hasse phrased it at the time, forms what would become the fruitful foundation of Hasse's research in various directions in the coming years. This methodological approach led to his work on hypercomplex number systems (that is, essentially normal simple algebras) over p-adic fields [Hasse, 1931]. Finally, it culminated in his joint work with Emmy Noether and Richard Brauer completed in the early 1930s on the structure theory of normal simple algebras over algebraic number fields [Brauer-Hasse-Noether, 1932].

A valuable source for Hasse's view on his own work at that time is a letter to Hermann Weyl, written on 15 December, 1931. Weyl had sent Hasse some words of admiration for the result of Brauer, Hasse, and Noether, and Hasse responded

> [t]hat I also, on my own, esteem the importance of my last result and am very happy on that account, I may say without presumption. But I have to ascribe a very considerable role to the elegant theory of E. Noether, as the formal algebraic foundation, to the newly revitalized p-adic methods of Hensel as a tool, and above all to the "idea" of the

---

[1] Hasse's thesis and *Habilitationsschrift* (see below) were never published in their original forms. [Hasse, 1923a] grew out of the thesis, while [Hasse, 1923b] grew out of his *Habilitationsschrift*.

p-adic inference from the Small to the Large, which emerged with Minkowski in all clarity [Hasse, 1931b, ETH].[2]

This chapter highlights both the mathematics and the mathematicians behind what came to be known as the Local-Global Principle. Although the chapter presents the work of several mathematicians, it primarily focuses, as the title suggests, on the contributions of Hermann Minkowski, Kurt Hensel, and Helmut Hasse. It opens with an introduction to the arithmetic theory of integral binary quadratic forms and the important question of how to determine if an integer can be represented by a binary quadratic form. It includes an overview of Gauss's notion of genus and its fundamental role in establishing congruence conditions that determine whether or not a given quadratic form represents an integer and in how many ways it does, if at all.

The focus then shifts to the title characters. First, Minkowski, with his influential training in the mathematical-physico seminar at the University of Königsberg, gave general conditions for determining if two rational quadratic forms are rationally equivalent. This result hinged in a decisive way on his previous work in 1882 on integral quadratic forms. Next, Hensel developed the concepts of *p*-adic numbers and their associated arithmetic theory. The valuation-theoretic point of view is also discussed as it appears in the later approach due to Kürschák [Kürschák, 1913a; 1913b]. Finally, Hasse, inspired by his adviser, Hensel, ultimately improved his thesis result and established that two quadratic forms are equivalent in the rationals if and only if they are equivalent in all fields $K(p)$ for p a prime or $p = \infty$ (Fundamental Theorem).[3] Hasse viewed his contribution as a simplification of Minkowski's result. More than a decade later, in a talk given in Königsberg in 1936, Hasse again considered his proof of the Fundamental Theorem and, in particular, the significance of his early work in the development of number theory. The chapter concludes with a discussion of what we learn about Hasse through his work on the Local-Global Principle, documenting thereby how the mathematics, as it were, provides valuable insight into the mathematician.

Toward an Arithmetic Theory of Quadratic Forms

Diophantine analysis pertains to the study of the solvability of polynomial equations in integers, or, alternatively, in the rationals. This subject has a long tradition. The treatise by Leonard E. Dickson on the history of the theory of numbers, published in three volumes between 1919 and 1923, gives an account of the early results in this field [Dickson, 1919–1923]. Initially, investigations of particular Diophantine equations involved methods of a more ad hoc nature. Only in relatively recent times have more coherent theories emerged. Still, the body of knowledge is less systematic than that in other more recently established branches of mathematics. The reason stems from the connection with the most basic of mathematical objects, namely, the rational integers.

The field $\mathbb{Q}$ of rational numbers is the field of fractions of the ring of integers. The study of solutions in the rationals is by far more systematic than that of solutions in integers. The former finds its natural place within the general field of

---

[2]The full German text and translation of this letter is given in the appendix to this chapter.

[3]Here, we use Hasse's notation $K(p)$ for the field $\mathbb{Q}_p$ of *p*-adic numbers if p is a prime or for the field $\mathbb{R} = \mathbb{Q}_\infty$ in the case of the infinite prime ∞.

algebraic geometry over $\mathbb{Q}$, whereas only for specific classes of polynomial equations in integers has a more or less satisfactory theory evolved over the last few centuries.

Quadratic forms over the integers and the corresponding arithmetic theory comprise one of these basic topics in Diophantine analysis. Joseph-Louis Lagrange unified various investigations of particular equations in two unknowns by Pierre de Fermat and Leonhard Euler in the systematic work, *Recherches d'arithmétique* [Lagrange, 1773–1775]. There, he laid the foundations of the theory of binary quadratic forms, that is, equations of the form

$$f(x,y) = ax^2 + bxy + cy^2$$

with integral coefficients a, b, c and variables x, y. It was the particular question of whether a given integer t can be represented by the quadratic form f—that is, do there exist integers x_0, y_0 so that $f(x_0, y_0) = t$—which played a decisive role in the formation of the theory. Lagrange's approach hinged on the notion of equivalence of two quadratic forms f and g and the fact that equivalent forms represent the same numbers.

Two binary quadratic forms f and g are said to be equivalent, if one can be transformed into the other by a substitution of the form

$$X = px + qy, \quad Y = rx + sy,$$

where p, q, r, s are integers with $ps - rq = \pm 1$. (In more modern terms, this relation is reflexive, symmetric, and transitive, that is, it is an equivalence relation.) An integral solution of the equation $f(x, y) = t$ gives rise to one for $g(X, Y) = t$ and vice versa.

Attached to a form f as above is the discriminant $D(f) = b^2 - 4ac$. This is related to the determinant $\Delta(f)$ by $D(f) = -4\Delta(f)$. We note first that $D(f) \equiv 0 \pmod 4$ if b is even, and $D(f) \equiv 1 \pmod 4$ if b is odd, and, second, that the relation $4af(x,y) = (2ax + by)^2 - D(f)y^2$ obtains. Thus, the values taken on by f are all of the same sign or zero if $D(f) < 0$; accordingly, f is called positive or negative definite. If $D(f) > 0$, then f takes values of both signs, and f is called indefinite. In what follows, we suppose that $D(f)$ is not a square and that $D(f)$ does not equal zero. A quadratic form f with $D(f)$ a square splits into a product of two linear factors.

A substitution $X = px + qy$, $Y = rx + sy$ as above does not alter the discriminant of f, hence, equivalent binary quadratic forms have the same discriminant. As Lagrange observed, a finite sequence of substitutions transforms a given form into an equivalent form f for which the estimates $|b| \leq |a|$ and $|b| \leq c$ hold [Lagrange, 1774]. A binary quadratic form for which these two conditions hold is said to be reduced. There are only finitely many triples (a, b, c) satisfying these conditions so that the discriminant $D(f) = b^2 - 4ac$ of f takes on a given value D. For if $|a| \leq |c|$ (which we may suppose), then $|D(f)|^2 \geq 3|a|^2$. Thus, the integral coefficients a, b, c of f must satisfy the conditions $|a| \leq (|D(f)|/3)^{1/2}$, $|b| \leq |a|$, and $c = (b^2 - D(f))/4a$, depending on the given D. This approach of Lagrange initiated the far-reaching theory of reduction for quadratic forms.

To illustrate what the phrase "if a given integer is represented by a quadratic form" involves, consider any binary quadratic form over the ring of integers $\mathbb{Z}$, say,

$$f = x^2 + y^2$$

and the corresponding Diophantine problem. First, it is hopeless to find an integral solution of, say, the equation, $x^2 + y^2 = -7$, since there is not even a real solution. This argument says that our knowledge about real solutions already gives us information about its solutions over $\mathbb{Z}$. In turn, solvability of the given equation over $\mathbb{R}$ is a necessary condition for the existence of an integral solution. Second, there are other necessary conditions, given by an analysis of the problem of finding its solutions modulo m, for any fixed modulus $m > 0$. Any integer x satisfies $x^2 \equiv 0$ or $1 \pmod 4$, thus if the integer t can be represented by the form f above (that is, as a sum of two squares), it must be the case that $t \not\equiv 3 \pmod 4$. It was Euler who showed in 1749 that these congruence considerations already imply the existence of a representation of a given integer by f as a sum of two squares [Euler, 1911, 1:295-327].[4] More precisely, the integer $t > 0$ can be represented as a sum of two squares if and only if no prime $p \equiv 3 \pmod 4$ divides t to an odd power. In particular, the equation $x^2 + y^2 = q$, for q a prime unequal to 2, has an integral solution if and only if $q \equiv 1 \pmod 4$.

This is a simple example. In general, congruence conditions and the solvability of the given equation over the reals are not sufficient to ensure the existence of a solution in the integers. For example, consider the quadratic form $g = 3x^2 + xy + 2y^2$. Obviously, the equation

$$3x^2 + xy + 2y^2 = 1$$

has a solution in the reals. More work is required to show that this equation has a solution modulo any modulus $m > 0$. However, there is clearly no solution in the integers.[5]

These simple examples already show that determining the integral solutions of such equations can be difficult. Although it is insufficient for a complete treatment, the examination of a given equation $f(x, y) = t$ modulo a modulus $m > 0$ is considerable help. In particular, finding its solutions in integers modulo m is a finite problem. The latter question, thanks to the Chinese remainder theorem, can be replaced by the analogous one of finding its solutions modulo the prime powers p^ν dividing m. Note that given an integral solution x_0, y_0 for the equation $f(x, y) = t$, there is a solution $\pmod{p^\nu}$, for any prime p and $\nu > 0$, by viewing x_0, y_0 as a pair of integers modulo p^ν. A guiding question is when does knowledge derived from congruence considerations for all prime numbers p, and from an analysis over $\mathbb{R}$, provide useful information regarding the solution of the integral problem, that is, the solvability over $\mathbb{Z}$.

A major part of Carl Friedrich Gauss's *Disquisitiones arithmeticæ*, published in 1801, deals with a systematic development of a theory of integral binary quadratic forms. Of particular importance, he gave a conclusive treatment of the representation problem alluded to above, that is, to determine the values taken on by a given form f and to find the number of possible ways in which a given number t can be represented by f [Gauss, 1801, arts. 180-181, 205, and 212].

Suppose that an equation $f(x, y) = t$ corresponding to a quadratic form f with discriminant $D(f)$ (not a square) has an integral solution x_0, y_0 so that $(x_0, y_0) = 1$,

---

[4]For a discussion of Fermat's contribution to this result, see [Hofmann, 1960] and [Weil, 1984, pp. 37-157].

[5]This follows by writing $3x^2 + xy + 2y^2 = (1/2)(x+y)^2 + (5/2)x^2 + (3/2)y^2$.

that is, t is properly represented by f. Gauss encoded his result in a set of necessary congruence conditions for the prime divisors of t modulo $2D(f)$ (respectively, $4D(f)$), that is, they have to reside in certain residue classes modulo these moduli.

As already illustrated by the example above, congruence conditions cannot be sufficient to ensure representability. However, Gauss showed that there is always a form f', in his terminology "in the same genus" of f, so that the given number t can be represented by f' [Gauss, 1801, arts. 228-233]. Forms which are integrally equivalent everywhere locally are said to be, by definition, in the same genus. The genus of f contains only a finite number of equivalence classes of quadratic forms over $\mathbb{Z}$, all determined by f and with the same discriminant $D(f)$. It is this notion of the genus of a form f which plays a decisive role. It gives the methodological framework within which one can set forth systematically the best possible description of how congruence conditions determine the question of representability in the integers. The concept of genus of forms is fundamental in further developments.

Mathematical Digression: Quadratic Forms over Rings

Consider now n-ary quadratic forms. For later use, we define the notion in some generality. Let A be a subring of a field F, for example, an integral domain contained in its field of fractions. For the sake of simplicity, suppose that the characteristic of A is not equal 2. An n-ary quadratic form f in the variables $x_1, \ldots, x_n$ over the ring A is a homogeneous polynomial of degree two of the form

$$f = \sum_{i,k=1}^{n} a_{ik} x_i x_k$$

with coefficients a_{ik} in A, where $a_{ik} = a_{ki}$, for all $i, k \in \mathbb{Z}^+$. The symmetric matrix $F = (a_{ik})$ is called the matrix of the quadratic form f. By definition, the determinant $\Delta(f)$ of the form f is the determinant $\det(F)$ of F. We say that f is regular or non-singular, if $\Delta(f) \neq 0$; otherwise, it is singular. We focus on the case of regular forms since the study of singular ones can be reduced to that of regular forms in a smaller number of variables, at least in the cases in which we are interested.

The form f represents an element $t \in F$ over A, if there are elements $\lambda_1, \ldots, \lambda_n \in A$ so that $f(\lambda_1, \ldots, \lambda_n) = \sum_{i,k} a_{ik} \lambda_i \lambda_k = t$. More generally, by a substitution of the variables

$$y_i = \sum_{j=1}^{n} \gamma_{ij} x_j, \text{ for } i = 1, \ldots, m, \, \gamma_{ij} \in A, \text{ and } m \leq n,$$

we obtain a new form. Any m-ary quadratic form with $m \leq n$ obtained in this way is said to be represented by f over A. We say that two forms f, g in the same number of variables are equivalent over A, if each represents the other over A. This notion defines an equivalence relation in the usual sense. Thus, we can speak of an A-equivalence class of quadratic forms in n variables. Note that equivalent forms represent the same elements in F. A quadratic form f is said to be isotropic over A, if it represents 0 non-trivially over A; otherwise, it is anisotropic.

This leads to a natural classification problem. For example, in the case $A = \mathbb{R} = F$, by Sylvester's so-called law of inertia, an n-ary quadratic form with non-vanishing determinant can be transformed by a substitution with real coefficients into a sum of $n - I$ positive and I negative squares [Sylvester, 1852]. The number

I is uniquely determined by f and is called the index of f. Thus, the $\mathbb{R}$-equivalence classes of n-ary quadratic forms over $\mathbb{R}$ are classified by this invariant.

In general, it is a much more difficult task to determine the equivalence classes in a given case and to characterize them in a convenient way in terms of sets of invariants attached to the forms in question. In the historical development up to the end of the nineteenth century, the classification problem in the case $A = \mathbb{Z}$, $F = \mathbb{Q}$ was a major theme.

Hermann Minkowski's Early Work

While still attending the Altstätische Gymnasium in Königsberg, Prussia, Hermann Minkowski had already made contact with Heinrich Weber, ordinary professor in mathematics from 1875 to 1882 at the Albertus University of Königsberg.[6] Minkowski graduated in April 1880 and went on to the Albertus University, where he studied mathematics principally with Weber, Johann Georg Rosenhain, and Woldemar Voigt. In 1834, the physicist Franz Neumann had established, together with Carl Jacobi the mathematical-physico seminar at the University of Königsberg. It was composed of the division for pure and applied mathematics and the division for mathematical physics. As the first official seminar in Prussia to incorporate mathematical methods in physics, this institution became the center of a school of mathematical physics.[7]

Beginning in 1876, Weber directed the mathematical division of the seminar, succeeding Jacobi and Friedrich Richelot in this office, while Voigt oversaw the physical division. One of Minkowski's fellow students was David Hilbert. Both attended the basic courses, given partly by Weber. His mathematical expertise and his instruction shaped Minkowski's mathematical development considerably. In 1883, however, the academic situation at the seminar changed. Weber moved to Berlin, and Voigt accepted an offer from Göttingen, becoming professor of theoretical physics.

Meanwhile, Minkowski was mainly attracted by a problem posed in 1881 by the *Académie des Sciences* in Paris as a competition for the *Grand Prix des Sciences Mathématiques* of 1882. The theory of the decomposition of integers as a sum of five squares was the problem. This question originated in some results by Gotthold Eisenstein announced without proof in 1847 [Eisenstein, 1847]. They contained precise formulas for the number of possible representations in this case. In his work on this question, Minkowski viewed it as a special case of the general problem of representations of positive definite quadratic forms with n variables by a quadratic form with a higher number of variables. Upon completion of his successful approach, Minkowski anonymously submitted a manuscript containing his results to the Academy on 29 May, 1882 [Minkowski, 1884]. There, he laid the foundations of a quite general theory of quadratic forms in n variables with integral coefficients. He extended the notion of the genus as developed by Gauss for binary quadratic forms to the case of forms with a higher number of variables. His main focus was on orders and genera of quadratic forms over $\mathbb{Z}$, their relationship to one another, and an in-depth analysis of possible sets of invariants characterizing these objects in a suitable arithmetical way. Eisenstein's assertions turned out to

---

[6] For biographical details on Minkowski, see [Schwermer, 1994; 1995].

[7] The foundation and evolution of the mathematical–physico seminar at the University of Königsberg in East Prussia are reconstructed in fine detail by Kathryn Olesko in [Olesko, 1991].

be easy consequences of Minkowski's results applied in a specific case. It seems that, at the time, Minkowski was unaware of the work of Henry Smith at Oxford on this topic. Smith had already published some investigations on the orders and genera of quadratic forms in 1867 [Smith, 1867; 1868]. Notably, these publications went unnoticed even by the Academy, so Smith had to write a more comprehensive memoir for submission developing his results anew [Smith, 1882]. On 2 April, 1883, the Academy awarded the prize to both Smith and Minkowski.

In the fall of 1883, Ferdinand Lindemann came to Königsberg as Weber's successor. At his side, Adolf Hurwitz worked as an extraordinary professor. These mathematicians played a central role in the mathematical training and personal development of both Hilbert and Minkowski. First, they gave lectures to the students on the foundations of and on more advanced material in analysis, the theory of functions, algebra, number theory, geometry, and mathematical physics. Second, the mathematical colloquium offered a forum for talks and scientific exchange at the research level on topics such as results of the participants and profound surveys of the work of others in mathematics and mathematical physics.[8]

In 1885, Minkowski received his doctorate and, owing to a lack of students at the University of Königsberg, moved to the west after his mandatory, year-long tour of duty in the Army. In February of 1887, Minkowski asked the Philosophical Faculty of the University of Bonn for admission to the *Habilitation*, and he submitted two papers, written in the fall of 1886, on finite groups of linear transformations with integral coefficients as his *Habilitationsschrift* [Minkowski, 1887a; 1887b].

One line of Minkowski's research in the preceding years pertained to the theory of reduction relating to positive definite quadratic forms with integral coefficients. He had seen in the work of Eisenstein, Gustav Peter Lejeune Dirichlet, Charles Hermite, and others the strong relationship between quadratic forms and problems in number theory. Closely related to this circle of ideas was the question of finding a qualitatively good estimate for the ratio of the minimum of a positive definite quadratic form in n variables to the nth root of its determinant by a constant depending only on n. This was the topic of the *Probevorlesung* he gave in Bonn on 15 March, 1887. Its contents proved decisive in Minkowski's approach to the theory of quadratic forms. There, he unfolded a geometric treatment of the arithmetically defined objects, that is, the quadratic forms in question. Now, the methods of investigation were steadily directed through geometrical concepts. This geometric interpretation ultimately led Minkowski to his geometry of numbers.[9]

Two quite different themes were central to Minkowski's research work in the years following his *Habilitation* in Bonn. On the one hand, as he wrote to Hilbert on 19 June, 1889, "I'm still totally stuck in mathematical physics [Ich stecke noch immer ganz in der mathematischen Physik]" [Minkowski, 1973, pp. 34-36]. Due to a lack of competent colleagues in mathematics, the young *Privatdozent* was more inclined to deepen his interest in physics, a notion implanted in his mind since the time of the mathematical-physico colloquium at Königsberg. Treatises of Voigt, Thomson, Helmholtz, and others had formed part of his studies. On the other

---

[8] For an attempt to trace their acquisition of knowledge and their evolving attitudes toward their field and its research practice, see [Schwermer, 2004].

[9] Minkowski's *Habilitation* is discussed and documented in detail in [Schwermer, 1991]. The manuscript of his *Probevorlesung* as well as the minutes of the corresponding faculty meeting are also reproduced there.

hand, he vigorously pursued his geometric approach to the arithmetic theory of quadratic forms and its relationship to questions in number theory. One of his fundamental achievements obtained in this period was the so-called lattice-point theorem.[10]

Hermann Minkowski's Letter to Adolf Hurwitz in 1890

Of interest for our objectives is a letter from Minkowski to Hurwitz, published in Crelle's *Journal* in 1890 in response to a specific question about rational transformations of ternary Diophantine equations posed by Hilbert and Hurwitz [Minkowski, 1890]. In it, Minkowski gave a complete solution of the classification problem of quadratic forms over the field $\mathbb{Q}$ of rational numbers. More precisely, he set up fully general conditions under which two rational quadratic forms are equivalent over $\mathbb{Q}$. Using the theory of integral quadratic forms he had worked out in 1882 [Minkowski, 1884], Minkowski attached to a given form f a set of invariants that characterizes the equivalence class of f over $\mathbb{Q}$. This set consists of the number n of variables in f, the determinant of f, its index I of inertia in the sense of Sylvester, and units $C_p = \pm 1$ associated with each prime p. Moreover, Minkowski determined under what conditions a given form represents zero non-trivially over $\mathbb{Q}$ (that is, when it is isotropic over $\mathbb{Q}$). He also stated and proved for which combinations of the invariants quadratic forms of corresponding type actually exist.

It is of interest to follow closely Minkowski's set-up for his definition of the units C_p, for p a prime, attached to a quadratic form f with rational coefficients. Given f (with non-vanishing determinant $\Delta(f)$) in the variables x_i, for $i = 1, \ldots, n$, the substitution $x_i = Ny_i$, for all i, where N denotes the general denominator of the coefficients of f, transforms this form into one with integral coefficients. Thus, we can suppose that f has integral coefficients. The index I of f remains the same under any transformation T with rational coefficients and non-vanishing determinant $\Delta(T)$, whereas its determinant $\Delta(f)$ is changed by the factor $\det(T)^2$. As a consequence, the family $\mathcal{P}$ of prime numbers dividing $\Delta(f)$ by an *odd* power is unchanged as well. This basic observation concerning the primes dividing $\Delta(f)$ under a rational transformation led Minkowski to take into account the product

$$A = (-1)^I \prod_{p \in \mathcal{P}} p$$

over all primes in $\mathcal{P}$ endowed with the sign of the determinant of f. If the set $\mathcal{P}$ is empty, the convention $A = (-1)^I$ is appropriate.

Furthermore, if p denotes an arbitrary prime, the reduction of f modulo sufficiently high powers of p gives rise to a construction of a unit, called C_p. By definition, it only takes the values ± 1 and is unchanged under a rational transformation of f. These invariants have the following two important properties. First, the way the C_p's are defined makes it apparent that for odd primes q not dividing the determinant $\Delta(f)$ or the denominator N, the corresponding values C_q are $+1$. Hence, only a finite number of the units C_p can take the value -1. Secondly, there is a product formula binding together the various C_p's and relating them to the invariants n, I, and A. If j denotes the number of primes in $\mathcal{P}$ which are congruent

---

[10]For a detailed discussion of the development of the reduction theory of quadratic forms, in particular Minkowski's early work, see [Schwermer, 2005; 2007].

to 3(mod 4), then one has

$$\prod_p C_p = \begin{cases} +1 & \text{for } n - 2I - j \equiv 0, 1, 6, 7 \pmod{8} \\ -1 & \text{for } n - 2I - j \equiv 2, 3, 4, 5 \pmod{8}. \end{cases}$$

It is this product relation that proved decisive in Minkowski's proof of the invariance property of the units C_p alluded to above.

The source for the very definition of the units C_p lies in Minkowski's own characterization of the genus of an integral quadratic form f as given in his first paper in 1882 [Minkowski, 1884]. There, he showed that two forms belong to the same genus, if their orders coincide and if their systems of characters equal to ± 1 (as defined in [Minkowski, 1884]) match. The latter are essentially given in terms of Legendre symbols, hence, they equal ± 1. The values taken are determined by certain Gauss sums.

Again, using certain results that he obtained in 1882 [Minkowski, 1884] independently from the earlier work of Henry Smith for ternary forms [Smith 1867; 1868], Minkowski proved that f can be rationally transformed into a form f' of determinant $A\beta^2$, the genus of which is completely determined by the units C_p, for p prime (respectively, n, I, and A). If B denotes the product of all odd primes q for which $C_q = -1$, then β is the quotient of B and the greatest common divisor of A and B [Minkowski, 1890, p. 8].

Thus, building on his deep results in the theory of genus for integral quadratic forms, Minkowski arrived at the following solution of the classification problem for rational quadratic forms: two rational quadratic forms with n variables and non-vanishing determinant can be rationally transformed into one another if and only if they have the same invariants I, A, and B [Minkowski, 1890, p. 8]. Minkowski noted, before stating this result, that the value B encodes all information concerning the values of the C_p's. The product formula serves as an indispensable arithmetic link between the finitely many C_p which take the value -1.

After more than thirty years, it was Helmut Hasse who used the p-adic methods of Hensel as developed at the turn of the century to reinterpret Minkowski's result in a more structural framework [Hasse, 1924b]. This approach made possible the generalization of the result to forms over algebraic number fields.

Hensel's p-adic Numbers: Series, Expansions, or Numbers as Functions

As we have seen, congruence considerations might be of considerable help in an analysis of the solvability over $\mathbb{Z}$ of a given equation $f(x, y) = t$ with integral coefficients. However, manipulations of congruences are exceedingly tiresome, essentially because the ring of integers modulo m may have zero divisors. The work of Leopold Kronecker [Kronecker, 1882] and, later on, of Hensel [Hensel, 1897] led to a more suitable formulation: an integral solution to such an equation (or, more generally, to any Diophantine equation) yields, for every prime p, a system of solutions modulo the prime powers p^ν, for $\nu \in \mathbb{Z}^+$. This system is compatible in the sense that reduction of the $(\nu+1)$th solution modulo p^ν gives the νth solution. Such a compatible system of solutions for a fixed prime p is, in the context of current mathematics, interpreted as an element in the ring $\mathbb{Z}_p$ of p-adic integers. This ring is defined as the projective limit

$$\mathbb{Z}_p := \varprojlim_n \mathbb{Z}/p^n\mathbb{Z}$$

of the finite rings $\mathbb{Z}/p^n\mathbb{Z}$. More precisely, we consider the direct product

$$\prod_{n=1}^{\infty} \mathbb{Z}/p^n\mathbb{Z} = \{(x_n)_{n\in\mathbb{N}} \mid x_n \in \mathbb{Z}/p^n\mathbb{Z}\}.$$

Then the projective limit $\mathbb{Z}_p$ is defined as

$$\mathbb{Z}_p := \left\{(x_n) \in \prod_{n=1}^{\infty} \mathbb{Z}/p^n\mathbb{Z} \mid \lambda_n(x_{n+1}) = x_n, n \in \mathbb{N}\right\},$$

where $\lambda_n : \mathbb{Z}/p^{n+1}\mathbb{Z} \longrightarrow \mathbb{Z}/p^n\mathbb{Z}$ denotes the canonical projection. It carries the structure of a ring in a natural way, namely, as a subring of the direct product. This ring is called the ring of p-adic integers. It can be considered as a topological ring, endowed with the profinite topology, that is, the topology that it inherits as a projective limit of the rings $\mathbb{Z}/p^\nu\mathbb{Z}$ given the discrete topology.

The ring $\mathbb{Z}_p$ of p-adic integers or, more generally, its field of fractions, the field $\mathbb{Q}_p$ of p-adic numbers is the completion of the field of rational numbers with respect to the p-adic valuation (see below). However, this modern point of view was not Hensel's when he introduced the p-adic numbers into mathematics at the end of the nineteenth century.[11] When he finally worked out his theory in detail in his book, *Zahlentheorie*, published in 1913, his formulation was different from the valuation-theoretic approach used nowadays based on the work of Josef Kürschák in 1912 [Kürschák, 1913a; 1913b]. Hensel aimed to introduce a methodological tool from the theory of functions. Consider the calculus of series expansions of a given function around any point c, that is, a series of the form

$$\sum_{\mu=N}^{\infty} a_\mu (z-c)^\mu,$$

for $c \in \mathbb{C}$, $N \in \mathbb{Z}$, and $a_\mu \in \mathbb{C}$, where $\mu \in \mathbb{Z}$ and $\mu \geq N$. (Here, $1/z$ instead of $z-c$ is taken as the local parameter of the expansion around the point at infinity.) This encodes almost complete information about the analytic nature of the function in question. Hensel suggested an analogous program in his investigations of the arithmetic nature of algebraic or transcendental numbers. He wrote: "If we want to deal with transcendental numbers in a similar way [and one] as simple as with the transcendental functions, we have to try to find, instead of the single representation by a decimal fraction, infinitely many others each of which throws new light on the relation of that number to a specific integer."[12]

Given a prime p, Hensel considered formal infinite series [Hensel, 1904]

$$\sum_{\nu=n}^{\infty} a_\nu p^\nu, n \in \mathbb{Z}$$

---

[11]The genesis of Hensel's theory of p-adic numbers is discussed from a historical point of view in [Ullrich, 1998]. The close analogy between function fields and algebraic number fields, as a motivation for this development in number theory, is described in [Ullrich, 1999]. In her forthcoming thesis (TU Darmstadt), Birgit Petri sheds new light on the influence of Kronecker's work on Hensel [Petri, to appear].

[12]Wollen wir ... die ... transzendenten Zahlen ähnlich einfach behandeln, wie die transzendenten Funktionen, so müssen wir versuchen, statt der einzigen Darstellung durch einen Dezimalbruch unendlich viele andere zu finden, von denen jede einzelne uns einen neuen Aufschluss über das Verhältnis jener Zahl zu einer bestimmten ganzen Zahl gewährt" [Hensel, 1905, p. 546].

with $a_\nu \in \{0, 1, ..., p-1\}$ for $\nu \in \mathbb{Z}$ and $\nu \geq n$, called p-adic numbers. In order to obtain the p-adic development of a rational number q, in analogy to the function-theoretic procedure, one has to use the fact that the residue classes in the finite ring $\mathbb{Z}/p^\nu \mathbb{Z}$ can be uniquely represented as

$$a_0 + a_1 p + a_2 p^2 + \cdots + a_{\nu-1} p^{\nu-1} (\bmod\ p^\nu),$$

where $0 \leq a_j \leq p-1$ and $j = 0, \ldots, \nu-1$. Thus, given a rational number $q = s/t$ whose denominator is not divisible by p, one obtains a system of residue classes

$$\bar{r}_\nu = q (\bmod\ p^\nu), \quad \nu = 1, 2, \ldots,$$

where $\bar{r}_1 = a_0 (\bmod\ p)$, $\bar{r}_2 = a_0 + a_1 p (\bmod\ p^2)$, $\bar{r}_3 = a_0 + a_1 p + a_2 p^2 (\bmod\ p^3), \ldots$ with uniquely determined coefficients $a_j \in \{0, 1, \ldots, p-1\}$. This compatible system gives rise to the p-adic number $\sum_{\nu=0}^{\infty} a_\nu p^\nu$, called the p-adic development of q. More generally, if q' is an arbitrary rational number, we write $q' = (s/t) p^{-n}$ with st prime to p. If $q := (s/t)$ has the p-adic development $\sum a_\nu p^\nu$, for $\nu = 0, 1, \ldots$, then we define

$$a_o p^{-n} + a_1 p^{-n+1} + \cdots + a_n + a_{n+1} p + a_{n+2} p^2 + \cdots$$

as the p-adic development of q'. This assignment establishes a map $\mathbb{Q} \longrightarrow K(p)$ from $\mathbb{Q}$ into the set $K(p)$ of all p-adic numbers. This map is injective. The image of the ring of integers $\mathbb{Z}$ is the set $R(p)$ of integral p-adic numbers given by the formal series $\sum a_\nu p^\nu$, for $\nu = 0, 1, \ldots$.

Note that an integral p-adic number given by the collection,

$$\{r_n\}_{n \in \mathbb{N}}, r_n = \sum_{\nu=0}^{n-1} a_\nu p^\nu$$

determines an element $(\bar{r}_n)_{n \in \mathbb{N}}$ in $\mathbb{Z}_p$, where $\bar{r}_n = r_n \in \mathbb{Z}/p^n \mathbb{Z}$ denotes the residue class of r_n. This gives rise to a bijection

$$R(p) \xrightarrow{\sim} \mathbb{Z}_p.$$

The ring structure on the righthand side induces one on the lefthand side.

The arithmetic significance of the theory of p-adic numbers lies in the fact that the solvability of a given Diophantine equation over the p-adic numbers is equivalent to satisfying a congruence to arbitrarily high powers of p. We enunciate this precisely as follows. Let $f \in \mathbb{Z}[X_1, \ldots, X_m]$ be a homogeneous polynomial with integral coefficients of degree $d \geq 1$. A necessary and sufficient condition for the existence of $x_1, \ldots, x_m \in \mathbb{Q}_p$ not all zero, such that $f(x_1, \ldots, x_m) = 0$ is that, for all natural numbers ν, there are integers $y_{1,\nu}, y_{2,\nu}, \ldots, y_{m,\nu}$ whose greatest common divisor is not divisible by p such that the congruence $f(X_1, \ldots, X_m) \equiv 0 \ (\bmod\ p^\nu)$ has $(y_{1,\nu}, \ldots, y_{m,\nu})$ as a non-trivial solution.

This result shows that the use of p-adic numbers, which at first sight seems quite artificial compared to congruence considerations, offers great technical advantages since $\mathbb{Q}_p$, the field of fractions of the ring $\mathbb{Z}_p$ of p-adic integers, is a field, whereas $\mathbb{Z}/p^\nu \mathbb{Z}$ is not even an integral domain.

Mathematical Digression: Valuations and p-adic Fields

The field $\mathbb{R}$ of real numbers is the completion of the field $\mathbb{Q}$ with respect to the ordinary absolute value $|\ |_\infty$. Similarly, the field $\mathbb{Q}_p$ contains $\mathbb{Q}$ and can be regarded as the topological field obtained by completing the field $\mathbb{Q}$ with respect to the p-adic absolute value $| \ |_p$ on $\mathbb{Q}$. It is defined as follows: if $x = (r/s)p^n \in \mathbb{Q}^*$ with $r, s, n \in \mathbb{Z}$ such that rs is not divisible by p, then

$$| \ x \ |_p = p^{-n}.$$

For $x = 0$, set $|0|_p = 0$. Thus, a rational integer has a small p-adic value, if it is divisible by a large power of p. In particular, the summands in a p-adic formal series $a_0 + a_1 p + a_2 p^2 + \cdots$ form a sequence converging to zero with respect to $| \ |_p$. The p-adic absolute value defined by $| \ |_p : \mathbb{Q} \longrightarrow \mathbb{R}$, $x \longmapsto |x|_p$ is a valuation, that is, it has the following three properties:

(1) $|x| \geq 0$ and $|x|_p = 0$ if and only if $x = 0$
(2) $|xy|_p \leq |x|_p |y|_p$, for all $x, y \in \mathbb{Q}$
(3) $|x + y|_p \leq |x|_p + |x|_p$, for all $x, y \in \mathbb{Q}$.

Note that not merely is condition (3) true, but $| \ |_p$ satisfies the stronger condition

$$|x + y|_p \leq \max\{|x|_p, |y|_p\}, \quad x, y \in \mathbb{Q}.$$

A valuation v on a field k induces a topology on k given by the metric defined by $d(x, y) := v(x - y)$, for $x, y \in k$. Endowed with the topology (induced by v), k is a topological field. Every field with a valuation can be embedded in a complete field. The completion of $\mathbb{Q}$ with respect to $| \ |_p$ is equal to the field $\mathbb{Q}_p$ of p-adic numbers. The absolute value $| \ |_p$ extends to a valuation on $\mathbb{Q}_p$ again denoted $| \ |_p$. The set $\{x \in \mathbb{Q}_p \mid |x|_p \leq 1\}$ forms a unique maximal compact subring of $\mathbb{Q}_p$. It coincides with the ring $\mathbb{Z}_p$ of p-adic integers. This is a principal ideal domain with unique nonzero prime ideal $p\mathbb{Z}_p$. The residue field $\mathbb{Z}_p/p\mathbb{Z}_p$ is isomorphic to the finite field $\mathbb{Z}/p\mathbb{Z}$. This is the valuation-theoretic interpretation of Hensel's field K_p of p-adic numbers, p a prime [Hensel, 1904]. They step up beside the completion $\mathbb{Q}_\infty = \mathbb{R}$ with respect to the absolute value $| \ |_\infty$, called "the infinite prime." The fields $\mathbb{Q}_p$, for $p \leq \infty$, are exactly (up to isomorphism) the locally compact non-discrete fields that contain $\mathbb{Q}$ as a dense subfield.

Let k be an algebraic number field, that is, a finite extension field of $\mathbb{Q}$. For each (normalized) valuation $| \ |_v$ of k, where v ranges over the set V of places of k (that is, the classes of inequivalent valuations of k), denote by k_v the completion of k with respect to the topology induced on k by $| \ |_v$. There are a finite number of so-called Archimedean completions for $k_v = \mathbb{R}$ or $\mathbb{C}$ corresponding to the infinite places given by embeddings of k into $\mathbb{R}$ or $\mathbb{C}$ (up to complex conjugation in the latter case). Then there are an infinite number of non-Archimedean completions. These correspond to the finite places v of k, one for each prime ideal in the ring of integers of k. If the prime ideal parametrized by v contains $p\mathbb{Z}$, then k_v is a p-adic field, that is, a finite extension of the field $\mathbb{Q}_p$ of p-adic numbers. The integral closure of $\mathbb{Z}_p$ in the field k_v is the unique maximal compact subring $\mathcal{O}_v$ of k_v. This ring is a principal ideal domain with a unique nonzero prime ideal.

Hasse's Thesis and *Habilitationsschrift*

Born in 1898, Helmut Hasse left the Fichte Gymnasium in Berlin in 1915 having passed the *Notabitur*, an anticipated form of the usual final examination during World War I.[13] While still a volunteer in the Navy, he enrolled at the Christian-Albrecht University in Kiel in 1917 to study mathematics. Otto Toeplitz was one

---

[13]For Hasse's biographical details, see [Frei, 1985].

of his teachers. In December 1918, after World War I, Hasse moved on to the Georg-August University in Göttingen to pursue his studies. There, he had the good fortune to attend lectures by Edmund Landau, David Hilbert, Erich Hecke, and Richard Courant. Beginning in 1915, Emmy Noether also worked in Göttingen, mostly with Klein and Hilbert. At that time, by law, women did not have the right to get their *Habilitation* in Prussia. Thus, despite Hilbert's support of Noether, the ministry twice denied her application for a *Habilitation*. Consequently, Noether had to announce her lecture courses—for example, on invariant theory—together with Hilbert. Finally, in 1919, she was formally entitled to complete the procedures for her *Habilitation*. This was connected with an appointment as a *Privatdozent*, still an unpaid position, but at least one involving the right to teach.[14]

Meanwhile, Hasse took a course in number theory from Erich Hecke in 1919, which might have played a crucial role in Hasse's early training. In 1920, Hecke accepted a position at the newly founded Hamburg University. Meanwhile, Hasse discovered Hensel's *Zahlentheorie* in a bookstore and found the ideas of his theory of p-adic numbers so attractive that he matriculated at Marburg University where Hensel held a professorship. As Hasse later explained:

> Though only being in my seventh semester of study, I obtained from Hensel after just a short time a theme for a *Staatsexamensarbeit* .... I was supposed to show that the necessary p-adic conditions for the representability of a rational integer by an integral binary quadratic form as derived in the last chapter of the number theory book alluded to are also sufficient and, if possible, also to carry out corresponding investigations for ternary, quaternary ... forms [Hasse, 1962, p. 3].[15]

Early in 1921, Hasse completed his thesis, solving the problem posed by Hensel. He achieved a quite general result pertaining to quadratic forms over the rationals and questions of representability. It is characterized by the structural insight, possibly conjectured by Hensel, that there is a relation between the behavior of f over $\mathbb{Q}$ and over the fields $K(p)$ of p-adic numbers, where p is a prime or $p = \infty$. In the final published version of his thesis, Hasse stated his result in the following terms [Hasse, 1923a, pp. 130-131]: "Fundamental Theorem: In order for a rational number m to be represented by a quadratic form f with rational coefficients, it is necessary and sufficient that m be represented by f in all $K(p)$."[16]

In remarks on the background of this result, Hasse emphasized how critical it is from a methodological point of view first to deal with the class of rational quadratic forms and questions of representability of numbers by elements in this family. In the

---

[14] Emmy Noether's *Habilitation*, as well as her earlier attempts to be admitted to it, are documented and discussed in detail in [Tollmien, 1990; 1991].

[15] "Obwohl erst im siebten Studiensemester stehend, erhielt ich von Hensel schon nach kurzer Zeit ein Thema für eine Staatsexamensarbeit .... Ich sollte die im Schlußkapitel des genannten Zahlentheoriebuches hergeleiteten notwendigen p-adischen Bedingungen für die Darstellbarkeit einer rationalen Zahl durch eine rationalzahlige binäre quadratische Form als hinreichend erweisen und wenn möglich entsprechende Untersuchungen auch für ternäre, quaternäre, ... Formen durchführen."

[16] "Fundamentalsatz: Damit eine rationale Zahl m durch eine quadratische Form f mit rationalen Koeffizienten rational darstellbar ist, ist notwendig und hinreichend, daß m durch f in allen $K(p)$ darstellbar ist."

long run, this can serve as a foundation for the more difficult, analogous questions in the arithmetic theory of integral quadratic forms.[17] It was Hasse's aim to make a fresh methodological start by freeing himself from the problems related to the notion of genus and the quite complicated investigations made by others such as Gauss, Smith, Minkowski, and Poincaré [Poincaré, 1882].

Hasse's "Fundamental Theorem," today called the "Strong Hasse Principle," and its methods of proof laid the foundation for another major theorem, obtained immediately after his thesis and submitted in December of 1921 as his *Habilitationsschrift*. In the latter, he dealt with the following question: "What are the necessary and sufficient conditions for two quadratic forms with rational coefficients to be related by an invertible linear substitution with rational coefficients?" [Hasse, 1923b, p. 205].[18] In working on this problem, Hasse was well aware of the result in which Minkowski had described a complete set of invariants characterizing the equivalence class of a quadratic form over $\mathbb{Q}$. However, Minkowski's approach hinged on the theory of integral quadratic forms he had worked out in 1882. It was Hasse's main objective—following his earlier lines of thought in relating the behavior of f over $\mathbb{Q}$ with that over the fields $K(p)$, for p a prime or $p = \infty$—"to substantiate the theory of equivalence in $K(1)$ independently of the theory in $R(1)$ and, thus, to avoid the quite complicated and involved way of Minkowski" [Hasse, 1923b, p. 208].[19] For each prime p, Hasse set up the existence of a complete set of invariants characterizing the equivalence class of a form f over $K(p)$. Binding together this information for each p, he obtained a complete set of invariants for the classification problem over $\mathbb{Q} = K(1)$. He showed that "[q]uadratic forms are equivalent in $K(1)$ if and only if they are equivalent in all $K(p)$" [Hasse, 1923b, p. 208].[20] As Hasse himself stated: "All these investigations, which have in part already been briefly carried out or indicated by Minkowski, gain in my systematic presentation uniformity and completeness. I obtain a specific simplification by using the Hilbert symbol in representing my [invariant] c_p. Thereby, a separate treatment of the prime 2 becomes redundant ..." [Hasse, 1923b, p. 208].[21] Concluding his investigations, Hasse described the precise relation between Minkowski's set of invariants and that given by himself. They turned out to be equivalent.

From the Small to the Large, or a Local-Global Principle

This result—giving a characterization of the equivalence class of a rational quadratic form, called the Weak Hasse Principle, as well as the former one, the

---

[17]Hasse gave a lecture of a quite elementary nature on this Fundamental Theorem on 17 November, 1927 [Hasse, 1927, NSUB].

[18]"Welches sind die notwendigen und hinreichenden Bedingungen, daß zwei quadratische Formen mit rationalen Koeffizienten durch eine umkehrbare, lineare Transformation mit rationalen Koeffizienten zusammenhängen?"

[19]"... eine von der Theorie in $R(1)$ unabhängige Begründung der Äquivalenztheorie in $K(1)$ zu geben und so den z.T. recht komplizierten und weit ausholenden Weg von Minkowski zu vermeiden."

[20]"Quadratische Formen sind dann und nur dann in $K(1)$ äquivalent, wenn sie in allen $K(p)$ äquivalent sind."

[21]"Alle diese Untersuchungen, die z.T. auch schon von Minkowski a.a.O. kurz ausgeführt oder angedeutet sind, gewinnen in meiner systematischen Darstellung an Einheitlichkeit und Vollständigkeit. Eine besondere Vereinfachung erziele ich durch Verwendung des Hilbertschen Symbols zur Darstellung meiner c_p. Hierdurch wird jede gesonderte Behandlung der Primzahl 2 überflüssig, ..."

Strong Hasse Principle—are well documented and elaborated upon in the modern literature.[22] Hasse proved obvious analogs of these theorems for any algebraic number field in addition to the field of rational numbers. He obtained these results in quick succession in two papers published in 1924 [Hasse, 1924a; 1924b]. However, the number theory involved in proving this generalization had to be on a different level from that required for the field $\mathbb{Q}$. The main ingredients in the result over the rationals included the Law of Quadratic Reciprocity (in the form of the product formula), Dirichlet's theorem about the existence of primes in arithmetic progressions, and the Hilbert symbol as a technical tool. Then-recent developments in the theory of the norm residue symbol as obtained by Hilbert [Hilbert, 1899] and Philipp Furtwängler [Furtwängler, 1912], Hilbert's quadratic reciprocity law for this norm residue symbol [Hilbert, 1899], and Weber's generalization of Dirichlet's theorem [Weber, 1886; 1887] were all crucial in Hasse's method of proof. In his publications on these questions [Hasse, 1923a; 1923b; 1924a; and 1924b], Hasse discussed in detail the different arithmetic tools he used. At the same time, he explained very carefully the intrinsic intention of his investigations, namely, to come up with a unifying approach based on Hensel's theory of p-adic numbers and, thereby, to open the way to possible generalizations. The basic insight in Hasse's work is what is today called the Local-Global Principle.

Hasse himself stressed in his related publications how important and fruitful Hensel's theory of p-adic numbers was for his work. Hasse ascribed a vague local-global concept in this respect to Hensel. In a short note published in 1962, Hasse commented (and even documented with a postcard written by Hensel in 1920) on the emergence of this principle [Hasse, 1962]. In working on the topic given to him by Hensel for this *Staatsexamensarbeit*, Hasse came to a point in dealing with ternary forms where his results concerning the representability of zero could not be immediately interpreted in a p-adic framework. Thus, Hasse asked Hensel for his advice. In his response, Hensel wrote: "I always have the idea that there is a certain question at the root. If I know of an analytic function that has rational character at all places ..., then it is rational. If I know the same of a number, that it is p-adic for the domain of each prime p and p_∞ [that is, with respect to the Archimedean valuation], then I do not yet know if it is a rational number. How should this be completed?" [Hasse, 1962].[23] It was this final hint, encoded in a question, that suggested that there might be a relation between the arithmetic nature of a problem over all p-adic fields $K(p)$ and its true nature over the field of rational numbers. Hasse shaped this into a fruitful concept leading to deep structural insight into the arithmetic theory of quadratic forms.

A Talk by Hasse in Königsberg in 1936

In May of 1936, Hasse gave a talk in Königsberg entitled "A Principal Theorem on Quadratic Forms with Rational Coefficients [Ein Hauptsatz über quadratische Formen mit rationalen Koeffizienten]" [Hasse, 1936, NSUB]. In the manuscript for this lecture, Hasse discussed in detail his approach to the Fundamental Theorem,

---

[22]See, for example, [Cassels, 1978, chap. 6] or [Kneser, 2002, chap. 6].

[23]"Ich habe immer die Idee, daß da eine ganz bestimmte Frage zu Grunde liegt. Wenn ich von einer analytischen Funktion weiß, daß sie an allen Stellen rationalen Charakter hat, so ist sie rational. Wenn ich bei einer Zahl dasselbe weiß, daß sie für den Bereich jeder Primzahl p und für p_∞ p-adisch ist, so weiß ich noch nicht, ob sie eine rationale Zahl ist. Wie wäre das zu ergänzen?"

its proof, and its relation to a previous result of Legendre in the ternary case. Based on Lagrange's reduction theory, Legendre had established conditions under which a ternary form f with rational coefficients gives a non-trivial representation of zero [Legendre, 1830, 1:§4]. Thus, Hasse described, in particular, the passage from these results to the p-adic framework, and vice versa, as he had carried it out in his thesis. He presented his transfer from the Small to the Large as the underlying key point which allowed him to prove the explicit criterion of Legendre. Hasse emphasized (in language resembling Hensel's lines on the postcard quoted above) that this method:

> ... is the simplest case of a principle which governs all of modern number theory, namely, the reduction of properties in the *Large* (for rational numbers) to properties in the *Small* or *local* properties (behavior as a congruence modulo the individual prime powers, behavior over the reals), totally analogous to the principle one has in complex function theory: a function is rational if and only if it has rational character at each finite place and at infinity. The individual prime numbers correspond to the finite places $z = a$ of the complex plane (the corresponding linear factors $z - a$ are the prime functions out of which the rational functions are composed), and the real corresponds to the infinite place (on the ball) [Hasse, 1936, NSUB, p. 3-4].[24]

This quote reflects Hasse's view in 1936 of his own early work in the theory of quadratic forms and of its significance in the development of number theory. At that time, the Local-Global Principle had already made its impact. Apart from the results discussed above, it had played a decisive role in the structure theory of normal simple algebras over algebraic number fields as completed in the early 1930s in the work of Brauer, Hasse, and Noether [Brauer-Hasse-Noether, 1932].[25]

With regard to the theory of quadratic forms with integral coefficients, Hasse pointed out in his talk that "[t]he corresponding theory of equivalence and the theory of representability under general restriction on integral coefficients and variables lie much deeper. One does not know a complete system of invariants in this case" [Hasse, 1936, NSUB, p. 13].[26] In conclusion, Hasse referred to the reception of his work by pointing toward a recent result of Carl Ludwig Siegel:

> In his deep papers on the theory of quadratic forms recently published in the *Annals of Mathematics*, Siegel has built below my purely qualitative principle of transfer from the individual places to the Large via a quantitative relation which formally has the same character as

---

[24]"... ist aber auch an sich von hohem theoretischem Interesse. Sie ist der einfachste Fall eines die gesamte moderne Zahlentheorie beherrschenden Prinzips, nämlich der Zurückführung von Eigenschaften im Großen (für rationale Zahlen) auf Eigenschaften im Kleinen oder lokale Eigenschaften (Verhalten als Kongruenz nach den einzelnen Primzahlpotenzen, Verhalten im Reellen), ganz analog wie man in der komplexen Funktionentheorie das Prinzip hat: Eine Funktion ist dann und nur dann rational, wenn sie an jeder endlichen Stelle und im unendlichen rationalen Charakter hat. Die einzelnen rationalen Primzahlen entsprechen den endlichen Stellen $z = a$ der komplexen Ebene (die zugehörigen Linearfaktoren $z - a$ sind die Primfunktionen, aus denen sich die rationalen Funktionen zusammensetzen), und das Reelle entspricht der unendlichen Stelle (auf der Kugel)."

[25]See [Hasse, 1931a, NSUB; 1932a, NSUB].

[26]"Die entsprechende Äquivalenztheorie und Darstellungstheorie bei durchgängiger Beschränkung auf ganzzahlige Koeffizienten und Variablenwerte liegt viel tiefer. Man kennt für sie kein vollständiges Invariantensystem."

Hilbert's product formula. It states that the suitably defined density of the quasi-integral representations by an integral quadratic form is equal to the product of the suitably defined representation densities for the individual p [Hasse, 1936, NSUB, p. 13].[27]

It is noteworthy that this talk was delivered by Hasse in Königsberg, the city where Minkowski had studied mathematics in the early 1880s and had held a professorship between 1893 and 1896. Furthermore, Richard Brauer had earned his *Habilitation* at the Albertus University in 1927 and had worked there in the years following as a *Privatdozent*. However, in 1934, Brauer (as well as Noether in Göttingen) were dismissed from their universities and emigrated to the United States.

Conclusion

Late in 1931, Richard Brauer, Helmut Hasse, and Emmy Noether established the principal theorem in the arithmetic theory of algebras, that is, the classification of normal simple algebras over algebraic number fields [Brauer-Hasse-Noether, 1932].[28] These investigations were partly inspired by questions in class field theory; the relations of these hypercomplex number systems to the commutative theory played an important role [Noether, 1932]. Later work of Claude Chevalley [Chevalley, 1933; 1940] and Hasse [Hasse, 1933] made evident the fact that the theory of normal simple algebras is an appropriate tool for the construction of local class field theory and the passage to global class field theory. Artin's reciprocity law found a natural place in this framework [Artin, 1927].

One of the decisive reduction steps in the proof of the principal theorem is the idea of the p-adic inference from the Small to the Large. This result, now called the Local-Global Principle for algebras, may be stated as follows. Let A be a normal simple algebra over an algebraic number field k. If the direct product algebra $A_v := A \otimes_k k_v$ splits everywhere locally (that is, for each place $v \in V$, A_v is isomorphic to a matrix algebra over k_v), then A splits over k, that is, A splits globally. Remarkably, Hasse attributed the source of this idea to Minkowski in a letter to Hermann Weyl, written on 15 December, 1931. In response to some words of admiration concerning the principal theorem which Weyl had sent on 8 December, 1931, Hasse wrote:

> Dear colleague!
> Accept my sincere thanks for your kind letter. The high words of praise from such a refined intellectual and visionary mathematical and philosophical man, as the author of "Space, Time and Matter,"

---

[27]"In seinen kürzlich in den Annals of Mathematics erschienenen tiefen Arbeiten zur Theorie der quadratischen Formen hat Siegel mein rein qualitatives Übertragungsprizip von den einzelnen Stellen aufs Große durch eine quantitative Relation unterbaut, die formal denselben Character hat, wie die Hilbertsche Produktformel. Sie sagt aus, daß die geeignet definierte Dichte der quasi-ganzen Darstellungen durch eine ganzzahlige quadratische Form gleich dem Produkt der geeignet definierten Darstellungsdichten für die einzelnen p ist." The papers of Siegel to which Hasse refers are [Siegel, 1935; 1936; 1937]. For an exposition of these results, see [Kneser, 1967, chap. 3] and [Kneser, 2002, chap. 10]. Notice, in particular, the concluding remarks in the latter reference, in which the emergence of Siegel's results and subsequent developments toward Tamagawa numbers of algebraic groups are discussed.

[28]In [Fenster and Schwermer, 2005], the authors discuss the development of this result as well as the other joint work linked with the proof of this theorem, namely, that of A. Adrian Albert and Hasse [Albert and Hasse, 1932]. See also Della Fenster's chapter in the present volume.

"The Idea of the Riemann Surface" and many other world renowned publications and treatises, are naturally a great honor and deep delight. That I also, on my own, esteem the importance of my last result and am very happy on that account, I may say without presumption. But I have to ascribe a very considerable role to the elegant theory of E. Noether, as the formal algebraic foundation, to the newly revitalized p-adic methods of Hensel as a tool, and above all to the "idea" of the p-adic inference from the Small to the Large, which emerged with Minkowski in all clarity [Hasse, 1931b, ETH].

This first paragraph of the letter provides profound insight into Hasse, the young thirty-three-year-old man.[29] The language Hasse used exposes his perspective. The words of thanks culminate in words of praise for Weyl and his work, almost deifying Weyl. Hasse continued, without presumption, by expressing his pride in his result and its importance.[30] At the same time, he acknowledged influences and gave credit to specific circles of ideas, namely, the algebraic theory of Noether, the p-adic methods of Hensel, and, last but not least, the first appearance of the p-adic inference from the Small to the Large in the work of Minkowski. No doubt, these are the sources for an "idea" that Hasse shaped into a powerful structural concept of still lasting influence in arithmetic and algebra.

Appendix: Helmut Hasse to Hermann Weyl, 15 December, 1931

Sehr verehrter Herr Kollege!

Nehmen Sie meinen herzlichsten Dank für Ihren freundlichen Brief. Die hohen Worte der Anerkennung eines so feingeistigen und weitblickenden mathematischen und philosophischen Menschen, wie des Autors von "Raum, Zeit und Materie," "Die Idee der Riemannschen Fläche" und vieler anderer weltberühmter Schriften und Abhandlungen, ist mir natürlich eine große Ehre und tiefe Freude. Daß ich auch von mir aus die Bedeutung meines letzten Resultats wohl zu schätzen weiß und darüber sehr glücklich bin, darf ich wohl ohne Überheblichkeit sagen. Einen sehr erheblichen Anteil muß ich aber der eleganten Theorie von E. Noether als dem formal algebraischen Unterbau, den zu neuem Leben erweckten p-adischen Methoden von Hensel als Werkzeug, und vor allem der zuerst bei Minkowski in aller Klarheit hervortretenden "Idee" des p-adischen Schlußes vom Kleinen aufs Große zuschreiben.

Auch ich erinnere mich sehr gut an Ihre ersten Worte zu mir anläßlich meines Vortrages über die erste explizite Reziprozitätsformel für höheren Exponenten in Innsbruck. Sie zweifelten damals ein wenig an der inneren Berechtigung solcher Untersuchungen, indem Sie ins Feld führten, es sei doch gerade Hilberts Verdienst, die Theorie des Reziprozitätsgesetzes von den expliziten Rechnungen früherer Forscher, insbesondere Kummers, befreit zu haben. Ich hoffe, inzwischen durch weitere Resultate und durch meine zusammenfassende Darstellung in Abschn. IV meines Berichts Teil II die Berechtigung solcher Untersuchungen auch für Augen wie die Ihrigen, die weniger auf Einzeltatsachen als auf die leitende "Idee" sehen, dargetan zu haben. Insbesondere dürfte jetzt klar sein, daß ich niemals einen Rückfall in die

---

[29]For the full text of the letter, see the appendix below.

[30]Note Hasse's use of the words "my last result" instead of "our last result" in the above quotation.

Vor-Hilbertschen Methoden gewollt habe, sondern nur einen Aufbau auf dem mit den formal-eleganten Hilbertschen Methoden begründeten Unterbau (Abschn. I-III). Ich kann aber natürlich gut verstehen, daß Dinge wie diese expliziten Reziprozitätsformeln einem Manne Ihrer hohen Geistes- und Geschmacksrichtung weniger zusagen, als mir, der ich durch die abstrakte Mathematik Dedekind-E. Noetherscher Art nie restlos befriedigt bin, ehe ich nicht zum mindesten auch eine explizite, formelmäßige konstruktive Behandlung daneben halten kann. Erst von der letzteren können sich die eleganten Methoden und schönen Ideen der ersteren wirklich vorteilhaft abheben.

Ich möchte die Gelegenheit benutzen, um Sie als den derzeitigen Vorsitzenden der Deutschen Mathematiker-Vereinigung wegen Ihrer Absichten zu Hensels 70. Geburtstag am 29. Dezember zu befragen. Bieberbach schreibt mir, daß bisher der Vorsitzende bei solchen Gelegenheiten einen Glückwunschbrief im Namen der Vereinigung geschrieben hat. Aber wäre es bei der Nähe Marburgs an Göttingen nicht sehr nett, wenn Sie persönlich herüberkämen? Ich würde mich jedenfalls sehr darüber freuen, von Hensel gar nicht zu reden. Ich selbst habe einen Festband des Crelleschen Journals zu überreichen.

Nochmals sehr herzlichen Dank für Ihren Brief.
Mit ergebensten Grüßen
Ihr

H. Hasse

Translation

Dear colleague!
Accept my sincere thanks for your kind letter. The high words of praise from such a refined intellectual and visionary mathematical and philosophical man, as the author of "Space, Time and Matter," "The Idea of the Riemann Surface" and many other world renowned publications and treatises are naturally a great honor and deep delight. That I also, on my own, esteem the importance of my last result and am very happy on that account I may say without presumption. But I have to ascribe a very considerable role to the elegant theory of E. Noether, as the formal algebraic foundation, to the newly revitalized p-adic methods of Hensel as a tool, and above all to the "idea" of the p-adic inference from the Small to the Large, which emerged with Minkowski in all clarity.

Also I recollect very well your first words to me on the occasion of my lecture in Innsbruck on the first explicit reciprocity formula for higher exponents. At that time you expressed some doubts about the essential justification of such investigations, by marshalling the argument that it is precisely Hilbert's achievement to have freed the theory of the reciprocity law from the explicit computations of former researchers, in particular, those of Kummer. I hope meanwhile to have made apparent by my further results and by my comprehensive exposition in section IV of part II of my report the justification of such investigations even in eyes such as yours, which see less the individual facts than the guiding "idea." In particular, it should now be clear that I never intended a relapse into the pre-Hilbertian methods but only a building upon the foundation which is established by the formally elegant Hilbertian methods (sections I-III). But, of course, I can well understand that things such as these explicit reciprocity formulas speak less to a man of your high intellect and taste than to me as I am never entirely satisfied by the abstract mathematics of the Dedekind-E. Noether kind, at least until I can also place next to it an explicit formula based on a constructive treatment. Only against the latter can the elegant methods and beautiful ideas of the former really profitably stand out.

I would like to take the opportunity to ask you, as the current chairman of the German Mathematical Association, about your intentions concerning Hensel's 70th birthday on 29 December. Bieberbach writes to me, that previously on such occasions the chairman has written a congratulatory letter on behalf of the association. But would it not be very nice, in view of the proximity of Marburg to Göttingen, if you would personally come? In any case I would be very pleased, to say nothing of Hensel. I myself have to present a special issue of Crelle's Journal.

Once again very sincere thanks for your letter.
With most devoted regard
Yours

H. Hasse

References

Archival Sources[31]

Hasse, Helmut. 1927. "Der Fundamentalsatz über quadratische Formen." Vortrag in der Leopoldina. 17 November, 1927. Niedersächsische Staats- und Universitätsbibliothek Göttingen [= NSUB], Handschriftenabteilung . Cod. Ms. H. Hasse 13:10 (13 pages).

———. 1931a. "Über Schiefkörper." Vortrag in der Mathematischen Gesellschaft Göttingen. NSUB. 13 January, 1931. Cod. Ms. H. Hasse 13:16 (17 pages).

———. 1931b. Letter to Hermann Weyl, 15 December, 1931. ETH–Bibliothek Zürich. Archiv [= ETH]. HS 91:591.

———. 1932a. [No title given]. Vortrag Zürich. November 1932. NSUB. Cod. Ms. H. Hasse 13:19 (4 pages).

———. 1936. "Ein Hauptsatz über quadratische Formen mit rationalen Koeffizienten." Vortrag in Königsberg. 13 May, 1936. NSUB. Cod. Ms. H. Hasse 13:28 (13 pages).

Printed Sources

Albert, A. Adrian and Hasse, Helmut. 1932. "A Determination of All Normal Division Algebras over an Algebraic Number Field." *Transactions of the American Mathematical Society* 34, 722-726.

Artin, Emil. 1927. "Beweis des allgemeinen Reziprozitätsgesetzes." *Abhandlungen Mathematisches Seminar, Universität Hamburg* 5, 353-363.

---

[31]I would like to thank Dr. Helmut Rohlfing for his permission to quote from the archives at the Niedersächsische Staats- und Universitätsbibliothek Göttingen. Similarly, I thank the ETH-Bibliothek, Archives, Zürich.

Brauer, Richard; Hasse, Helmut; and Noether, Emmy. 1932. "Beweis eines Hauptsatzes in der Theorie der Algebren." *Journal für die reine und angewandte Mathematik* 167, 399-404.

Cassels, John W. S. 1978. *Rational Quadratic Forms*. London/New York: Academic Press, Inc.

Chevalley, Claude. 1933. "Sur la théorie du corps de classes dans les corps finis et les corps locaux." *Journal of the Faculty of Sciences Tokyo University* 2, 365-476.

Chevalley, Claude. 1940. "La theorie du corps de classes." *Annals of Mathematics* 41, 398-418.

Dickson, Leonard Eugene. 1919–1923. *History of the Theory of Numbers*. 3 Vols. Washington D.C.: Carnegie Institution of Washington.

Eisenstein, Gotthold. 1847. "Note sur la représentation d'un nombre par la somme de cinq carrés." *Journal für die reine und angewandte Mathematik* 35, 368.

Euler, Leonhard. 1911. *Vollständige Anleitung zur Algebra mit den Zusätzen von J. L. Lagrange*. Ed. Heinrich Weber. In *Opera omnia*. 1st Ser. *Opera mathematica*. Vol. 1. Leipzig-Berlin: B. G. Teubner Verlag, 499-651.

Fenster, Della Dumbaugh and Schwermer, Joachim. 2005. "A Delicate Collaboration: Adrian Albert and Helmut Hasse and the Principal Theorem in Division Algebras in the Early 1930s." *Archive for History of Exact Sciences* 59, 349-379.

Frei, Günther. 1985. "Helmut Hasse (1898–1979)." *Expositiones Mathematicae* 3, 55-69.

_____. 2001. "How Hasse Was Led to the Theory of Quadratic Forms, the Local–Global Principle, the Theory of the Norm Residue Symbol, the Reciprocity Laws, and to Class Field Theory." In *Class Field Theory: Its Centenary and Prospect*. Ed. Katsuya Miyake. Advanced Studies in Pure Mathematics. Vol. 30. Tokyo: Mathematical Society of Japan, 31-62.

Furtwängler, Philipp. 1912. "Reziprozitätsgesetze für Potenzreste mit Primzahlexponenten in algebraischen Zahlkörpern II." *Mathematische Annalen* 72, 346-386.

Gauss, Carl Friedrich. 1801. *Disquisitiones arithmeticæ*. Leipzig: B. G. Teubner Verlag.

Hasse, Helmut. 1923a. "Über die Darstellbarkeit von Zahlen durch quadratische Formen im Körper der rationalen Zahlen." *Journal für die reine und angewandte Mathematik* 152, 129-148.

_____. 1923b. "Über die Äquivalenz quadratischer Formen im Körper der rationalen Zahlen." *Journal für die reine und angewandte Mathematik* 152, 205-244.

_____. 1924a. "Darstellbarkeit von Zahlen durch quadratische Formen in einem beliebigen algebraischen Zahlkörper." *Journal für die reine und angewandte Mathematik* 153, 113-130.

_____. 1924b. "Äquivalenz quadratischer Formen in einem beliebigen algebraischen Zahlkörper." *Journal für die reine und angewandte Mathematik* 153, 184-191.

_____. 1924c. "Zur Theorie des quadratischen Hilbertschen Normenrestsymbols in algebraischen Körpern." *Journal für die reine und angewandte Mathematik* 153, 76-93.

_____. 1931. "Über p-adische Schiefkörper und ihre Bedeutung für die Arithmetik hyperkomplexer Zahlsysteme." *Mathematische Annalen* 104, 495-534.

_____. 1933. "Die Struktur der R–Brauerschen Algebrenklassengruppe über einem algebraischen Zahlkörper." *Mathematische Annalen* 107, 731-760.

_____. 1962. "Kurt Hensels entscheidender Anstoß zur Entdeckung des Lokal-Global-Prinzips." *Journal für die reine und angewandte Mathematik* 209, 3-4.

Hensel, Kurt. 1897. "Über eine neue Begründung der Theorie der algebraischen Zahlen." *Jahresbericht der Deutschen Mathematiker-Vereinigung* 6, 83-88.

_____. 1904. "Neue Grundlagen der Arithmetik." *Journal für die reine und angewandte Mathmatik* 127, 51-84.

_____. 1905. "Über die arithmetischen Eigenschaften der algebraischen und transzendenten Zahlen." *Jahresbericht der Deutschen Mathematiker-Vereinigung* 14, 545-558.

_____. 1908. *Theorie der algebraischen Zahlen*. Leipzig and Berlin: B. G. Teubner Verlag.

_____. 1913. *Zahlentheorie*. Berlin and Leipzig: G. J. Göschen'sche Verlagshandlung.

Hilbert, David. 1899. "Über die Theorie des relativquadratischen Zahlkörpers." *Mathematische Annalen* 51, 1-127.

Hofmann, Josef E. 1960. "Über zahlentheoretische Methoden Fermats und Euler, ihre Zusammenhänge und ihre Bedeutung." *Archive for History of Exact Sciences* 1, 122-159.

Kneser, Martin. 1967. "Semi-simple Algebraic Groups." In *Algebraic Number Theory*. Ed. John W. S. Cassels and Albrecht Fröhlich. London/New York: Academic Press, Inc., 250-265.

_____. 2002. *Quadratische Formen*. Berlin/Heidelberg/New York: Springer-Verlag.

Kronecker, Leopold. 1882. "Grundzüge einer arithmetischen Theorie der algebraischen Grössen." *Journal für die reine und angewandte Mathematik* 92, 1-122.

Kürschák, Josef. 1913a. "Über Limesbildung und allgemeine Körpertheorie." In *Proceedings of the Fifth International Congress of Mathematicians (Cambridge, 22-28 August 1912)*. Ed. Ernest W. Hobson and Augustus E. H. Love. 2 Vols. Cambridge: Cambridge University Press, 1:285-289.

_____. 1913b. "Über Limesbildung und allgemeine Körpertheorie." *Journal für die reine und angewandte Mathematik* 142, 211-253.

Lagrange, Joseph-Louis. 1773–1775. "Recherches d'arithmétique." *Nouveaux mémoires de l'Académie royale des Sciences et Belles-Lettres de Berlin*. 1st Pt. 1773. 2nd Pt. 1775. Reprint Ed. Lagrange, Joseph-Louis. 1867–1892. *Oeuvres*. Ed. Joseph Serret and Gaston Darboux. 14 Vols. Paris: Gauthier-Villars, 3:693-795.

_____. 1774. "Additions à l'analyse indeterminée." In *Éléments d'algèbre par M. Leonard Euler.* 2 Vols. Lyon: Bryset, 2:369-664; Reprinted in [Euler, 1911].

Legendre, Adrien-Marie. 1830. *Théorie des nombres.* 3d Ed. 2 Vols. Paris: Firmin-Didot.

Leopoldt, Heinrich W. 1973. "Zum wissenschaftlichen Werk von Helmut Hasse." *Journal für die reine und angewandte Mathematik* 262/263, 1-17.

Minkowski, Hermann. 1884. "Grundlagen für eine Theorie der quadratischen Formen mit ganzzahligen Koeffizienten." Trans. *Mémoire sur la théorie des formes quadratiques.* Vol. 29. *Mémoires presentés par divers savants à l'Académie des Sciences de l'Institut national de France.* 180 pp., or [Minkowski, 1911, 1:3-144].

_____. 1886. "Über positive quadratische Formen." *Journal für die reine und angewandte Mathematik* 99, 1-9, or [Minkowski, 1911, 1:149-156].

_____. 1887a. "Über den arithmetischen Begriff der Äquivalenz und über die endlichen Gruppen linearer ganzzahliger Substitutionen." *Journal für die reine und angewandte Mathematik* 100, 449-458, or [Minkowski, 1911, 1:203-211].

_____. 1887b. "Zur Theorie der positiven quadratischen Formen." *Journal für die reine und angewandte Mathematik* 101, 196-202, or [Minkowski, 1911, 1:212-218].

_____. 1890. "Über die Bedingungen, unter welchen zwei quadratische Formen mit rationalen Koeffizienten ineinander rational transformiert werden können (Auszug aus einem von Herrn H. Minkowski an Herrn Adolf Hurwitz gerichteten Brief)." *Journal für die reine und angewandte Mathematik* 106, 5-26, or [Minkowski, 1911, 1:219-242].

_____. 1911. *Gesammelte Abhandlungen von Hermann Minkowski.* Ed. David Hilbert. 2 Vols. Berlin/Leipzig: B. G. Teubner Verlag.

_____. 1973. *Briefe an David Hilbert.* Berlin/Heidelberg/New York: Springer-Verlag.

Noether, Emmy. 1932. "Hyperkomplexe Systeme in ihren Beziehungen zur kommutativen Algebra und Zahlentheorie." In *Verhandlungen des Internationalen Mathematiker Kongresses Zürich 1932.* Ed. Walter Saxer, 2 Vols. Zürich/Leipzig: Orell Füssli, 1:189-194.

Olesko, Kathryn. 1991. *Physics as a Calling: Discipline and Practice in the Königsberg Seminar for Physics.* Ithaca and London: Cornell University Press.

Poincaré, Henri. 1882. "Sur les formes cubiques ternaires et quaternaires." *Journal de l'École polytechique* 51, 45-91.

Schwermer, Joachim. 1991. "Räumliche Anschauung und Minima positiv definiter quadratischer Formen: Zur Habilitation von Hermann Minkowski 1887 in Bonn." *Jahresbericht der Deutschen Mathematiker-Vereinigung* 93, 49-105.

_____. 1994. "H. Minkowski." In *Neue Deutsche Biographie.* Munich: Bayerische Akademie der Wissenschaften, 17:537-538.

_____. 1995. "Hermann Minkowski (1864–1909)." In *Die Albertus-Universität zu Königsberg und ihre Professoren.* Ed. Dietrich Rauschning *et al. Jahrbuch der Albertus-Universität zu Königsberg.* Vol. 29. Berlin: Duncker Humblot, pp. 553-560.

———. 2004. "Mathematics as Discipline and Practice in the Königsberg Seminar for Mathematics and Physics in the 1880s." Lecture at the University of Richmond. May 2004. In preparation.

———. 2007. "Reduction Theory of Quadratic Forms: Towards Räumliche Anschauung in Minkowski's Early Work." In *The Shaping of Arithmetic after C. F. Gauss's Disquisitiones Arithmeticae.* Ed. Catherine Goldstein *et al.* Berlin/Heidelberg/New York: Springer-Verlag, pp. 491-510.

Siegel, Carl-Ludwig. 1935. "Über die analytische Theorie der quadratischen Formen I." *Annals of Mathematics* 36, 527-606.

———. 1936. "Über die analytische Theorie der quadratischen Formen II." *Annals of Mathematics* 37, 230-263.

———. 1937. "Über die analytische Theorie der quadratischen Formen III." *Annals of Mathematics* 38, 212-291

Smith, Henry J. S. 1867. "On the Orders and Genera of Ternary Quadratic Forms." *Philosophical Transactions of the Royal Society of London* 157, 255-298.

———. 1868. "On the Orders and Genera of Quadratic Forms Containing More Than Three Indeterminates." *Proceedings of the Royal Society of London* 16, 197-208.

———. 1882. "Mémoire sur la représentation des nombres par des sommes de cinq carrés." In [Smith, 1894, pp. 623-680].

———. 1894. *The Collected Mathematical Papers.* Ed. James W. L. Glaisher. 2 Vols. Oxford: Clarendon Press.

Sylvester, James Joseph. 1852. "A Demonstration of the Theorem That Every Homogeneous Quadratic Polynomial Is Reducible by Real Orthogonal Substitutions to the Form of a Sum of Positive and Negative Squares." *Philosophical Magazine* 4, 138-142.

Tollmien, Cordula. 1990. " 'Sind wir doch der Meinung, dass ein weiblicher Kopf nur ganz ausnahmsweise in der Mathematik schöpferisch tätig sein kann ...': Emmy Noether 1882–1935." *Göttinger Jahrbuch* 38, 153-219.

———. 1991. "Die Habilitation von Emmy Noether an der Universität Göttingen." *NTM. Schriftenreihe für Geschichte der Naturwissenschaften, Technik und Medizin* 28, 13-32.

Ullrich, Peter. 1998. "The Genesis of Hensel's p-adic Numbers." In *Charlemagne and His Heritage: 1200 Years of Civilization and Science in Europe.* Vol. 2. *Mathematical Arts.* Ed. Paul L. Butzer *et al.* Turnhout: Brepols Publishers, 163-178.

Ullrich, Peter. 1999. "Die Entdeckung der Analogie zwischen Zahl- und Funktionenkörpern: der Ursprung der 'Dedekind–Ringe'." *Jahresbericht der Deutschen Mathematiker-Vereinigung* 101, 116-134.

Weber, Heinrich. 1886. "Theorie der abelschen Zahkörper I." *Acta Mathematica* 8, 193-263.

———. 1887. "Theorie der abelschen Zahlkörper II." *Acta Mathematica* 9, 105-130.

Weil, André. 1984. *Number Theory: An Approach through History from Hammurapi to Legendre.* Boston/Basel/Stuttgart: Birkhäuser Verlag.

CHAPTER 8

Research in Algebra at the University of Chicago: Leonard Eugene Dickson and A. Adrian Albert

Della Dumbaugh Fenster
University of Richmond, United States

Introduction

This chapter ostensibly focuses on three topics: Leonard Dickson, A. Adrian Albert, and the institutional context of the University of Chicago. At the same time, however, it is also concerned more broadly with American mathematicians working on the theory of algebras, with German algebraists and number-theorists working on algebras over algebraic number fields, and with the way these groups came together in connection with the solution of one of the important problems of twentieth-century mathematics.

A given historical study of mathematics might appear to have a narrow, well-defined focus, but the details are, in fact, rarely straightforward. Historical studies in mathematics often sprawl out like a winding road with its own unique combination of twists and turns. Historians try to follow the path pursued by others in another time and, sometimes, place. The history of mathematics owes its richness, in large part, to the individual historical actors behind the theorems or communities or fields of mathematics. This chapter, in particular, hinges on relationships: the interactions between adviser and Ph.D. student, the connections between faculty colleagues, and the competition among mathematicians pursuing related areas of research. Mathematics can create an inextricable link in each of these kinds of relationships. Consequently, these associations between people must, as they do in this chapter, assume a central place in historical analysis.

Leonard Dickson: Student

Born in 1874, Dickson attended the University of Texas for his undergraduate and master's education. He arrived at the University of Chicago in 1894 and went on to earn one of the first doctorates awarded in the Mathematics Department there.[1] E. H. Moore directed Dickson's 1896 thesis, entitled "The Analytic Representation of Substitutions on a Power of a Prime Number of Letters with a Discussion of the Linear Group" [Dickson, 1897]. Like his adviser, Dickson sought to extend the theory of finite fields and to establish further its connections with group theory. He

---

[1] Dickson and John Irwin Hutchinson (a student of Oskar Bolza) earned the first two doctorates in mathematics from Chicago in 1896. See [Parshall, 1992a, p. 43]. On the role of the Chicago Mathematics Department in the history of late nineteenth- and early twentieth-century American mathematics, see [Parshall and Rowe, 1994].

followed his dissertation with his first book, *Linear Groups with an Exposition of the Galois Field Theory*, published in 1901 [Dickson, 1901]. This text proved to be a primary source of information about finite simple groups for the next fifty years [Parshall, 1991].

Like many aspiring American mathematicians of his day, Dickson traveled to Europe for a year of study following the completion of his Ph.D. Given the focus of his dissertation, he chose, not surprisingly, to study with Sophus Lie in Leipzig and Camille Jordan, Émile Picard, and Charles Hermite in Paris.[2] On his return to America, and, most likely en route to his first job as an instructor at the University of California, Dickson spoke on his year abroad at the Chicago Mathematics Club on 30 July, 1897. His lecture, "The Influence of Galois on Recent Mathematics," displayed his potential as an expositor [Dickson, 10 July, 1897].[3] Although his talk focused on the study of continuous and discontinuous groups, and, in particular, on Lie's contributions to the study of the former, Dickson also discussed Galois's contributions in terms of the broader mathematical picture with a special emphasis on his own place in the heritage (a rather bold approach given his youth and his audience).

An obviously confident twenty-three-year-old Dickson brought his remarks to a close with a summary of the influence of Galois in various universities throughout the world. "It would thus appear," the young American concluded,

> that the brief products of the genius of Galois was still at work to unify and magnify the things that form the essence of many fields of mathematics. Along with these active forces at Paris, with both Picard and Jordan in the group lines, also at Göttingen with Klein and Hilbert, and at Berlin with Frobenius (whose recent work on solvable groups has attracted so much attention), I am proud to feel that in the investigation as well as the teaching of the fields of mathematics traceable back to the influence of Galois, Chicago holds a high place, with Prof. Moore in both the group field and the field of Algebraic and number Corpora and Professors Bolza and Maschke in the group and substitution field [Dickson, 10 July, 1897].

Dickson was one of the few people (at the time) who could compare the training provided by this new American university with that of the older, established European schools. His view, quite naturally perhaps, cast a strong vote in favor of his countrymen. The triumvirate at Chicago was capable of inspiring the highest level of research mathematics, and Dickson, at least, recognized himself as one of the beneficiaries.

Leonard Dickson: University of Chicago Faculty Member

After two years at the University of California and one year at his undergraduate *alma mater*, the University of Texas, Dickson returned to the University of Chicago in 1900 to join his former professors as a colleague. In his forty-year tenure at Chicago, Dickson authored more than 300 manuscripts and eighteen books. His extraordinary productivity reinforced and contributed to the broad view of

---

[2] For an account of Dickson's peregrinations, see [*University of Texas Record*, 1899].

[3] Dickson recorded his talk (by hand) in the Mathematical Club Records, misspellings and all.

an American mathematical community that he had adopted from Moore, Bolza, and Maschke.

Inspired by his adviser, Moore, Dickson pursued algebraic investigations which would ultimately define his research interests for more than forty years. Yet, this meant algebra in its broadest sense, for Dickson worked in group theory, invariant theory, finite field theory, number theory, and the theory of algebras. Moreover, Dickson advanced the Chicago algebraic school by imparting his standards for solid mathematics to the next generation of aspiring doctoral students [Fenster, 1997]. His sixty-seven Ph.D. students took Dickson's ideas with them, teaching at no fewer than forty-five academic institutions in at least twenty-two states and three foreign countries. Almost all of these first-generation Dickson students and second generation American students secured academic positions and passed on the hallmarks of both their adviser and their degree-granting institution [Fenster, 1997; 1998].[4]

With sixty-seven doctoral students, and countless others in his classroom, Dickson naturally imparted a significant amount of what I. C. Russell described as "that intangible something which is transmitted from person to person by association and contact, but can not [sic] be written or spoken" [Russell, 1904] to the next generation of American mathematicians. In his assessment of "Fifty Years of American Mathematics," written just a year before Dickson retired, George D. Birkhoff described Dickson as "the foremost American algebraist of the period [1888–1938]." "Moreover," he continued, "the influence which he has exerted through his students has been very considerable. In the height of his activity today, Dickson will always remain one of our great figures" [Birkhoff, 1938, p. 287]. Fifty years later, in conjunction with the centennial celebration of the American Mathematical Society (AMS), Saunders Mac Lane described Dickson as a "powerful and assertive" mathematician [Mac Lane, 1989, p. 133]. Relative to his students, Mac Lane remarked that "[o]ne can contemplate with amazement the wide influence exerted by Dickson" [Mac Lane, 1989, pp. 133-134]. Thus, when considering Dickson in the panorama of American mathematics, these reliable sources recall both his mathematical prowess and his mathematical investment in and through his students.

A. Adrian Albert and the Classification of Division Algebras

As Dickson began to formulate his ideas on the arithmetic of algebras and to publicize this "remarkable" theory internationally both in print and in person [Dickson, 1923, p. 141], a seventeen-year-old boy of immigrant parents arrived at the University of Chicago to begin his undergraduate studies.[5] In the fall of 1922, in

---

[4]For more on how Dickson and his students fit into the larger scheme of a Chicago "research school" in algebra, see [Parshall, 2005]. The accumulat*ing* tradition at Chicago, combined with the departmental hiring procedure (see below) also suggest a characterization of the University of Chicago algebraists as a strong mathematics program in terms of the guidelines offered by George David Birkhoff in 1926. "It is not the number [of mathematicians in the group] and the material equipment only that counts," Birkhoff determined after his on-site visits to the strongest mathematical centers in Europe. "An accumulated tradition, a wise policy in the choice of men, and an ability to work together as a group are factors of equal importance. George D. Birkhoff to Alexander Trowbridge, 8 September, 1926, as quoted in [Siegmund-Schultze, 2001, p. 268].

[5]For the biographical material in this paragraph on Albert's early life, see various materials contained in the Abraham Adrian Albert Papers held in the University of Chicago Archives as well as [Fenster, 1994, pp. 183-184].

his first quarter at his hometown university, A. Adrian Albert enrolled in a college algebra course. The winter term introduced him to plane analytic geometry with Mayme Logsdon and differential and integral calculus with Ernest Wilczynski. In the fall of 1924, he had his first class, "Theory of Equations," with Dickson. He took courses in the theory of numbers and continuous groups from Dickson the following autumn, and just before he received his B.S. in 1926, he heard Dickson lecture on one of his then-current mathematical interests, algebras and their arithmetics. Albert found algebra (and, one might conclude, its proponent at Chicago) so attractive that he joined Dickson's long line of graduate students and continued on for a Master's and Ph.D. He was the thirty-fifth student to work with Dickson; thirty-two others followed suit. By the time this student became Dr. A. Adrian Albert in 1928, he had already secured a place for himself at the center of activity in the field of linear associative algebras. In less than six years, the University of Chicago Mathematics Department, and especially Dickson, had contributed to the transformation of Albert (and others) from a student of college algebra to a first-rate algebraist.[6]

Albert's early research interests reflect the most obvious signs of Dickson's "considerable" influence on him [Kaplansky, 1988, p. 246]. Daniel Zelinsky, one of Albert's subsequent students, describes Dickson as "the one mainly responsible for steering Albert into the subject of algebras over fields, which is the subject that primarily concerned him throughout his career" [Zelinsky, 1973, p. 662].[7] Albert's master's thesis on "A Determination of All Associative Algebras in Two, Three, and Four Units [basis elements] over a Non-Modular Field F" and his dissertation, "Algebras and Their Radicals, and Division Algebras," represent his first investigations in this area. In the latter, he (almost) established the important result that every central division algebra of degree four (dimension sixteen) is always a crossed product.[8] He polished this work and presented it in his first major publication, "A Determination of All Normal Division Algebras in Sixteen Units" [A. Albert, 1929]. This very early work (Albert was only twenty-three at the time) at once "stamped him as one of the outstanding algebraists of his day" [Zelinsky, 1973, p. 662] and displayed "the hallmarks of his mathematical personality" [Kaplansky, 1988, p. 247]. As Irving Kaplansky described the situation: "Here was a tough problem that had defeated his predecessors; he attacked it with tenacity till it yielded" [Kaplansky, 1988, p. 247]. As for Dickson's view of Albert's result, "[o]ne can only imagine how delighted [he] must have been" [Kaplansky, 1988, p. 247].

---

[6]On Albert, see, for example, [Herstein, 1973], [Zelinsky, 1973], [Jacobson, 1974], and [Kaplansky, 1988]. Albert's daughter Nancy has recently published a personal reminiscence of her father in [N. Albert, 2005].

[7]Nathan Jacobson referred to Zelinsky as "[o]ne of the most distinguished of Albert's students" in [Jacobson, 1974, p. 1078].

[8]The notion of a crossed product arose in the work of Emmy Noether. See the next note. For the history of the notion of a crossed product, see [Scharlau, 1999]. In a colloquium given in January of 1931, Helmut Hasse claimed that there was a "certain confusion in the recent results of Albert pertaining to this problem [ist eine gewisse Verwirrung in den jeuen darauf bezüglichen Resultaten von Albert]" [Hasse, 13 January, 1931]. Albert proved that for $n = 4$, S is an abelian crossed product. In other words, degree four central division algebras are crossed products of abelian extension fields. This part was correct. It was Albert's assertion that the associated Galois groups were always of type $(2,2)$ that proved problematic for Hasse. It seems likely that this point led Hasse to contact Albert (see below) [Fenster and Schwermer, 2005, p. 353].

The problem Albert solved in his dissertation had its origins in the work of both Dickson and Joseph H. M. Wedderburn. In 1906, Dickson had introduced the concept of a cyclic algebra [Dickson, 1914].[9] In the final publication of this work, Dickson considered various conditions that led to the construction of new division algebras. Wedderburn's structure theorems of 1907 essentially reduced the study of associative algebras over a field to the classification of division algebras [Wedderburn, 1907]. In 1921, Wedderburn noted that "one may as well consider these [division algebras] as algebras over their centers" and found the dimension of any central division algebra over its base field to be a perfect square n^2 [Wedderburn, 1921]. Dickson had settled the "easy" $n = 2$ case and proved that a central division algebra of dimension four is cyclic. Wedderburn tackled the more difficult $n = 3$ case (also cyclic) in his 1921 paper, and Albert solved the still harder case in his dissertation by determining, in part, the structure of a dimension 16 ($n = 4$) central division algebra over its base field (see below).

The main goal of the structure theory of algebras in 1929–1932 was the determination and classification of finite-dimensional division algebras over Q or, equivalently, finite-dimensional central division algebras over number fields. As Albert had noted in the 1929 *Transactions* paper which had grown out of his dissertation, "[t]he chief outstanding problem in the theory of linear associative algebras over an infinite field F is the determination of all division algebras" [A. Albert, 1929, p. 253]. Others besides Albert had recognized the centrality of this problem, however. In Germany, Richard Brauer, Helmut Hasse, and Emmy Noether were also at work on it. Ultimately, late in 1931, this trio of Germans established the principal theorem that every central division algebra over an algebraic number field of finite degree is cyclic [Brauer-Hasse-Noether, 1932]. Two months later, Albert and Hasse published a joint work, entitled "A Determination of All Normal Division Algebras Over an Algebraic Number Field" [Albert and Hasse, 1932], which gave the history of the theorem and included an alternative proof of the principal theorem. What had happened? Newly uncovered archival materials provide further insight into this remarkable international development in the history of mathematics and, consequently, into Albert himself.

In a colloquium talk at Göttingen University on 13 January, 1931, Helmut Hasse called attention to the conjecture that every normal division algebra over an algebraic number field is cyclic. Around this same time, apparently, Hasse wrote to Albert via Dickson.[10] Hasse enclosed reprints of his work, including, at least, his

---

[9]In the opening footnote of [Dickson, 1914], Dickson remarked that he presented the first half of this work (which contained the construction of a cyclic algebra) to the American Mathematical Society on 14 April, 1906 and added the remaining sections in March of 1913. A cyclic algebra is a division algebra of dimension n^2 over the base field. These algebras contain a maximal subfield S which is cyclic over the ground field F, that is, they are Galois with $G(S/F)$ a cyclic group generated by a single element s. This generating element s, together with elements in the maximal subfield and base field, satisfy certain multiplicative relationships. This concept of a cyclic algebra was later generalized by Emmy Noether to the concept of a crossed product algebra. See the section entitled "Mathematical Digression: Crossed Products" in [Fenster and Schwermer, 2005, pp. 359-360]. A cyclic algebra is the special case in which the Galois extension has a cyclic Galois group. The crossed product is called abelian, or solvable, if the Galois group $G(S/F)$ is an abelian, or, respectively, solvable group.

For more on the consequences of research in this line, see Curtis's chapter below.

[10]The following discussion parallels that given in [Fenster and Schwermer, 2005, pp. 363-370], which relies, in part, on ten crucial letters written by Albert to Hasse between 6 February, 1931

paper published in the *Mathematische Annalen* on the determination of p-adic skew fields (division algebras) [Hasse, 1931]. Albert responded cordially on 6 February, 1931 from his position as an instructor at Columbia University and informed Hasse that he would assume his new position as an assistant professor of mathematics at the University of Chicago in August of 1931 [A. Albert, 6 February, 1931].

Albert's letter probably reached Hasse in mid-February, shortly before Hasse hosted a Congress at Marburg on 26-28 February, 1931 devoted to a discussion of known results on the conjecture regarding division algebras. The participants included, among others, Hasse, Brauer, and Noether. It seems that the joint efforts of Brauer, Hasse, and Noether to classify all divison algebras over an algebraic number field took shape during this Congress [Fenster and Schwermer, 2005, p. 362].

Around this same time, Hasse wrote to Albert and apparently raised a question regarding the existence of division algebras of order 16, the very subject of Albert's dissertation. Hasse inquired about the possibility of a more precise result. In particular, he wondered if it was possible to determine whether division algebras of order 16 were—or were not—cyclic?

In his reply, Albert pointed Hasse to one of his recent papers [A. Albert, 1930], in which he had considered the existence of noncyclic division algebras. In the end, however, Albert had to admit, "[t]he question seems to be a number-theoretic one and I see no way to get an algebraic hold on it. It seems to be a hopeless problem to me after more than a year's work on it" [A. Albert, 23 March, 1931]. Albert's statement hit at the heart of the classification of division algebras. That is, as he would later describe it himself, "the problems in the theory of normal division algebras are in a great measure number-theoretic as well as algebraic" [A. Albert, 11 May, 1931]. Although Albert recognized the importance of the arithmetic method, he was, to use Nathan Jacobson's description, "handicapped" in its use by the fact that he remained unaware until rather late of the powerful results in algebraic number theory which had been developed in Germany [Jacobson, 1974, p. 1080]. It was, apparently, Hasse's first letter (and enclosed reprints) that made Albert aware of these arithmetic methods.

Meanwhile, Hasse realized how little the German results were known in America. If Albert, one of the chief American mathematicians pursuing similar researches, did not know about his work, then who did? Consequently, Hasse submitted his "Theory of Cyclic Algebras over an Algebraic Number Field" to the *Transactions of the American Mathematical Society* in April of 1931. His introduction emphasized his motivation for writing.

> I present this paper for publication to an American journal and in English for the following reason:
> The theory of linear algebras has been greatly extended through the work of American mathematicians. Of late, German mathematicians have become active in this theory. In particular, they have succeeded in obtaining some apparently remarkable results by using the theory of algebraic numbers, ideals and abstract algebra, highly developed in Germany in recent decades. These results do not seem to be as well known in America as they should be on account of their

---

and 1 April, 1932. The whereabouts of the letters from Hasse to Albert, if they exist, are currently unknown.

importance. This fact is due, perhaps, to the language difference or to the unavailability of the widely scattered sources.

This paper develops a new application of the above mentioned theories to the theory of linear algebras. Of particular importance is the fact that purely algebraic results are obtained from deep-lying arithmetical theorems. In the middle part, an account is given of the fundamental algebraic basis for these arithmetical methods. This account is more extended than is necessary for this paper, and should obviate an extended study of several German papers [Hasse, 1932, p. 171].

Thus, Hasse aimed to bring German results, including his own, to an American audience. Although Hasse viewed the German results as "not as well known ... as they should be," it was also the other way around.

By May of 1931, Albert had begun to incorporate Hasse's theory on quadratic forms over algebraic number fields (global objects) and its relation to the theory of quadratic forms over p-adic fields (local objects) in his own work [A. Albert, 1931]. On 30 June, 1931, Albert wrote to Hasse to correct Hasse's impression that Albert could prove a result concerning the exponent and degree of a normal division algebra which was very close to the principal theorem. Albert casually concluded this letter promising to write at the end of summer, "when I may perhaps have more things of interest to you" [A. Albert, 30 June, 1931]. In a warm postscript, Albert added that "in perhaps two years I may visit Germany and there see you and discuss our beautiful subject, linear algebras" [A. Albert, 30 June, 1931]. In less than half a year, then, the correspondence between Albert and Hasse had penetrated to the very heart of their subject and had helped cultivate a cordial professional association (at least on paper) between the two mathematicians.

If Albert updated Hasse at the end of the summer, the letter does not survive. Albert did, however, reply with urgency to Hasse in early November. Hasse had apparently communicated to Albert that he could prove the principal theorem for abelian crossed products [Fenster and Schwermer, 2005, p. 366].[11] Albert immediately sent word to Hasse that this result allowed Albert to establish that:

> I. All normal division algebras of degree 2^e are cyclic algebras ...
> II. The exponent of any normal division algebra of degree n over an algebraic number field Ω is n ...
> III. Let A be a normal division algebra of degree p^e, p a prime, over an algebraic number field Ω. Then there exists an algebraic field Ω' over Ω (of degree r prime to p over Ω) such that $A' = A \times \Omega'$ is a cyclic normal division algebra of degree 2^e over Ω' ...
>
> This above [III] result says that while we don't know if A is cyclic we at least have this property *extensionally obtainable* while A' remains a divisional algebra. [A. Albert, 6 November, 1931 (Albert's emphasis)]

Hasse's result allowed Albert to come "very close to the principal theorem" [Albert and Hasse, p. 723]. Albert must have also realized how close Hasse was to a complete proof of the principal theorem, since he suggested that they publish their

---

[11]Recall from footnote 9 that a crossed product is called abelian, or solvable, if the Galois group $G(S/F)$ is an abelian group.

results in the same journal, or, even as a joint work. At this point, apparently, Albert was unaware of Hasse's collaborative work with Brauer and Noether. It would not be long, however, before he would realize his naïveté.

By the time Albert's reply reached Hasse, Brauer had established the "missing" reduction step in the Brauer-Hasse-Noether Theorem. Hasse submitted the Brauer-Hasse-Noether manuscript to himself on 11 November, 1931 in his role as editor of the upcoming issue of Crelle's *Journal* in honor of Kurt Hensel's seventieth birthday [Fenster and Schwermer, 2005, p. 367]. That same day, Hasse wrote Albert to inform him that they had established the principal theorem. Albert sent his congratulations to Hasse on 26 November, 1931. He also pointed out a difficulty in the proof Hasse had made available to him.[12] At the same time, Albert emphasized Hasse's crucial number-theoretic results. "In all my work on division algebras," Albert admitted, "the principal difficulty has been to somehow find a cyclic splitting field. This your p-adic method accomplishes" [A. Albert, 26 November, 1931].

At the same time, however, Albert was quick to call attention to his own contributions (which were overlooked, from his perspective). "The part of the proof of Theorem I which you attribute to Brauer and Noether," Albert asserted, "is however already in print" [A. Albert, 26 November, 1931]. Albert informed Hasse of his equivalent result which had just appeared in the *Bulletin of the American Mathematical Society* in October of 1931 [A. Albert, 1931]. "As my theorems have already been printed I believe that I may perhaps deserve some priority on that part of your proof" [A. Albert, 26 November, 1931]. Albert did not specify what he had in mind by "priority." Meanwhile, in two long letters containing corrections and suggestions for the final version of their paper, Emmy Noether called Hasse's attention to Albert's work [Fenster and Schwermer, 2005]. Ultimately, in the final corrections to the Brauer-Hasse-Noether paper, Hasse added a substantial footnote which gave the results Albert conveyed to Hasse in his letters of 6 and 26 November, 1931 [Brauer-Hasse-Noether, 1932].

Even given the tense topics Albert raised in his letter of 26 November, 1931, he brought the letter to a close by thanking Hasse "for communicating your wonderful theorem to me. I deeply appreciate it" [A. Albert, 26 November, 1931]. He also firmed up dates for his proposed trip to Germany in the fall of 1933. After Hasse sent Albert the original proof-sheets of the Brauer-Hasse-Noether paper, Albert tried once more to emphasize his more simplified approach [A. Albert, 9 December, 1931].

Albert's proof eventually made it into print in the joint paper he and Hasse submitted to the *Transactions of the American Mathematical Society* in January of 1932 [Albert and Hasse, 1932]. Albert updated Hasse on this joint project on 25 January, 1932 when he wrote:

> I have finally found time to write up the article by both of us "A Determination of all Normal Division Algebras over an Algebraic Number Field" for the Transactions. I gave a historical sketch of the proof, my short proof, and a slight revision (to make it more suitable for American readers) of your proof. I believe the presentation will be approved

---

[12]With this remark, Albert put his finger on a significant "gap" in the Brauer-Hasse-Noether proof which would take nearly twenty years to fill. See [Fenster and Schwermer, 2005, pp. 370-372] for the mathematical details behind this difficulty.

by you and, with a footnote to the effect that I undertook the writing of the article at your suggestion, I have presented the paper to the American Mathematical Society and will send it soon to the editors of the Trans. [A. Albert, 25 January, 1932]

This paper appeared in the April volume of the *Transactions*, along with Hasse's "Theory of Cyclic Algebras over an Algebraic Number Field" and Hasse's corresponding corrections [Hasse, 1932]. As the 25 January, 1932 letter and the footnote on the opening page of the published paper indicate, Albert wrote the joint paper at Hasse's suggestion. Moreover, although Hasse saw an initial draft of this manuscript, he did not see the proof-sheets. Albert completed the final corrections [Fenster and Schwermer, 2005, p. 369].

In the Albert-Hasse paper, the authors claimed that Brauer, Hasse, and Noether "used a reduction not as simple as the one by Albert already in print. The authors of the present article feel that it is desirable to show how the proof of the main theorem is an immediate consequence of Hasse's arithmetic and Albert's algebraic results (*first proof*). We shall also give a new proof (of the algebraic part) using the line of Albert's reasoning ..." [Albert and Hasse, 1932, p. 724 (their emphasis)]. Thus, in their proof of what they referred to as "*our* principal theorem," Albert and Hasse highlighted Albert's algebraic contributions and Hasse's number-theoretic results [Albert and Hasse, 1932, p. 722]. It seems unlikely that Hasse would have characterized his earlier work with Brauer and Noether in these less than favorable terms. "These comments may reflect Albert's initial writing of the paper and final correcting of the proofs. Or, perhaps, they represent some sort of concession on the part of Hasse" [Fenster and Schwermer, 2005, p. 370]. In any event, Hasse here in this joint publication agreed to present a proof of the principal theorem using Albert's algebra and his own number theory.[13]

In the context of the present chapter, these mathematical discussions and publications reveal that Hasse drew Albert into his work and vice versa. Moreover, the two mathematicians managed to maintain an extraordinary amount of diplomacy in their correspondence, even during the presumably tense months of November and December 1931. No doubt Albert would have preferred more credit than a lengthy footnote in eight-point type in the Brauer-Hasse-Noether manuscript. The Albert-Hasse paper at least gave Albert a chance to tell his version of the story to a largely American audience.[14]

The efforts to establish this theorem reveal a striking difference in the development of results in the United States and in Germany. Although he regularly communicated his results to Wedderburn, Albert primarily worked on his own. Hasse, on the other hand, "had access to a large number of distinguished European mathematicians right from the start" of his career [Fenster and Schwermer, 2005, p. 355]. Hasse presumably saw Emmy Noether when he gave a colloquium in Göttingen; he initiated an informative mathematics meeting on division algebras that apparently sent Brauer, Hasse, and Noether in pursuit of the principal

---

[13]Interestingly, Hasse did not include the Albert-Hasse paper in his collected works. The editors of Albert's collected works, however, did include it. Compare [Fenster and Schwermer, 2005, p. 373].

[14]For more on the later history of this theorem, see Curtis's chapter below.

theorem in the theory of algebras. Finally, Hasse, Brauer, and Noether—and particularly Hasse and Noether—remained in regular contact as they worked toward establishing this theorem.

This close-up look at an intense year of Albert's professional career shows how his dissertation work—born as it had been in Dickson's research on algebras—took him immediately to the center stage of algebra that had once been occupied by Dickson and Wedderburn alone. Only a few years later, however, center stage had become even more crowded. A number of distinguished German mathematicians—Brauer, Hasse, and Noether, in particular—had taken their places there as well.

This look at one central problem in division algebras over a brief time interval also elucidates the significance of the adviser-student association, the collaborative efforts of mathematicians, the heightened sense of urgency among mathematicians pursuing the same problem, and the (very human) desire for credit where credit is due. This is not a history of one mathematician's proof of a single theorem. Rather, it is a history that hinges on various associations within a broadening mathematical community that, in this case, spanned an ocean. Hasse needed the algebraic results of Brauer and Noether and an awareness of Albert's progress on the theorem. Albert needed the powerful number-theoretic techniques developed in Germany. As the Albert-Hasse correspondence reveals, Hasse was the only mathematician with a complete view of the developments related to this theorem. By the time Albert was fully aware of the advances made on the other side of the Atlantic, it was essentially too late. But all was not lost.

A. Adrian Albert: Professional Overview

As early as 1928–1929, Albert's algebraic researches had helped him secure a National Research Council Fellowship in 1928–1929. He spent the year at Princeton, attracted, apparently, by Wedderburn. While there, Albert met Solomon Lefschetz, who introduced him to the subject of Riemann matrices, and, in particular, interested him in a "major unsolved problem" in this area.[15]

The problem, on which Albert made critical breakthroughs in the mid 1930s, required a "classification of the algebraic correspondences of a Riemann surface (automorphisms of a complex curve). This had been reduced to the problem of finding the matrices that commute with a certain 'Riemann matrix' .... These commuting matrices form an algebra, and in the basic cases, a central simple algebra over the rational number field" [Zelinsky, 1973, p. 664]. This way of thinking about the problem was, as noted, precisely in Albert's area of expertise, and, according to Zelinsky, "he demolished it" [Zelinsky, 1973, p. 664]. The American Mathematical Society recognized the significance of this achievement by awarding Albert the Cole Prize in 1939. Like Dickson with his work at the interface of the theory of algebras and number theory, Albert had joined two areas—the theory of algebras and algebraic geometry—to solve a problem of key importance. Moreover, a fellowship, a colloquium talk, and lively conversations among mathematicians ultimately led to this valuable, definitive result in the theory of Riemann matrices.

Albert enjoyed similar success in 1933–1934 when he returned to Princeton as an invited member of the newly created Institute for Advanced Study. This year

---

[15]Kaplansky called this contact "fortunate" [Kaplansky, 1988, p. 247]. [Jacobson, 1974] provides the most detailed discussion of Albert's work in this field. For more on Solomon Lefschetz, consult [Lefschetz, 1989].

at the Institute led Albert to investigations in non-associative algebras that eventually "influenced a large number of mathematicians to break into this seemingly unpromising ground" [Zelinsky, 1973, p. 664]. In fact, at least half of Albert's thirty doctoral students selected a non-associative topic for their dissertation research.

Albert's main areas of research—associative algebras, Riemann matrices, and non-associative algebras—represent the most obvious components of his mathematical career. Yet his mathematical persona is not so easy to categorize. According to Jacobson, Albert possessed a "highly individualistic" mathematical style distinguished by

> the directness of his approach to a problem and his power and stamina to stick with it until he achieved a complete solution. He had a fantastic insight into what might be accomplished by intricate and subtle calculations of a highly original character. At times he could have obtained simpler proofs by using more sophisticated tools (e.g. representation theory), and one can almost always improve upon his arguments. However, this is of secondary importance compared to the first breakthrough which establishes a definitive result. It was in this that Albert really excelled [Jacobson, 1974, p. 1093].

Thus, Jacobson viewed Albert's strength as his ability to find that all-important initial approach to even the most sophisticated of problems rather than in his presentation of that result. From Jacobson's perspective, Albert possessed the far more valuable skill.

Moreover, Albert realized that mathematics extended far beyond the production and publication of mathematical results. In particular, he recognized the importance of committee and administrative work and he devoted time and energy to these activities alongside his mathematical researches. He, like Dickson and Moore, made significant contributions to both the University of Chicago and the American Mathematical Society. Relative to the former, Albert participated in a variety of committees, organized conferences, chaired the Mathematics Department and served as the dean of the Division of Physical Sciences. While chair, he "skillfully" found support to maintain a steady flow of visitors and research instructors [Kaplansky, 1988, pp. 251-252]. For example, he used his influence to persuade the University to donate an apartment building, affectionately known as "the compound," to house the visitors. Kaplansky claims that "the compound" became the "birthplace of many a fine theorem" [Kaplansky, 1988, p. 252].[16] It should come as no surprise that Albert strove to attract visitors to Chicago. He realized that a department which relied solely on its permanent faculty had the potential to become stale and narrow in its focus. An infiltration of new ideas frequently encouraged a fresh perspective on mathematics. From personal experience, Albert knew that mathematical progress often depended on just such an external spark.

Albert's career also reflected a strong commitment to the mathematical community at large. He served the AMS in a variety of capacities—as a committee member, as an editor of the *Bulletin* and *Transactions*, and, like Dickson and Moore, as President in 1965–1966. The concerns of American mathematicians in the middle two quarters of the twentieth century were, however, somewhat different from those in

---

[16]In particular, two visitors calling it home in 1960–1961, Walter Feit and John Thompson, were later to prove that all groups of odd order are solvable [Kaplansky, 1988, p. 252].

the early years when Moore and Dickson made their contributions, and Albert's service quite naturally addressed the changing needs of American mathematicians. In particular, Albert helped establish government research grants for mathematics comparable to those existing in other areas of science. He helped set the budget for mathematics of the National Science Foundation (NSF) (founded in 1950) and aided in the creation of NSF summer research institutes [Jacobson, 1974, p. 1077]. He apparently found satisfaction in this nationally oriented work for "[h]e was always pleased to use his influence in Washington to improve the status of mathematicians in general, and he was willing to do the same for individual mathematicians whom he considered worthy" [Zelinsky, 1973, p. 665]. This latter category surely included his students.

Beyond his service to Chicago, the AMS, and the mathematical community at large, Albert exerted considerable influence in mathematics through his students [Jacobson, 1974, p. 1078]. As I. N. Herstein observed, "Adrian was extremely good at working with students. This is attested by the 30 mathematicians who took their Ph.D.'s with him. In their number are many who are well known mathematicians today. His interest in his students—while they were students and forever afterwards—was known and appreciated by them" [Herstein, 1973, p. 186]. Zelinsky, in particular, described Albert as an adviser who treated his Ph.D. students "almost as members of his family" [Zelinsky, 1973, p. 663]. His students and colleagues regarded him warmly, a luxury, his own adviser had not often—if ever—known.

At certain times in his career, however, Albert did give explicit expressions of his admiration for Dickson. In the obituary he wrote for the *Bulletin of the American Mathematical Society*, he described his adviser as an "inspiring teacher .... He helped his students to get started in research after the Ph.D. and his books had a worldwide influence in stimulating research" [A. Albert, 1955, p. 331]. More specifically, in the preface to his *Structure of Algebras*, Albert stated that "I owe much to these expositions [*Algebras and Their Arithmetics* [Dickson, 1923] and *Algebren und ihre Zahlentheorie* [Dickson, 1927]] as well as to their author, who has been my teacher and the inspiration of all my research" [A. Albert, 1939, p. 5]. He had also valued Dickson's opinion in the early stages of his career enough to seek advice regarding employment options [Dickson, undated].

As noted above, the possibility of a faculty position at the University of Chicago had opened up for Albert not long after he had earned his doctoral degree there. On 25 November, 1930, Gilbert A. Bliss, then in the third year of a fourteen-year stint as chair of the Mathematics Department at Chicago, had this to say in his letter to the dean supporting Albert's appointment to an assistant professorship:

> He was one of the ablest students we have ever had in our Department. My only objection to him when he was here was a slight overconfidence in himself, which, however, may help to carry him steadily toward success in his mathematical career. His scientific activity has been unusual.... His work is in Professor Dickson's field, just now not adequately represented in our younger faculty group. Professor Dickson knows his work well and approves it highly [Bliss, 25 November, 1930].

Thus, the Mathematics Department at Chicago—Dickson, Bliss, and E. H. Moore, among others—recognized Albert's abilities and sought to secure him as a permanent faculty member. The following fall he began what would ultimately grow into his forty-one-year-long tenure at the University of Chicago.

Bliss's comments to the Dean point to a hiring procedure described later by Saunders Mac Lane as an "'inheritance principle': If X has been an outstanding professor in field F, appoint as his successor the best person in F, if possible the best student of X" [Mac Lane, 1989, p. 131]. This appointment procedure worked exceptionally well, as Moore, Dickson, and Albert established a strong algebraic research tradition at Chicago.[17]

Dickson and his "best" student ultimately spent eight years as colleagues on the Chicago mathematics faculty. During this time, the two rarely pursued similar topics of research. Dickson brought his career to a close with a thorough investigation of the Waring problem, while Albert launched his with studies in associative algebras and Riemann matrices. When Dickson retired from mathematics in 1939, he returned to Texas and, of his former students, allowed only Albert to see him [Duren, 10 December, 1992]. When Dickson died on 17 January, 1954, just five days shy of his eightieth birthday, Albert was thus the obvious choice to write the obituary of his adviser.

By the end of March 1954, Albert had, in fact, already agreed to write the obituary for the *Bulletin of the American Mathematical Society*. Detlev W. Bronk, President of the National Academy of Sciences at the time, also inquired whether he would "be willing to prepare the biographical memoir of Leonard Eugene Dickson for the Academy series?" [Bronk, 29 March, 1954]. Albert answered Bronk's request on 8 April, 1954. "I am somewhat puzzled as to how to reply to your letter asking me to write the Biographical Memoir on L. E. Dickson," Albert began. "I realize that I am the obvious victim but I have already agreed to write the obituary article for the Bulletin of the American Mathematical Society ..." [A. Albert, 8 April, 1954]. Albert eventually agreed to write a scientific article for the *Bulletin* and a more general article for the Academy. He wrote a terse, but cordial obituary of Dickson for the former, but the biography for the latter never appeared. Albert may have felt that he just did not have enough to say about Dickson to fill two such articles. He may also have been under pressure to complete other, unrelated projects. Whatever the case may be, Dickson, unlike Albert himself, was ultimately not memorialized in terms reflective of feelings of great warmth.

Still, Albert could not and did not deny the influence Dickson had on his early career. His early work in the theory of algebras, his joining of the theory of algebras with the study of Riemann matrices, and both his general and specific contributions to the theory of non-associative algebras, all reflect Dickson's interest in the theory of algebras in the early 1920s. Jacobson, Kaplansky, Zelinsky, and Albert himself concur that Dickson set the course for much of Albert's research.[18]

From a broader perspective, Albert in many ways fulfilled the role of an "heir" in that his mathematical career *looked* very much like Dickson's. Yet despite the

---

[17][Mac Lane, 1989, pp. 141-142] discusses how this "principle" worked in other fields at Chicago. Both Garrett Birkhoff [Birkhoff, 1976, p. 35] and Saunders Mac Lane [Mac Lane, 1989, p. 143; Mac Lane, 5-6 March, 1992] classify Albert as Dickson's "best" student.

[18]For their respective views, see [Jacobson, 1974, p. 1076], [Kaplansky, 1988, p. 246], [Zelinsky, 1973, p. 662], and [A. Albert, 1955, p. 331].

many similarities between their careers, as personalities Albert and Dickson hardly resembled one another at all. Dickson, apparently, had few close relationships within mathematics, whereas Albert enjoyed many close friends within the mathematical community as the four warm biographies written upon his death by Herstein, Jacobson, Kaplansky, and Zelinsky attest.

Conclusions

The University of Chicago was committed to establishing a strong German research tradition as an *American* research tradition, and, in particular, a Chicago research tradition. Whereas E. H. Moore and Dickson worked to gain the acceptance of the European mathematicians,[19] Albert found himself faced with very different issues right from the start. The Germans had begun to recognize the American contributions to algebra, partly as a consequence of Dickson's *Algebras and Their Arithmetics* and the corresponding German edition. These texts helped spread Wedderburn's structure theorems and Dickson's corresponding arithmetic [Birkhoff, 1938, p. 287]. In his remarks at a memorial for Emmy Noether in 1935, Hermann Weyl observed that the American algebraists may not have secured their deserved recognition. As he described it, Noether's "methods need not, however, be considered the only means of salvation. In addition to Artin and Hasse, who in some respects are akin to her, there are algebraists of a still more different stamp, such as I. Schur in Germany, Dickson and Wedderburn in America, whose achievements are certainly not behind hers in depth and significance. Perhaps her followers, in pardonable enthusiasm, have not always full recognized this fact" [Weyl, 1935, p. 218].

One can imagine how Weyl's remarks would have delighted Dickson. To put him in the same league with Emmy Noether, forty-three years after the University of Chicago had come into existence indirectly heralded the success of the University of Chicago Mathematics Department. Dickson had not only acquired his own graduate training there, but he had also guided the next generation, including Albert. Moreover, while still a very young mathematician, Albert had demonstrated his mathematical prowess and diplomatic finesse in his exchanges with another one of Weyl's all-stars, Helmut Hasse.

This chapter, however, has ranged far from the research of Dickson and Albert at the University of Chicago. It has explored the evolving association of a Ph.D. adviser and his student. It has highlighted the fresh approaches and insights brought to mathematicians by various forms of exchange—colloquia, visitors, and even ideas via the postal service. Perhaps most saliently, taking Albert as a case study, it has highlighted two very different collaborative initiatives in mathematics.

In Germany, Hasse employed a certain strategy to remain apprised of Albert's results without divulging his own work on the principal theorem with Noether and Brauer. Although Albert would not realize it until later, it was head-to-head competition from the start. At the same time, however, the mathematics discussed in the Albert-Hasse correspondence provided a certain initiative for Albert to continue his pursuit of the theorem. Both the Brauer-Hasse-Noether and the Albert-Hasse

---

[19]Witness the efforts of the Chicago Mathematicians to include Europeans and particularly Germans at the 1893 Chicago Mathematical Congress [Parshall and Rowe, 1994, pp. 295-330]; Felix Klein's Evanston Colloquium Lectures [Parshall and Rowe, 1994, pp. 331-361]; and Dickson's first book under a B. G. Teubner imprint [Parshall, 1991].

collaborative efforts show how deadlines and priority issues became tangled in the search for the classification of division algebras over number fields.

External funds, in the form of both philanthropy and, later, government grants also played a role, although literally a supporting one, in this development. Andrew Carnegie's Scottish trust, for example, financed Wedderburn's year at Chicago in 1903–1904 [Parshall, 1992b, p. 528]. More Carnegie funds, this time from the American-based Carnegie Institution of Washington, provided financial support for Dickson's *History of the Theory of Numbers* [Fenster, 2003]. Dickson's historical study of the theory of numbers, in turn, inspired his 1920 International Congress address in Strasbourg.[20] For this work, Dickson found himself compelled to come to terms with the arithmetic of the quaternions initially and with more general algebras later.[21] Dickson's mathematical interests included this research in the theory of algebras when Albert arrived at the University of Chicago.

Looking on the other side of the Atlantic, Rockefeller funds built the Courant Institute at the University of Göttingen, the university where Emmy Noether held a position during this particular development in the history of mathematics [Siegmund-Schultze, 2001]. Later, Rockefeller monies supported the successful emigration of German scientists to the U.S. Albert's postdoctoral fellowship reflected still other, new efforts to support American mathematics with government funds.

The standard biographies of Albert mention that in six short years A. Adrian Albert went from a course in college algebra to a Ph.D. that placed him at the center of international algebraic activity. While that is certainly accurate, a deeper look at the historical context reveals that those six years apparently gave Albert a strong foundation, in both mathematics *and* extra-mathematical ideals, on which to build a distinguished mathematical career. That career embodied many of the characteristics, including, in particular, the research interests of his adviser Dickson. But that career had its own unique attributes, too.

It would have been difficult for any graduate program or mathematician to prepare Albert for the intense competition, which eventually evolved into a collaborative effort, that he would face shortly after he received his Ph.D. As Kaplansky put it, Albert may have already displayed the "hallmarks of his mathematical personality" on finishing his Ph.D., but just a few years later, the year-long correspondence with Hasse would mold and shape that personality further. This time, Albert attacked an even more sophisticated problem until it yielded, but he needed results that lay outside his domain. He also apparently needed as much "tenacity" in his correspondence regarding the principal theorem in division algebras as he did in establishing the proof itself. That is, managing the association with Hasse, particularly after Albert knew about his joint efforts with Brauer and Noether, displays yet another aspect of Albert's mathematical personality.

This chapter also enlarges our understanding of Kaplansky's characterization of Albert as a "statesman and leader" [Kaplansky, 1988, p. 3] and Zelinsky's description of him as "a vigorous force for the advancement of mathematics, and a

---

[20][Fenster, 1999] provides a detailed account of the genesis of Dickson's *History of the Theory of Numbers*, the historiographic style revealed therein, and the mathematical contributions to the theory of algebras which arose out of it.

[21]For the details of this work, see [Fenster, 1998, pp. 136-152].

very warm and understanding human being" [Zelinsky, 1973, p. 661]. In particular, Albert's mathematical personality had a "vigorous force" of its own. Once Albert realized Hasse had reached a proof of the principal theorem with Brauer and Noether, for example, he immediately implored Hasse to give his own work the credit he felt it deserved. In the end, though, Albert could not control how Hasse moved forward with the Brauer-Hasse-Noether result. And he certainly could not undo what had been done in the preceding months; Albert *had* exchanged letters and results with Hasse without knowing of Hasse's work with Brauer and Noether. But Albert could tell his version of the story in the Albert-Hasse paper, and he could tell it in a way that highlighted *his* results. Albert authored the joint work and saw it through to almost immediate publication in the *Transactions of the American Mathematical Society* without Hasse seeing any version other than the initial draft.

Albert's strong personality or, perhaps, his clarity of purpose is also apparent in his discussions with the President of the National Academy of Sciences regarding the biographical memoir of Dickson shortly after Dickson's death. Albert did not disguise his dissatisfaction with Detlev Bronk's request for what Albert viewed as another biography of Dickson. Moreover, Albert did not suggest another former student or colleague. This event tells as much about Dickson as it does about Albert. When Dickson died just shy of his eightieth birthday and after an almost fifteen-year retirement, there were very few colleagues who could—or would?—write a biography of him. As noted, the situation was markedly different for Albert, who died at the age of sixty-six while still participating fully in the mathematical community.

Kaplansky describes the search for rational division algebras as "an unequal battle" with Albert "nosed out in a photo finish. In a joint paper with Hasse published in 1932 the full history of the matter was set out, and one can see how close Albert came to winning" [Kaplansky, 1988, p. 5]. With new insight into what took place in that "battle," we see that although it may have been "unequal" in terms of numbers of forces (three versus one), Albert showed his tremendous mathematical "power" [Herstein, 1973, p. 186] by nearly establishing the proof on his own. He certainly showed diplomacy on the front lines. Moreover, although Albert may not have "won" in Kaplansky's sense of the word, he certainly "won" in terms of leading a thriving mathematical career with significant mathematical results and influential personal contacts.

References

Archival Sources

Albert, A. Adrian. 8 April, 1954. Letter to Detle[v] Bronk. University of Chicago Archives. Department of Special Collections. A. A. Albert Papers.
———. 6 February, 1931. Letter to Helmut Hasse. University of Göttingen Archives. Helmut Hasse Nachlaß.
———. 23 March, 1931. Letter to Helmut Hasse. University of Göttingen Archives. Helmut Hasse Nachlaß.
———. 11 May, 1931. Letter to Helmut Hasse. University of Göttingen Archives. Helmut Hasse Nachlaß.

———. 30 June, 1931. Letter to Helmut Hasse. University of Göttingen Archives. Helmut Hasse Nachlaß.

———. 6 November, 1931. Letter to Helmut Hasse. University of Göttingen Archives. Helmut Hasse Nachlaß.

———. 26 November, 1931. Letter to Helmut Hasse. University of Göttingen Archives. Helmut Hasse Nachlaß.

———. 9 December, 1931. Letter to Helmut Hasse. University of Göttingen Archives. Helmut Hasse Nachlaß.

———. 25 January, 1932. Letter to Helmut Hasse. University of Göttingen Archives. Helmut Hasse Nachlaß.

Bliss, Gilbert A. 25 November, 1930. Letter to Dean Gale. University of Chicago Archives. Department of Special Collections. Presidents' Papers. Appointments and Budgets, 1925–1940. Box 31. Folder 6.

Bronk, Detlev W. 29 March, 1954. Letter to A. Adrian Albert. University of Chicago Archives. Department of Special Collections. A. A. Albert Papers.

Dickson, Leonard E. Undated. Letter to A. Adrian Albert. University of Chicago Archives. Department of Special Collections. A. A. Albert Papers.

———. 10 July, 1897. "The Influence of Galois on Recent Mathematics." University of Chicago. Department of Special Collections. Mathematical Club Records 1893–1921.

Duren, William L. 10 December, 1992. Interview by the author. Charlottesville, Virginia. Tape recording.

Hasse, Helmut. 13 January, 1931. "Über Schiefkörper." Vortrag in der Mathematischen Geselleschafte Göttingen. Handschriftenabteilung der Niedersächsische Staats- und Universitätsbibliothek (NSUB). Cod. Ms H. Hasse 13:16.

Mac Lane, Saunders. 5-6 March, 1992. Interview by the author. Charlottesville, Virginia. Tape recording.

Printed Sources

Albert, A. Adrian. 1929. "A Determination of All Normal Division Algebras in Sixteen Units." *Transactions of the American Mathematical Society* 31, 253-260.

———. 1930. "New Results in the Theory of Normal Division Algebras." *Transactions of the American Mathematical Society* 32, 171-195.

———. 1931. "Division Algebras over an Algebraic Number Field." *Bulletin of the American Mathematical Society* 37, 777-784.

———. 1939. *Structure of Algebras*. American Mathematical Society Colloquium Publications. Vol. 24. New York: American Mathematical Society.

———. 1955. "Leonard Eugene Dickson: 1874–1954." *Bulletin of the American Mathematical Society* 61, 331-345.

Albert, A. Adrian and Hasse, Helmut. 1932. "A Determination of All Normal Division Algebras Over an Algebraic Number Field." *Transactions of the American Mathematical Society* 34, 722-726.

Albert, Nancy. 2005. A^3 *& His Algebra: How a Boy form Chicago's West Side Became a Force in American Mathematics*. Lincoln, NE: iUniverse.

Archibald, Raymond C. 1938. *A Semicentennial History of the American Mathematical Society, 1888-1938.* New York: American Mathematical Society.

Birkhoff, Garrett. 1976. "Some Leaders in American Mathematics: 1891–1941." In *The Bicentennial Tribute to American Mathematics, 1776–1976.* Ed. Dalton Tarwater. N.p.: The Mathematical Association of America, pp. 25-78.

Birkhoff, George D. 1938. "Fifty Years of American Mathematics." In *Semicentennial Addresses of the American Mathematical Society 1888-1938.* Ed. Raymond C. Archibald. New York: American Mathematical Society, pp. 270-315.

Brauer, Richard; Hasse, Helmut; and Noether, Emmy. 1932. "Beweis eines Hauptsatzes in der Theorie der Algebren." *Journal für die reine und angewandte Mathematik* 167, 399-404.

Dickson, Leonard E. 1897. "The Analytic Representation of Substitutions on a Power of a Prime Number of Letters with a Discussion of the Linear Group." *Annals of Mathematics* 11, 65-143.

_____. 1901. *Linear Groups with an Exposition of the Galois Field Theory.* Leipzig: B. G. Teubner Verlag. Reprint Ed. 1958. New York: Dover Publications, Inc.

_____. 1914. "Linear Associative Algebras and Abelian Equations." *Transactions of the American Mathematical Society* 15, 31-46.

_____. 1923. *Algebras and Their Arithmetics.* Chicago: University of Chicago Press.

_____. 1927. *Algebren und ihre Zahlentheorie.* Zürich: Orell-Füssli.

Fenster, Della D. 1994. "Leonard Eugene Dickson and His Work in the Theory of Algebras." Unpublished Doctoral Dissertation: University of Virginia.

_____. 1997. "Role Modeling in Mathematics: The Case of Leonard Eugene Dickson (1874–1954)." *Historia Mathematica* 24, 7-24.

_____. 1998. "Leonard Eugene Dickson and His Work in the Arithmetics of Algebras." *Archive for History of Exact Sciences* 52, 119-159.

_____. 1999. "Leonard Dickson's History of the Theory of Numbers: An Historical Study with Mathematical Implications." *Revue d'histoire des mathématiques* 5, 159-179.

_____. 2003. "Funds for Mathematics: Carnegie Institution of Washington Support for Mathematics from 1902–1921." *Historia Mathematica* 30, 195-216.

Fenster, Della D. and Schwermer, Joachim. 2005. "A Delicate Collaboration: A. Adrian Albert and Helmut Hasse and the Principal Theorem in Division Algebras in the Early 1930s." *Archive for History of Exact Sciences* 59, 349-379.

Hasse, Helmut. 1931. "Über p-adische Schiefkörper und ihre Bedeutung für die Arithmetik hyperkomplexer Zahlensysteme." *Mathematische Annalen* 104, 495-534.

_____. 1932. "Theory of Cyclic Algebras over an Algebraic Number Field." *Transactions of the American Mathematical Society* 34, 171-214.

Herstein, Israel N. 1973. "A. Adrian Albert." *Scripta Mathematica* 29, 185-189.

Jacobson, Nathan. 1974. "Abraham Adrian Albert: 1905–1972." *Bulletin of the American Mathematical Society* 80, 1075-1100.

REFERENCES

Kaplansky, Irving. 1988. "Abraham Adrian Albert." In *A Century of Mathematics in America–Part I*. Ed. Peter L. Duren *et al.* Providence: American Mathematical Society, pp. 244-264.

Lefschetz, Solomon. 1989. "Reminiscences of a Mathematical Immigrant in the U.S." In *A Century of Mathematics in America–Part I*. Ed. Peter L. Duren *et al.* Providence: American Mathematical Society, pp. 201-207.

Mac Lane, Saunders. 1989. "Mathematics at the University of Chicago: A Brief History." In *A Century of Mathematics in America–Part II*. Ed. Peter L. Duren *et al.* Providence: American Mathematical Society, pp. 127-154.

Parshall, Karen Hunger. 1991. "A Study in Group Theory: Leonard Eugene Dickson's *Linear Groups*." *The Mathematical Intelligencer* 13, 7-11.

_____. 1992a. "The 100th Anniversary of Mathematics at the University of Chicago." *The Mathematical Intelligencer* 14, 39-44.

_____. 1992b. "New Light on the Life and Work of Joseph Henry Maclagan Wedderburn (1882–1948)." In *Amphora: Festschrift für Hans Wussing zu seinem 65. Geburtstag*. Ed. Menso Folkerts *et al.* Basel/Boston/Berlin: Birkhäuser Verlag, pp. 523-537.

_____. 2005. "Defining a Mathematical Research School: The Case of Algebra at Chicago, 1892–1945." *Historia Mathematica* 31, 263-278.

Parshall, Karen Hunger and Rowe, David E. 1994. *The Emergence of an American Mathematical Research Community, 1876–1900: J. J. Sylvester, Felix Klein, and E. H. Moore*. Providence: American Mathematical Society and London: London Mathematical Society.

Russell, I. C. 1904. "Research in State Universities." *Science* 19, 383.

Scharlau, Winfried. 1999. "Emmy Noether's Contributions to the Theory of Algebras." In *The Heritage of Emmy Noether: Proceedings of the Conference, Bar-Ilan University, Ramat-Gan, Israel, December 2-3, 1996*. Ed. Mina Teicher. Israeli Mathematics Conference Proceedings. Vol. 12. Ramat-Gan: Bar-Ilan Univ., The Emmy Noether Research Institute of Mathematics, pp. 39-55.

Siegmund-Schultze, Reinhard. 2001. *Rockefeller and the Internationalization of Mathematics Between the Two World Wars*. Basel/Boston/Berlin: Birkhäuser Verlag.

The University of Texas Record. 1899. Vol. 1. No. 3. August. University of Texas. The Center for American History. University of Texas Memorabilia Collection. James Benjamin Clark file.

Wedderburn, Joseph H. M. 1907. "On Hypercomplex Numbers." *Proceedings of the London Mathematical Society* 6, 77-118.

_____. 1921. "On Division Algebras." *Transactions of the American Mathematical Society* 22, 129-135.

Weyl, Hermann. 1935. "Emmy Noether." *Scripta Mathematica* 3, 201-220.

Zelinsky, Daniel. 1973. "A. A. Albert." *American Mathematical Monthly* 80, 661-665.

CHAPTER 9

Emmy Noether's 1932 ICM Lecture on Noncommutative Methods in Algebraic Number Theory

Charles W. Curtis
University of Oregon, United States

Introduction

This chapter provides an historical sketch covering roughly the eight-year period from 1926 to 1934 of the pursuit by Emmy Noether and others of the idea of using noncommutative methods—especially the theory of central simple algebras—in commutative algebra and, in particular, in algebraic number theory. This idea was the theme of Noether's plenary lecture, on 7 September, 1932, at the International Congress of Mathematicians (ICM) held in Zürich, entitled "Hyperkomplexe Systeme in ihren Beziehungen zur kommutativen Algebra und der Zahlentheorie [Hypercomplex Systems and Their Connections to Commutative Algebra and Number Theory]" [Noether, 1932]. The main points in her lecture, and related contributions by Noether and others, will be the focus of the discussion to follow. The principals, besides Noether herself, were Richard Brauer (in Königsberg) and Helmut Hasse (in Halle and Marburg), both of them more than fifteen years younger than Noether.[1]

At the time of her lecture, Noether held an honorary or unofficial professorship (as "nicht-beamteter ausserordentlicher Professor mit Lehrauftrag") at the University of Göttingen, a position she had held since soon after she became a Privatdozent.[2] By the mid-1920s, she had published her great papers on the foundations of commutative algebra and had turned to noncommutative algebras and representations of finite groups beginning with the course on "Gruppentheorie und hyperkomplexe Zahlen" she gave at Göttingen in 1924–1925. In the years immediately following, she gave courses regularly on different aspects of noncommutative algebra. A major work [Noether, 1929a] on the structure of hypercomplex systems and representation theory, based on her lectures during the Winter Semester 1927–1928 and written up for publication by Bartel L. van der Waerden, was published in 1929. By 1927, she had begun to collaborate with Brauer and Hasse (see below) and, a few years later, joined them as one of the invited speakers at a conference on algebras

---

[1]References to their correspondence are based on letters from Noether to Brauer [AMCCL], and letters to Hasse from Brauer and Noether [NSUB].

[2]Her position was no more than a title, with no obligations and no salary, and was accompanied by a Lehrauftrag for algebra, providing a small salary.

held at Marburg, 26-28 February, 1931.[3] The title of her lecture at the Marburg conference was "Hyperkomplexe Struktursätze mit zahltheoretischen Anwendungen [Hypercomplex Structure Theorems with Number-theoretical Applications]." The following year, she shared with Emil Artin the "Alfred-Ackermann-Teubner Gedächtnispreis zur Förderung der mathematischen wissenschaftlichen Leistungen [Alfred Ackermann-Teubner Memorial Prize for the Advancement of the Mathematical Sciences]," awarded by the Deutsche Mathematiker-Vereinigung (DMV) for the totality of her scientific achievements. The prize, and the international recognition represented by the invitation to give one of the major addresses at the Zürich Congress, were events occurring in the year she celebrated her fiftieth birthday.

Her 1932 Congress lecture began:

> During the past few years the theory of hypercomplex systems, or algebras, has had a strong upswing, but it is only very recently that the significance of this theory for commutative problems has become clear. Today I should like to comment on the significance of the noncommutative for the commutative, and in particular I want to do this in the light of two classical problems going back to Gauss, the principal genus theorem and the closely related norm principle. The formulation of these problems has undergone continuous change. With Gauss they appear as the conclusion of his theory of quadratic forms; then they play an essential role in the characterization of relative cyclic and abelian number fields via class field theory;[4] and finally they manifest themselves as theorems on automorphisms and on the splitting of algebras, and this last formulation allows the theorems to be extended to arbitrary Galois extensions of number fields [Noether, 1932, p. 189].[5]

The last point expressed the hope that the noncommutative methods would play a part in extending class field theory to the case of Galois extensions of algebraic number fields with nonabelian Galois groups. She described it as the "Prinzip der Anwendung des Nichtkommutativen auf das Kommutative [principle of the application of the noncommutative to the commutative]": "... By means of the theory of algebras, one tries to obtain invariant and simple formulations of known facts

---

[3]Mathematisches Kolloquium an der Universität Marburg: "Vortragsreihe über hyperkomplexe Systeme." For the program of the meeting, see *Jahresbericht der Deutschen Mathematiker-Vereinigung* 41 (1932), 16.

[4]It is worthwhile to recall the meaning of some of the terms. A "number field" or "algebraic number field" is a finite extension field of the field of rational numbers. "Class field theory" in what follows is the description of finite abelian extensions of an algebraic number field K (Galois extensions having an abelian Galois group) in terms of the arithmetic of the field K.

[5]"Die Theorie der hyperkomplexen Systeme, der Algebren, hat in den letzen Jahren einen starken Aufschwung genommen; aber erst in allerneuester Zeit ist die Bedeutung dieser Theorie für kommutative Fragestellungen klar geworden. Über diese Bedeutung des Nichtkommutativen für das Kommutative möchte ich heute berichten: und zwar will ich das im einzelnen verfolgen an zwei klassischen, auf Gauss zurückgehenden Fragestellungen, dem Hauptgeschlechtsatz und dem eng damit verbundenen Normensatz. Diese Fragestellungen haben sich im Laufe der Zeit in ihrer Formulierung immer weider gewandelt: bei Gauss treten sie auf als Abschluss seiner Theorie der quadratischen Formen; dann spielen sie eine wesentliche Rolle in der Charakerisierung der relativ zyklischen und abelschen Zahlkörper durch die Klassenkörpertheorie, und schliesslich lassen sie sich ausprechen als Sätzes über Automorphismen und über das Zerfallen von Algebren, und diese letzere Formulierung gibt dann zugleich eine Übertragung der Sätze auf beliebige relativ galoissche Zahlkörper." Christina M. Mynhardt's translation.

on quadratic forms or cyclic fields, i.e. that depend only on structural properties of the algebras. Once one has obtained these formulations—as in the case of the examples mentioned above—these facts automatically carry over to arbitrary Galois extension fields."[6] An important part of the background of the lecture was research on the theory of central simple algebras by Brauer and Noether, first done independently, and then in a collaboration that brought Hasse into their circle of ideas. This came about in the following way.

Richard Brauer's investigation of the structure of simple algebras was based on representation theory and his theory of factor sets and was published in his *Habilitationsschrift* [Brauer, 1928] at the University of Königsberg and related papers, beginning in 1926. Noether developed her own direct approach to central simple algebras and began to correspond with Brauer. In 1927, they realized that they had independently arrived at the concept of a splitting field of a central simple algebra. One of Noether's statements in a letter to Brauer was incorrect, and at a meeting of the Deutsche Mathematiker-Vereinigung held at Kissingen in September of 1927, Brauer showed her a counterexample. They decided to publish a joint paper on the subject [Brauer and Noether, 1927]. The paper contained examples to show that the degrees of minimal splitting fields of a central simple algebra may be unbounded. For these examples, they used results from algebraic number theory supplied by Hasse.[7]

Following their interaction in connection with the Brauer-Noether paper, the three of them began an intense period of collaborative research, culminating in a proof in 1931 of the theorem that all central division algebras over algebraic number fields are cyclic algebras [Brauer-Hasse-Noether, 1932]. Cyclic algebras, or Dickson algebras, as Joseph Wedderburn called them, were first defined by Leonard Dickson at Chicago in an announcement in 1906. The Brauer-Hasse-Noether Theorem solved what was considered at the time to be a central open problem in the theory of algebras: the classification of rational division algebras. A former Ph.D. student of Dickson at Chicago, A. Adrian Albert, had also been working towards a proof of the result and in a note added to their paper, Brauer, Hasse, and Noether acknowledged [Brauer-Hasse-Noether, 1932, p. 400] that Albert had independently obtained results which gave him "a share [ein unabhängiger Anteil]" of the proof of the main theorem (see also [Albert and Hasse, 1932]).

The proof of the Brauer-Hasse-Noether Theorem contained several fundamental connections between the theory of central simple algebras and algebraic number theory. Upon learning about them, Emil Artin wrote to Hasse that "... you cannot imagine how ever so pleased I was about the proof, finally successful, for the cyclic systems. This is the greatest advance in number theory of the last years" [NSUB (Roquette's translation)].[8]

---

[6]"Man sucht vermöge der Theorie der Algebren invariante und einfache Formulierungen für bekannte Tatsachen über quadratische Formen oder zyklische Körper zu gewinnen, d. h. solche Formulierungen, die nur von Struktureigenschaften der Algebren abhängen. Hat man einmal diese invarianten Formulierungen bewiesen—und das ist in den oben angegebenen Beispielen der Fall—so is damit von selbst eine Übertragung dieser Tatsachen auf beliebige galoissche Körper gewonnen." See [Noether, 1932, p. 189].

[7]See [Curtis, 1999, pp. 226-228], and [Roquette, 2004, pp. 51-54] for discussions of their correspondence related to the Brauer-Noether paper.

[8]The letter is believed to have been sent in November of 1931.

Noether's vision of the applicability of noncommutative methods to commutative algebra included more than the connections with algebraic number theory in the Brauer-Hasse-Noether paper. For example, in 1930, she received a preprint of a paper of Brauer [Brauer, 1932], in which he showed that from a "Galois theory" of simple algebras, he was able to give a new proof of the basic results in the Galois theory of fields. When Noether received the preprint, she wrote:

> I have enjoyed very much your noncommutative proof of commutative Galois theory. I believe anyhow that the noncommutative methods will with time penetrate more and more into the theory of commutative algebras. In this direction, Deuring [Noether's doctoral student], whose theory of class fields of algebraic functions I reported on in our discussions in Königsberg and who worked out the material of my lecture, has given a noncommutative foundation for the theory of norm residues. ...Only with such direct noncommutative proofs [referring to Brauer's results] can one really speak of subordination of the commutative to the noncommutative.[9]

In a letter to Hasse, Noether wrote, "What I know about the connection between hypercomplex algebra and class field theory is very modest and entirely formal."[10] Nevertheless the letter contained a sketch of some ideas about a general form of a principal genus theorem, which she was eventually able to complete and include as one of the main points in her Congress lecture and in her paper [Noether, 1933]. The ideas underlying her statement about the norm principle in her lecture evolved from a course she gave at Göttingen in the Winter Semester 1929–1930, in which she presented her theory of crossed product algebras, and in her joint work with Brauer and Hasse leading to their paper [Brauer-Hasse-Noether, 1932]. I consider now some of the mathematics involved in these developments, beginning with the background that led to the Brauer-Noether paper.

Brauer's Factor Sets, the Brauer Group, and Crossed Products

Brauer's approach to the theory of central simple algebras was based on a theory of factor sets associated with absolutely irreducible matrix groups over a field.[11] Here, we give only a brief outline.[12] Brauer's starting point [Brauer, 1928] was a continuation of an article by Schur [Schur, 1909] in which his theory of the Schur index for representations of finite groups [Schur, 1906] was extended to what he called irreducible *matrix groups*. In their terminology, a matrix group $\mathfrak{H}$ was

---

[9] "Ihr nichtkomm. Beweis der komm. Galois'schen Theorie hat mir viel Vergnügen gemacht. Ich glaube überhaupt, dass mit der Zeit die nichtkomm. Methoden immer stärker auch in der kommutative Algebren eindringen werden. So hat Deuring—über dessen Klassenk örpertheorie der algebr. Funkt. ich in Königsberg in der diskussion berichtete, und der die Vorlesung ausgearbeitet hat—die Normenresttheorie ...nichtkommutativ begr ündet. ...Erst bei solchen direkt nichtkomm. Beweisen kann man eigentlich von Unterordnung des Komm. unter das Nichtkomm. sprechen." See [AMCCL, 26 October, 1930].

[10] "Vorerst—was ich über den Zusammenhang von hyperkomplexer Algebra und Klassenkörpertheorie weiss, ist sehr bescheiden und ganz formal ...." See [NSUB, 2 October, 1929].

[11] In this chapter, fields are assumed to be subfields of the field of complex numbers. In his research, Brauer considered algebras over perfect fields.

[12] A full account, with references and proofs of the main results can be found in [Curtis, 1999, pp. 228-234].

a set of matrices with complex entries such that the product of any two matrices in $\mathfrak{H}$ belongs to $\mathfrak{H}$, in other words, a monoid consisting of matrices, as we would say today. Irreducibility and equivalence were defined as for representations of finite groups. A field L containing a base field K is called a *splitting field* for an absolutely irreducible matrix group $\mathfrak{H}$, if $\mathfrak{H}$ is L-rational, in the sense that $\mathfrak{H}$ is equivalent to a matrix group consisting of matrices all of whose entries belong to L. A splitting field L of $\mathfrak{H}$ contains the field $K(\chi)$ generated by the character values $\chi(a) = \text{Trace}(a)$, for $a \in \mathfrak{H}$. The *Schur index* of an absolutely irreducible matrix group $\mathfrak{H}$ with respect to a field K is the minimum dimension $(L : K(\chi))$, for all splitting fields L of $\mathfrak{H}$ [Schur, 1909, p. 168].[13]

Central simple algebras arose in Brauer's set-up as follows. Let K be a field, and let $\mathfrak{H}$ be an absolutely irreducible matrix group with the property that K contains all the character values $\chi(a)$, for $a \in \mathfrak{H}$. Then the set of all K-linear combinations of elements of $\mathfrak{H}$ is a central simple algebra over K, that is, a simple algebra whose center consists of the scalar multiples of the identity by elements of K [Brauer, 1929, p. 94].

Noether approached central simple algebras and splitting fields more directly. Let A be a central simple algebra over K. An extension field L of K was called a splitting field of A whenever the algebra $A_L = L \otimes_K A$ obtained by extension of the field from K to L is isomorphic over L to a total matrix algebra $M_d(L)$ for some d. The representation theory of central simple algebras and Brauer's theorem stated above show that this definition is consistent with the notion of a splitting field for an absolutely irreducible matrix group introduced previously.

One of the main results in the joint paper, [Brauer and Noether, 1927, p. 223], can be stated as follows:

THEOREM 1. *(i) Let D be a central division algebra over a field K. Then the dimension $(D : K) = m^2$ for some positive integer m. Moreover, m is the degree of the unique absolutely irreducible representation of D, and is the Schur index of this absolutely irreducible representation with respect to K.*

(ii) A central division algebra D, of dimension m^2 over K, has splitting fields of dimension m over K; these are the splitting fields of D of minimal degree over K and are realized as maximal subfields of D. The Schur index divides the dimension $(L : K)$ of an arbitrary splitting field L of D.

(iii) The splitting fields of D are the same as the splitting fields of central simple algebras of the form $M_r(D)$, for positive integers r.

With the intervention of Hasse, they were able to give examples to show that the dimensions over the base field of minimal splitting fields of certain central division algebras are unbounded [Brauer and Noether, 1927, p. 225].

The next step in these developments was Brauer's discovery of the abelian group of classes of central simple algebras: the *Brauer group*. In his *Habilitationsschrift* [Brauer, 1928], Brauer introduced the concept of a factor set c_{ijk} associated with a central simple algebra A (or an absolutely irreducible matrix group). The elements of a factor set were complex numbers in the Galois closure of a fixed splitting field L of A, and indexed by integers i, j, k between 1 and n, where n is the dimension of the Galois closure. Brauer showed that the elements of a factor set satisfy what we would call today a cocycle condition. Using them, he was able to describe the

---

[13]See also [Schur, 1906, p. 174].

structure of the algebra A, and he planned to use factor sets to obtain information about the Schur index of A. He proved that if A and B are central simple algebras over K, with factor sets c_{ijk} and d_{ijk} associated with them, then $A \otimes B$ is a central simple algebra, associated with the factor set $c_{ijk}d_{ijk}$. Moreover, letting A be a central simple algebra, associated with the factor set c_{ijk}, the opposite algebra A' is a central simple algebra associated with the factor set c_{ijk}^{-1} and $A \otimes A' \cong M_d(K)$, for some integer d, a total matrix algebra over K [Brauer, 1929, p. 103].

Using these results, Brauer was able to introduce the structure of an abelian group on the set of equivalence classes of central simple algebras over a field K. Two central simple algebras A and B over K are said to belong to the same class if the division algebras associated with them by Wedderburn's theorem are isomorphic; this is clearly an equivalence relation. Brauer defined the product of two classes $\{A\}$ and $\{B\}$ by $\{A\}\{B\} = \{A \otimes B\}$, and noted that the set of classes of central simple algebras over K, which we shall denote by $B(K)$, is an abelian group [Brauer, 1932, p. 243], generally known soon after he defined it as the Brauer group. The identity element in $B(K)$ is the class of total matrix algebras over K, and the inverse of a class $\{A\}$ is given by the class of the opposite algebra $\{A'\}$, by the results mentioned above.

We shall use the notation (from the Brauer-Hasse-Noether paper) $A \sim B$ to mean that $\{A\} = \{B\}$ in $B(K)$. The Schur index of a central simple algebra A coincides with the Schur index of the division algebra in the class of A and is the same for all central simple algebras in the class of A. Using the theory of factor sets, Brauer proved the kind of result he had hoped to find connecting the theory of factor sets to the Schur index, namely, that each element $\{A\}$ in the Brauer group $B(K)$ has finite order dividing the Schur index of A [Brauer, 1932, p. 243]. The order of an element $\{A\}$ in the Brauer group $B(K)$ is called the exponent of A, so the result stated above due to Brauer and Noether states that the exponent of a central simple algebra divides the Schur index.

In the Winter Semester 1929–1930, Noether gave a course at Göttingen with the title "Algebra der hyperkomplexen Grössen [Algebra of Hypercomplex Numbers]" [Noether, 1929]. Notes of the course were written up by her student, Max Deuring, and were distributed at the time. The last two chapters of the lecture notes contained a survey of Brauer's theory of factor sets, followed by her own theory of "crossed products [verschränkte Produkte]." The first published version of Noether's theory of crossed products appeared in 1932, with Noether's permission, in a paper by Hasse published in English in the *Transactions of the American Mathematical Society* in 1932 and entitled "Theory of Cyclic Algebras over an Algebraic Number Field" [Hasse, 1932]. In the introduction to the paper, Hasse explained why he published the paper in English in an American journal: "The theory of linear algebras has been greatly extended through the work of American mathematicians. Of late German mathematicians have been active in this theory. In particular they have succeeded in obtaining some apparently remarkable results by using the theory of algebraic numbers, ideals, and abstract algebra, highly developed in Germany in recent decades. These results do not seem to be as well known in America as they should be on account of their importance. This fact is due, perhaps, to the language difference or to the unavailability of the widely scattered sources" [Hasse, 1932, p. 171].

The starting point in the construction of Brauer's factor sets was an arbitrary splitting field $K(\theta)$ of a central simple algebra A over K. In her theory of crossed products, Noether showed [Noether, 1929, pp. 39-41] that if one starts instead with a Galois extension field L of K with Galois group $\mathcal{G}$, one is led to a clearer description of central simple algebras A having L as a splitting field. The action of elements of $\mathcal{G}$ on L, given by $\ell \to \ell^S$ for $\ell \in L$ and $S \in \mathcal{G}$, is realized by inner automorphisms of A, and the multiplicative structure of A is determined by a new kind of factor set, which she called a "small factor set [kleines Faktorensystem]."

In more detail, consider a vector space A over L with a basis $\{u_S\}$ over L parametrized by the elements $S \in \mathcal{G}$. Noether defined a multiplication on A by setting

$$\ell u_S = u_S \ell^S \ (\text{or } u_S^{-1} \ell u_S = \ell^S), \ u_S u_T = u_{ST} a_{S,T},$$

for $\ell \in L$, $S, T \in \mathcal{G}$, and for some nonzero elements $a_{S,T} \in L$. She observed that A becomes an associative algebra over K provided that $u_R(u_S u_T) = (u_R u_S) u_T$, or that the elements of the factor set $a_{S,T}$ satisfy what we recognize today as another cocycle condition, $a_{R,ST} a_{S,T} = a_{RS,T} a_{R,S}^T$, for all $R, S, T \in \mathcal{G}$ [Noether, 1929, p. 45]. The following basic result was proved in Hasse's paper. In the crossed product algebra $A = (a, L)$ defined above, the field L is a maximal subfield of A. Moreover, A is a central simple algebra, and L is a splitting field of A [Hasse, 1932, p. 184].

The basis elements u_S in a crossed product are not uniquely determined. If v_S is another set of elements of A such that $v_S^{-1} \ell v_S = \ell^S$, for all $\ell \in L$, then $u_S^{-1} v_S$ commutes with all the elements of L, so that by the previous theorem, $v_S = u_S c_S$ for some nonzero elements $c_S \in L$. Then $v_S v_T = v_{ST} b_{S,T}$ with $b_{S,T} \in L$, for all $S, T \in \mathcal{G}$. The second factor set $b_{S,T}$ is related to the first by the conditions

$$b_{S,T} = a_{S,T} \frac{c_S^T c_T}{c_{ST}},$$

for all $S, T \in \mathcal{G}$. The relation stated above is an equivalence relation on the factor sets associated with central simple algebras with splitting field L [Hasse, 1932, p. 183].

In her lectures, Noether proved that the tensor product (called "Produkt" in [Noether, 1929, p. 44]) $A \otimes B$ of two crossed product algebras $A = (a, L)$ and $B = (b, L)$, both with splitting field L, is again a crossed product algebra with a factor set given by $a_{S,T} b_{S,T}$ for $S, T \in \mathcal{G}$. It was then not difficult to define the structure of an abelian group on equivalence classes of factor sets, and to prove the following theorem, stated in the lecture notes from her course and in her Zürich ICM lecture, and proved in [Hasse, 1932, p. 194]:

THEOREM 2. *The classes of central simple algebras over K having the property that L is a splitting field form a subgroup $B(L/K)$ of the Brauer group $B(K)$. Moreover, this subgroup is isomorphic to the group of equivalence classes of factor sets associated with crossed product algebras (a, L) with splitting field L.*

This theorem, in modern terminology, states that $B(L/K) \cong H^2(G, L^*)$, where the latter is the second cohomology group of $\mathcal{G}$ with coefficients in the multiplicative group L^* of L. It is one of the sources, along with Hilbert's Theorem 90 [Hilbert, 1897, 1:149] and Noether's Principal Genus Theorem, also presented in her ICM lecture, of the subject known today as Galois cohomology (see below).

Cyclic Algebras and the Albert-Brauer-Hasse-Noether Theorem

In an announcement [Dickson, 1906] and in a subsequent paper [Dickson, 1914], Dickson defined a special kind of central simple algebra, called a cyclic algebra, as follows. Let K be a field, and L a finite Galois extension of K with a cyclic Galois group $\mathcal{G}$ of order n, generated by S. A cyclic algebra $A = (\alpha, L, S)$ is a vector space over L of dimension n^2, generated by elements $1, u, u^2, \ldots$ and satisfying the relations $u^n = \alpha \in K$ and $\ell u = u \ell^S$, for all $\ell \in L$. He stated [Dickson, 1914, p. 32], and Wedderburn proved at about the same time [Wedderburn, 1914, p. 166] that a cyclic algebra is a division algebra if α^n is the smallest power of α which is the norm of an element of L. He proved that central division algebras over K of dimension n^2 are cyclic, if $n = 2$ [Dickson, 1914, p. 32], and Wedderburn proved the corresponding result if $n = 3$ [Wedderburn, 1921]. Albert handled the more difficult case $n = 4$ in his dissertation at Chicago [Albert, 1932, p. 372]. Dickson gave a detailed account of what was known about cyclic algebras at the time in his book, *Algebras and Their Arithmetics* [Dickson, 1923]. A German translation of the book published in 1927 had an immediate impact among European mathematicians working in the theory of algebras and was cited in Noether's Zürich lecture [Noether, 1932, p. 189] as a source of the notion of crossed product algebra. There was general agreement that a central problem in the theory of algebras was the classification of rational division algebras, or division algebras of finite dimension over the field of rational numbers. The center of such an algebra is an algebraic number field, so the problem is equivalent to the problem of classifying central division algebras over number fields.[14]

In their paper, [Brauer-Hasse-Noether, 1932], Brauer, Hasse, and Noether proved the theorem that central division algebras over algebraic number fields are cyclic algebras, with Albert hard on their heels.[15] The proof of the theorem was based on p-adic methods in algebraic number theory, pioneered by Kurt Hensel, and was an application of what is called today the Hasse Principle, or the Local-Global Principle for algebras over algebraic number fields (see below for a precise statement).[16] The Local-Global Principle for cyclic algebras, stated in the form of a criterion for a cyclic algebra to represent the unit class in the Brauer group [Hasse, 1932, p. 179 (Theorem 3)], was one of the main results in Hasse's *Transactions* paper. There, he also showed that for a central simple cyclic algebra, the exponent and index coincide [Hasse, 1932, p. 179 (Theorem 5)].

The Brauer-Hasse-Noether paper was dedicated to Hensel and was written by Hasse in time to appear in a volume of the *Journal für die reine und angewandte Mathematik* (Crelle's *Journal*) to be presented to Hensel on his seventieth birthday. Emmy Noether commented on the text of the dedication of their paper to Hensel in a letter to Hasse: "Of course I agree with the bow to Hensel. My methods are useful and conceptual methods and therefore have anonymously penetrated

---

[14]The importance of cyclic algebras for this problem was clearly understood from the beginning. They provided the only known examples of central division algebras over algebraic number fields. In his paper [Wedderburn, 1914, p. 162], Wedderburn wrote: "So far as I am aware no algebra other than these two [the cases $n = 2$ and $n = 3$ mentioned above] and fields has been proved to be primitive [a division algebra]; hence it is of considerable interest to find that for any value of n, θ, and g [S and α in the definition of cyclic algebras given above] can be chosen to make Dickson's algebra primitive."

[15]For this story, see Fenster's chapter above.

[16]Compare Schwermer's chapter above for the early history of this theorem.

everywhere."[17] In his historical analysis of the Brauer-Hasse-Noether Theorem, Peter Roquette wrote:

> The second sentence of this comment has become famous in the Noether literature. It puts into evidence that she was very sure about the power and success of "her methods" which she describes quite to the point. But why did she write the sentence just here, while discussing the dedication text for Hensel? The answer which suggests itself is that, on the one hand, Noether wishes to express to Hasse that, after all, "her methods" (as distinguished from Hensel's p-adic methods) were equally responsible for their success. On the other hand, she does not care whether this is publicly acknowledged or not.
>
> In the present connection, "her methods" means two things. First, she had insisted that the classical representation theory be done in the framework of the abstract theory of algebras (or hypercomplex systems in her terminology) instead of matrix groups and semi-groups as Schur had started it. Second, she had strongly proposed that the noncommutative theory of algebras should be used for a better understanding of commutative algebraic number theory.
>
> Perhaps we should add a third aspect of "her methods": the power to transmit her ideas and concepts to people around her. In this way she had decisively influenced Richard Brauer's and Helmut Hasse's way of thinking: Brauer investigated division algebras and Hasse did noncommutative arithmetic [Roquette, 2004, p. 12].

It was not immediately clear that central division algebras over number fields are cyclic algebras. Brauer had already published an example of a noncyclic central division algebra, but the center was not an algebraic number field [Brauer, 1930, p. 745]. In a letter to Hasse dated 19 December, 1930 [NSUB], Noether wrote that she believed that some recent work of Albert might lead to a central simple algebra whose exponent was less than the index and that such an algebra could not be cyclic, by a result in Hasse's *Transactions* paper. In another letter written five days later, on Christmas Eve, evidently responding to a message from Hasse that she was wrong, Noether explained that she had misinterpreted Albert's result, and said she now believed it entirely likely that for a central simple algebra A over an algebraic number field, the Schur index and the exponent of the class of A in the Brauer group should coincide [NSUB].[18]

The two main steps in the proof of the theorem that all central division algebras over number fields are cyclic were the Local-Global Principle for algebras, and an existence theorem for cyclic extension fields of an algebraic number field with certain properties, known today as the Grunwald-Wang Theorem. Let A be a central simple algebra over an algebraic number field K. For each prime $\mathfrak{p}$ of K, finite or infinite, let $K_\mathfrak{p}$ be the completion of K at $\mathfrak{p}$. Then $K_\mathfrak{p}$ is an extension field of K, and it was known in general that the algebra $A_{K_\mathfrak{p}} = K_\mathfrak{p} \otimes A$ obtained by extension of the base

---

[17]"Mit der Verbeugung vor Hensel bin ich selbstverständlich einverstanden. Meine Methoden sind Arbeits—und Auffassungsmethoden, und daher anonym überall eingedrungen." See [NSUB, 12 November, 1931].

[18]For commentary on the Brauer-Hasse-Noether correspondence as they were working on their joint paper, see [Roquette, 2004].

field from K to $K_\mathfrak{p}$, and denoted in what follows by $A_\mathfrak{p}$, is a central simple algebra over $K_\mathfrak{p}$.

THEOREM 3 (Local-Global Principle for algebras). *A central simple algebra A over K splits, that is, $A \sim K$ in the Brauer group, if and only if it splits everywhere locally, that is, $A_\mathfrak{p} \sim K_\mathfrak{p}$ for all primes $\mathfrak{p}$, finite or infinite, in K.*

In their paper, [Brauer-Hasse-Noether, 1932, p. 399], Brauer, Hasse, and Noether proved the Local-Global Principle for arbitrary central simple algebras by a series of reduction steps to the case of a cyclic algebra, where the result had already been proved by Hasse in his *Transactions* paper mentioned earlier. These steps had essentially been obtained independently by Albert and had been communicated in a letter to Hasse, as Brauer, Hasse, and Noether explained in a footnote to their paper [Brauer-Hasse-Noether, 1932, p. 400]. At this point, both parties knew something of each others' efforts. But communication overseas was slow, and each party was unaware of important progress by the other at certain stages in the investigation. An attempt to sort out the communicaton, or lack of it, and to present a summary of the joint work of all four on the classification of rational division algebras, was published in 1932 in a joint paper by Albert and Hasse [Albert and Hasse, 1932] in the same volume of the *Transactions* as Hasse's paper [Hasse, 1932].[19]

The proof of the Local-Global Principle for cyclic algebras required a preliminary result, which is stated in two parts. Let A be a cyclic algebra (α, L, S) over K. Then for each prime $\mathfrak{p}$ in K, and for some prime $\mathfrak{P}$ in L extending $\mathfrak{p}$, $L_\mathfrak{P}$ is a cyclic extension of $K_\mathfrak{p}$. Moreover, $A_\mathfrak{p}$ is a cyclic algebra $(\alpha, L_\mathfrak{P}, \sigma)$, where σ is a generator of the Galois group of $L_\mathfrak{P}$ over $K_\mathfrak{p}$. The second part is the assertion that a cyclic algebra (α, L, S) over K is split over K, that is, $(\alpha, L, S) \sim K$, if and only if $\alpha = N_{L/K}(\ell)$ for some nonzero element $\ell \in L$. The corresponding result holds for the cyclic algebra $(\alpha, L_\mathfrak{P}, \sigma)$ over $K_\mathfrak{p}$.

By the preceding discussion, the Local-Global Principle for cyclic algebras is equivalent to the following result, proved in class field theory:

THEOREM 4 (Hasse Norm Theorem). *Let L be a finite cyclic extension of an algebraic number field K. For each prime $\mathfrak{p}$ of K, let $\mathfrak{P}$ be a prime of L extending $\mathfrak{p}$. Then $\alpha \in N_{L/K}(L)$, if and only if $\alpha \in N_{L_\mathfrak{P}/K_\mathfrak{p}}(L_\mathfrak{P})$, for each $\mathfrak{p}$.*

At the meeting of the DMV in September, 1925, Hasse gave a report on class field theory. A written version of the report was commissioned by the DMV at Hilbert's suggestion, and was undertaken by Hasse as a follow-up to Hilbert's monumental report [Hilbert, 1897], which had also been commissioned by the DMV.[20] In his *Transactions* paper containing a proof of the Local-Global Principle for cyclic algebras, Hasse wrote that in Part II of his report on class field theory, published in 1930, he had been able to prove the Norm Theorem only for cyclic extensions of prime order, but that, "inspired by the important application in the theory of cyclic algebras developed in this paper, I succeeded recently in proving the result for general degree n" [Hasse, 1932, p. 175 (footnote)].

The "norm principle" announced in Noether's ICM talk in Zürich was simply the remarkable fact that the Local-Global Principle for central simple algebras was equivalent to Hasse's Norm Theorem for cyclic extensions of algebraic number fields.

---

[19]Compare the discussion in Fenster's chapter above.
[20]See [Roquette, 2001, p. 555].

The latter statement, she wrote, "is the norm principle proved in class field theory by using well-known analytical tools. And the proof of the general theorem on splitting algebras can be obtained from this cyclic special case by pure algebraic-arithmetic considerations."[21]

The proof of the theorem that central division algebras over algebraic number fields are cyclic algebras also gives the corresponding theorem for central simple algebras over number fields, and is today credited to Albert, Brauer, Hasse, and Noether. Here is an outline of the proof. The theory of crossed product algebras showed that an arbitrary central simple algebra A over a number field K is a cyclic algebra if it has a splitting field which is a finite cyclic extension of K whose degree is the Schur index of A with respect to K. The Local-Global Principle for algebras gave a numerical criterion for an extension field L of K to be a splitting field, namely, that for each prime $\mathfrak{p}$ of K, the Schur index $m_\mathfrak{p}$ of the algebra $A_\mathfrak{p}$ divides the degree $(L_\mathfrak{P} : K_\mathfrak{p})$ for each prime $\mathfrak{P}$ of L extending $\mathfrak{p}$.

Now let A be a central simple algebra over a number field K. The proof of the theorem that A is a cyclic algebra came down to an existence problem, namely, whether there exists a cyclic extension field L of K satisfying the numerical criteria for L to be a splitting field for A and whose degree $(L : K)$ is the Schur index of A with respect to K. In their paper [Albert and Hasse, 1932, p. 723], the authors stated: "At the time (April, 1931) when Hasse presented his paper [Hasse, 1932] to these Transactions he had also outlined a proof of the following existence theorem. Let A be a normal simple algebra of degree[22] n over F. Then there exists a cyclic field C of the same degree n over F such that A_C splits everywhere." The existence theorem stated above, from the Albert-Hasse paper, combined with the Local-Global Principle for algebras, implied the main theorem. At the time, the Brauer-Hasse-Noether paper was submitted, however, the required existence theorem had been proved in a sharper form by a Ph.D. student of Hasse, Wilhelm Grunwald, based on his dissertation [Grunwald, 1932].[23] The existence theorem completed the proof of the main result.

THEOREM 5 (Albert-Brauer-Hasse-Noether). *Let A be a central simple algebra over an algebraic number field K. Then A is a cyclic algebra.*

But there was a further development. In 1948, a graduate student, Shianghaw Wang, at Princeton University, writing his dissertation under the supervision of Emil Artin, published a short note in Volume 49 of the *Annals of Mathematics* which opened: "In recent research, the author of this note found that Grunwald's Theorem on the existence of cyclic extensions over an algebraic number field was not always true. On the other hand he has obtained a proof of the theorem under appropriate restrictions. Here a counter-example will be given and existing proofs analyzed" [Wang, 1948, p. 1008]. John Tate, a participant along with Wang in

---

[21]"Dieser letztere Fassung ist aber der Normensatz, der in der Klassen körpertheorie bewiesen wird, unter Benutzung der bekannten analytischen Hilfsmittel. Und der Beweis des allgemeinen Satzes über zerfallende Algebren lässt sich aus diesem zyklischen Spezialfall durch rein algebraisch—arithmetische Betrachtungen gewinnen." See [Noether, 1932, p. 193]. On the last point, Peter Roquette wrote (e-mail to the author 22 April, 2003) that the last phrase "by pure algebraic-arithmetic arguments" is to be interpreted today as "by cohomology."

[22]By "degree," Albert and Hasse meant "Schur index."

[23]See [Roquette, 2004, pp. 29 and 30].

Artin's seminar on class field theory at Princeton in 1947-1948, has recounted[24] that late in the Spring term Bill Mills was reporting on George Whaples's paper [Whaples, 1942] containing a new proof of Grunwald's theorem when Wang found the mistake.[25]

Fortunately, Wang's version [Wang, 1950] of Grunwald's existence theorem, known today as the Grunwald-Wang Theorem, was sufficient to produce the required cyclic splitting field for each central simple algebra over an algebraic number field, and filled the gap in the proof of the Albert-Brauer-Hasse-Noether Theorem.

The Principal Genus Theorem

As a second illustration of noncommutative methods in algebraic number theory, Noether gave an outline in her ICM lecture of her version of a principal genus theorem in the abelian group of classes of fractional ideals in a finite Galois extension L of an algebraic number field K. A full account was published the following year [Noether, 1933]. Concerning it, Albrecht Fröhlich wrote:

> The main aim of [Noether, 1933] is a new proof, based on the theory of algebras, and a far reaching generalization of the principal genus theorem of class field theory, which deals with cyclic extensions and which in turn had its roots in Gauss's principal genus theorem on rational quadratic forms. The basic program for Noether's work on this topic, contained in her Congress lecture [Noether, 1932], is to reformulate results of this type in terms of simple algebras, i.e. of crossed products where the statement makes sense for arbitrary rather than just cyclic Galois groups, and then to prove them in this much greater generality. This outlook puts Noether well ahead of her time. It was one of the main successes of the cohomological theory in later years to show systematically that part of class field theory really deals with arbitrary Galois extensions rather than just abelian ones [Fröhlich, 1981, p. 160].

In her Congress lecture [Noether, 1932, p. 189], the "principle of the application on the noncommutative to the commutative" quoted in the introduction to this chapter above and described by Fröhlich in the preceding remarks, expressed Noether's plan for extending at least parts of class field theory to the case of general Galois extensions of number fields. In his report on a nonabelian generalization of the reciprocity law of local class field theory known as the Local Langlands Correspondence, Jonathan Rogawski wrote that "[o]ne gets a sense of the difficulty of the problem [of developing a nonabelian class field theory] from a remark of Weil ([Weil, 1979, 3:457] written in 1971) to the effect that E. Noether, E. Artin, and H. Hasse had hoped in vain that their theory of simple algebras would lead to a nonabelian theory but that by 1947 Artin confided that he was no longer sure that such a theory existed" [Rogawski, 2000, p. 36].

The concepts of genus, and principal genus, were first introduced by Gauss in Section V of his *Disquisitiones arithmeticæ*, where he divided classes of rational

---

[24]E-mail to the author, 24 March, 2003.

[25]The author of this chapter was a graduate student at Yale at the time and remembers very well the sensation in the mathematical community caused by Wang's announcement.

quadratic forms with a given determinant into genera using his notion of character and proved that a class belonged to the principal genus containing the unit class, if and only if it is obtained by what he called duplication, so that it is a square in the abelian group of classes of quadratic forms with a given determinant.

After recalling the background from Gauss's theory of quadratic forms, Noether, in [Noether, 1933, p. 414], presented her first result, a generalization of Hilbert's Theorem 90 from the report [Hilbert, 1897]. Hilbert's Theorem 90 states that in case L is a cyclic extension field of K, and S is a generator of the Galois group of L over K, then $N_{L/K}(a) = 1$ for an element a in the multiplicative group L^* of L, if and only if $a = b^{1-S}$, for some element $b \in L^*$. Her generalization, which she called "Hauptgeschlechtsatz im Minimalen [Principal Genus Theorem–Minimal Version]," was a purely formal algebraic result requiring nothing from the arithmetic of number fields. Letting c_S be a set of elements from the multiplicative group L^* of a finite Galois extension L of an algebraic number field K with Galois group $\mathcal{G}$ parametrized by the elements $S \in \mathcal{G}$, and satisfying the condition $c_R^S c_S / c_{RS} = 1$, for all $R, S \in \mathcal{G}$, the theorem states that there exists an element $b \in L^*$ such that $c_S = b^{1-S}$, for all $S \in \mathcal{G}$.

The equations $c_R^S c_S = c_{RS}$ are known as the Noether equations, so the point of the theorem was to give a solution for them. As Fröhlich said, this theorem was later interpreted in cohomological terms, namely, that the first cohomology group $H^1(\mathcal{G}, L^*)$ is zero [Fröhlich, 1981, p. 161].

The second point in her Zürich lecture [Noether, 1932, p. 193], and the main theorem in [Noether, 1933, p. 417], was what she called a principal genus theorem for the group of fractional ideals $\mathfrak{J}$ in a finite Galois extension L of an algebraic number field K with Galois group $\mathcal{G}$, in which the arithmetic and, in particular, the Local-Global Principle for algebras would be involved. The first step was to extend the notion of factor set used to define a crossed product algebra to the concept of an ideal factor system, namely, a system of fractional ideals $\mathfrak{a}_{S,T}$ in L parametrized by pairs S, T from $\mathcal{G}$, and satisfying the conditions

$$\mathfrak{a}_{R,ST}\mathfrak{a}_{S,T} = \mathfrak{a}_{RS,T}\mathfrak{a}_{R,S}^T,$$

for all R, S, T in $\mathcal{G}$. Equivalence of two ideal factor systems was defined by means of a set of ideals $\mathfrak{c}_S$ as before. She needed the following preliminary result, that is, the Principal Genus Theorem for ideals: For a set of ideals $\mathfrak{c}_S$, one has $\mathfrak{c}_R^S \mathfrak{c}_S / \mathfrak{c}_{RS} = 1$ in $\mathfrak{J}$, if and only if there exists an ideal $\mathfrak{b} \in \mathfrak{J}$ such that $\mathfrak{b}^{1-S} = \mathfrak{c}_S$, for all $S \in \mathcal{G}$ [Noether, 1933, p. 417].

The proof is essentially the same as for the "minimal version" and is interpreted today as a proof of the statement that the first cohomology group $H^1(\mathcal{G}, \mathfrak{J})$ vanishes. Let $\mathfrak{b} = \sum_R \mathfrak{c}_R$. Then

$$\mathfrak{b}^S \mathfrak{c}_S = \sum_R \mathfrak{c}_R^S \mathfrak{c}_S = \sum \mathfrak{c}_{RS} = \mathfrak{b},$$

and the result follows.

Noether defined the "principal class [Hauptklasse]" in the group of classes of ideal factor systems to be those ideal factor systems given by principal ideals $\mathfrak{a}_{S,T} = (a_{S,T})$, with $a_{S,T}$ a factor set in L having the property that the corresponding crossed product algebra $A = (a, L)$ is split at all ramified primes $\mathfrak{p}$ of K, that is, $A_{\mathfrak{p}} \sim K_{\mathfrak{p}}$. Her main result was:

THEOREM 6 (Principal Genus Theorem). *Let an ideal factor system of the form $\mathfrak{c}_R^S \mathfrak{c}_S / \mathfrak{c}_{RS}$ belong to the principal class. Then there exists a fractional ideal $\mathfrak{b} \in \mathfrak{J}$ such that $\mathfrak{b}^{1-S}$ belongs to the same ideal class as $\mathfrak{c}_S$, for all $S \in \mathcal{G}$.*

In case the hypothesis of the theorem holds for a set of fractional ideals $\mathfrak{c}_S$, Noether interpreted the set of ideals $\mathfrak{c}_S$ as belonging to the principal genus. One of the main steps in the proof was to show that, for a set of ideals $\mathfrak{c}_S$ belonging to the principal genus, with

$$\mathfrak{c}_R^S \mathfrak{c}_S / \mathfrak{c}_{RS} = (a_{R,S})$$

and the corresponding crossed product algebra $A = (a, L)$ split at all ramified primes, then indeed A splits at all primes, finite or infinite, so that A is a total matrix algebra, by the Local-Global Principle for algebras (Theorem 3).

In comments on Noether's paper [Noether, 1933], Peter Roquette wrote that "[t]hat paper is not easy to understand because she does not have the proper notions to deal with the problem. This is a clear example that here we have reached the limit of Noether's idea to translate everything into the language of algebras. With the formalism of cohomology we can at least understand what she has in mind."[26] Fröhlich and, more recently, Lemmermeyer gave modern versions of Noether's Principal Genus Theorem and its proof in terms of cohomology sequences involving idèle groups ([Fröhlich, 1981] and [Lemmermeyer, 2003]).

In concluding her ICM talk, Noether made the following observation showing how precisely her Principal Genus Theorem for ideals, stated above, can be viewed as an extension of Gauss's principal genus result for quadratic forms.

> In this way, however, the principal genus defined here goes over into the Gaussian because the ideal classes [in quadratic extensions of the rationals] correspond to the quadratic forms, and the norms of the classes to the numbers representable by the forms. That the unit class generates the algebras which split at the ramified places of L can therefore be expressed by saying that these representable numbers are quadratic residues at the ramified places; the associated forms thus possess the total character of the principal form, hence form the Gaussian principal genus. It is already known that the $(1-S)$th power goes over into the square.[27]

Applications to Algebraic Number Theory by Hasse and Chevalley

In a paper entitled "Die Struktur der R. Brauerschen Algebrenklassengruppe über einem algebraischen Zahlkörper. Inbesondere Begründung der Theorie des Normrestsymbols und die Herleitung des Reziprozitätsgesetzes mit nichtkommutativen Hilfsmitteln [The Structure of R. Brauer's Algebraic Class Groups over an Algebraic Number Field. In Particular the Establishment of the Theory of the Norm

---

[26] E-mail to the author, 22 April, 2003.

[27] "Damit geht aber das hier definierte Hauptgeschlecht für quadratische Körper in das Gausssche über; denn den Idealklassen entsprechen die quadratischen Formen, den Normen der Klassen die durch die Formen darstellbaren Zahlen. Dass die Einsklasse die an den Verzweigungsstellen von K zerfallenden Algebren erzeugt, heisst also dass diese darstellbaren Zahlen an den Verzweigungsstellen quadratische Reste werden; die zugehörigen Formen besitzen somit den Totalcharakter der Hauptform, bilden also das Gausssche Hauptgeschlecht. Dass die $(1-S)$te symbolische Potenz in die Duplikation übergeht, ist bekannt." See [Noether, 1932, p. 194].

Residue Symbol and the Derivation of the Reciprocity Theorem by Noncommutative Means]" [Hasse, 1933], dedicated to Emmy Noether on her fiftieth birthday, Hasse gave a criterion for a central simple algebra over an algebraic number field K to belong to a given class in the Brauer group $B(K)$ in terms of arithmetical invariants, today called Hasse invariants, and used his theory of invariants to give a new proof based on the theory of central simple algebras of Artin's reciprocity law in class field theory.

In the introduction to the paper, Hasse stated that "I will in the following develop a new proof of the reciprocity law from the arithmetic of noncommutative algebras. This fulfills a wish Emmy Noether has stated to me."[28] In her biography of Noether, Auguste Dick wrote that "[i]n the same year [1932] the Göttingen algebraists celebrated Emmy Noether's fiftieth birthday. In spite of her modesty, this sign of awareness from her colleagues did make her happy. What she enjoyed most was the paper Helmut Hasse dedicated to her in the *Mathematische Annalen*. It contained a noncommutative derivation of the law of reciprocity and confirms Noether's stated thought that the theory of noncommutative algebras is governed by simpler laws than is the theory of commutative algebras" [Dick, 1981, p. 73].

In his paper, Hasse defined the now so-called Hasse invariant $\left(\frac{A}{\mathfrak{p}}\right)$ of a central simple algebra A over an algebraic number field K, for each prime $\mathfrak{p}$ in K. The invariant $\left(\frac{A}{\mathfrak{p}}\right)$ is a rational number taken modulo 1 and was defined in terms of arithmetic properties of the p-adic division algebra associated with the central simple algebra $A_{\mathfrak{p}}$ over $K_{\mathfrak{p}}$ by Wedderburn's theory, using the fact that a p-adic division algebra was known to be a cyclic algebra. Hasse proved that the Brauer class of a central simple algebra was determined by the Hasse invariants [Hasse, 1933, p. 750 (6.51)]:

THEOREM 7. *The central simple algebras A and B over an algebraic number field K belong to the same class in the Brauer group $B(K)$, if and only if*

$$\left(\frac{A}{\mathfrak{p}}\right) \equiv \left(\frac{B}{\mathfrak{p}}\right) \pmod{1}$$

for all primes $\mathfrak{p}$ of K.

Hasse's noncommutative proof of Artin's reciprocity law was based on the following theorem [Hasse, 1933, p. 750 (6.54)].[29]

---

[28] "Ich will im folgenden aus dieser einfachen Arithmetik der nichtkommutativen Algebren einen neuen Beweis des Reziprozitätsgesetzes entwickeln. Damit erfülle ich einen mir schon vor einiger Zeit von Emmy Noether ausgesprochenen Wunsch." See [Hasse, 1933, p. 731].

[29] On this point, Peter Roquette wrote in an e-mail message to the author on 22 April, 2003, that "the essential breakthrough came already in an earlier Annalen paper of Hasse (cited as Hasse [5] in the references for [Hasse, 1933]). There Hasse had shown that locally every central simple algebra admits an unramified splitting field of minimal degree. Based on this, he could define, in his second Annalen paper [Hasse, 1933], the local norm residue symbol for cyclic local extensions by means of the structure theory of (local) algebras. This idea was in fact Emmy Noether's, as Hasse himself says in [Hasse, 1933]. This theorem is, to my knowledge the only contribution where the structure theory of algebras (and not only their cohomological description) could contribute essentially to class field theory–and we should attribute this not only to Hasse but also to Emmy Noether."

THEOREM 8. *Let A be a central simple algebra over an algebraic number field K. Then*
$$\sum_{\mathfrak{p}} \left(\frac{A}{\mathfrak{p}}\right) \equiv 0 \pmod{1}.$$

Claude Chevalley worked on his thesis on class field theory during a trip to Hamburg in the early 1930s, visited Göttingen during that time, and became familiar with the noncommutative methods of Hasse and Noether in number theory. He made a contribution to this area of research himself, in a paper on the norm residue symbol in local class field theory [Chevalley, 1933]. Hasse had already developed a theory of the norm residue symbol in local class field theory, but had derived his results using the global theory. Chevalley's aim was to obtain the theory of the norm residue symbol in local class field theory directly, without using the global theory. His approach was based on the theory of central simple algebras, and in particular, on Noether's theory of crossed product algebras. He stated that the idea of the definition in question was obtained from a remark of Noether in her mimeographed notes on the theory of algebras. As editor of the *Journal für die reine und angewandte Mathematik*, Hasse had accepted Chevalley's paper [Roquette, 2001, p. 597] and referred to Chevalley's approach to the norm residue symbol in connection with his definition of Hasse invariants and the local norm residue symbol in his paper [Hasse, 1933, p. 732].

In a continuation of the letter from Artin to Hasse quoted above, Artin said that "[a]t present I am lecturing on class field theory, and next semester I will continue and will become hypercomplex" [NSUB]. In other words, he was going to lecture on the theory of algebras. This raises the question as to whether Artin included in his lectures of 1931–1932 or later applications of the theory of algebras to algebraic number theory. The published contents[30] show that the 1931–1932 lectures contained the theory of splitting fields, crossed product algebras, and the Brauer group, but do not list applications to number theory. At Peter Roquette's suggestion I asked John Tate about Artin's seminar on class field theory mentioned in the previous section, and, in particular, whether the theory of algebras was used in it. Tate replied:

> The Princeton seminar you refer to was in Artin's second year there, 1947–48. He was not using cohomological methods then, but I don't recall much use of algebras either. I think he was more or less using the methods of Chevalley's paper—the one where he did class field theory without using analytic methods: 1st inequality, 2nd inequality, Artin's original proof of reciprocity, Chevalley's method for the existence theorem. I don't recall any mention of Brauer groups. ... A few years later, when we redid things with cohomological methods, there was more emphasis on the Brauer groups as 2 dimensional Galois cohomology groups, but with the appreciation of the fact that the cohomological interpretation was more flexible than the one with simple algebras, because $H^2(Gal, M)$ made sense for an arbitrary Galois module M, and not only for M the multiplicative group of a Galois extension. In particular $H^2(Gal(K/F), J_K)$,[31] which is critical for the

---

[30] See the Nachlaß Artin in the Archiv der UB Hamburg.
[31] The notation J_K denotes the idèle group of K.

global theory, is not a Brauer group, although its local counterpart is.[32]

Noether's idea, expressed in her ICM lecture, pointed towards the subsequent use of cohomology theory as a powerful tool in algebra and number theory, as discussed in the work of Artin and Tate ([Artin and Tate, 1967] and [Tate, 1967]) and in the papers of Fröhlich and Lemmermeyer cited in the previous section. As Roquette sees it, "[s]he did this somewhat unconsciously—after all, cohomology in the algebraic sense had not yet been formulated. Hence she used the language of algebras where she was used to 2-cocycles, and similarly group extensions. But in fact what she was talking about were Galois modules and their cohomology. With her intuition she foresaw that those methods would become important in the future."[33]

Conclusion

In 1930, Noether's student Max Deuring completed his dissertation and received his Ph.D. at the age of twenty-three. At about this time, Noether was asked by the editors of the series *Ergebnisse der Mathematik* to write a volume in the series on the theory of algebras. She decided not to do it herself, but arranged in 1931 for Deuring to be commissioned to write it. The result, entitled *Algebren*, was a masterful report [Deuring, 1935], containing a crisp presentation of the work of Albert, Brauer, and Noether on simple algebras, the theories of factor systems of Brauer and Noether, and the work on the arithmetic of algebras by Dickson, Brandt, Speiser, and Artin. The last chapter contained a report on the connections between the theory of algebras and the arithmetic of algebraic number fields obtained by Hasse and Noether, and the new results of Chevalley and Deuring himself on the theory of norm residues. In a letter to the author, Peter Roquette wrote that "[i]t can be said that Deuring's book represents to a large extent Emmy Noether's vision of the theory of algebras."[34]

In addition to the work surveyed above, Deuring included in his report an account of the Ph.D. thesis of Käte Hey, who received her degree with Artin at Hamburg in 1929. Her thesis [Hey, 1929], which remains unpublished, developed the analytic number theory and, in particular, the theory of the zeta function of a central simple algebra over the field of rational numbers. "I think Artin had hopes in the thirties," John Tate wrote, "that the zeta functions of simple algebras might contain a clue to nonabelian class field theory, but as Käte Hey's thesis showed, they did not seem to give much that was new."[35]

The collaboration of Brauer, Hasse, and Noether ended abruptly in 1933 when Hitler came to power in Germany. The newly installed Nazi government dismissed Jewish faculty members in German universities, including Brauer and Noether, from their positions. The following year, Brauer and Noether moved to the United States, Brauer to the University of Kentucky and Noether to Bryn Mawr College. Brauer changed his field and began his development of the modular representation theory of finite groups. Noether returned to Göttingen in the summer of 1934 "to visit her brother Fritz (for the last time) before his emigration to Siberia, to visit

---

[32]Tate to the author, 24 March, 2003.

[33]Roquette to the author, 1 May, 2003.

[34]For a survey of Deuring's research and the impact of the book *Algebren*, see [Roquette, 1989].

[35]Tate to the author, 24 March, 2003.

old friends and to break up her small household" [Dick, 1981, p. 82]. On 15 April, 1934, she wrote to Brauer at Lexington about her travel plans, and thanked him for helping to arrange a position for her student, Hans Fitting: "Fitting wrote to me that he is going to Königsberg for the S. S. (Summer Semester) 34; many thanks for your mediation."[36] Towards the end of the letter she added that

> Hasse has been proposed in the first place as Weyl's successor; besides him also Blaschke, also in the first place, but apparently with much less emphasis. ... Hasse expects the call from the ministry of the 'Reichs'-culture—which is supposed to exist—to come only in the autumn. Hasse is able—generalizing complex multiplication—to construct the class fields by division equations of abelian functions; I am not sure whether all of them or only in the case of totally imaginary base field. At the same time he now proves the Riemann hypothesis generally in the case of function fields with Galois field coefficients! He is steadily becoming more brilliant.[37]

In 1934, Hasse was indeed appointed professor and director of the Mathematical Institute at the University of Göttingen as successor to Hermann Weyl. Weyl had left Göttingen in 1933 to become one of the original faculty members, along with Albert Einstein, at the newly established Institute for Advanced Study in Princeton. In a note to Brauer before she left for Germany, on 13 May, 1934, Noether wrote from Bryn Mawr that "I heard in Princeton that all your family is now together; I am very glad about that. Next year, in a seminar at Princeton, I shall read (or have it read) class field theory following Chevalley's presentation."[38]

Brauer was appointed assistant to Herman Weyl at the Institute for Advanced Study in Princeton for the academic year 1934–1935. During that year, Noether held a contract with the Institute for Advanced Study to give a two-hour lecture each week. The subject was class field theory, as she said in her letter to Brauer. In April of 1935, she took a recess from her lectures to undergo surgery. Unable to recover from the operation, she died on 14 April. On the Wednesday following Noether's death, a simple service was held at the home of President Marion Edwards Park of Bryn Mawr College with talks by Hermann Weyl and Richard Brauer, both of whom had come over from Princeton, and by Anna Pell Wheeler and Olga Taussky at Bryn Mawr. The following day, Marguerite Lehr, a faculty member at Bryn Mawr College, spoke in the Bryn Mawr College Chapel. Referring to Brauer's talk the preceding day, she recalled that "Professor Brauer, in speaking yesterday of Miss Noether's powerful influence professionally and personally among the young scholars who surrounded her at Göttingen, said that they were called the Noether

---

[36] "Fitting schreib mir, dass er einstweilen für SS 34 (Sommersemester) — nach Königsberg geht; besten Dank für Ihre Vermittlung." See [AMCCL].

[37] "Hasse ist in erster Stelle als Weyls Nachfolger vorgeschlagen, ausserdem Blaschke—auch an erster Stelle, aber anschleinend viel weniger betont. ... Hasse vermutet, dass er ruf erst zum Herbst kommt, vom dann wohl existierenden Reichs-Kultusministerium. Hasse kann—in Verallgemeinerung der komplexen Multiplikation— die Klassenkörper durch Teilungsgleichungen Abelscher Funktionen konstruieren; ich weiss nicht genau, ob alle oder nur bei ... imaginärem Grundkörper. Zugleich beweist er die Riemann'sche Vermutung bei Funkionenkörpern mit Galoisfeld- Koeffizienten jetzt allgemein: Er wird immer glänzender!" See [AMCCL].

[38] "In Princeton höre ich dass Sie jetzt alle zusammen sind, was mich sehr freut. Ich will nächstes Jahr in Princeton die Klassenkörpertheorie in der Darstellung von Chevalley im Seminar lesen oder vortragen lassen." See [AMCCL].

family, and that when she had to leave Göttingen, she dreamed of building again somewhere what was destroyed there. We realize now with pride and thankfulness that we saw the beginning of a new 'Noether family' here" [Kimberling, 1981, p. 38].

In her 1932 lecture, Noether stated important connections between the theory of central simple algebras and algebraic number theory and announced her version of the principal genus theorem which she published the following year. These results perhaps fell short of the breakthrough she had hoped for, as envisioned in her "principle of the application of the noncommutative to the commutative." The force of her idea of the significance of the noncommutative for the commutative was realized only many years later in the uses of cohomology theory which are standard today in algebraic number theory, commutative algebra, and, in particular, class field theory. Nevertheless, algebras have not disappeared from the scene completely. For example, André Weil, in his book *Basic Number Theory* [Weil, 1967], gave a modern treatment of class field theory based on the theory of central simple algebras.

Acknowledgments

The author wishes to express his thanks to Burton Fein, Gerald J. Janusz, Peter Roquette, and John Tate for e-mail messages containing comments on the manuscript and information about items discussed in the text.

The author also gratefully acknowledges permission from Routledge/Taylor & Francis Group LLC to publish the following quotations (copyright 1981) from *Emmy Noether: A Tribute to Her Life and Work*, ed. James W. Brewer and Martha K. Smith: on p. 200-201 (the two quotations from the translation from Noether's ICM lecture), on p. 210 (the quotation from Fröhlich 1981), and on pp. 216-217 (the quotation from Kimberling's article).

References

Archival Sources

AMCCL. Letters from Emmy Noether to Richard Brauer. Archives of the Mariam Coffin Canaday Library. Bryn Mawr College. Bryn Mawr, PA, USA.

NSUB. Letters from Emil Artin, Richard Brauer, and Emmy Noether to Helmut Hasse. Niedersächsische Staats-und Universitätsbibliothek Göttingen. Handschriftenabteilung. Göttingen, Germany.

Printed Sources

Albert, A. Adrian. 1932. "Normal Division Algebras of Degree Four over an Algebraic Field." *Transactions of the American Mathematical Society* 34, 363-372.

Albert, A. Adrian and Hasse, Helmut. 1932. "A Determination of all Normal Division Algebras over an Algebraic Number Field." *Transactions of the American Mathematical Society* 34, 722-726.

Artin, Emil and Tate, John T. 1967. *Class Field Theory*. New York: Benjamin.

Brauer, Richard D. 1928. "Untersuchungen über die arithmetischen Eigenschaften von Gruppen linearer Substitutionen I." *Mathematische Zeitschrift* 28, 677-696, or [Brauer, 1980, 1:20-39].

———. 1929. "Über Systeme hyperkomplexer Zahlen." *Mathematische Zeitschrift* 30, 79-107, or [Brauer, 1980, 1:40-68].

———. 1930. "Untersuchungen über die arithmetischen Eigenschaften von Gruppen linearer Substitutionen II." *Mathematische Zeitschrift* 31, 733-747, or [Brauer, 1980, 1:88-102].

———. 1932. "Über die algebraische Struktur der Schiefkörpern." *Journal für die reine und angewandte Mathematik* 166, 241-252, or [Brauer, 1980, 1:103-114].

———. 1980. *Richard Brauer: Collected Papers*. Ed. Paul Fong and Warren J. Wong. 3 Vols. Cambridge, MA: The MIT Press.

Brauer, Richard D. and Noether, Emmy. 1927. "Über minimale Zerfällungskörper irreduzibler Darstellungen." *Sitzungsberichte der Königlich Preussischen Akademie der Wissenschaften zu Berlin*, 221-228, or [Brauer, 1980, 1:12-19].

Brauer, Richard D.; Hasse, Helmut; and Noether, Emmy. 1932. "Beweis eines Hauptsatzes in der Theorie der Algebren," *Journal für die reine und angewandte Mathematik* 168, 399-404, or [Brauer, 1980, 1:115-120].

Chevalley, Claude. 1933. "La théorie du symbole de restes normiques." *Journal für die reine und angewandte Mathematik* 169, 141-157.

Curtis, Charles W. 1999. *Pioneers of Representation Theory: Frobenius, Burnside, Schur, and Brauer*. History of Mathematics. Vol. 15. Providence: American Mathematical Society and London: London Mathematical Society.

Deuring, Max. 1935. *Algebren*. Ergebnisse der Mathematik und ihre Grenzgebiete. Vol. 4. Berlin: Springer-Verlag.

Dick, Auguste. 1981. *Emmy Noether 1882–1935*, Trans. H. I. Blocher. Boston/Basel/Stuttgart: Birkhäuser Verlag.

Dickson, Leonard E. 1906. "Linear Associative Algebras and Abelian Equations. Abstract." *Bulletin of the American Mathematical Society* 12, 442.

———. 1914. "Linear Associative Algebras and Abelian Equations." *Transactions of the American Mathematical Society* 15, 31-46.

———. 1923. *Algebras and Their Arithmetics*. Chicago: University of Chicago Press.

Fröhlich, Albrecht. 1981. "Algebraic Number Theory." In *Emmy Noether: A Tribute to her Life and Work*. New York: Marcel Dekker Inc., pp. 157-163.

Grunwald, Wilhelm. 1932. "Charakterisierung des Normrestsymbols durch die p-Stetigkeit den vorderen Zerlegungssatz und die Produktformel." *Mathematische Annalen* 107, 145-164.

Hasse, Helmut. 1932. "Theory of Cyclic Algebras over an Algebraic Number Field." *Transactions of the American Mathematical Society* 34, 171-214.

———. 1933. "Die Struktur der R. Brauerschen Algebrenklassengruppe über einem algebraischen Zahlkörper. Inbesondere Begründung der Theorie des Normrestsymbols und die Herleitung des Reziprozitätsgesetzes mit nichtkommutativen Hilfsmitteln." *Mathematische Annalen* 107, 731-760, or Hasse, Helmut. 1975. *Mathematische Abhandlungen*. Ed. Heinrich

W. Leopoldt and Peter Roquette. 3 Vols. Berlin: G. de Gruyter, 1:501-530.

Hey, Käte. 1929. "Analytische Zahlentheorie in Systemen hyperkomplexer Zahlen." Unpublished Doctoral Dissertation: Hamburg University.

Hilbert, David. 1897. "Die Theorie der algebraischen Zahlkörper." *Jahresbericht der Deutschen Mathematiker-Vereinigung* 4, 175-546, or Hilbert, David. 1932–1935. *Gesammelte Abhhandlungen*. 3 Vols. Berlin: Springer-Verlag, 1:63-434.

Kimberling, Clark. 1981. "Emmy Noether and Her Influence." In *Emmy Noether: A Tribute to her Life and Work*, Ed. James W. Brewer and Martha K. Smith. New York: Marcel Dekker Inc., pp. 3-61.

Lemmermeyer, Franz. 2003. "The Development of the Principal Genus Theorem." Preprint.

Noether, Emmy. 1929. *Algebra der Hyperkomplexen Grössen, Vorlesung von Prof. E. Noether, W. S. 1929/30, Ausgearbeitet von Prof. M. Deuring*. In [Noether, 1983, pp. 711-763].

──────. 1929a. "Hyperkomplexe Grössen und Darstellungstheorie." *Mathematische Zeitschrift* 30, 641-692, or [Noether, 1983, pp. 563-614].

──────. 1932. "Hyperkomplexe Systeme in ihren Beziehungen zur kommutativen Algebra und der Zahlentheorie" *Verhandlungen des Internationalen Mathematiker-Kongresses Zürich*. 1932. Ed. Walter Saxer. 2 Vols. Reprint Ed. 1967. Nendeln/Liechtenstein: Kraus Reprint Limited, 1:189-194, or [Noether, 1983, pp. 636-641].

──────. 1933. "Der Hauptgeschlechtssatz für relativ-galoische Zahlkörper." *Mathematische Annalen* 108, 411-419, or [Noether, 1983, pp. 670-678].

──────. 1983. *Emmy Noether Gesammelte Abhandlungen*. Ed. Nathan Jacobson. Berlin: Springer-Verlag.

Rogawski, Jonathan. 2000. "The Nonabelian Reciprocity Law for Local Fields." *Notices of the American Mathematical Society* 47, 35-41.

Roquette, Peter. 1989. "Über die algebraisch-zahlentheoretischen Arbeiten von Max Deuring." *Jahresbericht der Deutschen Mathematiker-Vereinigung* 91, 109-125.

──────. 2001. "Class Field Theory in Characteristic p, Its Origin and Development." In *Class Field Theory: Its Centenary and Prospect*. Advanced Studies in Pure Mathematics 30, pp. 549-631.

──────. 2004. "The Brauer-Hasse Noether Theorem in Historical Perspective." *Schriften der Mathematisch-naturwissenschaftlicher Klasse der Heidelberger Akademie der Wissenschaften*. No. 15. Berlin/Heidelberg: Springer-Verlag, pp. 1-92.

Schur, Issai. 1906. "Arithmetische Untersuchungen über endliche Gruppen linearer Substitutionen." *Sitzungsberichte der Preussischen Akademie der Wissenschaften, Physikalisch-Mathematische Klasse*, 164-184, or [Schur, 1973, 1:177-197].

──────. 1909. "Beiträge zur Theorie der Gruppen linearer homogener Substitutionen." *Transactions of the American mathematical Society* 10, 159-175, or [Schur, 1973 1:95-111].

──────. 1973. *Gesammelte Abhandlungen*. Ed. Alfred Brauer and Hans Rohrbach. 3 Vols. Berlin: Springer-Verlag.

Tate, John T. 1967. "Global Class Field Theory." In *Algebraic Number Theory: Proceedings of the Brighton Conference.* Ed. John W. S. Cassels and Albrecht Fröhlich. New York: Academic Press, Inc., pp. 163-203.

Wang Shianghaw. 1948. "A Counter-example to Grunwald's Theorem." *Annals of Mathematics* 49, 1008-1009.

──────────. 1950. "On Grunwald's Theorem." *Annals of Mathematics* 51, 471-484.

Wedderburn, Joseph H. M. 1914. "A Type of Primitive Algebra." *Transactions of the American Mathematical Society* 15, 162-166.

──────────. 1921. "On Division Algebras." *Transactions of the American Mathematical Society* 22, 129-135.

Weil, André. 1967. *Basic Number Theory.* New York: Springer-Verlag.

──────────. 1979. *Oeuvres scientifiques.* 3 Vols. New York: Springer-Verlag.

Whaples, George. 1942. "Non-analytic Class Field Theory and Grunwald's Theorem." *Duke Mathematical Journal* 9, 455-473.

CHAPTER 10

From *Algebra* (1895) to *Moderne Algebra* (1930): Changing Conceptions of a Discipline–A Guided Tour Using the *Jahrbuch über die Fortschritte der Mathematik*

Leo Corry
Tel Aviv University, Israel

Introduction

The discipline of algebra underwent significant changes between the last third of the nineteenth century and the first third of the twentieth. These changes comprised not only the addition of important new results, new concepts, and new techniques, but also fundamental shifts in the very way that the aims and scope of the discipline were conceived by its practitioners. During the nineteenth century, algebraic research had meant mainly research on the theory of polynomial equations and the theory of polynomial forms, including algebraic invariants. The ideas implied by Évariste Galois's works had become increasingly visible and central after their publication by Joseph Liouville in 1846. Together with important progress in the theory of fields of algebraic numbers, especially in the hands of Leopold Kronecker and Richard Dedekind, they gave rise to an increased interest in new concepts such as groups, fields, and modules.

A very popular textbook of algebra from the middle of the century was the *Cours d'algèbre supérieure* by Joseph Serret, which went through three editions in 1849, 1854, and 1866, respectively [Serret, 1849]. In these successive editions, this book gradually incorporated the techniques introduced by Galois, and in the third, it became the first university textbook to publish a full exposition of the theory. Still, it continued to formulate the main results of Galois theory in the traditional language of solvability dating back to the works of Lagrange and Abel at the beginning of the century. Thus, it did not even include a separate discussion of the concept of group. A second important, contemporary textbook was Camille Jordan's *Traité des substitutions et des équations algébriques* [Jordan, 1870], which already included a more elaborate presentation of the theory of groups, but which still treated this theory as subsidiary to the main tasks of algebra, and, above all, to the elucidation of solvability conditions for polynomial equations.

Towards the end of the century, Heinrich Weber published a three-volume textbook, *Lehrbuch der Algebra* [Weber, 1895] that incorporated an entire body of new ideas and techniques developed in the nineteenth century, thereby providing a full picture of what the body of algebraic knowledge looked like at the time. Concurrently, it implicitly embodied in the most elaborate and detailed way to date the disciplinary conception of algebra over the century. It laid down the main aims of this discipline, stressed the most relevant questions that practitioners had and should address, and presented the main available techniques available to do so successfully. In spite of the great amount of specific knowledge it added over books like Serret's or Jordan's, Weber's *Lehrbuch* did not embody an essentially different conception of the discipline from theirs;[1] algebra is seen as the discipline of polynomial equations and polynomial forms. Abstract concepts such as groups, in so far as they appear in the book, are subordinate to the main classical tasks of algebra. And, most importantly, all the results are based on the assumption of a thorough knowledge of the basic properties of the systems of rational and real numbers; these systems are conceived as conceptually prior to algebra. Whatever is said about polynomials or about factorization properties of algebraic numbers is based on what is known about the various systems of numbers.

The first two decades of the twentieth century were ripe with new algebraic ideas. Toward the end of the 1920s, one finds a growing number of works that can be identified with only recently consolidated theories, usually aimed at investigating the properties of abstractly defined mathematical entities now seen as the focus of interest in algebraic research: groups, fields, ideals, rings, and others. Like many other important textbooks, *Moderne Algebra*, written by the young Bartel L. van der Waerden, appeared in 1930 at a time when the need was felt for a comprehensive synthesis of what had been achieved since the publication of its predecessor, in this case Weber's *Lehrbuch*. It presented ideas that had been developed earlier by Emmy Noether and Emil Artin—whose courses van der Waerden had recently attended in Göttingen and Hamburg, respectively—and also by other algebraists, such as Ernst Steinitz, whose works van der Waerden had also studied under their guidance. Van der Waerden masterfully incorporated a great deal of the important innovations accumulated over the early decades of the twentieth century at the level of the body of algebraic knowledge. But the originality and importance of this book is best recognized by focusing on its totally new way of conceiving of the discipline. Van der Waerden presented systematically those mathematical branches then related to algebra, deriving all the relevant results from a single, unified perspective, and using similar concepts and methods for all those branches. This original perspective, which turned out to be enormously fruitful over the next decades of research—and not only in algebra, but in mathematics at large—is what I will call here the structural image of algebra.

The structural image of algebra as put forward in van der Waerden's textbook is based on the realization that a certain family of notions (that is, groups, ideals, rings, fields, etc.) are, in fact, individual instances of one and the same underlying idea, namely, the general idea of an algebraic structure, and that the aim of research in algebra is the full elucidation of those notions. None of these notions, to be sure, appeared as such for the first time in this book. Groups, as noted, had appeared in mainstream textbooks on algebra as early as 1866, in the third edition of

---

[1]Throughout this chapter, I will refer to the distinction between "body" and "images" of mathematical knowledge, on which I have elaborated in greater detail in [Corry, 2001; 2003]. Roughly stated, answers to questions directly related to the subject matter of any given discipline constitute the body of knowledge of that discipline, whereas claims and knowledge *about* that discipline constitute their images of knowledge. The images of knowledge help in discussing questions arising from the body of knowledge that are, in general, not part of, and cannot be settled within, the body of knowledge itself.

Serret's *Cours*. Ideals and fields, in turn, had been introduced in 1871 by Dedekind in his elaboration of Ernst Edward Kummer's factorization theory of algebraic numbers. But the unified treatment they were accorded in *Moderne Algebra*, the single methodological approach adopted to define and study each and all of them, and the compelling, new picture it provided of a variety of domains that were formerly seen as only vaguely related, all these implied a striking and original innovation.

This is not the place to describe in detail the aims and contents of van der Waerden's book.[2] One fundamental innovation implied by his approach, however, that merits stressing here is the redefinition of the conceptual hierarchy underlying the discipline of algebra. Rational and real numbers no longer have conceptual priority over, say, polynomials. Rather, they are defined as particular cases of abstract algebraic constructs. Thus, for instance, van der Waerden introduced the concept of a field of fractions for integral domains in general, and then obtained the rational numbers as a particular case of this kind of construction, namely, as the field of quotients of the ring of integers. His definition of the system of real numbers in purely algebraic terms was based on the concept of a "real field," recently elaborated by Artin and Otto Schreier [Artin and Schreier, 1926], whose seminars van der Waerden had attended in Hamburg.

The task of finding the real and complex roots of an algebraic equation, which was the classical main core of algebra in the previous century, was relegated in van der Waerden's book for the first time to a subsidiary role. Three short sections in his chapter on Galois theory dealt with this specific application of the theory, and they assume no previous knowledge of the properties of real numbers. In this way, two central concepts of classical algebra (rational and real numbers) are presented here merely as final products of a series of successive algebraic constructs, the "structure" of which was gradually elucidated. On the other hand, additional, non-algebraic properties such as continuity and density were not considered at all by van der Waerden as part of his discussion of those systems.

Another important innovation implied by the book concerns the particular way in which the advantages of the axiomatic method are exploited in conjunction with all other components of the structural image of algebra, such as those mentioned above. Once one has realized that the basic notions of algebra (groups, rings, fields, etc.) are, in fact, different varieties of a same species ("varieties" and "species" understood here in a "biological," and not in a mathematical sense), namely, different kinds of algebraic structures, the abstract axiomatic formulation of the concepts becomes, in a natural way, the most appropriate one. The central disciplinary concern of algebra becomes, in this conception, the systematic study of those different varieties through a common approach. In fact, this fundamental recognition appears in *Moderne Algebra* not only implicitly, but rather explicitly and even didactically epitomized in the *Leitfaden* that appears in the introduction to the book, and that pictures the hierarchical, structural interrelation between the various concepts investigated in the book (see the next page).

Obviously, the new image of algebra presented by van der Waerden reflected the then-current state of development of the body of algebraic knowledge. However, the important point is that the image was *not a necessary* outcome of the body,

---

[2]For that, see [Corry, 2003, pp. 43-54].

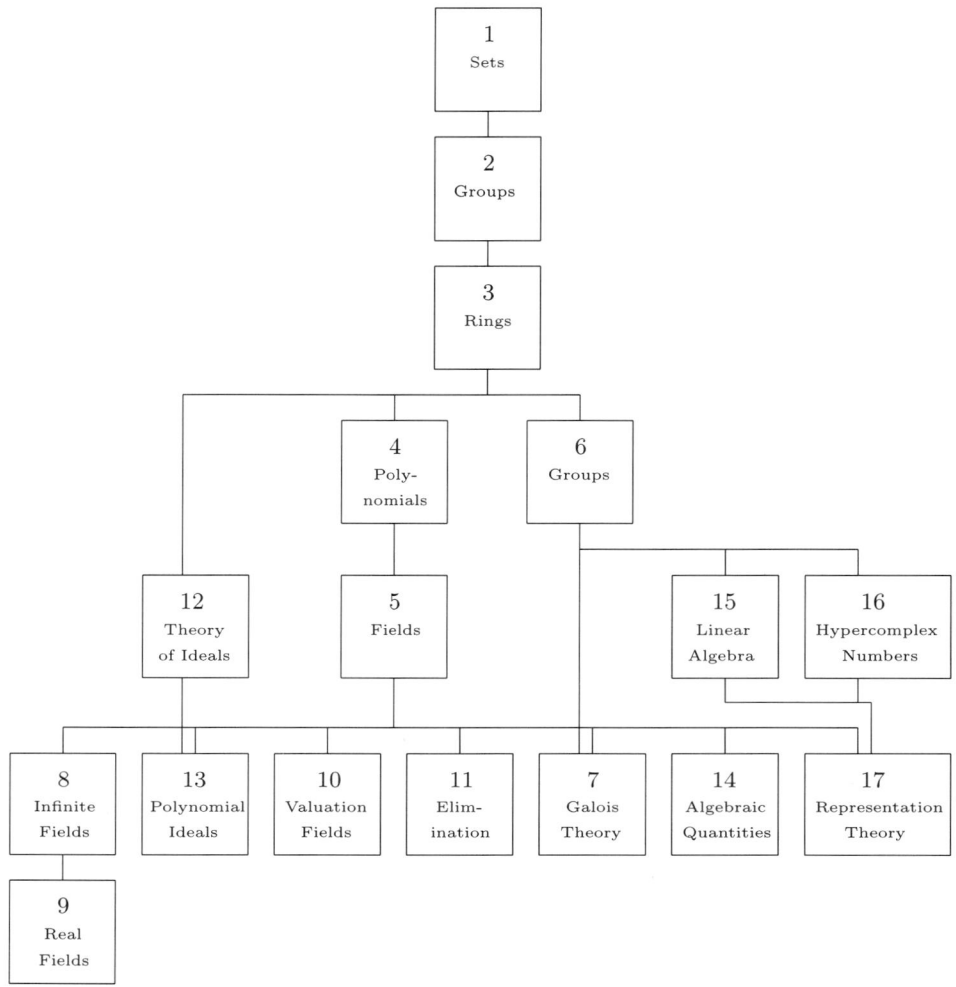

but rather an independent development of intrinsic value. This becomes clear when we notice that parallel to van der Waerden's, several other textbooks on algebra were published which also contained most of the latest developments in the body of knowledge, but which essentially preserved the classical image of algebra. Examples of these are Leonard Eugene Dickson's *Modern Algebraic Theories* [Dickson, 1926], Helmut Hasse's *Höhere Algebra* [Hasse, 1926], and Otto Haupt's *Einführung in die Algebra* [1929]. But perhaps the most interesting example in this direction is provided by Robert Fricke's *Lehrbuch der Algebra*, published in 1924, with the revealing subtitle: "Based on Heinrich Weber's Homonymous Book [*Verfasst mit Benutzung vom Heinrich Webers gleichnamigem Buche*]" [Haupt, 1924]. These books were by no means of secondary importance. Dickson's, for instance, became after its publication the most advanced algebra text available in the United States, and it was not until 1941 that a new one, better adapted to recent developments in algebra and closer to the spirit of *Moderne Algebra*, was published there: *A Survey of Modern Algebra* by Garrett Birkhoff and Saunders Mac Lane [Birkhoff and Mac Lane, 1941].

Weber's *Lehrbuch* and van der Waerden's *Moderne Algebra*, then, embody in their respective presentations—thirty years apart—two very different images of the discipline they discuss. Faced with this fact, the question becomes how to account for the historical process that led from the former to the latter. The most immediate, and perhaps necessary, way to tackle this is to look at the most prominent works that, acting as milestones, progressively produced the main concepts, theorems, and techniques that came to stand at the center of algebraic research as it was practiced in the 1920s, and to explain how they helped shape the new images of algebra.[3] Parallel to this perspective, however, one may also look for additional hints that clarify how the practitioners of the discipline interpreted this progressive evolution and how their image of algebra changed accordingly. One way of illuminating this is to look at the leading, German review journal of the period, the *Jahrbuch über die Fortschritte der Mathematik* and, particularly, at the changing classificatory schemes adopted by the journal to account for the current situation at various, important crossroads of this story. As will be seen below, this perspective sheds interesting light on our understanding of the sometimes tortuous path from "Algebra" to "Modern Algebra."

The *Jahrbuch über die Fortschritte der Mathematik*

The *Jahrbuch über die Fortschritte der Mathematik* was established in 1868 by Carl Ohrtmann and Felix Müller, two Berlin Gymnasium teachers. It soon became the world's leading review journal of mathematics, to be eclipsed only after 1931 with the foundation in Germany of the *Zentralblatt für Mathematik und ihre Grenzgebiete* and later of the *Mathematical Reviews* in the United States beginning in 1940. The *Jahrbuch* published its last volume in 1945.[4] Other contemporary, but much less visible and influential review journals were the French *Bulletin des sciences mathématiques et astronomiques* founded in 1895 by Gaston Darboux, and the Dutch *Revue semestrielle des publications mathématiques* founded in 1897 [Siegmund-Schultze, 1993, pp. 14-20].

Although the advent of the *Jahrbuch* was enthusiastically welcomed by many in the German mathematical community, the degree of actual collaboration from that community's leading representatives was rather negligible. Some prominent German mathematicians did occasionally participate in writing reviews, but the truly outstanding names can hardly be found in the list.[5] The Berlin triumvirate of Kummer, Kronecker, and Karl Weierstrass, as well as their colleagues Carl Wilhelm Borchardt, Lazarus Fuchs, and Friedrich Schottky never contributed to the journal; Georg Frobenius and Hermann Amandus Schwarz did so only rarely. Felix Klein and David Hilbert, likewise, contributed very little, and they did so only before their great Göttingen years. Nor were any of the editors of the *Jahrbuch* truly first-rate mathematicians [Siegmund-Schultze, 1993, pp. 15-16 and 201-203].

---

[3]This was the main topic of Part I of [Corry, 2003].

[4]For the history of the *Jahrbuch* and other mathematical review journals in the twentieth century, see [Siegmund-Schultze, 1993; 1994].

[5]Between 1900 and 1920, the quality of the reviewers considerably improved [Siegmund-Schultze, 1993, pp. 25-26].

The *Jahrbuch* was conceived as a yearly publication summarizing the relevant research activity of that period of time. It usually took several years, in some cases up to seven, before a given volume was completed and finally published. By the 1930s, this had become one of the journal's main drawbacks when compared to its competitors, the *Zentralblatt* and the *Mathematical Reviews* [Siegmund-Schultze, 1993, pp. 16-17].

It should come as no surprise that defining the classification schemes to be used for the mathematical works reviewed in the journal proved a challenging task from its inception. As Emil Lampe, co-editor of the *Jahrbuch* from 1885 until his death, explained "certainly there is no exhaustive classification of the mathematical disciplines, and although many groups can be easily demarcated on a coarse scale, it is extraordinarily hard to divide all the mathematical fields according to a precise uniform scheme. It is also quite easy for one to find fault in a division of too many categories with much too fine, puzzling principles of classification" [Lampe 1903, p. 5].[6] To state this in the terminology introduced above, adopting any given classification implies spelling out a certain image of mathematical knowledge that by nature is implicit, somewhat unstable, and only tacitly shared by some—but not necessarily all—members of the community. Moreover, specifying a classification scheme involves an attempt to freeze a conception that is essentially dynamic and that will certainly change as the body of knowledge changes. The editors of the *Jahrbuch* were clearly aware of this problem, and, over time, the tension was evident in their attempts, on the one hand, to preserve the existing schemes so as to make it easier for a reader to find articles and, on the other hand, to keep abreast of current developments in the body of knowledge and how they affect the images of mathematics.[7]

The use of the successive classificatory schemes of the *Jahrbuch* for algebra between 1900 and 1930 as an expression of the then-currently accepted images of the discipline must be set in the correct context by considering the background provided in the above account. For one thing, the schemes reflect the images of the editors of the journal at any given point in time, and there is no reason to believe that they were universally shared. For another, it is not absolutely clear how the schemes and the changes introduced in them were decided upon, whether by the editors alone, or by the editors in consultation with other mathematicians. Still, it would seem evident that the schemes attempted to express in the most coherent way possible what would appear to be an existing consensus in this regard, and thus they do express a certain common denominator that may have been shared by many contemporary mathematicians. More specifically, since the focus here is the gradual emergence of the structural conception of algebra and of the idea of an algebraic structure as a unifying principle across different algebraic subdisciplines, the absence of this idea as a leading classificatory component in the journal until the late 1920s may be taken, in my opinion, as a reliable criterion for its actual absence in the existing conceptions of the discipline, rather than only in the eyes of the editors of the *Jahrbuch*.

---

[6]The English translation is quoted from [Despeaux, 2002, p. 298]. I thank Sloan Despeaux for providing me with a copy of her unpublished Ph.D. dissertation and for allowing me to quote from it.

[7]Compare [Despeaux, 2002, p. 299]. In her chapter 7, Despeaux charts the changes in the general classification schemes (not only in algebra) between 1868 and 1900.

Algebra by the Turn of the Century: The *Jahrbuch* in 1900

The image of algebra reflected in the classification scheme of the *Jahrbuch* at the beginning of the twentieth century is as close as it can be to that embodied in Weber's *Lehrbuch der Algebra*, as described above. Before presenting detailed evidence for this claim, however, it is illuminating to discuss briefly a preliminary instance, namely, an important and well-known article published by the same Weber in 1893 on "The General Foundations of Galois Theory" [Weber, 1893], and to see how it was classified in the relevant volume of the *Jahrbuch*.

In many respects, Weber's article represents the first truly modern published presentation of Galois theory, wherein the latter appears not just as an analysis of the problem of solvability, but rather as a more general examination of the interplay between specific groups and certain well-defined fields. In particular, Weber focused on establishing an isomorphism between the group of permutations of the roots of the equation and the group of automorphisms of the splitting field that leave the elements of the base field invariant. This approach leads in a natural way to adopting abstract formulations of the central concepts involved—group and field—while stressing the interplay between what we may in retrospect call their "structural" properties.

Weber, to be sure, was not the first to define groups abstractly, but this is, indeed, the first place where fields appear as an extension of the concept of group, obtained by adding a second operation to the already existing one. This permitted finite and infinite fields to be subsumed under a single, general definition, although, significantly, Weber did not consider the problem of the characteristic of the field [Corry, 2003, pp. 35-43].

The point that concerns us here is that the "structural" potentialities implied by the basic formulation of the theory in Weber's article were never fully exploited by him, and, in particular, his 1895 *Lehrbuch* contains no trace of them. Rather, Weber's book elaborated the classical, nineteenth-century approach to algebra in greater detail and comprehensiveness than ever before. This was, obviously, his true conception of the discipline. The incipient structural character present in his article strongly resonates because of subsequent developments, but it was of lesser importance for, and has less direct impact on, Weber himself and his contemporaries.

And, indeed, the *Jahrbuch* classifies this article in its 1893 volume according to classical standards, namely, as "Algebra: Equations (General Theory. Special Algebraic and Transcendental Equations)." In fact, all articles on Galois theory appeared in this section until much later, according to the classical conception that the theory is an auxiliary tool for dealing with the question of solvability of polynomial equations.

Consider now how algebraic works were classified in the *Jahrbuch* at the turn of the century, starting with volume 31 (1900).[8] This volume was published in 1902 and the editors were Emil Lampe and Georg Wallenberg. The two sections of particular interest here are Section II (Algebra) and Section III (Elementary and Higher Arithmetic). They are divided into subsections as follows:

---

[8]Very recently a web-based database comprising all the reviews published in the *Jahrbuch* was established as an "Electronic Research Archive for Mathematics" at http://www.emis.de/projects/JFM/JFM.html.

Section II: Algebra
- Ch. 1: Equations: General Theory. Special Algebraic and Transcendental Equations
- Ch. 2: Theory of Forms (Theory of Invariants)
- Ch. 3: Substitutions and Group Theory. Determinants, Elimination and Symmetric Functions

Section III: Elementary and Higher Arithmetic
- Ch. 1: Elementary Arithmetic
- Ch. 2: Number Theory
    - A. General
    - B. Theory of Forms
- Ch. 3: Continued Fractions (*Kettenbrüche*)

As in the *Lehrbuch*, this scheme acknowledges group theory specifically, but it is seen as closely related to substitutions and determinants, rather than, say, to fields, rings, or algebras. On the other hand, articles specifically dealing with "Galois Theory" all appear under the heading of "Special Algebraic and Transcendental Equations," since the theory is seen as simply one among several existing tools for dealing with the theory of equations, rather than as an autonomous topic or as one ancillary to the theory of groups.

Before the 1905 volume, there is only one noticeable change in the scheme adopted at the beginning of the century, and that appears in volume 34 (1903): the subsection on the theory of forms is divided there into (A) Theory of Algebraic Forms and (B) Differential Invariants. Representative of the articles included under the latter heading is a series of works by the Italian Ernesto Pascal [Pascal, 1903]. Pascal had published many similar articles prior to that date (for example, [Pascal, 1902]), and these were typically reviewed in the *Jahrbuch* in the section on differential and integral calculus under the sub-heading: "Partial Differential Equations." Thus, although the origins of problems in the theory of differential invariants lay in the domain of differential equations, the editors seem to have acknowledged in 1903 that the approach used in dealing with these problems was essentially similar to that used for algebraic invariants. Still, this is a rather minor change without overall implications for the disciplinary conception of algebra.

The classification scheme adopted in 1900 by the *Jahrbuch*, as already stressed, followed closely the image of algebraic knowledge put forward in Weber's *Lehrbuch*. Still, it was certainly possible at the time to follow that same image, while organizing algebraic knowledge slightly differently. Indeed, just at the turn of the century, the first chapters of Felix Klein's *Encyklopädie der mathematischen Wissenschaften mit Einschluss ihrer Anwendungen* began to appear in print [Gispert, 2001]. The articles composing the first volume of this collection, dealing with "Arithmetic and Algebra," were all published between 1898 and 1900, and it is illuminating to see how the then-current state of knowledge in algebra was distributed there among the various sections and articles. The volume was divided into three major sections on "Arithmetic," "Algebra," and "Number Theory," respectively, and these were further subdivided as follows:

A. Arithmetic
- Foundations of arithmetic (that is, elementary operations with numbers)
- Combinatorics
- Irrational numbers and convergence
- "Higher complex numbers" (that is, quaternions and hypercomplex systems)
- Set theory
- Finite groups

B. Algebra
- Rational functions of one variable
- Rational functions of several variables
- Algebraic forms. Arithmetic theory of algebraic magnitudes
- Algebraic invariants
- Separation and approximation of roots
- Rational functions of roots: symmetric functions
- Galois theory and applications
- Systems of equations
- Finite groups of linear substitutions

C. Number theory
- Elementary number theory
- Arithmetic theory of forms (that is, quadratic and bilinear forms, etc.)
- Analytic number theory
- Theory of algebraic number fields
- Cyclotomy fields
- Complex multiplication (that is, class field theory)

What better way to express implicitly the idea that algebra is above all the discipline dealing with the theory of polynomial equations and polynomial forms? The theory of algebraic number fields, for all its use of "algebraic" techniques and concepts, is seen here as part of a different (if neighboring) discipline, with stronger connections to analytic number theory than to, say, Galois theory. In turn, this latter theory is associated with the same mathematical family as the pursuit of analytical methods for approximating roots. There is no hint, of course, of the possibility of unifying under a separate, common heading works on groups, on fields or rings, or on associative algebras and hypercomplex systems. Not only do all of these concepts appear dispersed across the various subsections, but even the elementary theory of linear groups and the theory of linear groups appear under two different sections. It is also pertinent to notice that most of the chapters of the *Encyklopädie* were also individually reviewed as articles in the *Jahrbuch*, and they were often classified there under very different headings. Thus, for instance, the chapter on "Algebraic forms. Arithmetic theory of algebraic magnitudes" [Landsberg, 1899], that in the index of the *Encyklopädie* appears twice, once under "Algebra" and once under "Number Theory," is classified in the *Jahrbuch* as "Algebra: General Theory. Special Algebraic and Transcendental Equations."

Some Tentative Changes: 1905–1915

The first truly significant change in the classification scheme of the *Jahrbuch* for algebra after 1900 appeared in volume 36 (1905). The section on "Equations" became "Equations, Universal Algebra and Vector Analysis" and was further divided into (A) Equations and (B) Universal Algebra and Vector Analysis. While the first of these two subsections remained basically as before, the second includes topics that had never been covered under algebra, at least in any consistent way. Prominent in this subsection are works related to quaternions and vectors, among them, expository books such as Charles J. Joly's *A Manual of Quaternions* [Joly, 1905], and (in volume 37 (1906)) the second edition of Alexander MacFarlane's *Vector Analysis and Quaternions* [MacFarlane, 1906]. In earlier volumes, works of this kind had appeared variously under "Analytic Geometry: Textbooks, Coordinates" (for example, [Kelland and Tait, 1904]), "Differential and Integral Calculus: Determinate Equations" (for example, [Joly, 1903a]) or, in many cases, "Algebra: Equations" (for example, [Bucherer, 1903] and [Joly, 1903b]).

In order to understand more precisely the context and the difficulties faced by the editors with the choice of this, or any other scheme, however, it must be stressed that even with the availability of a new, special subsection for works dealing with quaternions and vectors, the 1905 volume still reflects a certain ambivalence concerning the most adequate way to classify them. Thus, for instance, while an article on "Hamilton's Quaternion Vector Analysis" [Knott, 1905] is classified in the new section on "Universal Algebra and Vector Analysis," one on "Quaternion Number-Systems" [Hawkes, 1905] is still classified as "Number Theory" and one on "Quaternion Products" [Stringham, 1905] under the more traditional heading of "Analytic Geometry: Coordinates." Only gradually were all articles on vectors and quaternions considered as belonging naturally to a single category, which is itself a subsection of "Algebra."

This change in the classificatory scheme of the *Jahrbuch* corresponds to parallel developments in the related body of knowledge. The last decade of the nineteenth century and the early years of the twentieth witnessed intense activity in the somewhat diverging approaches provided by quaternions and vectors to the same general set of topics. On the one hand, the "International Association for Promoting the Study of Quaternions and Allied Systems of Mathematics" published several issues of its *Bulletin* between 1900 and 1913 [Crowe, 1967, p. 218]. On the other hand, influential books that adopted the vectorial approach, especially through its application to physical theories, led to an eventual dominance of this approach and to a unification of all existing, related languages. This was the case with [Bucherer, 1903], [Gans, 1905], and [Marcolongo and Burali-Forti, 1907], as well as with influential articles in Klein's *Encyklopädie*, such as Abraham's article on the mechanics of deformable bodies [Abraham, 1901] or Lorentz's article on electron theory [Lorentz, 1904]. After 1905, the modern system of vector analysis had essentially been absorbed into the mainstream treatment of physical theories [Crowe, 1975, pp. 238–242], and, as the *Jahrbuch* makes clear, the more purely algebraic spirit implied by the approach became increasingly evident.

But besides quaternions and vectors, and perhaps more significantly for the purposes of the present account, among the articles covered under the new heading of "Universal Algebra" were the groundbreaking contributions of Joseph H. M. Wedderburn and Leonard E. Dickson to the study of finite algebras and hypercomplex systems, for example, [Wedderburn, 1905] and [Dickson, 1905a; 1905b].[9] Wedderburn's previous articles on related topics had been variously classified as "Algebra:

---

[9]Compare [Parshall, 1983].

Determinants" [Wedderburn, 1902] or "Algebra: Equations" [Wedderburn, 1903]. Dickson, on the other hand, had published an astonishingly large amount of work on group theory, especially linear groups, and some of his articles more closely related to algebras and hypercomplex systems had also appeared under the same heading of group theory (for example, [Dickson, 1903a; 1903b]), or sometimes simply under "Algebra: Equations" (for example, [Dickson, 1903c]). Another interesting example is provided by works of Issai Schur on the theory of matrices; in 1902 they were classified as "Algebra: Group Theory, Determinants," whereas in 1905 they already appeared under "Universal Algebra."

A different perspective on the meaning of the classification scheme is provided by works that would later be strongly identified with the modern, structural approach to algebra, for instance, Emanuel Lasker's important 1905 article on the factorization of polynomial ideals [Lasker, 1905].[10] Later generalized by the work of Macaulay (see below), Lasker's main results defined one of the central pillars, together with Dedekind's factorization theorems for algebraic integers, of Emmy Noether's abstract theory of rings. Lasker elaborated on ideas that had appeared in previous works of Hilbert on the theory of polynomial invariants. The article was reviewed in the chapter on the "Theory of Forms," but not in the section on "Algebra," rather in the section on "Elementary and Higher Arithmetic."

More significant is the case of Ernst Steinitz's seminal article on the abstract theory of fields [Steinitz, 1910]. The importance of this article for the rise of the structural approach in algebra can hardly be overstated. This is the first place where we find an analysis of the kind that can be dubbed "structural" with full justification of an algebraic entity that is defined in purely abstract terms. A purely abstract definition of field, as already mentioned above, had already been formulated in 1893 by Weber. However, as Steinitz pointed out in his article, beyond the definition itself, no truly abstract investigation of the theory implied by this concept had ever been put forward. Steinitz used for the first time truly set-theoretical considerations in a work of this kind (including a thoughtful application of the axiom of choice only at the right place, where it was truly necessary). He also based his whole analysis on a systematic discussion of the possible cases of the most basic kinds of fields (prime fields), and a study of the kinds of properties that are passed over from these basic fields to any extensions or subentities thereof.[11] One could even say that in writing his textbook of 1930, van der Waerden was actually extending to the whole of algebra the paradigm embraced by Steinitz in his study of fields.

It is thus all the more surprising to see where this article was reviewed in the 1910 volume of the *Jahrbuch*: it is not even classified in the section on algebra, but rather under "Function Theory: General." One wonders what criteria resulted in classifying the article this way. Or perhaps it was a simple technical mistake? Looking at the existing classification scheme of the journal at this time, however, it is unclear into what section it would have fit in a completely natural way. Such a subsection appeared only in 1916, following important developments in algebra for which Steinitz's article itself was a main driving force. This shift eventually resulted in a reclassification of a later reprint of Steinitz's article.

---

[10]Compare [Corry, 2003, pp. 214-219].

[11]See [Corry, 2003, pp. 192-196] for details.

Transactions of the American Mathematical Society: 1910

Additional insight may be gained at this point by looking at the few contemporary classificatory schemes used in venues other than the *Jahrbuch*. In the United States, for instance, there was no specialized review journal at the time; the *Mathematical Reviews* would only be established in 1940. Still, the index of the first ten volumes of the *Transactions of the American Mathematical Society* was published as an appendix to the 1909 volume of this leading American journal. There, the articles are classified according to topics and subtopics, and, in this sense, it is useful to compare it with the *Jahrbuch*'s scheme, in spite of its limited scope; it only covered works published in the journal itself.

The first section in the index refers to the "Logical Analysis of Mathematical Disciplines," which includes a relatively long list of works associated with the distinctly American tradition of "Postulational Analysis." This tradition derived from Hilbert's work on the foundations of geometry at the turn of the century, but took a twist of its own following the lead of Eliakim Hastings Moore. The idea was to take sets of postulates defining various mathematical entities pertaining to different disciplines and to analyze them using the tools developed by Hilbert in the *Grundlagen der Geometrie*. The focus was on the sets of postulates themselves, rather than on the specific disciplines to which they referred, and the expected results of works of this kind were the formulation of a minimal set of independent postulates underlying the various mathematical theories. This trend was very active in the United States until 1920, and involved contributions by the leading American mathematicians, such as E. H. Moore, Dickson, Robert L. Moore, and Edward Huntington. In Germany, on the other hand, it received scant attention.[12] It is therefore not surprising that a separate section in the index of the *Transactions* was specifically devoted to it. Nothing similar to this appears until 1916 in the *Jahrbuch*, that continued meanwhile to classify articles on postulational analysis according to the discipline whose postulates were analyzed, varying from "Differential and Integral Calculus" [Huntington, 1902], to "Philosophy" [Huntington, 1904], to "Groups" [Huntington, 1905] and [Dickson, 1905b], to "Universal Algebra" [Carstens, 1906], to "Mechanics" [Carmichael, 1912], to "Geometry" [Huntington, 1913].

Our main interest here, though, is the section on algebra in the index of the *Transactions*. This section is quite different from its counterpart in the *Jahrbuch*, reflecting once again more typically American conceptions of the discipline at the time. It is divided into four subsections, as follows:

B1. Rational Functions. Theory of Equations. Determinants. Symmetric Functions
B2. Algebraic Forms
B3. Linear Associative Algebra. Hypercomplex Systems. Fields
B4. Algebra of Logic

Especially worthy of attention are the two last subsections. Linear associative algebras and hypercomplex systems had traditionally been a focal point of interest for American mathematicians at least since the related works of Benjamin Peirce in 1881, with deeper roots in the British tradition originating with William R. Hamilton and having James Joseph Sylvester as the bridge between the two continents

---

[12]Compare [Corry, 2003, pp. 214-219] for details.

[Parshall, 1985, pp. 226-261]. It is thus no wonder to find these topics here in a special section. That section also offers, of course, a natural location for [Wedderburn, 1905], the important article on finite algebras, and also could have offered the natural classification of his seminal article on the structure of algebras [Wedderburn, 1907] had it been published in this journal.[13] In the *Jahrbuch*, these two articles of Wedderburn appeared under the new subsection on "Universal Algebra and Vector Analysis." Considering the algebra of logic as part of algebra occurred only much later in German classificatory schemes, although works specifically devoted to such topics had existed in German at least as early as 1890 with Ernst Schröder's *Vorlesungen über die Algebra der Logik* [Schröder, 1890]. This work had been reviewed in the 1890 volume of the *Jahrbuch* in the section on "History and Philosophy: Philosophy and Pedagogy." This classification was preserved for decades and included articles such as Huntington's postulational analysis of this discipline [Huntington, 1904], which in the index of the *Transactions* appeared both under "Logical Analysis of Mathematical Disciplines" and "Algebra: Algebra of Logic."

The theory of numbers received in the index of the *Transactions* a section separate from algebra. So did the theory of groups, in which intense activity traditionally existed in the United States. This latter section was further divided into three subsections dealing, respectively, with "Discrete Groups in General," "Linear Groups in Arbitrary or Special Fields," and "Continuous Groups." A last section related to algebraic topics, but grouped under a different heading in the index of the *Transactions*, concerned "Algebraic Geometry," which appears here not as part of "Algebra," but rather as part of "Geometry." Interestingly enough, articles belonging to what we would associate in retrospect with the discipline of "Algebraic Geometry," both in its German and Italian traditions, appear in the various volumes of the *Jahrbuch* under different chapters of the section on "Geometry," and more often than not under "Analytic Geometry."

It is worth mentioning, to conclude this section, that the index of the next ten volumes of the *Transactions*, published as an appendix to its 1919 volume, repeated essentially the 1909 scheme. Then, in the 1928 volume, where the next ten volumes were indexed, no classificatory scheme whatsoever was used, and the articles appeared simply in alphabetical order by author.

The *Jahrbuch* after 1916

The next important change in the classificatory scheme of the *Jahrbuch* appeared in its volume 46, published in 1923 under the editorship of Leon Lichtenstein and containing reviews of the mathematical activity of the years 1916–1917.[14] The classificatory scheme contained many innovations for all the topics covered, and this was particularly the case for the two topics, algebra and arithmetic, which appeared now unified as part of a single section. This section, "Algebra and Arithmetic," was further subdivided into the following nine chapters:

---

[13] Compare [Parshall, 1985, pp. 309–331].

[14] After Lampe's death in 1918, Arthur Korn edited a single issue of the journal, and then the job was immediately taken over by Lichtenstein. He was perhaps the most proficient of the mathematicians who held this position. Compare [Siegmund-Schultze, 1993, p. 202].

Ch. 1 Foundations of Arithmetic and Algebra. General
Ch. 2 Elementary Arithmetic and Algebra. Combinatorics
Ch. 3 Theory of Polynomials and Algebraic Equations
Ch. 4 Theory of Forms. Determinants. Theory of Invariants
Ch. 5 Group Theory. Abstract Theory of Fields and Modules
Ch. 6 Elementary Theory of Numbers. Additive Number Theory
Ch. 7 Arithmetic Theory of Forms
Ch. 8 Algebraic Number Theory. Analytic Number Theory
Ch. 9 Transcendental Numbers

The chapter on foundations of algebra and arithmetic represents the first reference, as a separate category, for articles in the tradition of "postulational analysis." As noted, this trend did not attract the same kind of attention in Germany that it did in the United States. Perhaps the first work to introduce it in the German literature was a relatively unknown textbook on algebra by Alfred Loewy [Loewy, 1915].[15] Loewy had published several related articles in the *Transactions* during the first decade of the century and, at the same time, had been the main reviewer for the *Jahrbuch* of articles connected with this trend. Among those who read Loewy's book and were influenced by him was his nephew Abraham Fraenkel, who in 1912 published an analysis of a set of postulates for defining Hensel's system of p-adic numbers. This article, in turn, eventually evolved into a series of works that mark the starting point of the abstract theory of rings. This may have been among the factors that triggered the addition of a separate section for articles of this kind. At any rate, typical of the works that were reviewed under this heading is a series of articles by the Berkeley mathematician, Benjamin Abram Bernstein. Whereas his article on the postulates of Boolean algebra [Bernstein, 1916] was now seen as dealing with the foundations of algebra, his earlier ones [Bernstein, 1911; 1913; 1914] had been reviewed in the sections on probability, elementary arithmetic, and philosophy, respectively.

The chapter on "Elementary Arithmetic and Algebra. Combinatorics" included works on elementary methods for calculating logarithms or roots, but also on combinatorics. In previous volumes, articles on combinatorics had appeared in the section on probabilities.

The most striking innovation implied by the new classification, and the one that matters most for the purposes of the present account, is the inclusion of a unified chapter for group theory and the "Abstract Theory of Fields and Modules." For the first time, the classification of algebraic topics adopted by the *Jahrbuch* implicitly indicated that domains of inquiry associated with the concepts of groups, fields, and modules, which had originated within separate contexts and which had been hitherto considered as basically different in their nature and aims, were best understood if seen as different manifestations of one and the same underlying general idea, the idea that we will later identify as that of algebraic structure. However, this is just an early indication of an ongoing process which at this point in time still showed many signs of continuity with previous conceptions. Thus, for instance, Galois theory was still included as a special subsection of chapter 3 on the "Theory of Polynomials and Algebraic Equations." The more modern conception of this theory attained its definitive form in the works of Artin in the late 1920s; Artin conceived

---

[15]See [Corry, 2003, pp. 196-201] for details.

it as the study of the interrelation among certain fields and their algebraic extensions, on the other hand, and their associated Galois groups with their subgroups, on the other hand. This became one of the prominent hallmarks of the structural conception of algebra as manifest in van der Waerden's book, but, clearly, here it was yet to be fully developed. Likewise, some works on the theory of matrices still appear in this volume of the *Jahrbuch* as part of the chapter on "Theory of Forms: Determinants" rather than as connected with rings or even vector spaces.

At any rate, one wonders what was the direct motivation behind the decision to introduce this new chapter on "Abstract Theory of Fields and Modules" at this point in time, since the absolute majority of the articles reviewed in it, at least in the 1916–1917 volume are, in fact, articles on group theory. Among the few that are not was one of Fraenkel's early expositions of the essentials of an abstract theory of rings [Fraenkel, 1916]. This was a natural continuation of Fraenkel's dissertation [Fraenkel, 1914], and, in fact, he had submitted it as his *Habilitationsschrift*. In the former work, Fraenkel had followed the paradigm of Steinitz's 1910 work on fields and applied it to a new mathematical domain, that of rings, similar to fields but also containing zero divisors. Thus, Fraenkel had proved that the investigation of the algebraic properties of any "separable ring" (which he defined according to the existing definition of separable fields) may be reduced to that of "simpler rings," namely, rings that in essence contain only one prime zero divisor. In 1916, Fraenkel attempted to extend to rings the full range of questions addressed previously by Steinitz for fields and, in particular, the question of how to characterize all possible, algebraic and transcendental extensions of a given ring thereof.[16] In spite of their close mutual connection, these two works of Fraenkel were reviewed in two volumes of the *Jahrbuch* that used different classification schemes, and thus, while the earlier one was classified as "Higher and Elementary Arithmetic: Number Theory," the second appeared under "Algebra: Group Theory. Abstract Theory of Fields and Modules (Systems of Hypercomplex Numbers)."

Some additional works reviewed in this same section on abstract algebraic theories help illustrate the import of the gradual change in the images of algebra introduced in 1916 in the *Jahrbuch*, for instance Macaulay's 1916 important tract on factorization of polynomials [Macaulay, 1916]. An earlier article in which Macaulay had dealt with similar issues, extending Lasker's results to what later became known as the Lasker-Macaulay Theorem, appeared in [Macaulay, 1913]. This article had been reviewed in the *Jahrbuch*, like Lasker's, in the category "Higher and Elementary Arithmetic: Number Theory." The 1921–1922 volume, published in 1925 and still using the same classification scheme, included in this section works of Schur and of Wolfgang Krull on the abstract theory of rings [Krull, 1922; Schur, 1922], and, of course, the path-breaking work of Emmy Noether on factorization theorems in this theory [Noether, 1921].[17] Also in 1921, Fraenkel published an additional work on the same topic [Fraenkel, 1921]. The following year two important works of Dickson on algebras, that could have previously been reviewed in the section on "Arithmetic," appeared now under the new general heading [Dickson, 1923a; 1923b].

All of these works are interesting for the purposes of the present article, and not only for the way in which they were classified in the *Jahrbuch*. In fact, their

---

[16]See [Corry, 2003, pp. 201-213] for additional details.

[17]See [Corry, 2003, pp. 225-237].

importance lies in how they actively contributed to the increasing realization of the deep change in process in the conceptions of the various algebraic domains, as well as in the interrelations among them. The important results proved in these works—and the fruitful way in which they implemented abstract formulations of concepts and increasingly structural research methods—turned them into harbingers of the emerging, structural image of algebra. Thus, in volume 51 of 1925, the subsection of algebra under consideration here, "Group Theory: Abstract Theory of Fields and Modules. Group Theory. System of Hypercomplex Numbers," turned into simply "Group Theory. Abstract Algebra." In retrospect, one may wonder why such a classification appeared only in 1925. After all, abstract formulations of central concepts had been known for many decades in algebra. As this account has stressed, however, the real change in the conception of the discipline occurred only slowly, as manifested in the *Jahrbuch*'s classification. There, it was only in 1925 that a separate part of algebra was that part dealing with several theories, all of which were defined by similar, abstract methods and of all which covered, abstractly formulated, structural questions.

On the other hand, the 1925 classification of algebraic topics in the *Jahrbuch* contains besides the already mentioned chapter on groups and abstract algebra, an additional one on the "Theory of Ideals," including works by Hasse and Masaso Sono [Hasse, 1924; Sono, 1924]. Once articles on abstract rings and abstract fields were classified under a common heading, it might have been the case that works on the abstract theory of factorization in terms of ideals would also fall into that category. That this was not the case can be taken, in my view, as further evidence of how slowly the full import of the structural conception of algebra, as embodied in van der Waerden's book, was understood. The year 1930 is, of course, important in this story as it is the year of publication of van der Waerden's *Moderne Algebra*. The stage was adequately set for its appearance, in terms of the classification scheme in the *Jahrbuch*, and, indeed, the book was reviewed in the section on "Algebra and Arithmetic" in the appropriate chapter: "Group Theory. Abstract Algebra." The same section also provided a natural framework for an important expository article on the latest developments in algebra, published that year by Helmut Hasse under the title "Die moderne algebraische Methode [Modern Algebraic Methods]" [Hasse, 1930]. The article appeared in the *Jahresbericht der Deutscher Mathematiker-Vereingung*, following in a tradition of publication of comprehensive reviews of recent work in this journal.[18] And a further work of interest reviewed in the same section that year is a new edition of Steinitz's work on fields, published with comments by Reinhold Baer and the same Hasse [Steinitz, 1930]. The *Jahrbuch*'s review of this article reproduced the introduction written by Baer and Hasse, which emphasizes the importance of the new image of algebra and the seminal role played in 1910 by the article in bringing about its consolidation. The review stressed that the article represented the starting point of much important research in algebra, and that it had became not only a milestone in the development of this discipline but also "an excellent and absolutely essential introduction for anyone intent on devoting himself to the study of the new algebra."

Moderne Algebra, as noted, played a decisive role in extending these ideas to the whole of algebra, thus crystallizing and helping to spread among a broad audience

---

[18]See the editorial remarks in the *Jahresbericht der Deutscher Mathematiker-Vereinigung* 1 (1891), 12.

in the mathematical community the new image of the discipline at the center of which stood the idea of an algebraic structure. The classification schemes of the *Jahrbuch* show, however, that there was still room for uncertainty as to the details of this image, hence the slight changes in this scheme in subsequent years. In volume 61 of 1935, the chapter on "Abstract Algebra" is further subdivided into a first section on groups and a second on rings and fields, whereas in volume 63 of 1937, the second section includes lattices together with groups and fields. The idea of a lattice had made its initial appearance in the late nineteenth century in the works of Schröder and Dedekind, but an elaborate theory built around this concept in its abstract formulation developed only after 1935 with the works of Garrett Birkhoff [Birkhoff, 1935] and Oystein Øre [Øre, 1935; 1936], and certainly under the new structural spirit promoted by van der Waerden's book.[19] The first works of this kind were classified in the *Jahrbuch* together with rings and fields, but it gradually became apparent to the editors that it would be adequate to point out explicitly that the section reviewed dealt not only with works on rings and fields but also with works on lattices.

The 1939 volume of the *Jahrbuch*, published in 1941, presented a final version of the classification scheme for the section on arithmetic and algebra, which reflected, in fact, the view of the discipline that would remain standard for decades to come. It comprised the following sections:

Ch. 1 General and Combinatorics
Ch. 2 Linear Algebra. Theory of Invariants
Ch. 3 Polynomials and Algebraic Equations
Ch. 4 Group Theory
Ch. 5 Abstract Theory of Lattices, Rings and Fields
Ch. 6 Fields of Numbers and Functions
Ch. 7 Number Theory
Ch. 8 Diophantine Approximations and Transcendental Problems

Concluding Remarks

To conclude this overview of the development of the images of algebra between 1900 and 1930 as reflected in the classification schemes of the *Jahrbuch*, it is illuminating to describe briefly the schemes adopted by the two new mathematical review journals, the *Zentralblatt* and the *Mathematical Reviews* in their early volumes. These two journals soon superseded the *Jahrbuch* and, in many respects, they may be considered as portraying more modern images and conceptions of mathematics than their predecessor [Siegmund-Schultze, 1994]. Yet, at least inasmuch as algebra is concerned, the two new reviewing journals were slow fully to adopt the structural image of the discipline. Consider, for instance, the classification scheme of the first volume of the *Zentralblatt*, published in 1931 under the editorship of Otto Neugebauer. Works in algebra appear under the major heading of "Arithmetic, Algebra and Group Theory," which comprises the following subsections:

- Foundations of Arithmetic and Algebra
- Linear Algebra. Determinants. Bilinear and Quadratic Forms
- Algebraic Equations
- Group Theory

---

[19]See [Corry, 2003, pp. 259-268] for additional details.

- Algebraic Numbers, Field Theory, Galois Theory, Ideal Theory in Fields of Numbers and of Functions
- Abstract Theory of Rings. Hypercomplex Numbers
- Invariant Theory. Elimination of Polynomial Ideals
- Number Theory
- Analytic Number Theory. Dirichlet Series, Diophantine approximations

Thus, at a time when the *Jahrbuch* was already reflecting the new, structural image of algebra, as described above, in a much more consistent fashion, the *Zentralblatt* still hesitated on this issue. On the one hand, rings and hypercomplex systems were conceptually associated there, but on the other hand, they were still separated from both field theory and the abstract theory of ideals.

The first issue of the *Mathematical Reviews* appeared in January 1941. The main driving force behind the new American reviewing journal was, once again, Neugebauer, who had emigrated to America, establishing himself at Brown University. Issues of the journal appeared monthly and were collected as a yearly volume, of which an index was compiled and articles classified retrospectively. Thus, although some kind of division into topics is found in this index, it appears more as an *a posteriori* organization of what was published than as a preconceived idea of how algebra, as well as all other mathematical disciplines, should be organized and subdivided. The classification of algebraic issues was far from systematic and changed from issue to issue in this first volume. The second volume of the journal seems to have been compiled under a more systematic and preconceived classification scheme, and this scheme gives a clearer idea of how algebra was conceived in the initial stages of the *Reviews*. Thus, articles on algebra are classified according to three major categories, namely, "Abstract Algebra," "Linear Algebra," and "Equations." These are further subdivided as follows:

Abstract Algebra
- Lattices and Boolean Algebras
- Rings and Ideal Theory
- Fields and Algebras
- Galois Theory
- p-adic Theories
- Function Fields

Algebra: Equations
- Symmetric Functions
- Zeros
- Classical Galois Theory
- Systems of Equations, Elimination
- Special Equations

Linear Algebra
- Matrices, Determinants, General Theory
- Special Matrices, Determinants
- Hypercomplex Systems
- Linear Forms and Equations
- Quadratic and Bilinear Forms
- Forms of Higher Degree
- Characteristic Values, Elementary Divisors

Beyond these basic categories, there was also a major section on the theory of groups, which is subdivided into the following: "Finite," "Abelian," "Abstract Representations," "Characters," "Continuous Topological," "Lie," "Crystallography," "Generalized." In addition, there were shorter sections on topics related to algebra but not seen as part of the hard core of the discipline such as "Algebraic Functions," "Algebraic Geometry," "Algebraic Invariants" (as part of a more general section on invariants), "Algebra of Logic," and "Algebraic Number Theory" (as part of a more general section on number theory).

Although very close to the conception of algebra embodied in van der Waerden's book, there are still interesting differences such as the separate status accorded to group theory in all its manifestations. The year of publication of the second volume of the *Mathematical Reviews*, 1941, was also the year of publication of Birkhoff and Mac Lane's *Survey*, a textbook whose widespread adoption in American universities helped bring about the adoption of the new image of algebra in the burgeoning community of algebraists in the United States [Birkhoff and Mac Lane, 1941].

References

Abraham, Max. 1901. "Mechanik der deformierbaren Körper. Geometrische Grundbegriffe." *Encyklopädie der mathematischen Wissenschaften mit Einschluss ihrer Anwendungen* 4 (2), 3-47.

Artin, Emil and Schreier, Oscar. 1926. "Algebraische Konstruktion reeller Körper." *Abhandlungen aus dem mathematisches Seminar der Hamburgischen Universität* 5, 85-99.

Bernstein, Benjamin A. 1912. "On an Algebra of Probability." *Bulletin of the American Mathematical Society* 18, 450.

_____. 1913. "A Set of Postulates for the Algebra of Positive Rational Numbers with Zero." *Bulletin of the American Mathematical Society* 9, 517.

_____. 1914. "A Complete Set of Postulates for the Logic of Classes Expressed in Terms of the Operation 'Exception' and a Proof of the Independence of a Set of Postulates due to Del Re." *Bulletin of the American Mathematical Society* 21, 103.

_____. 1915. "A Set of Four Independent Postulates for Boolean Algebras." *Bulletin of the American Mathematical Society* 22, 6.

_____. 1916. "A Set of Four Independent Postulates for Boolean Algebras." *Transactions of the American Mathematical Society* 17, 50-52.

Birkhoff, Garrett. 1935. "On the Lattice Theory of Ideals." *Bulletin of the American Mathematical Society* 40, 613-619.

Birkhoff, Garrett and Mac Lane, Saunders. 1941. *A Survey of Modern Algebra*. New York: MacMillan.

Bucherer, Alfred. 1903. *Elemente der Vektor-Analysis*. Leipzig: B. G. Teubner Verlag.

Carmichael, Robert D. 1912. "On the Theory of Relativity: Analysis of the Postulates." *Physical Reviews* 35, 153-176.

Carstens, R. L. 1906. "A Definition of Quaternions by Independent Postulates." *Bulletin of the American Mathematical Society* (2) 12, 392-394.

Corry, Leo. 2001. " Mathematical Structures from Hilbert to Bourbaki: The Evolution of an Image of Mathematics." In *Changing Images in Mathematics: From the French Revolution to the New Millennium.* Ed. Umberto Bottazzini and Amy Dahan-Dalmedico. London and New York: Routledge, 167-186.

―――. 2003. *Modern Algebra and the Rise of Mathematical Structures.* 2d Rev. Ed. Boston and Basel: Birkhäuser Verlag. (1st Ed. 1996).

Crowe, Michael. 1967. *A History of Vector Analysis.* Notre Dame: Notre Dame University Press.

Despeaux, Sloan. 2002 "The Development of a Publication Community: Nineteenth-Century Mathematics in British Scientific Journals." Unpublished Ph.D. Dissertation: University of Virginia.

Dickson, Leonard E. 1903a. "Fields Whose Elements Are Linear Differential Expressions." *Bulletin of the American Mathematical Society* 10, 30-31.

―――. 1903b. "Definitions of a Field by Independent Postulates." *Transactions of the American Mathematical Society* 4, 13-20.

―――. 1903c. "Definitions of a Linear Associative Algebra by Independent Postulates." *Transactions of the American Mathematical Society* 4, 21-26.

―――. 1905a. "On Finite Algebras." *Göttingen Nachrichten*, 358-393.

―――. 1905b. "Definitions of a Group and a Field by Independent Postulates." *Transactions of the American Mathematical Society* 6, 198-204.

―――. 1923a. *Algebras and their Arithmetics.* Chicago: University of Chicago Press.

―――. 1923b. "General Theory of Hypercomplex Integers." *Bulletin of the American Mathematical Society* 29, 200.

―――. 1926. *Algebraic Theories.* Chicago: Benjamin H. Sanborn.

Fraenkel, Abraham H. 1912. "Axiomatische Begründung von Hensels *p*-adischen Zahlen." *Journal für die reine und angewandte Mathematik* 141, 43-76.

―――. 1914. "Über die Teiler der Null und die Zerlegung von Ringen." *Journal für die reine und angewandte Mathematik* 145, 139-176.

―――. 1916. *Über gewisse Teilbereiche und Erweiterungen von Ringen.* Leipzig: B. G. Teubner Verlag.

―――. 1921. "Über einfache Erweiterungen zerlegbarer Ringe." *Journal für die reine und angewandte Mathematik* 151, 121-166.

Fricke, Robert. 1924. *Lehrbuch der Algebra - verfasst mit Benutzung vom Heinrich Webers gleichnamigem Buche.* Vol. 1. Braunschweig: Vieweg Verlag.

Gans, Richard. 1905. *Einführung in die Vektoranalysis mit Anwendungen auf die mathematische Physik.* Leipzig: B. G. Teubner Verlag.

Gispert, Hélène. 2001. "The German and French Editions of the Klein-Molk Encyclopedia: Contrasted Images." In *Changing Images in Mathematics: From the French Revolution to the New Millennium.* Ed. Umberto Bottazzini and Amy Dahan-Dalmedico. London and New York: Routledge, 93-112.

Hasse, Helmut. 1925. "Über das allgemeine Reziprozitätsgesetz in algebraischen Zahlkörpern." *Jahresbericht der Deutschen Mathematiker Vereinigung* 33, 97-101.

―――. 1926. *Höhere Algebra.* Berlin: Sammlung Göschen.

---------. 1930. "Die moderne algebraische Methode." *Jahresbericht der Deutschen Mathematiker Vereinigung* 39, 22-34.

Hawkes, Herbert E. 1905. "On Quaternion Number-Systems." *Mathematische Annalen* 60, 437-447.

Huntington, Edward. 1902. "A Complete Set of Postulates for the Theory of Absolute Continuous Magnitude." *Transactions of the American Mathematical Society* 3, 264-279.

---------. 1904. "Sets of Independent Postulates for the Algebra of Logic." *Transactions of the American Mathematical Society* 5, 288-309.

---------. 1905. "Note on the Definitions of Abstract Groups and Fields by Sets of Independent Postulates." *Transactions of the American Mathematical Society* 6, 181-197.

---------. 1913. "A Set of Postulates for Abstract Geometry, Expressed in Terms of the Simple Relation of Inclusion." *Bulletin of the American Mathematical Society* 19, 171-172.

Joly, Charles J. 1903a. "Integrals Depending on a Single Quaternion Variable." *Proceedings of the Royal Dublin Society* 8, 6-20.

---------. 1903b. "The Multilinear Quaternion Function." *Proceedings of the Royal Dublin Society* 8, 47-52.

---------. 1905. *Manual of Quaternions.* London: Macmillan.

Jordan, Camille. 1870. *Traité des substitutions et des équations algébriques.* Paris: Gauthier-Villars.

Kelland, Phillip and Tait, Peter G. 1906. *Introduction to Quaternions.* Prepared by Cargill G. Knott. London: Macmillan and Co.

Knott, Cargill G. 1905. "Hamilton's Quaternion Vector Analysis." *Jahresbericht der Deutschen Mathematiker Vereinigung* 14, 167-171.

Krull, Wolfgang. 1922. "Algebraische Theorie der Ringe. I." *Mathematische Annalen* 88, 80-122.

Lampe, Emil. 1903. "Das *Jahrbuch* über die Fortschritte der Mathematik: Rückblick und Ausblick." *Jahrbuch über die Fortschritte der Mathematik* 33, 1-5.

Landsberg, Georg. 1899. "Algebraische Gebilde: Arithmetische Theorie algebraischer Grössen." *Encyklopädie der mathematischen Wissenschaften mit Einschluss ihrer Anwendungen,* I.1:283-319.

Lasker, Emanuel. 1905. "Zur Theorie der Moduln und Ideale." *Mathematische Annalen* 60, 20-115.

Loewy, Alfred. 1915. *Lehrbuch der Algebra. Erster Teil: Grundlagen der Arithmetik.* Berlin: Veit.

Lorentz, Hendrik A. 1904. "Weiterbildung der Maxwellschen Theorie. Elektronentheorie." *Encyklopädie der mathematischen Wissenschaften mit Einschluss ihrer Anwendungen,* V.1-2:145-280.

Macaulay, Francis S. 1913. "On the Resolution of a Given Modular System into Primary Systems including Some Properties of Hilbert Numbers." *Mathematische Annalen* 74, 66-121.

---------. 1916. *The Algebraic Theory of Modular Systems.* Cambridge: Cambridge University Press.

MacFarlane, Alexander. 1906. *Vector Analysis and Quaternions.* 2d Ed. New York: Wiley.

Marcolongo, Roberto and Burali-Forti, Cesare. 1907. "Per l'unificazione delle notazioni vettoriali. I, II, III." *Rendiconti del Circolo matematico di Palermo* 23, 324-328; 24, 65-80, 318-332.

Noether, Emmy. 1921. "Idealtheorie in Ringbereichen." *Mathematische Annalen* 83, 24-66.

———. 1926. "Abstrakter Aufbau der Idealtheorie in algebraischen Zahl und Funktionenkörpern." *Mathematische Annalen* 96, 26-61.

Øre, Oystein. 1935. "On the Foundation of Abstract Algebra. I." *Annals of Mathematics* 36, 406-437.

———. 1936. "On the Foundation of Abstract Algebra. II." *Annals of Mathematics* 37, 265-292.

Parshall, Karen H. 1983. "In Pursuit of the Finite Division Algebra Theorem and Beyond: Joseph H. M. Wedderburn, Leonard E. Dickson, and Oswald Veblen." *Archives internationales d'histoire des sciences* 33, 223-349.

———. 1985. "Joseph H. M. Wedderburn and the Structure Theory of Algebras." *Archive for History of Exact Sciences* 32, 223-349.

Pascal, Ernesto. 1902. "Introduzione alla teoria invariantiva delle equazioni di tipo generale ai differenziali totali di secondo ordine (Memoria I)." *Annali di matematica* 7, 1-38.

———. 1903. "Introduzione alla teoria delle forme differenzialle di ordine qualunque." *Atti della Academia nazionale dei Lincei (Roma), Rendiconti*, 12, 325-332.

Schröder, Ernst. 1890. *Vorlesungen über die Algebra der Logik. (Exacte Logik).* I. Leipzig: B. G. Teubner Verlag.

Schur, Issai. 1902. "Über einen Satz aus der Theorie der vertauschbaren Matrizen." *Königlich preussische Akademie der Wissenchaften (Berlin) Sitzungsberichte*, 120-125.

———. 1905. "Zur Theorie der vertauschbaren Matrizen." *Journal für die reine und angewandte Mathematik* 130, 66-76.

———. 1922. "Über Ringbereiche im Gebiete der ganzzahligen linearen Substitutionen." *Königlich preussische Akademie der Wissenchaften (Berlin) Sitzungsberichte*, 145-168.

Serret, Joseph. 1849. *Cours d'algèbre supérieure*. Paris: Gauthier-Villars. (2d Ed. 1854; 3d Ed. 1866).

Siegmund-Schultze, Reinhard. 1993. *Mathematische Berichterstattung in Deutschland: Der Niedergang des "Jahrbuchs über die Fortschritte der Mathematik."* Göttingen: Vandenhoeck & Ruprecht.

———. 1994. "'Scientific Control' in Mathematical Reviewing and German-US-American Relations between the Two World Wars." *Historia Mathematica* 21, 306-329.

Sono, Masaso. 1924. "On the Reduction of Ideals." *Memoirs Kyoto* (A) 7 (1924), 191-204.

Steinitz, Ernst. 1910. "Algebraische Theorie der Körper." *Journal für die reine und angewandte Mathematik* 137, 167-309.

———. 1930. *Algebraische Theorie der Körper: Neu herausgegeben, mit Erläuterungen und einem Anhang: Abriss der Galoisschen Theorie versehen von R. Baer und H. Hasse*. Berlin: W. de Gruyter & Co.

Stringham, Irving. 1905. "A Geometric Construction for Quaternion Products." *Bulletin of the American Mathematical Society* 2, 437-439.

van der Waerden, Bartel L. 1930. *Moderne Algebra*. 2 Vols. Berlin: Springer-Verlag.

Weber, Heinrich. 1893. "Die allgemeinen Grundlagen der Galoisschen Gleichungstheorie." *Mathematische Annalen* 43, 521-549.

———. 1895 *Lehrbuch der Algebra*. Braunschweig: F. Vieweg und Sohn.

Wedderburn, Joseph H. M. 1903. "On the General Scalar Function of a Vector." *Proceedings of the Edinburgh Royal Society* 24, 409-412.

———. 1904. "Note on the Linear Matrix Equation." *Proceedings of the Edinburgh Mathematical Society* 22, 49-53.

———. 1905. "A Theorem on Finite Algebras." *Transactions of the American Mathematical Society* 6, 349-352.

———. 1907. "On Hypercomplex Numbers." *Proceedings of the London Mathematical Society* 6, 77-118.

CHAPTER 11

A Historical Sketch of B. L. van der Waerden's Work in Algebraic Geometry: 1926–1946

Norbert Schappacher
Université Louis Pasteur, France

> I am simply not a Platonist. For me mathematics is not a contemplation of essences but intellectual construction. The *Tetragonizein te kai parateinein kai prostithenai* that Plato speaks of so contemptuously in *Republic 527A* is my element.[1]

Introduction

Algebraic geometry might be defined as the treatment of geometrical objects and problems by algebraic methods. According to this *ad hoc* definition,[2] what algebraic geometry is at a given point in history will naturally depend on the kind of geometrical objects and problems accepted at the time, and even more on the contemporary state of algebra. For instance, in Descartes's early seventeenth century, "algebraic geometry" (in the sense just defined) consisted primarily in applying the new algebra of the time to problems of geometrical constructions inherited mostly from antiquity. In other words, the "algebraic geometry" of early modern times was the so-called analytic art of Descartes, Viète, and others.[3]

The discipline which is called algebraic geometry today is much younger. It was first created by a process of gradual dissociation from analysis after the Riemannian revolution in geometry. Bernhard Riemann had opened the door to new objects that eventually gave rise to the various sorts of varieties—topological, differentiable, analytic, algebraic, etc.—which happily populate geometry today. After a

---

[1] "Ich bin halt doch kein Platoniker. Für mich ist Mathematik keine Betrachtung von Seiendem, sondern Konstruieren im Geiste. Das *Tetragonizein te kai parateinein kai prostithenai*, von dem Platon im Staat 527A so verächtlich redet, ist mein Element." Postscript of Bartel L. van der Waerden's letter to Hellmuth Kneser dated Zürich, 10 July, 1966, [NSUB, Cod. Ms. H. Kneser A 93, Blatt 19]. Van der Waerden begs to differ with the following passage of Plato's *Republic* (as it appears in Benjamin Jowett's translation): "Yet anybody who has the least acquaintance with geometry will not deny that such a conception of the science is in flat contradiction to the ordinary language of geometricians.—How so?—They have in view practice only, and are always speaking in a narrow and ridiculous manner, of *squaring and extending and applying* and the like—they confuse the necessities of geometry with those of daily life; whereas knowledge is the real object of the whole science." The italicized words are quoted in Greek by van der Waerden.

[2] This definition was suggested to me by Catherine Goldstein several years ago to fix ideas in the course of a discussion.

[3] Compare [Bos, 2001].

strong initial contribution by Alfred Clebsch and Max Noether as well as Alexander von Brill and Paul Gordan, the main development—important foreign influence notwithstanding, for instance by the Frenchman Émile Picard—came at the hands of Italian mathematicians such as the two leading figures in the classification of algebraic surfaces, Guido Castelnuovo and Federigo Enriques, as well as Eugenio Bertini, Pasquale del Pezzo, Corrado Segre, Beppo Levi, Ruggiero Torelli, and Carlo Rosati in his earlier works. This—I am tempted to say—golden period of Italian algebraic geometry may be argued to have more or less ended with World War I.[4] Yet, some of the authors, like Rosati, continued to be active and were joined by younger colleagues like Beniamino Segre. The strongest and most visible element of continuity of Italian algebraic geometry, after World War I and into the 1950s, however, was the towering figure of Francesco Severi, whose long and active life connects the golden first period with the following second period. At the end of this second period, Italian algebraic geometry essentially ceased to exist as a school identifiable by its method and production.

Meanwhile on an international scale, the discipline of algebraic geometry underwent a major methodological upheaval in the 1930s and 1940s, which today tends to be principally associated with the names of André Weil and Oscar Zariski. Subsequently, another rewriting occurred under Alexander Grothendieck's influence as of the early 1960s. Both of these twentieth-century upheavals redefined algebraic geometry, changing its methods and creating new types of mathematical practice. The second rewriting, at the hands of Grothendieck, also clearly changed the realm of objects; algebraic geometry became the theory of schemes in the 1960s. In contrast to this, the relevance of new objects for the rewriting of algebraic geometry in the 1930s and 1940s is less marked and depends in part on the authors and papers considered. At any rate, both rewritings appear to have preserved both the objects and the big problems studied in the previous incarnations of algebraic geometry. For example, the resolution of singularities for higher-dimensional algebraic varieties was prominent in Italian algebraic geometry, which claimed to have solved it up to dimension 2, and it continues to arouse interest even today. But new problems were added at the crossroads of history, either inherited from other traditions which had formerly not belonged to algebraic geometry—for instance, the analog of the Riemann Hypothesis for (function fields of) curves over finite fields—or created by the new methods—like Grothendieck's so-called "Standard Conjectures."

In this chapter, I discuss Bartel Leendert van der Waerden's contributions to algebraic geometry in the 1920s and 1930s (as well as a few later articles) with a view to an historical assessment of the process by which a new type of algebraic geometry was established during the 1930s and 1940s. The simultaneous decline of Italian algebraic geometry, its causes and the way it happened, is at best a side issue of the present chapter.[5] However, the relationship between new and old algebraic geometry in the 1930s and 1940s is at the heart of the discussion here, in part because of the interesting way in which van der Waerden's position with respect

---

[4]This point of view is also taken in [Brigaglia and Ciliberto, 1995].

[5]I plan to treat this in greater detail elsewhere. In fact, the present chapter on van der Waerden sketches only one slice of a larger project to study the history of algebraic and arithmetic geometry between 1919 (Noether's report on the arithmetic theory of algebraic functions in one variable) and 1954 (Weil's well-prepared coup against Severi at the International Congress of Mathematicians (ICM) in Amsterdam), that is, before the advent of cohomological methods in algebraic geometry.

to Italian algebraic geometry evolved in the 1930s (see the section on the years 1933 to 1939 below), but mostly because any historical account of the rewriting of algebraic geometry must answer the question of how the old and new practices related to each other.

A first explanation of this historical process could interpret the dramatic changes of the 1930s and 1940s as the natural consequence of the profound remodeling of algebra in the first third of the twentieth century; such an interpretation is perhaps suggested by the *ad hoc* definition in terms of objects, problems, and methods of algebraic geometry given above and by the fact that this rewriting essentially meant to preserve the objects and problems treated by the Italian authors. In this view, new, powerful algebra was being brought to bear on algebraic geometry, transforming this field so as to bring it closer to the algebraic taste of the day. The decline of Italian algebraic geometry around the same time might then simply express the failure on the part of the Italians to adopt that new way of doing algebra. Within this historical scheme, one would still wish to have a more specific explanation of why the Italian algebraic geometers failed to adapt to the new ways of algebra between the wars; for instance, some thought that algebraic geometry was a discipline separated from the rest of mathematics by a special sort of intuition needed to give evidence to its insights.[6] But even in the absence of this kind of a more detailed analysis, a plain historical mechanism—the adoption of a new algebraic methodology, the roots of which could be studied independently[7]—would be used to account for the rewriting of algebraic geometry in the 1930s and 1940s.

This first scheme of historical explanation would seem *a priori* to be particularly well adapted to an analysis of van der Waerden's contributions because the remodeling of algebra to which we have alluded was epitomized in his emblematic textbook *Moderne Algebra* [van der Waerden, 1930–1931]. Even though its author was but the skillful compiler and presenter of lectures by Emil Artin and Emmy Noether, he would obviously appear to have been particularly well placed to play an important role when it came to injecting modern algebra into algebraic geometry. As we will see in the next section, he appears to have set out to do precisely that. Moreover, main actors of the then modern and new development of algebra were aware of its potential usefulness for recasting algebraic geometry. This applies in the first place to Emmy Noether. As early as 1918, she had written a report for the *Deutsche Mathematiker-Vereinigung* (DMV) on the arithmetic theory of algebraic functions of one variable and its relation, especially, to the theory of algebraic number fields [Noether, 1919] and, in so doing, had complemented the earlier

---

[6] See, for example, [Weil, 1979, p. 555], where he states that "On the subject of algebraic geometry, some confusion still reigned. A growing number of mathematicians, and among them the adepts of Bourbaki, had convinced themselves of the necessity of founding all of mathematics on the theory of sets; others doubted that that would be possible. Exception was taken for probability, ..., differential geometry, algebraic geometry; it was held that they needed autonomous foundations, or even (confounding the needs of invention with those of logic) that the constant intervention of a mysterious intuition was required [Au sujet de la géométrie algébrique, il régnait encore quelque confusion dans les esprits. Un nombre croissant de mathématiciens, et parmi eux les adeptes de Bourbaki, s'étaient convaincus de la nécessité de fonder sur la théorie des ensembles toutes les mathématiques; d'autres doutaient que cela fût possible. On nous objectait le calcul des probabilités, ..., la géométrie différentielle, la géométrie algébrique; on soutenait qu'il leur fallait des fondations autonomes, ou même (confondant en cela les nécessités de l'invention avec celles de la logique) qu'il y fallait l'intervention constante d'une mystérieuse intuition]."

[7] These *have* been studied independently. See, for instance, [Corry, 1996/2003].

report of 1892–1893 by Alexander Brill and her father Max Noether [Brill and Noether, 1892–1893]. Noether had also actively helped introduce ideal-theoretic methods into algebraic geometry in the 1920s, in particular via her rewriting of Hentzelt's dissertation [Noether, 1923a] and her article on "Eliminationstheorie und allgemeine Idealtheorie [Elimination Theory and General Ideal Theory]" [Noether, 1923b], which inspired the young van der Waerden's first publication on algebraic geometry.

As we shall see below, however, this first scheme of explanation, according to which modern algebra is the principal motor of the process, does not suffice to account for van der Waerden's changing relationship with Italian algebraic geometry, let alone serve as an historical model for the whole rewriting of algebraic geometry in the 1930s and 1940s. Not only is the notion of applying modern algebra to algebraic geometry too vague as it stands, but following the first scheme carries the risk of missing the gossamer fabric of motivations, movements, and authors which renders the historiography of the first rewriting of algebraic geometry in the twentieth century so challenging and instructive.

Another explanation of this historical process, several variants of which are widespread among mathematicians, is implicit in the following quote by David Mumford from the preface to Carol Parikh's biography of Oscar Zariski:

> The Italian school of algebraic geometry was created in the late 19th century by a half dozen geniuses who were hugely gifted and who thought deeply and nearly always correctly about their field. ... But they found the geometric ideas much more seductive than the formal details of the proofs .... So, in the twenties and thirties, they began to go astray. It was Zariski and, at about the same time, Weil who set about to tame their intuition, to find the principles and techniques that could truly express the geometry while embodying the rigor without which mathematics eventually must degenerate to fantasy [Parikh 1991, pp. xxv–xxvi].

According to this view, the principal origin of the process lay in the lack of rigor on the part of the Italians; the injection of new algebraic techniques into algebraic geometry was simply necessary in order "truly" to bring out what the Italians had been trying to do with their inadequate methodology. Aside from the fact that no human mathematical formulation of a problem or phenomenon can ever reasonably be called the "true" one, Mumford's last sentence above is especially difficult to reconcile with the historical facts because of the considerable variety of ways to rewrite algebraic geometry which were under discussion in the 1930s and 1940s (compare the section on the years from 1933 to 1946 below).

The first part of Mumford's account, which isolates the Italians' lack of rigor as the principal motivation behind the development and interprets the rewriting of algebraic geometry as a reaction to it, has its origin in the experience of many mathematicians trying to work their way through the Italian literature on algebraic geometry. We shall see van der Waerden, too, was occasionally exasperated with the Italian sources. But there are two reasons why such an explanation of what happened in the 1930s and 1940s is insufficient. On the one hand, I will show on another occasion that these difficulties were not just due to a lack of rigor on the

Italian side, but can best be described as a clash of cultures of scientific publishing.[8] On the other hand, I shall sketch below—and this will show the need to correct both schemes of explanations discussed so far—how the rewriting of algebraic geometry was a much more complicated process in which several different mathematicians or mathematical schools, with different goals and methods, interacted, each in a different way, with Italian algebraic geometry. Political factors will be seen to play a non-negligible part in this dynamic. At the end of the day, Weil and Zariski indeed stand out as those who accomplished the decisive shift after which the practice of algebraic geometry could no longer resemble that of the Italian school.

Note, incidentally, that van der Waerden is not mentioned by Mumford as one of those who put algebraic geometry back on the right track. I am in no way pointing this out to suggest that Mumford did not want to give van der Waerden his due—in fact, he does mention him in a similar context in an article which is also reproduced in Parikh's biography of Zariski—but it seems to me that van der Waerden's sinuous path between algebra and geometry, which I will outline in this chapter, simply does not suggest Mumford's claim about "the principles and techniques that could truly express the geometry while embodying the rigor" [Parikh, 1991, p. 204]. Zariski's and Weil's (different!) algebraic reconstructions of algebraic geometry, on the other hand, may indeed convey the impression of justifying it because of the way in which these latter authors presented their findings. My main claim, then, which will be developed in this chapter at least as far as van der Waerden is concerned, is that the difference, especially between van der Waerden and Weil, is less a matter of mathematical substance than of style.

Indeed, compared to Weil's momentous treatise *Foundations of Algebraic Geometry* [Weil, 1946a], van der Waerden's articles on algebraic geometry may appear piecemeal, even though they do add up to an impressive body of theory,[9] most, if unfortunately not all,[10] of which has been assembled in [van der Waerden, 1983]. This piecemeal appearance may be related to van der Waerden's "non-platonic" way of doing mathematics as he described it to Hellmuth Kneser in the postscript chosen as the epigraph of this chapter. Van der Waerden was quite happy to develop bit by bit the minimum techniques needed to algebraize algebraic geometry, but he left the more essentialist discourse to others. Later in his life, he would feel that he was world-famous for the wrong reason—namely, for his book on algebra—whereas his more original contributions, especially those he had made to algebraic geometry, were largely forgotten.[11]

---

[8]It therefore goes without saying that I do not go along with the caricature of Italian algebraic geometry presented in [de Boer, 1994].

[9]Elements of this body continue to be used today in research to great advantage. For instance, Chow coordinates have had a kind of renaissance recently in Arakelov theory as seen, for example, in [Philippon, 1991–1995], and transcendence techniques have been improved using multi-homogeneous techniques first developed by van der Waerden. See, for instance, the reference to [van der Waerden, 1928c] in [Rémond, 2001, p. 57].

[10]Van der Waerden's papers sadly and surprisingly missing from the volume [van der Waerden, 1983] include: [van der Waerden, 1926b; 1928b; 1928c; 1941; 1946; 1947b; 1948; 1950a; 1950b; 1956a; 1956b; and 1958].

[11]Compare Hirzebruch's *Geleitwort* to the volume [van der Waerden, 1983, p. iii].

1925: Algebraizing Algebraic Geometry à la Emmy Noether

On 21 October, 1924, Luitzen Egbertus Jan Brouwer from Laren (Nord-Holland) wrote a letter to Hellmuth Kneser, then assistant to Richard Courant in Göttingen, announcing the arrival of Bartel Leendert van der Waerden:

> In a few days, a student of mine (or actually rather of Weitzenböck's) will come to Göttingen for the winter term. His name is van der Waerden, he is very bright and has already published things (especially about invariant theory). I do not know whether the formalities a foreigner has to go through in order to register at the University are difficult at the moment; at any rate, it would be very valuable for van der Waerden if he could find help and guidance. May he then contact you? Many thanks in advance for this.[12]

About ten months after his arrival in Göttingen, on 14 August, 1925, the twenty-two-year-old van der Waerden submitted his first paper on algebraic geometry to the *Mathematische Annalen* with the help of Emmy Noether: "Zur Nullstellentheorie der Polynomideale" [van der Waerden, 1926a]. Its immediate reference point was [Noether, 1923b], and its opening sentences sound like a vindication of the thesis indicated above that the development of algebraic geometry reflects the state of algebra at a given time. This interpretation was also endorsed by the author himself when he looked back on it forty-five years later: "Thus, armed with the powerful tools of Modern Algebra, I returned to my main problem: to give algebraic geometry a solid foundation."[13]

Van der Waerden opened his article in no uncertain terms:

> The rigorous foundation of the theory of algebraic varieties in n-dimensional spaces can only be given in terms of ideal theory because the definition of an algebraic variety itself leads immediately to polynomial ideals. Indeed, a variety is called algebraic, if it is given by algebraic equations in the n coordinates, and the lefthand sides of all equations that follow from the given ones form a polynomial ideal.
>
> However, this foundation can be formulated more simply than it has been so far, without the help of elimination theory, on the sole basis of field theory and of the general theory of ideals in ring domains.[14]

---

[12]"In einigen Tagen kommt ein Schüler von mir (oder eigentlich mehr von Weitzenböck) nach Göttingen zum Wintersemester. Er heisst van der Waerden, ist sehr gescheit und hat schon einiges publiziert (namentlich über Invariantentheorie). Ich weiss nicht, ob für einen Ausländer, der sich immatrikulieren will, die zu erfüllenden Formalitäten momentan schwierig sind; jedenfalls wäre es für van der Waerden von hohem Wert, wenn er dort etwas Hilfe und Führung fände. Darf er dann vielleicht einmal bei Ihnen vorsprechen? Vielen Dank im Voraus dafür" [NSUB, Cod. Ms. H. Kneser].

[13]See [van der Waerden, 1971, p. 172]. This passage goes on to recount the genesis and the main idea of [van der Waerden, 1926a].

[14]"Die exakte Begründung der Theorie der algebraischen Mannigfaltigkeiten in n-dimensionalen Räumen kann nur mit den Hilfsmitteln der Idealtheorie geschehen, weil schon die Definition einer algebraischen Mannigfaltigkeit unmittelbar auf Polynomideale führt. Eine Mannigfaltigkeit heißt ja algebraisch, wenn sie durch algebraische Gleichungen in den n Koordinaten bestimmt wird, und die linken Seiten aller Geichungen, die aus diesen Gleichungen folgen, bilden ein Polynomideal.

As we shall soon see, van der Waerden would change his discourse about the usefulness—let alone the necessity—of ideal theory for algebraic geometry quickly and radically. Looking back, he wrote on 13 January, 1955 in a letter to Wolfgang Gröbner (who, contrary to van der Waerden, adhered almost dogmatically to ideal theory as the royal road to algebraic geometry practically until his death): "Should one sacrifice this whole comprehensive theory only because one wants to stick to the ideal-theoretic definition of multiplicity? The common love of our youth, ideal theory, is fortunately not a living person, but a tool, which one drops as soon as one finds a better one."[15]

This statement belongs to a debate about the correct definition of intersection multiplicities (a first stage of which will be discussed in the next section). But one might actually wonder whether van der Waerden *ever* fully embraced the first sentence of his paper [van der Waerden, 1926a] about the necessity of ideal theory as the foundation of algebraic geometry. In all probability, in fact, the young author did not write the introduction. As van der Waerden states in his obituary for Emmy Noether, it was her habit with papers of her young students to write their introductions for them. In that way, she could highlight their main ideas, something they often could not do themselves [van der Waerden, 1935, p. 474]. Also, the fact that he felt or kept a certain distance from her can be gathered from remarks that van der Waerden made at different times. For instance, in a letter written on 26 April, 1926 to Hellmuth Kneser (then absent from Göttingen), van der Waerden wrote: "But you may be able to imagine that I value a conversation with you more highly than the one with Emmy Noether, which I am now facing every day (in complete recognition of Emmy's kindheartedness and mathematical capacities)."[16] And the obituary for his Jewish teacher—while in itself an act of courage in Nazi Germany, considering, in particular, the difficulties that local party officials at Leipzig created for van der Waerden then and afterwards[17]—insisted so strongly on how very special and different from ordinary mathematicians, and therefore also

---

"Die Begründung kann nur einfacher gestaltet werden als es bisher geschehen ist, nämlich ohne Hilfe der Eliminationstheorie, ausschließlich auf dem Boden der Körpertheorie und der allgemeinen Idealtheorie in Ringbereichen" [van der Waerden, 1926a, p. 183].

[15] "Soll man nun diese ganze umfassende Theorie opfern, nur weil man an der idealtheoretischen Multiplizität festzuhalten wünscht? Unsere gemeinsame Jugendliebe, die Idealtheorie, ist zum Glück kein lebender Mensch, sondern ein Werkzeug, das man aus der Hand legt, sobald man ein besseres findet" [ETHZ, Nachlass van der Waerden, HS 652:3107]. I thank Silke Slembek, who first pointed out this correspondence to me.

[16] "Dennoch werden Sie sich vielleicht vorstellen können[,] daß ich Ihre Unterhaltung höher schätze als diejenige Emmy Noethers, die mir jetzt täglich wartet (mit vollständiger Anerkennung von Emmy's Herzensgüte und mathematische Kapazitäten)" [NSUB, Cod. Ms. H. Kneser A 93, Blatt 3].

[17] Van der Waerden's personal file in the University Archives at Leipzig [UAL, Film 513] records political difficulties he had especially with local Nazis. After initial problems with Nazi students in May of 1933 and after the refusal of the ministry in Dresden to let him accept an invitation to Princeton for the winter term of 1933–1934, an incident occurred in a faculty meeting on 8 May, 1935 (that is, less than a month after Emmy Noether's death and slightly more than a month before van der Waerden submitted his obituary to the *Mathematische Annalen*).

Van der Waerden and the physicists Heisenberg and Hund inquired critically about the government's decision to dismiss four "non-Aryan" colleagues in spite of the fact that they were covered by the exceptional clause for World War I Frontline Fighters of the law of 7 April, 1933, and van der Waerden went so far as to suggest that these dismissals amounted to a disregard of the law on the part of the government. Even though he took this back seconds afterwards when attacked by a colleague, an investigation into this affair ensued which produced evidence that

from him, she had been that it makes her appear almost outlandish. Consider, for instance, the following passage (in which the gothic letters alluded to were at the time the usual symbols to denote ideals):

> It is true that her thinking differs in several respects from that of most other mathematicians. We all rely so happily on figures and formulæ. For her these utilities were worthless, even bothersome. She cared for concepts only, not for intuition or computation. The gothic letters which she hastily jotted on the blackboard or the paper in a characteristically simplified shape, represented concepts for her, not objects of a more or less mechanical computation.[18]

Regardless of van der Waerden's later opinions on the general relevance of ideal theory, in his first paper on algebraic geometry [van der Waerden, 1926a], he applied ideal theory to the very first steps of the theory of algebraic varieties. In so doing, he all but stripped it of elimination theory with which it was still intimately linked via Noether's immediately preceding works. More precisely, van der Waerden reduced to that of a mere tool the role of elimination theory in algebraic geometry, whereas ever since Kronecker, elimination theory had been an essential ingredient in the arithmetico-algebraic treatment of it. As van der Waerden put it: "Elimination theory in this setting is only left with the task to investigate how one can find in finitely many steps the variety of zeros of an ideal (once its basis is given) and the bases of its corresponding prime and primary ideals." He later repeated this move, as noted above, with respect to ideal theory.[19]

The key observation of the paper, which introduced one of the most fundamental notions into the new algebraic geometry, is today at the level of things taught in a standard algebra course. Paraphrasing §3 of [van der Waerden, 1926a], if $\Omega = \mathbf{P}(\xi_1, \ldots, \xi_n)$ is a finitely generated extension of fields, then all the polynomials f in $R = \mathbf{P}[x_1, \ldots, x_n]$, for which one has $f(\xi_1, \ldots, \xi_n) = 0$, form a prime ideal $\mathfrak{p}$ in R, and Ω is isomorphic to the field of quotients Π of the integral domain $R/\mathfrak{p}$, the isomorphism sending $\xi_1, \ldots, \xi_n$ to $x_1, \ldots, x_n$. Conversely, given a prime ideal $\mathfrak{p}$ in R (and distinct from R), there exists an extension field $\Omega = \mathbf{P}(\xi_1, \ldots, \xi_n)$ of finite type such that $\mathfrak{p}$ consists precisely of the polynomials f in $R = \mathbf{P}[x_1, \ldots, x_n]$

---

local Nazis thought him politically dangerous, citing also his behavior at the Bad Pyrmont meeting of the DMV in the fall of 1934. Van der Waerden continued not to be authorized to attend scientific events abroad to which he was invited; he was allowed neither to attend the ICM in Oslo (1936) nor events in Italy (1939, 1942). The Nazi *Dozentenbund* in April of 1940 considered van der Waerden unacceptable as a representative of "German Science," and thought him to be "downright philosemitic." I sincerely thank Birgit Petri who took the trouble to consult this file in detail.

[18] "Ihr Denken weicht in der Tat in einigen Hinsichten von dem der meisten anderen Mathematiker ab. Wir stützen uns doch alle so gerne auf Figuren und Formeln. Für sie waren diese Hilfsmittel wertlos, eher störend. Es war ihr ausschließlich um Begriffe zu tun, nicht um Anschauung oder Rechnung. Die deutschen Buchstaben, die sie in typisch-vereinfachter Form hastig an die Tafel oder auf das Papier warf, waren für sie Repräsentanten von Begriffen, nicht Objekte einer mehr oder weniger mechanischen Rechnung" [van der Waerden, 1935, p. 474].

[19] "Die Eliminationstheorie hat in diesem Schema nur die Aufgabe, zu untersuchen, wie man (bei gegebener Idealbasis) in endlichvielen Schritten die Nullstellenmannigfaltigkeit eines Ideals und die Basis der zugehörigen Primideale und Primärideale finden kann" [van der Waerden 1926a, pp. 183-184]. We do not discuss here the gradual shift from elimination to ideals from Kronecker, via König, Macaulay, and others, to Emmy Noether and her Dedekindian background. This history will, however, be treated for our larger project.

for which one has $f(\xi_1, \ldots, \xi_n) = 0$; indeed, it suffices to take $\xi_i = x_i \pmod{\mathfrak{p}}$ in $R/\mathfrak{p}$.

These constructions suggest a crucial generalization of the notion of zero, and thereby of the notion of point of an algebraic variety. The field Ω associated with $\mathfrak{p}$, which is unique up to isomorphism, "is called the *field of zeros* of $\mathfrak{p}$. The system of elements $\{\xi_1, \ldots, \xi_n\}$ is called a *generic zero* of $\mathfrak{p}$.[20] A *zero* (without further qualification) of an ideal $\mathfrak{m}$ is by definition any system of elements $\{\eta_1, \ldots, \eta_n\}$ of an extension field of $\mathbf{P}$, such that $f(\eta_1, \ldots, \eta_n) = 0$ whenever $f \equiv 0 \, (\mathfrak{p})$. A zero which is not generic is called *special*."[21] In a footnote to this passage, van der Waerden noted the analogy with the terminology of generic points used by (algebraic) geometers. He further developed this point in geometric language in §4, with reference to an affine algebraic variety M in affine n-space $C_n(\mathbf{P})$ over an algebraically closed field $\mathbf{P}$, defined by the ideal $\mathfrak{m}$:

> If M is irreducible, so that $\mathfrak{m}$ is prime, then every generic zero of the ideal $\mathfrak{m}$ is called a *generic point of the variety* M. This terminology agrees with the meaning that the words generic and special have in geometry. Indeed, by generic point of a variety, one usually means, even if this is not always clearly explained, a point which satisfies no special equation, except those equations which are met at every point. For a specific point of M, this is of course impossible to fulfil, and so one has to consider points that depend on sufficiently many parameters, that is, points that lie in a space $C_n(\Omega)$, where Ω is a transcendental extension of $\mathbf{P}$. But requiring of a point of $C_n(\Omega)$ that it be a zero of all those and only those polynomials of $\mathbf{P}[x_1, \ldots, x_n]$ that vanish at all points of the variety M yields precisely our definition of a generic point of the variety M.[22]

---

[20]Literally, van der Waerden speaks of "allgemeine Nullstelle," that is, "general zero," and continues to use the adjective "general" throughout. Our translation takes its cue from the English terminology which was later firmly established, in particular by Weil, and which echoes the Italian "punto generico."

[21]"Der nach **3** für jedes von R verschiedene Primideal $\mathfrak{p}$ konstruierbare, nach **1** auch nur für Primideale existierende, nach **2** bis auf Isomorphie eindeutig bestimmte Körper $\Omega = \mathbf{P}(\xi_1, \ldots, \xi_n)$, dessen Erzeugende ξ_i die Eigenschaft haben, daß $f(\xi_1, \ldots, \xi_n) = 0$ dann und nur dann, wenn $f \equiv 0 \, (\mathfrak{p})$, heißt Nullstellenkörper von $\mathfrak{p}$; das Elementsystem $\{\xi_1, \ldots, \xi_n\}$ heißt *allgemeine Nullstelle* von $\mathfrak{p}$. Unter *Nullstelle* schlechthin eines Ideals $\mathfrak{m}$ verstehen wir jedes Elementsystem $\{\eta_1, \ldots, \eta_n\}$ eines Erweiterungskörpers von $\mathbf{P}$, so daß $f(\eta_1, \ldots, \eta_n) = 0$, wenn $f \equiv 0 \, (\mathfrak{p})$. Jede nicht allgemeine Nullstelle heißt *speziell*" [van der Waerden, 1926a, p. 192].

[22]"Ist M irreduzibel, also $\mathfrak{m}$ prim, so heißt jede allgemeine Nullstelle des Ideals $\mathfrak{m}$ *allgemeiner Punkt der Mannigfaltigkeit* M. Diese Bezeichnung ist in Übereinstimmung mit der in der Geometrie geläufigen Bedeutung der Wörter allgemein und speziell. Man versteht doch meistens, wenn es auch nicht immer deutlich gesagt wird, unter einem allgemeinen Punkt einer Mannigfaltigkeit einen solchen Punkt, der keiner einzigen speziellen Gleichung genügt, außer denjenigen Gleichungen, die in allen Punkten erfüllt sind. Diese Forderung kann natürlich ein bestimmter Punkt von M niemals erfüllen, und so ist man genötigt, Punkte zu betrachten, die von hinreichend vielen Parametern abhängen, d.h. in einem Raum $C_n(\Omega)$ liegen, wo Ω eine transzendente Erweiterung von $\mathbf{P}$ ist. Fordert man aber von einem Punkt von $C_n(\Omega)$, daß er Nullstelle ist für alle die und nur die Polynome von $\mathbf{P}[x_1, \ldots, x_n]$, die in allen Punkten der Mannigfaltigkeit M verschwinden, so kommt man gerade auf unsere Definition eines allgemeinen Punktes der Mannigfaltigkeit M" [van der Waerden, 1926a, p. 197].

This builds a very elegant bridge from the classical to the new usage of the word. The meaning of "generic," however, was not formally defined, as van der Waerden himself remarked, in terms of parameters, even though objects depending on parameters are fairly ubiquitous in the geometric literature.[23] The word appears to have been considered as already understood, and therefore in no need of definition. Still, it is to the more philosophically minded Federigo Enriques that we owe a textbook explanation of what a generic point is that does not agree with van der Waerden's interpretation:

> The notion of a *generic* "point" or "element" of a variety, that is, the distinction between properties that pertain *in general* to the points of a variety and properties that only pertain to *exceptional* points, now takes on a precise meaning for all algebraic varieties.
>
> A property is said to pertain in general to the points of a variety V_n, of dimension n, if the points of V_n not satisfying it form—inside V_n—a variety of less than n dimensions.[24]

Contrary to van der Waerden's notion of generic points, Enriques's "points" are always points with complex coordinates, and genericity has to do with negligible exceptional sets, not with introducing parameters. This provides a first measure for the *modification* of basic notions that the rewriting of algebraic geometry entailed; defining a generic point as van der Waerden did brought out the aspect that he explained so well, but is quite different from Enriques's narrower notion of point. At the same time, the new framework of ideal theory barred all notions of (classical, analytic) continuity as, for example, in the variation of parameters; it made sense over arbitrary abstract fields.

The modest *ersatz* for classical continuity offered by the Zariski topology[25] was partially introduced in [van der Waerden, 1926a, p. 25], where the author defined the "*algebraische Abschließung*"[26] of a finite set of points to be what we would call their Zariski closure. He appended an optimistic footnote, in which he said, in

---

[23]To cite an example at random from the Italian literature, Severi's *Trattato* [Severi, 1926], which appeared in the same year as van der Waerden's paper under discussion, opened with a chapter on linear systems of plane curves. In the chapter's second section, the discussion of algebraic conditions imposed on curves in a linear system quickly turned to the case [Severi, 1926, p. 23] where the conditions vary (continuously), giving rise to the distinction between particular and general positions of the condition. The context there, as well as in many other texts of the period, was the foundation of enumerative geometry, a problem in which van der Waerden was especially interested. Compare the section on the years 1933–1939 below.

[24]"La nozione di 'punto' o 'elemento' *generico* di una varietà, cioè la distinzione fra proprietà spettanti *in generale* ai punti d'una varietà e proprietà che spettano solo a punti eccezionali, acquista ora un significato preciso per tutte le varietà algebriche.

"Si dice che una proprietà spetta in generale ai punti d'una varietà V_n, ad n dimensioni, se i punti di V_n per cui essa non è soddisfatta formano—entro V_n—una varietà a meno di n dimensioni" [Enriques and Chisini, 1915, p. 139].

[25]This is, of course, our modern terminology, not van der Waerden's in 1926. As is well known, it was actually Zariski who formally introduced this topology on his "Riemann manifolds" of function fields (the points of which are general valuations of the field) in [Zariski, 1944].

[26]The only reasonable translation of this would be "algebraic closure." However, van der Waerden used a participle of the verb "to close" instead of the noun "closure," presumably in order to avoid confusion with the algebraic closure (*algebraischer Abschluß*) of a field.

particular, that "as far as algebra is concerned, the algebraic closure is a perfect substitute for the topological closure."[27]

Finally, the *dimension* of a prime ideal $\mathfrak{p}$ (notations as above) was defined by van der Waerden, in classical geometrical style, to be the transcendence degree of the corresponding function field Ω over $\mathbf{P}$. Emmy Noether had given her "arithmetical version of the notion of dimension" via the maximum length of chains of prime ideals in §4 of [Noether, 1923b] under slightly more restrictive hypotheses, and van der Waerden generalized her results to his setting in [van der Waerden, 1926a, pp. 193-195]. He added in proof a footnote which sounded a word of caution against using chains for the notion of dimension in arbitrary rings. As is well known, this step was taken by Wolfgang Krull more than ten years later in [Krull, 1937].

As the section title just quoted from [Noether, 1923b] shows, and as repeatedly used in [van der Waerden, 1926a], developments using ideal theory were called *arithmetic* by Emmy Noether and her circle.[28] In this sense, van der Waerden's first paper on algebraic geometry provides an *arithmetization* of some of its basic notions. This terminology was made more precise by Krull, who reserved it for methods having to do with the multiplicative decomposition of ideals or valuations,[29] and from there it was adopted by Zariski for his way of rewriting the foundations of algebraic geometry as of 1938. It sounds out of place today; we would rather speak of *algebraization*. But taking the old terminology seriously and using it to a certain extent actually helps the historic analysis.

More precisely, van der Waerden's first contribution to the rewriting of algebraic geometry announced a transition from the *arithmetization* to the *algebraization* of algebraic geometry. The methods he used were undoubtedly called arithmetical at the time and place where the paper was written. The basic new notions that he brought to algebraic geometry, above all the notion of generic point, however, did not appeal to the more properly arithmetic aspects of ideal theory (like prime or primary decomposition), that is, they did not appeal to those aspects which are nowadays treated under the heading of "commutative algebra." With the success of "modern algebra," the general theory of fields as it was first presented by Steinitz, which was still considered an arithmetic theory in the 1920s, would simply be incorporated into algebra, as most of it became preparatory material for the modern treatment of the resolution of algebraic equations. Since I will describe van der Waerden's later contributions to algebraic geometry as a specific form of *algebraization*, the article [van der Waerden, 1926a] can be considered with hindsight as a first step in the direction that he would take, increasingly freeing himself from a more specifically arithmetic heritage.

---

[27]"Die algebraische Abschließung kann aber für die Algebra die Stelle der topologischen Abschließung vollständig vertreten" [van der Waerden, 1926a, pp. 197-198 (note 15)].

[28]It would be very interesting to study Emmy Noether's usage of the word "arithmetic" in detail. One might be able to argue that she tended to use the word as a synonym of "conceptual," taken in the sense that those coming after Emmy Noether have used to characterize her approach. A rather extreme example of such a characterization appeared in the passage from van der Waerden's obituary quoted above.

[29]See, in particular, [Krull, 1937, p. 745 (note 2)]: "Unter Sätzen von ausgesprochen 'arithmetischem' Charakter verstehe ich Sätze, die in den Gedankenkreis der 'multiplikativen', an Dedekind anknüpfenden Richtung der Idealtheorie und der Bewertungstheorie gehören ...."

1927–1932: Forays into Intersection Theory

It is probably not known what high or conflicting intentions the parents of H. C. H. Schubert had, in the proud town of Potsdam back in the turbulent year of 1848, when they christened their son Hermann Caesar Hannibal, but he who was thus named created a theory—the calculus of enumerative geometry—which, had it not been created, should have to be invented for the sake of historians of mathematics. For, like no other purely mathematical theory of the late nineteenth century, the so-called Schubert calculus can be regarded as an expression, in the realm of pure mathematics, of the mindset of contemporaneous industrialization. Consequently, later criticism of this theory—for what were viewed as its shaky foundations and/or for the occasional malfunctioning of its machinery at the hands of its practitioners—would eventually be cast in terms of metaphors of cultural critique.

Since the focus here, however, is on van der Waerden, I will not go into the history of the Schubert calculus. Suffice it to say that the precise goal of the theory was effectively to determine the number (not the nature!) of all the geometric objects satisfying a set of conditions, which, taken together, admit but finitely many solutions. Examples include: "(1) to find the number of circles tangent to 3 given circles, which Apollonius investigated about 200 B.C.; (2) to find the number of arbitrary conics—ellipses, parabolas and hyperbolas, as well as circles—tangent to 5 conics, which Steiner proposed in 1848 as a natural generalization of the problem of Apollonius; (3) to find the number of twisted cubics tangent to 12 quadratic surfaces, whose remarkable solution, published only in the book [Schubert, 1879] (culminating on p. 184), won Schubert the gold medal in 1875 from the Royal Danish Academy."[30] (Steiner thought the solution to (2) was $6^5 = 7776$, but was corrected by Chasles in 1864 who came up with the right answer of 3264. The prizeworthy number of solutions to (3) that Schubert found is 5,819,539,783,680.) Schubert constructed his theory as a special kind of propositional calculus—influenced by Ernst Schröder's logic, that is, by the continental counterpart of British developments in the algebra of logic—for geometric conditions. A key ingredient in building this effective calculus was Schubert's "principle of the conservation of number," which postulates the invariance—as long as the total number of solutions remains finite—of the number of solutions (always counted with multiplicities), when the constants in the equations of the geometric conditions vary.

The calculus works well and produces enormous numbers, digesting amazingly complicated situations. Its theoretical justification remained problematic, though, and in a very prominent way: David Hilbert's 15th problem in his famous 1900 ICM address called for the "rigorous foundation of Schubert's enumerative calculus," and, following artfully constructed counterexamples to Schubert's principle proposed as of 1903 by Gustav Kohn, Eduard Study, and Karl Rohn, even Francesco Severi admitted that the desire to secure the exact range of applicability of Schubert's principle was "something more than just a scruple about exaggerated rigor."[31] Severi, in the paper just quoted, reformulated the problem in terms of algebraic

---

[30] Quoted from Kleiman's concise introduction to the centennial reprint of [Schubert, 1979, p. 5]. This may also serve as a first orientation about the history of Schubert calculus.

[31] "Comunque, in questo caso si tratta di qualcosa più che un semplice scrupolo di eccessivo rigore; e la critica non è poi troppo esigente se richiede sia circoscritto con precisione il campo di validità del principio" [Severi, 1912, p. 313].

correspondences,[32] thereby providing one of the many reasons for the importance and increasing impetus of this subject in the algebraic geometry of the first half of the twentieth century. During World War I, Study's critique became more bitter, probably reflecting the fact that large numbers, without regard for the individuals in the masses that were counted, were acquiring a bad taste at that time.[33]

Van der Waerden first became acquainted with Schubert calculus, and indeed with algebraic geometry, in a course on enumerative geometry given by Hendrik de Vries at the University of Amsterdam, before he went to Göttingen.[34] He returned to this subject—apparently influenced by discussions with Emmy Noether[35]—in a paper that he submitted to the *Mathematische Annalen* just as [van der Waerden, 1926a] appeared. It is in this second paper on algebraic geometry, [van der Waerden, 1927], that one finds explicitly for the first time the other key ingredient, besides generic points, which characterized van der Waerden's rewriting of algebraic geometry, namely, what he called "*relationstreue Spezialisierung* [relation-preserving

---

[32] "I first observe, what is also implicit in Schubert's statement, that every variable condition [also of dimension less than k] imposed on the objects Γ of an algebraic variety V, ∞^k, translates into an algebraic correspondence between the elements Γ of V and the elements Γ' of another algebraic variety V' whose dimension k' has *as priori* nothing to do with k. Fixing one of the elements Γ', the elements Γ in correspondence with the given Γ' are those which satisfy a *specialization* of the variable condition.

"Thus, for example, the condition imposed on a line Γ in space to trisect an algebraic curve Γ' of given order n translates into an algebraic correspondence between the variety V_4 of all lines Γ and the algebraic variety V' (which in general is reducible and even consists of parts of different dimensions) of the curves Γ' of order n, by letting a line Γ and a curve Γ' be in correspondence if Γ trisects Γ' [Comincio dall'osservare che, come del resto è implicito nell'enunciato di Schubert, ogni condizione variabile [anche di dimensione inferiore a k] imposta agli enti Γ d'una varietà algebrica V, ∞^k, si traduce in una corrispondenza algebrica tra gli elementi Γ di V e gli elementi Γ' di un'altra varietà algebrica V', la cui dimensione k' non ha a priori alcuna relazione con k. Fissando uno degli elementi Γ', i Γ omologhi del dato Γ', son quelli che soddisfanno ad una *particolarizzazione* della condizione variabile.

"Così per esempio la condizione imposta ad una retta Γ dello spazio di trisecare una curva algebrica Γ' di dato ordine n, si traduce in una corrispondenza algebrica tra la varietà V_4 delle rette Γ e la varietà algebrica V' (generalmente riducibile e costituita anche da parti di diverse dimensioni) delle curve Γ' di ordine n, assumendosi omologhe una retta Γ ed una curva Γ', quando Γ triseca Γ']" [Severi, 1912, p. 314f].

[33] "In the case at hand, what is at issue is not only the massive figures produced by some representatives of the enumerative geometry, which one may or may not find interesting, but the methodology of algebraic geometry itself. ... The said 'principle' has also been applied in places where the usual means of algebra, applied in a thorough effort, would not only have been sufficient, but would have yielded *much more*. When one is interested in such and such 'results,' any method is welcome which appears to produce them as quickly and abundantly as possible [Im vorliegenden Fall handelt es sich nicht nur um die von einzelnen Vertretern der abzählenden Geometrie produzierten gewaltigen Zahlen, für die man sich interessieren mag oder nicht, sondern um die Methodik der algebraischen Geometrie überhaupt. ... Man hat das in Rede stehende 'Prinzip' auch da angewendet, wo, bei eingehenderer Bemühung, die gewöhnlichen Mittel der Algebra nicht nur ausgereicht, sondern auch sehr viel mehr geleistet haben würden. Man interessiert sich für diese oder jene 'Resultate', jede Methode ist willkommen, die sie möglichst geschwind und reichlich zu liefern scheint]" [Study, 1916, p. 65-66].

[34] In 1936, de Vries published a textbook in Dutch, *Introduction to Enumerative Geometry*, which van der Waerden reviewed very briefly for *Zentralblatt* (15, p. 368-369), writing in particular that, according to his own experience, there was no better way to learn geometry than to study Schubert's *Kalkül der abzählenden Geometrie*.

[35] Compare [van der Waerden, 1927 (note 5)].

specialization]." André Weil would later, in his *Foundations of Algebraic Geometry*, simply write "specialization."[36]

There is, however, a slight technical difference between the basic notion of specialization *à la* Weil—replacing one affine point ξ with coordinates in some extension field of the fixed ground field, which we call **P** as before, by another one η in such a way that every polynomial relation with coefficients in **P** involving the coordinates of ξ also holds for the coordinates of η—and the concept that van der Waerden introduced in his 1927 paper. Van der Waerden worked with multi-homogeneous coordinates in order to control the simultaneous specialization of a finite number of projective points (which will be taken to be all the generic solutions of an enumerative problem). More precisely,[37] starting from the ground field **P** and adjoining h unknowns (parameters) $\lambda_1, \ldots, \lambda_h$, he worked in some fixed algebraically closed extension field Ω of $\mathbf{P}(\lambda_1, \ldots, \lambda_h)$. Given q points

$$X^{(1)} = (\xi_0^{(1)} : \cdots : \xi_n^{(1)}), \ \ldots \ , \ X^{(q)} = (\xi_0^{(q)} : \cdots : \xi_n^{(q)})$$

in projective n-space over the algebraic closure $\overline{\mathbf{P}(\lambda_1, \ldots, \lambda_h)}$ inside Ω, a "*relationstreue Spezialisierung*" of $X^{(1)}, \ldots, X^{(q)}$ for the parameter values $\mu_1, \ldots, \mu_h \in \Omega$ is a set of q points

$$Y^{(1)} = (\eta_0^{(1)} : \cdots : \eta_n^{(1)}), \ \ldots \ , \ Y^{(q)} = (\eta_0^{(q)} : \cdots : \eta_n^{(q)})$$

in projective n-space over Ω such that, for any polynomial g in the variables $x_0^{(1)}, \ldots, x_n^{(1)}; x_0^{(2)}, \ldots, x_n^{(2)}; \ldots; x_0^{(q)}, \ldots, x_n^{(q)}; \lambda_1; \ldots; \lambda_h$ with coefficients in **P** which is homogeneous in each of the packets of variables separated by semicolons, and such that when

$$g(\xi_0^{(1)}, \ldots, \xi_n^{(1)}; \ldots; \xi_0^{(q)}, \ldots, \xi_n^{(q)}; \lambda_1; \ldots; \lambda_h) = 0,$$

one also has

$$g(\eta_0^{(1)}, \ldots, \eta_n^{(1)}; \ldots; \eta_0^{(q)}, \ldots, \eta_n^{(q)}; \mu_1; \ldots; \mu_h) = 0.$$

Van der Waerden uses this notion to analyze problems with Schubert's principle of the conservation of number in a way vaguely reminiscent of the avoidance of Russell's paradox by a theory of types; in order to make sense of the number of solutions which will be conserved, one has to specify the generic problem from which the given problem is considered to have been derived via specialization of parameters. Just as in the case of the theory of types, the prescribed diet makes it a little hard to survive. Thus, van der Waerden mentioned the example of the multiplicity of an intersection point of an r-dimensional with an $(n-r)$-dimensional subvariety in projective n-space, which, according to his analysis, is not well-defined (if none of the subvarieties is linear) as long as one has not specified the more general

---

[36]See [Weil, 1946a, Chap. II, §1]. In the introduction to this book, Weil acknowledged that "[t]he notion of specialization, the properties of which are the main subject of Chap. II, and (in a form adapted to our language and purposes) the theorem on the extension of a specialization ... will of course be recognized as coming from van der Waerden" [Weil, 1946a, p. x].

[37]Here, I am paraphrasing the beginning of §3 in [van der Waerden, 1927].

algebraic sets of which the given subvarieties are considered to be specializations.[38] We will soon encounter this example again.

On the positive side, given the reference to a generic problem, van der Waerden could simply define the multiplicity of a specialized solution to be the number of times it occurs among the specializations of all generic solutions. (This multiplicity can be zero, for generic solutions that do not specialize; see [van der Waerden, 1927, p. 765].) In this way, the "conservation of number" was verified by construction, and van der Waerden managed to solve a certain number of problems from enumerative geometry by interpreting them as specializations of generic problems which are completely under control. For instance, in the final §8, he demonstrated his method for lines on a (possibly singular) cubic surface over a base field of arbitrary characteristic.[39]

The technical heart of [van der Waerden, 1927] is the proof of the possibility and unicity (under suitable conditions) of extending (*"ergänzen"*) a specialization from a smaller to a larger finite set of points. It is for this that van der Waerden resorted to elimination theory (systems of resultants). The necessary results had been established in [van der Waerden, 1926b] which, as noted above, is strangely missing from [van der Waerden, 1983]. It is part of well-known folklore in algebraic geometry that André Weil in his *Foundations* would "finally eliminate ... the last traces of elimination theory" [Weil, 1946a, p. 31 (note)], at least from this part of the theory, using a trick of Chevalley's. As of the fourth edition of 1959, van der Waerden also dropped the chapter on elimination theory from the second volume of his algebra book. In the papers by van der Waerden to which I now turn, however, algebraic techniques become even more diverse, but this will be short-lived, for he ultimately settled on his own sort of minimal algebraization of algebraic geometry (see the next section).

Having seen how van der Waerden reduced the problem of Schubert's principle to that of a good definition of intersection multiplicity, it is not surprising to find him working on Bezout's Theorem in two papers the next year: the long article [van der Waerden, 1928a] as well as the note [van der Waerden, 1928c]. (This note is also not contained in [van der Waerden, 1983].) In the simplest case, Bezout's Theorem says that two plane projective curves of degree n, respectively m, intersect in precisely $m \cdot n$ points of the complex projective plane, provided one counts these points with the right multiplicities. In the introduction to [van der Waerden, 1928a], van der Waerden first recalled a "Theorem of Bézout in modern garb" following Macaulay, to the effect that the sum of multiplicities of the points of intersection of n algebraic hypersurfaces $f_i = 0$ in projective n-space equals the product of the degrees $\deg f_i$, provided the number of points of intersection remains finite. Here, the multiplicities are defined in terms of the decomposition into linear

---

[38]"The principle of specifying the generic problem has often been violated. For instance, one talks without definition of the multiplicity of the point of intersection of two varieties, of dimensions r and $n-r$ in the projective space P_n. But the generic sets of which M_r and M_{n-r} are considered to be specializations are not given [Gegen diesen Grundsatz ist oft verstoßen worden. Man redet z.B. ohne Definition von der Multiplizität eines Schnittpunktes zweier Mannigfaltigkeiten der Dimension r und $n - r$ im projektiven Raum P_n. Es wird dabei nicht angegeben, aus welchen allgemeineren Gebilden man die M_r und die M_{n-r} durch Spezialisierung entstanden denkt]" [van der Waerden, 1927, p. 766].

[39]In this paper, van der Waerden called hypersurfaces "principal varieties" because their corresponding ideals are principal. In a funny footnote [van der Waerden, 1927, p. 768], he even proposed to call them simply *"Häupter,"* that is, "heads."

forms of the so-called u-resultant of the system of hypersurfaces, that is, of the resultant of $(f_1, \ldots, f_n, \sum u_k x_k)$, where the u_k are unknowns and $x_0, \ldots, x_n$ are the projective coordinates. This entailed the "conservation of number" in the sense of the article discussed above, namely, the sum of multiplicities in each special case equals the number of solutions in the generic case (when the coefficients of the f_i are unknowns). Van der Waerden preserved this property as a guiding principle for generalizing Bezout's Theorem. As a consequence, for every application of the theorem, he had to define the "generic case" that is to be taken as reference.

Van der Waerden mentioned the general problem already encountered in [van der Waerden, 1927]: to define the multiplicity of the intersection of an r-dimensional subvariety and an $(n-r)$-dimensional subvariety in projective n-space. Again, he criticized earlier attempts to generalize Bezout's Theorem to this situation for their failure to make the notion of multiplicity precise. He solved the problem using a method which went back to Kronecker, and which used the wealth of automorphisms of projective space: transform the two subvarieties which we want to intersect via a sufficiently general matrix U of rank $n-r+1$, so that they are in general position to each other. Re-specializing U to the identity matrix will then realize the original problem as a special case of the generic one. Bezout's Theorem then states that the number of generic intersection points is just the product of the degrees of the two subvarieties (the degree of a k-dimensional subvariety being defined as the number of intersection points with a generic $(n-k)$-dimensional linear subspace).

The technical panoply employed in [van der Waerden, 1928a] was rich and varied: more Noetherian (and Noether-Hentzeltian) ideal theory than in the parsimonious [van der Waerden, 1926a], Macaulay's homogeneous ideals, David Hilbert's and Emmanuel Lasker's results about dimension theory with "Hilbert's Function,"[40] and linear transformations. Incidentally, van der Waerden performed all the constructions of §6 of the paper in what Weil later called a *universal domain* Ω, that is, an algebraically closed field of infinite transcendence degree over the base field:

> Ω then has the property that every time when, in the course of the investigation, finitely many quantities have been used, there will still be arbitrarily many unknowns left which are independent of those quantities. Fixing this field Ω once and for all saves us adjoining new unknowns time and again, and all constructions of algebraic extensions. If in the sequel at any point "unknowns from Ω" are introduced, it will be understood that they are unknowns which are algebraically independent of all quantities used up to that point.[41]

In spite of the considerable algebraic apparatus that van der Waerden brought to bear on the problems of intersection theory, his results remained unsatisfactory:

---

[40] Compare also the slightly later [van der Waerden, 1928c] in which another case of Bezout's Theorem was established, concerning the intersection of a subvariety with a hypersurface in projective space.

[41] "Ω hat dann die Eigenschaft, daß es immer, wenn im Laufe der Untersuchung endlichviele Größen aus Ω verwendet worden sind, noch beliebig viele neue, von diesen Größen unabhängige Unbestimmte in Ω gibt. Die Zugrundelegung des ein für allemal konstruierten Körpers Ω erspart uns also die immer erneute Adjunktion von Unbestimmten und alle Konstruktionen von algebraischen Erweiterungskörpern. Wenn im Folgenden an irgendeiner Stelle 'Unbestimmte aus Ω' eingeführt werden, so sind damit immer gemeint solche Unbestimmte von Ω, die von allen bis dahin verwendeten Größen aus Ω algebraisch-unabhängig sind" [van der Waerden, 1928a, p. 518].

As far as it went, the algebraic method had a greater generality than any analytic one, since it was applicable to arbitrary abstract geometries (belonging to abstract fields). But in transferring the methods to varieties of lines and the like, the proofs encountered ever mounting difficulties, and for ambient varieties which do not admit a transitive group of transformations like projective space, the transfer of the above notion of multiplicity is altogether excluded.[42]

Thus, van der Waerden changed horses:

But topology has a notion of multiplicity: the notion of index of a point of intersection of two complexes, which has already been applied with success by Lefschetz [1924] to the theory of algebraic surfaces as well as to correspondences on algebraic curves.

...

But topology achieves even more than making a useful definition of multiplicity possible. At the same time it provides plenty of means to determine in a simple manner the sum of indices of all the intersection points, or the "intersection number," the determination of which is the goal of all enumerative methods. For it shows that this sum of indices depends only on the homology classes of the varieties that are being intersected, and for the determination of the homology classes, it puts at our disposal the whole apparatus of "combinatorial topology."[43]

Van der Waerden was not the only mathematician involved in algebraic geometry to be tempted by Solomon Lefschetz's topology. Oscar Zariski's topological period around this same time, for instance, was brought about by immediate contact with Lefschetz and lasted roughly from 1928 until 1935. Interestingly, Lefschetz was skeptical of algebraic geometry, but did not so much bemoan its lack of rigor as deplore the amount of special training needed to practice this discipline in the traditional way. His idea was to incorporate algebraic geometry into more accessible mainstream mathematics, that is, into analysis in a broad sense. As he wrote to Hermann Weyl:

---

[42] "Soweit sie reichte, hatte die algebraische Methode eine größere Allgemeinheit als jede analytische, da sie auf beliebige abstrakte Geometrien (die zu abstrakten Körpern gehören) anwendbar war. Aber bei der Übertragung der Methoden auf Varietäten von Geraden u.dgl. stieß die Durchführung der Beweise auf immer wachsende Schwierigkeiten, und für solche Gebilde, die nicht wie der Projektive Raum eine transitive Gruppe von Transformationen in sich gestatten, ist die Übertragung der obigen Multiplizitätsdefinition ganz ausgeschlossen" [van der Waerden, 1929, p. 338].

[43] "Aber die Topologie besitzt einen Multiplizitätsbegriff: den Begriff des Schnittpunktes von zwei Komplexen, der schon von Lefschetz [1924] mit Erfolg auf die Theorie der algebraischen Flächen sowie auf Korrespondenzen auf algebraischen Kurven angewandt wurde. ... Die Topologie leistet aber noch mehr als die Ermöglichung einer brauchbaren Multiplizitätsdefinition. Sie verschafft zugleich eine Fülle von Mitteln, die Indexsumme aller Schnittpunkte oder 'Schnittpunktzahl', deren Bestimmung das Ziel aller abzählenden Methoden ist, in einfacher Weise zu bestimmen, indem sie zeigt, daß diese Indexsumme nur von den Homologieklassen der zum Schnitt gebrachten Varietäten abhängt, und indem sie für die Bestimmung der Homologieklassen den ganzen Apparat der 'kombinatorischen Topologie' zur Verfügung stellt" [van der Waerden, 1929, pp. 339-340].

> I was greatly interested in your "Randbemerkungen zu Hauptproblemen..." and especially in its opening sentence.[44] For any sincere mathematical or scientific worker it is a very difficult and heartsearching question. What about the young who are coming up? There is a great need to unify mathematics and cast off to the wind all unnecessary parts leaving only a skeleton that an average mathematician may more or less absorb. Methods that are extremely special should be avoided. Thus if I live long enough I shall endeavor to bring the theory of Algebraic Surfaces under the fold of Analysis and An.[alysis] Situs as indicated in Ch. 4 of my Monograph. The structure built by Castelnuovo, Enriques, Severi is no doubt magnificent but tremendously special and requires a terrible 'entraînement.' It is significant that since 1909 little has been done in that direction even in Italy. I think a parallel edifice can be built up within the grasp of an average analyst.[45]

Van der Waerden was apparently the first to realize Schubert's formal identities in the homology ring of the ambient variety:

> In general, each homology relation between algebraic varieties gives a symbolic equation in Schubert's sense, and these equations may be added and multiplied *ad libitum*, just as in Schubert's calculus. And the existence of a finite basis for the homologies in every closed manifold implies furthermore the solvability of Schubert's 'characteristics problems' in general.
> ...
>   I hope to give on a later occasion applications to concrete enumerative problems of the methods which are about to be developed here.[46]

---

[44]This refers to [Weyl, 1924, p. 131]: "Next to such works, which—exploding in all directions and therefore followed with a lively interest by only a few—explore new scientific territory, reflections like those presented here—which care less for augmenting than for clearing up and reformulating in a way as simple and adequate as possible results already obtained earlier—also have their right, if they focus on main problems that are of interest to all mathematicians who deserve to be called by this name [Neben solchen Arbeiten, die—in alle Richtungen sich zersplitternd und darum jeweils auch nur von wenigen mit lebhafterem Interesse verfolgt—in wissenschaftliches Neuland vorstoßen, haben wohl auch Betrachtungen wie die hier vorgelegten, in denen es sich weniger um Mehrung als um Klärung, um möglichst einfache und sachgemäße Fassung des schon Gewonnenen handelt, ihre Berechtigung, wenn sie sich auf Hauptprobleme richten, an denen alle Mathematiker, die überhaupt diesen Namen verdienen, ungefähr in gleicher Weise interessiert sind]."

[45]From page 4 of a long letter by Solomon Lefschetz to Hermann Weyl, dated 30 November, 1926 [ETHZ, HS 91:659]. Hearty thanks to David Rowe for pointing out this magnificent quote to me.

[46]"Allgemein ergibt jede Homologierelation zwischen algebraischen Varietäten eine symbolische Gleichung im Schubertschen Sinn, und man darf diese Gleichungeen unbeschränkt addieren und multiplizieren, wie es im Schubertschen Kalkül geschieht. Aus der Existentz einer endlichen Basis für die Homologien in jeder geschlossenen Mannigfaltigkeit ergbt sich weiter allgemein die Lösbarkeit der Schubertsche 'Charakteristikenprobleme.' ...Anwendungen der hier zu entwickelnden Methoden auf konkrete abzählende Probleme hoffe ich später zu geben" [van der Waerden, 1929, p. 340]. An example of such a concrete application is contained in the paper "Zur algebraischen Geometrie IV": [van der Waerden, 1983, pp. 156-161].

The article was written in the midst of the active development of topology. For example, in a note added in proof, van der Waerden put to immediate use van Kampen's thesis, which had just been completed.[47]

The whole topological approach, of course, only works over the complex (or real) numbers; it does not work in what was called at the time "abstract" algebraic geometry, over an arbitrary (algebraically closed) field, let alone over one of characteristic $p \neq 0$. There is, however, no reason to discard this work from the history of algebraic geometry simply because it seems to lead us away from a purely algebraic or arithmetic rewriting of it. Both Zariski and van der Waerden took the topological road for a while; and Italian algebraic geometry had never done without analytical or continuity arguments when needed. In fact (as a smiling Richard Pink once pointed out to me), algebraic topology meets the *ad hoc* definition of algebraic geometry with which this chapter opened: the treatment of geometrical objects and problems by algebraic methods.

Clearly, van der Waerden held no dogmatic views about arithmetic or algebraic approaches. He had tried the algebraic muscle on the problem of defining intersection multiplicities as generally as possible, and the result had not been conclusive. The fact that I have anticipated here and there how André Weil picked up van der Waerden's most basic ideas in his *Foundations of Algebraic Geometry* (1946) must, of course, not create the impression of an internal sense of direction for the history of algebraic geometry. At the end of the 1920s, that history remained wide open, full of different options, and—to anticipate once more—in the 1950s, topological (Hirzebruch) and analytical (Kodaira and Spencer) methods would make their strong reappearance in a discipline which had just been thoroughly algebraized.

History must also have seemed particularly open from the personal point of view of the young, brilliant van der Waerden, who, newly married, had started his first professorship in 1928 at Groningen, and had become Otto Hölder's successor in Leipzig in May of 1931. He had plenty of different interests. He was most attracted to Leipzig because of the prospect of contact with the physicists Heisenberg and Hund. While his *Moderne Algebra* appeared in 1930 (vol. I) and 1931 (vol. II), the following year of 1932 saw the publication of his book on group-theoretic methods in quantum mechanics. Within another five years, he had added statistics to his active research interests, and had even started to publish on the ancient history of mathematics.

Nevertheless, algebraic geometry, including topological methods when necessary, remained one of his chief research interests. Thus, following a tiny, four-page paper emending an oversight of Brill and Noether[48] and obviously confident that he had already explored and secured the methodological foundations for broad research in the field, van der Waerden launched in 1933 (paper submitted on 12 July, 1932) his series "Zur algebraischen Geometrie," or ZAG for short, coming back in the first installment to the problem of defining multiplicities, with a relatively light use of algebra, this time in the special case where one of the intersected varieties is a hypersurface.[49] This ZAG series, which appeared in the *Mathematische Annalen*

---

[47]See [van der Waerden, 1929, p. 118 (note 20)]. I will not go into the technical details of van der Waerden's topological work here.

[48][van der Waerden 1931] was submitted on 19 November, 1930. Severi later scolded van der Waerden for criticizing his elders. See the final footnote in [Severi, 1933, p. 364 (note 31)], respectively, [Severi, 1980, p. 129].

[49]See [van der Waerden, 1933].

and which was incorporated in the volume [van der Waerden, 1983], ran from the article ZAG I (1933) just mentioned, all the way to ZAG 20 which appeared in 1971. (Although it is only fair to say that the penultimate paper of the series, ZAG 19, had appeared in 1958.) Van der Waerden opened the series this way: "In three preceding articles in the *Annalen*, I have developed several algebraic and topological notions and methods upon which higher dimensional algebraic geometry may be based. The purpose of the present series of papers 'On algebraic geometry' is to demonstrate the applicability of these methods to various problems from algebraic geometry."[50]

We shall skip over the details of this paper as well as over the quick succession of ZAG II (submitted 27 July, 1932/appeared 1933), ZAG III (27 October, 1932/1933), ZAG IV (27 October 1932/1933), and ZAG V (8 October, 1933/1934), in order to get to the historically more significant encounter of van der Waerden with the Italian school of algebraic geometry, and the corresponding ripples in the mathematical literature.

1933–1939: When in Rome ... ?

The following remarkably dry account, taken from [van der Waerden, 1971, p. 176], is surely an understatement of what actually happened during and after that meeting between the twenty-nine-year-old Bartel L. van der Waerden and the impressive and impulsive fifty-three-year-old Francesco Severi:

> At the Zürich International Congress in 1932 I met Severi, and I asked him whether he could give me a good algebraic definition of the multiplicity of a point of intersection of two varieties A and B, of dimensions d and $n-d$, on a variety U of dimension n, on which the point in question is simple. The next day he gave me the answer, and he published it in the *Hamburger Abhandlungen* in 1933. He gave several equivalent definitions . . . .

In the absence of any first-hand documentary evidence about their relationship in the thirties,[51] one can only say that Severi's presence effectively confronted van der Waerden with the reality of Italian algebraic geometry for the first time in his life. This confrontation had an attractive and a repellent aspect. The attraction is clearly reflected in van der Waerden's desire to spend some time in Rome. In fact, just about a month before he had to abandon his function as director of the Göttingen Mathematics Institute, Richard Courant wrote a letter to Wilbur E. Tisdale at the Rockefeller Foundation in Paris in which he explained that

---

[50] "In drei früheren Annalenarbeiten habe ich einige algebraische und topologische Begriffe und Methoden entwickelt, die der mehrdimensionalen algebraischen Geometrie zugrunde gelegt werden können. Der Zweck der jetzigen Serie von Abhandlungen 'Zur Algebraischen Geometrie' ist, die Anwendbarkeit dieser Methoden auf verschiedene algebraisch-geometrische Probleme darzutun" [van der Waerden 1933].

[51] All of van der Waerden's correspondence before December 1943 seems to have burned with his Leipzig home in an air raid. On the other hand, Italian historian colleagues have assured me that, in spite of years of searching, they have never found any non-political correspondence of Severi's—except for those letters that were kept by the correspondents. A fair amount of later correspondence between Severi and van der Waerden, in particular in the long, emotional aftermath of the events at the 1954 ICM in Amsterdam, is conserved at ETHZ.

> Prof. Dr. B. L. van der Waerden, at present full professor at the University of Leipzig, about 30 or 31 years old, former Rockefeller fellow, has asked me to sound out whether the Rockefeller Foundation could arrange a prolonged sojourn in Italy for him.
>
> In spite of his great youth, van der Waerden is today one of the outstanding mathematicians in Europe. He was one of the three candidates of the Faculty for Hilbert's successor. For a few years now, van der Waerden has started to study the problems of algebraic geometry, and he seriously intends to promote the cultivation of this domain in Germany. As a matter of fact, the geometric-algebraic tradition is all but dead in Germany whereas it has come to full blossom in Italy over the past few decades. Several young mathematicians, for instance Dr. Fenchel and Dr. Kähler have spent time in Italy on a Rockefeller grant and have successfully studied algebraic geometry there. But for the advancement of science, it would be effective on quite a different scale, if such an outstanding man as van der Waerden could establish the necessary link on a broad basis.
>
> It is for these scientific reasons that van der Waerden has developed the wish to work for some time especially with Prof. Severi in Rome, and to then transplant the results back to Germany.[52]

In fact, van der Waerden did not get the Rockefeller grant, and he traveled neither to Italy nor to the United States in the 1930s, at least in part because of the travel restrictions that the Nazi Regime imposed on him.[53]

As to the repellent side of the encounter with Severi, Leonard Roth (who had spent the 1930–1931 academic year in Rome) left this analysis in his obituary of Severi. He explained that "[p]ersonal relationships with Severi, however complicated in appearance, were always reducible to two basically simple situations: either he had just taken offence or else he was in the process of giving it—and quite often genuinely unaware that he was doing so. Paradoxically, endowed as he was with even more wit than most of his fellow Tuscans, he showed a childlike incapacity

---

[52]"Prof. Dr. B. L. van der Waerden, gegenwärtig Ordinarius an der Universität Leipzig, etwa 30 oder 31 Jahre alt, früherer Rockefeller fellow, hat mich darum gebeten, die Möglichkeit zu sondieren, ob ihm von der Rockefeller Foundation ein längerer Aufenthalt in Italien ermöglicht werden kann.

"van der Waerden ist trotz seiner grossen Jugend einer der hervorragenden Mathematiker, die es augenblicklich in Europa gibt. Er war bei der Neubesetzung des Hilbertschen Lehrstuhls einer der drei Kandidaten der Fakultät. Nun hat van der Waerden seit einigen Jahren erfolgreich begonnen, sich mit den Problemen der algebraischen Geometrie zu beschäftigen, und es ist sein sehr ernstes Bestreben, die Pflege dieses Gebietes in Deutschland wirklich zu betreiben. Tatsächlich ist die geometrisch-algebraische Tradition in Deutschland fast ausgestorben, während sie in Italien im Laufe der letzten Jahrzehnte zu hoher Blüte gelangt ist. Schon mehrere junge Mathematiker, z.B. Dr. Fenchel und Dr. Kähler sind mit einem Rockefellerstipendium in Italien gewesen und haben dort erfolgreich algebraische Geometrie studiert. Aber es würde für die wissenschaftliche Entwicklung von ganz anderer Wirksamkeit sein, wenn ein so hervorragender Mann wie van der Waerden die notwendige Verbindung auf einer breiteren Front herstellen könnte.

"Aus solchen sachlichen Erwägungen ist van der Waerdens Wunsch entstanden, insbesondere in Kontakt mit Prof. Severi in Rom eine gewisse Zeit zu arbeiten und dann das Gewonnene hier nach Deutschland zu verpflanzen" (my translation). The letter is dated 2 March, 1933. Compare [Siegmund-Schultze, 2001, pp. 112-113]. I thank Reinhard Siegmund-Schultze for providing me with the original German text of the letter

[53]Recall the discussion of this point in note 17 above.

either for self-criticism or for cool judgement" [Roth, 1963, p. 307]. At the same time, such psychological observations must not obscure the fact that Severi wielded real academic power in the fascist Italy of the thirties, after having turned his back on his former socialist convictions and anti-fascist declarations when the possibility arose to take Enriques's seat at the Academy in Rome. For example, beginning in 1929 and in concert with the regime's philosopher Giovanni Gentile, Severi was actively preparing the transformation (which became effective in August of 1931) of the traditional professors' oath of allegiance into an oath to the fascist regime.[54]

The papers of van der Waerden that appeared before 1934 contain only very occasional references to Italian literature, and only one to Severi [van der Waerden, 1931, p. 475 (note 6)]. Severi's irritated reaction to this—and more generally to the content of van der Waerden's series of papers on algebraic geometry—shows clearly through the sometimes barely polite formulations in his German paper [Severi, 1933]. As Hellmuth Kneser nicely put it in his *Jahrbuch* review of this article, "[g]eneral and personal remarks scattered throughout the article impart even to the non-initiated reader a lively impression of the peculiarity and the achievements of the author and the Italian school."[55] Severi's overall vision of algebraic geometry and its relationship to neighboring disciplines is made clear straight away in the introductory remarks:

> I claimed that all the elements required to define the notion of "intersection multiplicity" completely rigorously and in the most general cases have been around, more or less well developed, for a long time in algebraic geometry, and that the proof of the principle of the conservation of number that I gave in 1912 is perfectly general. In order to lay the foundation for those concepts in a way covered against all criticism, it is therefore not necessary, as Mr. van der Waerden and Mr. Lefschetz think, to resort to topology as a means that would be particularly adapted to the question. Lefschetz's theorems ... and van der Waerden's applications thereof ... are undoubtedly of great interest already in that they demonstrate conclusively that fundamental algebraic facts have their deep and almost exclusive foundation in pure and simple continuity. ... As I already said in my ICM talk, it is rather topology that has learned from algebra and algebraic geometry than the other way around, because these two disciplines have served topology as examples and inspiration.[56]

---

[54] See [Guerraggio and Nastasi, 1993, pp. 76-83 and 211-213].

[55] "Allgemeine und persönliche Bemerkungen, die durch die Abhandlung verstreut sind, vermitteln auch dem Fernerstehenden einen lebhaften Eindruck von der Eigenart und den Leistungen des Verf. und der italienischen Schule."

[56] "... behauptete ich, daß sich in der algebraischen Geometrie schon seit längerer Zeit in mehr oder weniger entwickelter Form alle Elemente vorfinden, die den Begriff 'Schnittmultiplizität' mit aller Strenge und in den allgemeinsten Fällen zu definieren erlauben; und dass ferner der von mir 1912 gegebene Beweis für das Prinzip der Erhaltung der Anzahlen vollkommen allgemein ist. Es ist demnach nicht nötig, wie die Herren van der Waerden und Lefschetz meinen, zur Topologie als dem der Frage vor allem angemessenen Hilfsmittel zu greifen, um eine gegen alle Einwände gedeckte Begründung jener Begriffe zu geben. Die Sätze von Lefschetz ... und die Anwendungen, die Herr van der Waerden davon ... gemacht hat, bieten unzweifelhaft grosses Interesse, schon weil sie in erschöpfender Weise zeigen, daß fundamentale algebraische Tatsachen ihren tiefen und fast ausschließlichen Grund in der reinen und einfachen Kontinuität finden. ... Wie ich bereits

Mathematically, Severi's construction for the intersection multiplicity amounts to the following.[57] He wanted to define the intersection multiplicity of the two irreducible (for simplicity) subvarieties V_k (indices indicate dimensions) and W_{r-k} of a variety M_r, which, in turn, is embedded in projective d-space S_d at a point P of their intersection which is simple on M. Then Severi chose a generic linear projective subspace S_{d-r-1} in S_d, and took the corresponding cone N_{d-r+k} over V_k projected from S_{d-r-1}. Writing the intersection cycle $N \cap M = V + V'$ and observing that V' does not pass through P, he then defined the intersection multiplicity of V, W at P to be the intersection multiplicity of N, W at P. This thus reduced the problem to the intersection of subvarieties of complementary dimensions in projective d-space, where he argued with generic members of a family containing N, or alternatively, of a family on M containing $V + V'$. The definition was then supplemented by showing its independence of choices, within suitable equivalence classes.[58]

We have used here, for the convenience of the modern reader, the word "cycle" (instead of "variety") to denote a linear combination of irreducible varieties. Such a distinction was absent from the terminology of the thirties, and was only introduced in Weil's *Foundations*. Still, even if the word is anachronistic relative to the early thirties, the concept is not. Severi had just opened up a whole "new field of research" in 1932, which today would be described as the theory of rational equivalence of 0-cycles.[59] It is important to underscore Severi's amazing mathematical productivity during those years, and even later, lest one get a wrong picture about what it meant to *re*write algebraic geometry at the time.

Van der Waerden's reaction to Severi's explanations and critique was twofold: he was annoyed, but he heeded the advice. Both reactions are evident in his paper ZAG VI, that is, [van der Waerden, 1934]. Mathematically, van der Waerden reconstructed here a good deal of Severi's theory of correspondences and of the

---

in meinem [ICM-] Vortrag sagte, hat eher die Topologie von der Algebra und der algebraischen Geometrie gelernt als umgekehrt" [Severi, 1933, p. 335].

It is instructive to compare this passage to Dieudonné's account of the history of intersection theory. See [Dieudonné, 1974, pp. 132-133], where he says that "[t]he works of Severi and Lefschetz bring to light the essentially topological nature of the foundations of classical algebraic geometry; in order to be able to develop in the same manner algebraic geometry over any field whatsoever, it will be necessary to create purely algebraic tools which will be able to substitute for the topological notions.... It is to van der Waerden that the credit goes for having, beginning in 1926, placed the essential markers for this path [Les travaux de Severi et de Lefschetz mettaient donc en évidence la nature essentiellement topologique des fondements de la Géométrie algébrique classique; pour pouvoir développer de la même manière la Géométrie algébrique sur un corps quelconque, il fallait créer des outils purement algébriques qui puissent se substituer aux notions topologiques.... C'est à van der Waerden que revient le mérite d'avoir, à partir de 1926, posé les jalons essentiels dans cette voie]." Although globally correct, this analysis leaves Severi back in 1912 and glosses over van der Waerden's multifarious methods.

[57]We paraphrase [Severi, 1933, no. 8].

[58]In the endnote Severi added to his 1933 article in 1950 obviously under the influence of Weil's *Foundations* (see [Severi, 1980, pp. 129-131]), Severi observed (which he had not done explicitly in 1933) that the intersection multiplicity he defined was symmetric in the intersecting subvarieties. He went on to comment on Weil's definition of intersection multiplicity, in the same way as in many other papers of his from the 1950s, calling it "static" rather than dynamic.

[59]Since Severi is not the main focus of this article, I shall not go into this here. I refer the reader instead to the best available study of this aspect of Severi's work: [Brigaglia, Ciliberto, and Pedrini, 2004, pp. 325-333]. Compare also van der Waerden's account in [van der Waerden, 1970].

"principle of conservation of number" with his own, *mild algebraic* methods (that is, without elimination or other fancy ideal theory, but also without topology). The paper digests substantial mathematical input coming more or less directly from Severi (not only from Severi's article just discussed) and sticks again to exclusively algebraic techniques.

As for the annoyance, the first paragraph of the introduction announced a surprising change of orientation with political overtones which could not have been suspected after all his previous papers on algebraic geometry:

> The goal of the series of my articles "On Algebraic Geometry" (ZAG) is not only to establish new theorems but also to make the far-reaching methods and conceptions of the Italian geometric school accessible with a rigorous algebraic foundation to the circle of readers of the Math. Annalen. If I then perhaps prove again something here which has already been proved more or less properly elsewhere, this has two reasons. Firstly, the Italian geometers presuppose in their proofs a whole universe of ideas and a way of geometric reasoning with which, for instance, the German man of today is not immediately familiar. But secondly, it is impossible for me to search, for each theorem, through all the proofs in the literature in order to check whether there is one among them which is flawless. I rather formulate and prove the theorems my own way. Thus, if I occasionally indicate deficiencies in the most widely circulated literature, I do not claim in any way that I am the first who now presents things really rigorously.[60]

The fairly aggressive wording in this passage may not quite show in the English translation, but the other element of linguistic taint of the time, namely, the fact that the readers of the *Mathematische Annalen* are represented by "*der Deutsche von heute,*" gives a distinctly national vocation to the international journal and is obvious enough. In order to understand this peculiar twist of van der Waerden's anger, one may recall that in October of 1933, when the paper was submitted, the Berlin–Rome axis was still a long way in the future, and Italy's foreign politics looked potentially threatening to German interests, not only in Austria. Thus, van der Waerden, momentarily forgetting that he was himself a foreigner in Germany, having been criticized by a famous Italian colleague, comfortably used for his own sake the favorite discourse of the day: that Germany had to concentrate on herself to be fortified against attacks from abroad.

---

[60] "Das Ziel der Serie meiner Abhandlungen 'Zur Algebraischen Geometrie' (ZAG) ist nicht nur, neue Sätze aufzustellen, sondern auch, die weitreichenden Methoden und Begriffsbildungen der italienischen geometrischen Schule in exakter algebraischer Begründung dem Leserkreis der Math. Annalen näherzubringen. Wenn ich dabei vielleicht einiges, was schon mehr oder weniger einwandfrei bewiesen vorliegt, hier wieder beweise, so hat das einen doppelten Grund. Erstens setzen die italienischen Geometer in ihren Beweisen meistens eine ganze Begriffswelt, eine Art geometrischen Denkens, voraus, mit der z.B. der Deutsche von heute nicht von vornherein vertraut ist. Zweitens aber ist es mir unmöglich, bei jedem Satz alle in der Literatur vorhandenen Beweise dahin nachzuprüfen, ob sich ein völlig einwandfreier darunter befindet, sondern ich ziehe es vor, die Sätze in meiner eigenen Art zu formulieren und zu beweisen. Wenn ich also hin und wieder eimal auf Unzulänglichkeiten in den verbreitetsten Darstellungen hinweisen werde, so erhebe ich damit keineswegs den Anspruch, der erste zu sein, der die Sachen nun wirklich exakt darstellt" [van der Waerden, 1934, p. 168].

I emphasize here that van der Waerden somewhat surprisingly does *not* insist in the introduction to [van der Waerden, 1934] on the extra generality achieved by his methods. After all, Italian geometers had never proved (nor wanted to prove) a single theorem valid over a field of characteristic p. The whole presentation of this article—in which van der Waerden begins to develop his treatment of some of the most central notions of Italian geometry, like correspondences and linear systems—seems remarkably close in style to the Italian literature, much more so than the previous articles we have discussed. For instance, the field over which constructions are performed is hardly ever made explicit.

At the end of the introduction to this article, van der Waerden stated that "[t]he methods of proof of the present study consist firstly in an application of *relationstreue Spezialisierung* over and over again, and secondly in supplementing arbitrary subvarieties of an ambient variety $\mathfrak{M}$ to complete intersections of $\mathfrak{M}$ by adding residual intersections which do not contain a given point.[61] This second method I got from Severi [1933]."[62] The first and the last sentences of this introduction, taken together, can well serve as a motto for almost all of van der Waerden's ZAG articles in the 1930s, more precisely, for ZAG VI–ZAG XV with the exception of ZAG IX. The author enriched his own motivations and resources by Italian problems and ideas, and he wrote up his proofs with the mildest possible use of modern algebra, essentially only using generic points and specializations to translate classical constructions. A particularly striking illustration of this is ZAG XIV of 1938 [van der Waerden, 1983, pp. 273-296]. There, van der Waerden returned to intersection theory and managed to translate not only Severi's construction of 1933 but also a good deal of the latter's theory of equivalence families into his purely algebraic setting, while, at the same time, excising all of the fancier ideal theory of his earlier papers [van der Waerden, 1927] and [van der Waerden, 1928a].

There is, however, one fundamentally new ingredient, which I have not yet mentioned, that enters in the mathematical technology of ZAG XIV. It is due to the one article excluded above, namely, the brilliantly original and important ZAG IX written jointly with Wei-Liang Chow [Chow and van der Waerden, 1937]. As Serge Lang concisely described this work:

> To each projective variety, Chow saw how to associate a homogeneous polynomial in such a way that the association extends to a homomorphism from the additive monoid of effective cycles in projective space to the multiplicative monoid of homogeneous polynomials, and ..., if one cycle is a specialization of another, then the associated Chow form is also a specialization. Thus varieties of given degree in a given projective space decompose into a finite number of algebraic families, called Chow families. The coefficients of the Chow form are called the Chow coordinates of the cycle or of the variety. ... He was to use them all his life in various contexts dealing with algebraic families.

---

[61]These "residual subvarieties" are like the cycle V' in our sketch of Severi's argument above. Adding them is all that is meant here by obtaining a "complete intersection."

[62]Die Beweismethoden der vorliegenden Untersuchung bestehen erstens in einer immer wiederholten Anwendung der 'relationstreuen Spezialisierung' und zweitens der Ergänzung beliebiger Teilmannigfaltigkeiten einer Mannigfaltigkeit $\mathfrak{M}$ zu vollständigen Schnitten von $\mathfrak{M}$ durch Hinzunahme von Restschnitten, welche einen vorgegebenen Punkt nicht enthalten. Die zweite Methode habe ich von Severi [1933] übernommen" [van der Waerden, 1934, p. 137].

In Grothendieck's development of algebraic geometry, Chow coordinates were bypassed by Grothendieck's construction of Hilbert schemes whereby two schemes are in the same family whenever they have the same Hilbert polynomial. The Hilbert schemes can be used more advantageously than the Chow families in some cases. However, as frequently happens in mathematics, neither is a substitute for the other in all cases [Lang, 1996, pp. 1120-1121].

Wei-Liang Chow, born in Shanghai, was van der Waerden's doctoral student in Leipzig (although he was actually more often to be found in Hamburg). He submitted his dissertation [Chow, 1937] in May of 1936. In it, he gave a highly original—in some ways amazing—example of rewriting algebraic geometry in van der Waerden's way (including the so-called "Chow forms" and a subtle sharpening of Bertini's Theorem). The thesis reproved the whole theory of algebraic functions of one variable—the theory of algebraic curves—over a perfect ground field of arbitrary characteristic, and it did so all the way to the Riemann-Roch Theorem, following for much of the way Severi's so-called "*metodo rapido*."[63] This may seem like a modest goal to achieve. However, Chow got there without ever using differential forms. As van der Waerden wrote in the evaluation of this work, contrasting its algebraic-geometric approach with the approach via function field arithmetic by Friedrich Karl Schmidt, "[a]ltogether, this has established a very beautiful, self-contained and methodologically pure construction of the theory."[64]

These examples should suffice to convey the general picture of van der Waerden's algebraization of algebraic geometry in his Leipzig years. It produced often brilliantly original, and always viable and verifiable, theorems about exciting questions in algebraic geometry with a modicum of algebra. And even the algebra that was used no longer looked particularly modern at the time: just polynomials, fields, generic points, and specializations.

This *modest algebraization* of algebraic geometry, as it may be styled, did a lot to restore harmony with the Italian school. In 1939, van der Waerden published his textbook *Einführung in die algebraische Geometrie*, which digested a great deal of classical material from old algebraic geometry, but also included the results of a number of his articles of the thirties. The style is particularly pedagogical, going from linear subspaces of projective space to quadrics, etc., from curves to higher dimensional varieties, from the complex numbers to more general ground fields. In his preface, van der Waerden stated that "[i]n choosing the material, what mattered were not aesthetic considerations, but only the distinction: necessary–dispensable. Everything that absolutely has to be counted among the 'elements,' I hope to have taken in. Ideal theory, which guided me in my earlier investigations, has proved dispensable for the foundations; its place has been taken by the methods of the Italian school which go further."[65] The echo from Rome was very encouraging:

---

[63]This presentation of the theory of algebraic curves goes back to [Severi, 1920], and Severi himself returned to it several times. See, in particular, [Severi, 1926, pp. 145-169] and [Severi, 1952]. On a later occasion, I hope to publish a detailed comparison of Severi's method with other treatments from the 1930s, in particular André Weil's. See [Weil, 1938b], and compare [van der Waerden, 1959, chapter 19].

[64]"Insgesamt ist so ein sehr schöner, in sich geschlossener und methodisch reiner Aufbau der Theorie entstanden" [UAL, Phil. Fak. Prom. 1272, Blatt 2].

[65]"Bei der Auswahl des Stoffes waren nicht ästhetische Gesichtspunkte, sondern ausschliesslich die Unterscheidung: notwendig–entbehrlich maßgebend. Alles das, was unbedingt zu

> This volume, devoted to an introduction to algebraic geometry, shows some of the well-known characteristics of the works of its author, namely, the clarity of exposition, the conciseness of the treatment, kept within the limits of a severe economy, and the constant aspiration for rigor and transparency in the foundations. However, one does not find that dense game of abstract concepts which is so typical of the "Modern Algebra," and renders the latter so hard to read without extensive preliminary preparation. ... This remarkable book of van der Waerden will undoubtedly facilitate learning the methods of the Italian school, and contribute to a mutual understanding between the Italian geometers and the German algebraists, thus fulfilling a task of great importance.[66]

A letter from 1950 of van der Waerden to Severi (the latter had invited van der Waerden to come to Rome for a conference and to give a talk on abstract algebra) rings like an echo both of Conforto's words about van der Waerden's algebraic geometry and of Weil's recollection (recall the introductory section):

> I do not think I can give a really interesting talk on abstract algebra. The enthusiasm would be lacking. One knows me as an algebraist, but I much prefer geometry.
>
> In algebra, not much is marvelous. One reasons with signs that one has created oneself, one deduces consequences from arbitrary axioms: there is nothing to wonder about.
>
> But how marvelous geometry is! There is a preestablished harmony between algebra and geometry, between intuition and reason, between nature and man! What is a point? Can one see it? No. Can one define it? No. Can one dissolve it into arbitrary conventions, like the axioms of a ring? No, No, No! There is always a mysterious and divine remainder which escapes both reason and the senses. It is from this divine harmony that a talk on geometry derives its inspiration.
>
> This is why I ask you to let me talk on:
> 1) The principle of the conservation of number (historic overview)
>   or else
> 2) The theory of birational invariants, based on invariant notions.

---

den 'Elementen' gerechnet werden muß, hoffe ich, aufgenommen zu haben. Die Idealtheorie, die mich bei meinen früheren Untersuchungen leitete, hat sich für die Grundlegung als entbehrlich herausgestellt; an ihre Stelle sind die weitertragenden Methoden der italienischen Schule getreten" [van der Waerden, 1939, p. v].

[66] "Questo volume, dedicato ad un'introduzione alla geometria algebrica, presenta alcune delle ben note caratteristiche delle opere del suo Autore, e precisamente la nitidezza dell'esposizione, la rapidità e compattezza della trattazione, tenuta nei limiti di una severa economia, e la costante aspirazione al rigore ed alla chiarezza nei fondamenti. Non si trova invece quel serrato giuoco di concetto astratti, così caratteristico della 'Moderne Algebra,' che rende quest'ultima di difficile lettura per chi non abbia un'ampia preparazione preliminare. ... il notevole libro di van der Waerden agevolerà senza dubbio la conoscenza dei metodi della scuola italiana e coopererà ad una reciproca comprensione tra i geometri italiani e gli algebristi tedeschi, assolvendo così un compito di grande importanza." This passage is taken from the review of the book by Fabio Conforto (Rome) in *Zentralblatt* 21, 250.

I found this very recently, stimulated by a discussion with you at Liège.[67]

1933–1946: The Construction Site of Algebraic Geometry

Having traced the development of van der Waerden's research in algebraic geometry, the issue now becomes to attempt to situate his contributions with respect to other contemporaneous agendas in the area. This more global picture must of necessity remain sketchy here and will highlight only a few of the other relevant actors.[68] Among them, however, as we saw in the previous section, the Italians figure prominently; Fabio Conforto underscored this relationship in his review of van der Waerden's 1939 textbook on algebraic geometry, by referring to it as a contribution "to a mutual understanding between the Italian geometers and the German algebraists, thus fulfilling a task of great importance." Moreover, once the "Axis Berlin–Rome," as Mussolini termed it, was in place—that is, after the summer of 1936—it could also provide at least a metaphorical background for and justification of official invitations attempting to promote scientific exchange between Germany and Italy. The related activities on the German side actually constitute an interesting prelude to the war attempts to set up a European scientific policy under German domination.[69]

Van der Waerden's position in this miniature replica of a great political game was certainly handicapped by the hurdles that local Nazi officials created for him in Leipzig. Even if this had not been the case, however, that is, even if he could have engaged in direct contact at will, the strategy he followed after 1933 with respect to Italian algebraic geometry might have done him a disservice. As intellectually flexible as he was, he managed to present his rewritten algebraic geometry in a way that outwardly conformed, to a large extent, to the Italian model. It may have been his personal mathematical temperament, as reflected in the epigram with which this chapter opened, that made him place more emphasis on the rich geometric ideas and techniques than on the radically new kind of theory in which he was executing his constructions. He made it very easy for the Italians to consider him almost as a disciple, and as the later letters between him and Severi show, he never betrayed his loyalty to the Italian master. For instance, at one of the crisis points in their correspondence (after the 1954 ICM), Severi accused van der

---

[67]"Mon trés cher collègue. Je ne crois pas que je puisse présenter une conférence vraiment intéressante sur l'Algèbre abstraite. Il y manquera l'enthousiasme. On me connaît comme algébriste, mais j'aime la géométrie beaucoup plus. — Dans l'algèbre, il n'y a que peu de merveilleux. On raisonne sur des signes qu'on a créé[s] soi-même, on déduit des conséquences d'axiomes arbitraires: il n'y a pas de quoi s'étonner. — Mais la géométrie, quel[le] merveille! Il y a une harmonie préétabli[e] entre l'algèbre et la géométrie, entre l'intuition et la raison, entre la nature et l'homme! Qu'est-ce que c'est un point? Peut-on le voir? Non. Peut-on le définir? Non. Peut-on le résoudre en des conventions arbitraires, comme les axiomes d'un anneau? Non, non, non! Il y a toujours un reste mystérieux et divin, qui échappe à la raison comme aux sens. C'est de cette harmonie divine que s'inspire une conférence géométrique. — C'est pourquoi je vous propose de me laisser parler sur: 1) Le principe de la conservation du nombre (aperçu historique), ou bien: 2) La théorie des invariants biration[n]els basée sur des notions invariant[e?]s. J'ai trouvé cela tout récemment, stimulé par une discussion avec vous à Liège" [ETHZ, Nachlass van der Waerden, HS 652:11960]. This is a draft of a letter from van der Waerden to Severi, dated 15 February, 1950.

[68]I plan to return to this matter in the context of my larger research project.

[69]Compare [Siegmund-Schultze, 1986] and [Remmert, 2004].

Waerden of not sufficiently acknowledging the priority and accuracy of his ideas. For his part, however, van der Waerden was ready to plead with Severi, by pointing out that he had documented his complete confidence in Severi's approach as early as 1937.[70] Being the younger of the two, van der Waerden could appear as a junior partner, rewriting algebraic geometry; thus, at the beginning of his long review of van der Waerden's *Introduction to Algebraic Geometry* in volume sixty-five of the *Jahrbuch über die Fortschritte der Mathematik* for 1939, Harald Geppert attributed the fact that the foundations of algebraic geometry had now finally attained the necessary degree of rigor, mainly "to the works of Severi and of" van der Waerden.[71] Bearing this in mind, let us now consider some of the other mathematicians busy at the construction site.

Helmut Hasse and his school of function field arithmetic developed an increasing demand for ideas from algebraic geometry after Max Deuring had the idea, in the spring of 1936, to use the theory of correspondences in order to generalize Hasse's proof of the analog of the Riemann hypothesis for (function fields of) curves over finite fields from genus one to higher genera. Hasse organized a little conference on algebraic geometry in Göttingen on 6-8 January, 1937, with expository talks by Jung, van der Waerden, Geppert, and Deuring. The politically prestigious bicentennial celebration of Göttingen University in June of 1937 next provided the opportunity for Hasse and Severi to meet, and the mathematical and personal contact between them grew more intense from then on.

A few days after the Munich summit on the Bohemian crisis—the summit where Mussolini had used his unexpected role as a mediator to favor Hitler—Hasse wrote an amazing letter to Severi in which a political part, thanking "your incomparable Duce" for what he has done for the Germans, is followed by a plea for a corresponding mathematical axis. In particular, he mentioned a plan to start a German-Italian series of monographs in algebra and geometry with the goal of synchronizing the two schools.[72] Hasse and his school had a much more definite methodological paradigm than van der Waerden, however; they foresaw an arithmetic theory of function fields in the tradition of Dedekind and Weber, Hensel and Landsberg, etc. Translating ideas from classical algebraic geometry into this framework could not be presented as a relatively smooth transition as in van der Waerden's case. The "axis" between the schools of Hasse and Severi therefore took the form of expository work on function field arithmetic sent to or delivered in Italy, and published in Italian, as well as lists of bibliographical references about the classical theory of correspondences going the other way.

---

[70]As van der Waerden put it: "As far as I am concerned, I already wrote (ZAG XIV, Math. Annalen **115**, p. 642) with complete confidence in 1937: 'The calculus of intersection multiplicities can be used for the foundation of Severi's theory of equivalence families on algebraic varieties.' This means that I stressed the importance of your fundamental ideas and developed at the same time an algebraic apparatus to make them precise in an irrefutable manner [Quant à moi j'ai écrit déjà en 1937 (ZAG XIV, Math. Annalen **115**, p. 642) avec confiance complet [sic]: 'Der Kalkül der Schnittmannigfaltigkeiten kann zur Begündung der Severischen Theorie der Äquivalenzscharen auf algebraischen Mannigfaltigkeiten verwendet werden.' Cela veut dire que j'ai souligné l'importance de vos idées fondamentales et en même temps développé un apparat algébrique pour les préciser d'une manière irréfutable]" [ETHZ, Nachlass van der Waerden, HS 652:8394, page 3]. This is a draft of a letter from van der Waerden to Severi, dated "'Mars 1955."

[71]Es ist hauptsächlich den Arbeiten Severis und des Verf. zu danken, dass heute in den Grundlagen die erforderliche Exaktheit erreicht ist.

[72]See the appendix for the text (and a translation) of this remarkable archive.

In spite of the small Göttingen meeting mentioned above, collaboration inside Germany between van der Waerden and the Hasse group remained scant. A revealing exception to this occurred in the last few days of 1941, when van der Waerden sat down and worked out, in his way of doing algebraic geometry, the proofs of three theorems in [Hasse, 1942] that Hasse had been unable to prove in his set-up. Hasse was overjoyed[73] and asked van der Waerden to publish his proofs alongside his article. Van der Waerden only published them in 1947, however.[74] This was in another mathematical world, one in which Hasse, ever since his dismissal from Göttingen by the British military authorities in 1945, no longer had much institutional power. Van der Waerden was thus free[75] to criticize what he considered Hasse's inadequate approach. His criticism not only showed the distance between van der Waerden and Hasse when it came to algebraic geometry, but confirmed once more van der Waerden's dogmatically conservative attitude with respect to fundamental notions of algebraic geometry.[76] The episode suggests that the war and political or personal factors—that made effective collaboration between the two German groups difficult—mixed with differences of mathematical appreciation in an intricate web of relations which is not always easy to untwine.

We have seen that van der Waerden had been on very good terms with Hellmuth Kneser. In the short note [Kneser, 1935], the latter *very* barely sketched a proof of the Local Uniformization Theorem for algebraic varieties of arbitrary dimension, in the complex analytic setting. Van der Waerden reacted immediately in a letter, inviting Kneser to publish a full account of the argument in the *Mathematische Annalen* and pointing out its importance by comparing it with Walker's analytic

---

[73]"Your letter was a great joy for me. You will not believe how happy I am that the statements I came up with are not only meaningful and correct, but that you taught me a method to attack these and similar questions. I am convinced that I will make substantial progress with this method, provided I one day have the time to take up my mathematical research work again with full sails [Mit Ihrem Brief haben Sie mir eine grosse Freude gemacht. Sie glauben gar nicht wie glücklich ich bin, nicht nur dass die von mir ausgesprochenen Behauptungen überhaupt sinnvoll und richtig sind, sondern dass ich durch Sie eine Methode gelernt habe, wie man diese und dann auch ähnliche Fragen angreifen kann. Ich bin überzeugt, dass ich mit dieser Methode in meinem Programm erheblich weiterkommen werde, wenn ich einmal die Zeit habe, die mathematische Forschungsarbeit wieder mit vollen Segeln aufzunehmen" [UAG Cod. Ms. H. Hasse 1:1794, van der Waerden, Bartel Leendert; Hasse to van der Waerden, 9 January, 1942].

[74]See [van der Waerden, 1947a].

[75]A letter to H. Braun dated Leipzig, 3 May, 1944 [ETHZ, HS 652 : 10 552] shows that van der Waerden, conscious of his political difficulties at Leipzig, tried—apparently in vain—during World War II to get help from Hasse as well as Wilhelm Süss.

[76]In evidence of this, consider, for example, the following critique of Hasse's notion of a point: "Calling these homomorphisms 'points' fits badly with the terminology of algebraic geometry. A point in algebraic geometry is not a homomorphism but a sequence of homogeneous coordinates or something which is uniquely determined by such a sequence, and so many other notions and notations hinge on this concept of 'point,' that it is impossible to use the same word in another meaning. What Hasse calls 'point' is, in our terminology, a *relationstreue Spezialisierung* $\zeta \to z$, i.e., the transition from a generic to a special point of an algebraic variety [Zu der Terminologie der algebraischen Geometrie paßt die Bezeichnung dieser Homomorphismen als 'Punkte' nicht. Ein Punkt ist in der algebraischen Geometrie kein Homomorphismus, sondern eine Reihe von homogenen Koordinaten oder etwas, was durch eine solche Reihe eindeutig bestimmt ist, und an diesem Begriff 'Punkt' hängen soviele andere Begriffe und Bezeichnungen, daß man dasselbe Wort unmöglich in einer anderen Bedeutung verwenden kann. Was bei Hasse 'Punkt' heißt, ist in unserer Bezeichnungsweise eine *relationstreue Spezialisierung* $\zeta \to z$, der Übergang von einem allgemeinen zu einem speziellen Punkt einer algebraischen Mannigfaltigkeit" [van der Waerden, 1947a, p. 346].

proof [Walker, 1935] of the resolution of singularities of algebraic surfaces.[77] Kneser did not comply. As a result, when van der Waerden reported on 23 October, 1941 at the meeting in Jena of the DMV about "recent American investigations," that is, about Oscar Zariski's arithmetization of local uniformization and resolution of singularities of algebraic surfaces [van der Waerden, 1942], and when he mentioned Kneser's work as a balm for his German audience, he was promptly criticized in a review by Claude Chevalley because that proof had never been published in detail.[78]

Zariski's stupendous accomplishments in the rewriting of algebraic geometry—which between 1939 and 1944 included not only the basic "arithmetic" theory of algebraic varieties but also a good deal of the theory of normal varieties (a terminology introduced by Zariski) as well as the resolution of singularities for two- and three-dimensional varieties—were based on Wolfgang Krull's general theory of valuations much more than on van der Waerden's approach. This heavier algebro-arithmetic packaging visibly separated Zariski's approach from the Italian style in which he had been brought up. The independence of the mature Zariski from his mathematical origins gave him a distinct confidence in dealing with Severi after World War II. For example, it was Zariski who suggested inviting Severi to the algebraic geometry symposium held at the Amsterdam ICM and organized by Kloosterman and van der Waerden.[79]

As noted above, van der Waerden's basic ideas for an algebraic reformulation of algebraic geometry—his generic points and specializations—account for a good deal of the technical backbone of André Weil's *Foundations of Algebraic Geometry*. Moreover, van der Waerden's success in rewriting much of algebraic geometry with these modest methods had, of course, informed Weil's undertaking. In trying to pin down the most important differences between the contributions of van der Waerden and Weil to the rewriting of algebraic geometry, then, the mathematical chronicler must first isolate innovations that Weil brought to the subject and that

---

[77]See van der Waerden to Kneser, 23 March, 1936 [NSUB, Cod. Ms. H. Kneser A 93, Blatt 10].

[78]See *Mathematical Reviews* 5 (1944), 11. "A previous solution of the problem [of local uniformization] is credited to Kneser [Jber. Deutsch. Math. Verein. 45, 76 (1935)]. This attribution of priority seems unfair. Kneser published only a short note in which he outlined the idea of a proof of the local uniformization theorem. Considering the great importance of the result the fact that Kneser never came back to the question makes it seem probable that he ran into serious difficulties in trying to write down the missing details of his proof."

Totally outside of the context of the resolution of singularities, but as another interesting illustration of the variety of approaches to algebraic geometry that were in the air in the 1930s and 1940s, we mention in passing Teichmüller's sketch [Tecihmüller, 1942] of how to derive the theory of complex algebraic functions of one variable from the uniformization theory of Riemann surfaces. This paper is probably both an attempt to promote his research program towards what is today called Teichmüller Theory, and an expression of Teichmüller's ideas about adequate methods in complex geometry. For the latter aspect, compare the somewhat ideological discussion of relative merits of various methods of proof, and in particular the preference for "geometric" reasonings, in [Teichmüller, 1944, §6].

[79]See the correspondence between Zariski and Severi in [HUA, HUG 69.10, Box 2, 'Serre - Szegö']. In a letter to Kloosterman dated 15 January, 1954 [HUA, HUG 69.10, Box 2, 'Zariski (pers.)'] Zariski wrote: "I am particularly worried by the omission of the name of Severi. I think that Severi deserves a place of honor in any gathering of algebraic geometers as long as he is able and willing to attend such a gathering. We must try to avoid hurting the feelings of a man who has done so much for algebraic geometry. He is still mentally alert, despite his age, and his participation can only have a stimulating effect. I think he should be invited to participate."

went beyond what he found in his predecessors, namely, the local definition of intersection multiplicities, the proof of the Riemann Hypothesis, the formulation of the general Weil Conjectures, the use of abstract varieties, etc. But as in Zariski's case, where the valuation-theoretic language immediately created a sense of independence from predecessors or competitors (an independence, however, which would probably be considered pointless if it were not accompanied by mathematical success), Weil produced the same effect via the *style* of his *Foundations*. What struck many contemporaries (who had no notion yet of Bourbaki's texts) as a book full of mannerisms, effectively imposed a practice of doing algebraic geometry *à la* Weil.

Keeping both of these aspects in mind—the novelty of mathematical notions and the new style—is essential for a reasonable discussion of Weil's role in re-shaping algebraic geometry. For instance, pointing to the fact that Weil's *Foundations* get most of their mileage out of van der Waerden's basic notions, as does Serge Lang, does *not* suffice to invalidate Michel Raynaud's claim, quoted by Lang, that Weil's *Foundations* mark "a break (*rupture*) with respect to the works of his predecessors—B. L. van der Waerden and the German school" [Lang, 2002, p. 52]. In other words, Weil's book is a startling example showing how a history of mathematics that only looks at "mathematical content" easily misses an essential part of the story.

To fix ideas, consider the year 1947. A spectrum of five disciplinary practices of algebraic geometry exist:

(1) the classical Italian way,
(2) van der Waerden's way,
(3) the method of Weil's *Foundations*,
(4) Zariski's valuation-based arithmetization, and
(5) (only for the case of curves) the practice of function field arithmetic.

Given the force of the discourse about the lack of rigor in (1) compared to existing algebraic or arithmetic alternatives, and given the dimension-restriction of (5), the real competition took place between (2), (3), and (4). Then, the superficial resemblance between (2) and (1), on the one hand, and the fact, on the other hand, that the basic mathematical concepts of (2) are absorbed in (3), clearly left the finish between (3) and (4). This was precisely the constellation that Pierre Samuel described in the lovely beginning of the introduction to his thesis [Samuel, 1951, pp. 1-2] and with respect to which he opted for the more varied method of (4). A more precise analysis of the mathematical practice of each of the alternatives will yield interesting insights into one of the most spectacular developments in the history of pure mathematics in the twentieth century, but this chapter, it is to be hoped, represents at least a start down this historical path.

Appendix: Extract from a Letter from Hasse to Severi

Ew. Exzellenz und Hochverehrter Herr Kollege,

Es ist mir ein tiefes Bedürfnis, Ihnen heute endlich einen Brief zu schreiben, den ich eigentlich gleich im Anschluss an die Tagung in Baden-Baden schreiben wollte. Die grossen Ereignisse, die inzwischen eingetreten sind, rechtfertigen es wohl, wenn ich zunächst ein paar Worte an Sie als hervorragenden Vertreter Ihres Landes richte, ehe ich zu Ihnen als Mathematiker und Kollegen spreche. Uns Deutsche bewegt in diesen Tagen ein Gefühl tiefster Dankbarkeit für die Treue und Entschlossenheit, mit der Ihr unvergleichlicher Duce zu unserem Führer gestanden hat, und ebenso für die Einmütigkeit und Verbundenheit, mit der sich das ganze italienische Volk zu der Sache unseres Volkes bekannt hat. Es ist wohl auch dem letzten von uns in diesen Tagen klar geworden, dass wir das gesteckte Ziel, die Befreiung der Sudetendeutschen, niemals erreicht hätten, wenn nicht der unbeugsame Wille unseres Führers und unseres Volkes diese kräftige und entschlossene Stütze durch den anderen Pol unserer Axe gehabt hätte. Sie haben ja aus dem Munde unseres Führers gehört, wie er dies anerkennt und wie er bereit ist, auch seinerseits zu seinem Freunde, dem Duce zu stehen, sollte es einmal nötig sein. Sie dürfen überzeugt sein, dass auch hinter diesem Wort das ganze deutsche Volk aus innerster Überzeugung steht.

Dazu, dass auch in unserem Bezirk, der Mathematik, der herzliche Wunsch und das eifrige Bestreben besteht, das Fundament der politischen Axe auf kulturellem Boden zu unterbauen und zu festigen, hätte es wohl des kräftigen Anstosses der letzten Wochen schon gar nicht mehr bedurft. Ich hoffe, dass Sie in Baden-Baden gefühlt haben, wie wir deutschen Mathematiker in dieser Richtung denken und zu arbeiten gewillt sind. Ganz besonders habe ich mich gefreut, dort von dem Plan zu hören, durch eine Reihe von Monographien das gegenseitige Verstehen und die Gleichrichtung der beiderseitigen Schulen in der Algebra und Geometrie zu fördern. ...[80]

Translation

Your excellency, venerated colleague:

It is my deep-felt need at last to write you a letter today, which I had originally wanted to write just after the conference in Baden-Baden. The big events that have occurred in the meantime surely justify my addressing you first as an eminent representative of your country, before talking to you as a mathematician and colleague. All Germans are moved these days by the resolute faithfulness with which your incomparable *Duce* has stood beside our *Führer*, and by the united solidarity which the Italian people have acknowledged in the interest of our people. Down to the last one among us we have realized these days that the intended goal: the liberalization of the *Sudeten*-Germans, would never have been attained, if the unfaltering will of our *Führer* and our people had not enjoyed this strong and resolute support by the other pole of our axis. You have heard it from the mouth of our *Führer*, how he acknowledges this and how he is prepared also to stand by the side of his friend, the *Duce*, if ever this should prove necessary. You may be assured that the German people also stand behind this word with innermost conviction.

In order that also in our domain, mathematics, the heartfelt desire and arduous quest exist to underpin and stabilize the foundation of the political axis in the cultural terrain, the forceful impetus of the past weeks would not even have been necessary. I hope that you will have felt in Baden-Baden [at a meeting of the DMV where Severi had given an invited talk] how we, the German mathematicians, think and are willing to work. I was particularly glad to hear of the plan to enhance the mutual understanding and the synchronization of the schools on both sides in algebra and geometry. ...

****

---

[80]See [UAG Cod. Ms. H. Hasse 1:1585, Severi, Francesco; Hasse to Severi, 3 October, 1938].

Nothing more about this planned series of monographs is known, yet Severi's answer to the spirit of Hasse's letter may be found in the conclusions of his Baden-Baden lecture. There, Severi expressed the "hope that the important progress that Germany has realized in modern algebra will enable your magnificent mathematicians to penetrate ever more profoundly into algebraic geometry, which has been cultivated in Italy over the last 40 years, and that the ties between German and Italian science which have already been so close in this area at the times of our masters will grow every day more intimate, as they are today in the political and general cultural domain" [Severi, 1939, p. 389].[81]

Acknowledgments

Much of the material developed in this article was tried out in my Zürich lectures during the winter term of 2003–2004, while I was enjoying the wonderful hospitality of the *Collegium Helveticum* as well as of the mathematicians at ETH. Birgit Petri took notes of my lectures there and wrote them up afterwards, often adding extra material and developing further insights. This very substantial work of hers entered into my present write-up in so many places that I can only signal my indebtedness to her by this global acknowledgment and expression of profound gratitude.

References

Archival Sources

NSUB = Handschriftenabteilung der Staats- und Universitätsbibliothek Göttingen.
HUA = Harvard University Archives, Cambridge MA.
UAL = Universitätsarchiv Leipzig.
ETHZ = Archiv der ETH Zürich.

Printed Sources

Brigaglia, Aldo and Ciliberto, Ciro. 1995. *Italian Algebraic Geometry between the Two World Wars.* Vol. 100. Queen's Papers in Pure and Applied Mathematics. Kingston: Queen's University.

Brigaglia, Aldo; Ciliberto, Ciro; and Pedrini, Claudio. 2004. "The Italian School of Algebraic Geometry and Abel's Legacy." In *The Legacy of Niels Henrik Abel (The Abel Bicentennial, Oslo 2002).* Ed. Olav A. Laudal and Ragni Piene. Berlin/Heidelberg: Springer-Verlag, pp. 295-347.

de Boer, Jan Hendrik. 1994. "Van der Waerden's Foundations of Algebraic Geometry." *Nieuw Archief voor Wiskunde* 12, 159-168.

Bos, Henk J. M. 2001. *Redefining Geometrical Exactness: Descartes' Transformation of the Early Modern Concept of Construction.* Berlin/Heidelberg/New York: Springer-Verlag.

---

[81] "Spero che i progressi tanto importanti che la Germania ha conseguiti nell'algebra moderna, consentiranno ai suoi magnifici matematici di penetrare sempre più a fondo nella geometria algebrica, quale è stata coltivata in Italia negli ultimi 40 anni; e che i legami fra la scienza tedesca e la scienza italiana, che furono già tanto stretti in questo dominio ai tempi dei nostri Maestri, divengano ogni giorno più intimi, come lo sono oggi sul terreno politico e culturale generale."

Brill, Alexander von and Noether, Max. 1892–1893. "Bericht über die Entwicklung der Theorie der algebraischen Functionen in älterer und neuerer Zeit." *Jahresbericht der Deutschen Mathematiker-Vereinigung* 3, 107-565 (this actually appeared in 1894).

Chow, Wei-Liang. 1937. "Die geometrische Theorie der algebraischen Funktionen für beliebige vollkommene Körper." *Mathematische Annalen* 114, 655-682; Reprinted in [Chow, 2002, pp. 14-41].

_____. 2002. *The Collected Papers of Wei-Liang Chow*. Ed. Shiing Shen Chern and Vyacheslav V. Shokurov. London/Singapore/Hong Kong: World Scientific.

Chow, Wei-Liang and van der Waerden, Bartel L. 1937. "Zur algebraischen Geometrie IX: Über zugeordnete Formen und algebraische Systeme von algebraischen Mannigfaltigkeiten." *Mathematische Annalen* 113, 692-704; Reprinted in [van der Waerden, 1983, pp. 212-224] (a totally inadequate and incomplete English translation of this paper (but not the German original) appeared in [Chow, 2002, pp. 1-13]).

Corry, Leo. 1996/2003. *Modern Algebra and the Rise of Mathematical Structures*, Science Networks, Vol. 17; 1st Ed. 1996. 2nd Ed. 2003. Basel/Boston: Birkhäuser Verlag.

Dieudonné, Jean. 1974. *Cours de géométrie algébrique*. Vol. 1. *Aperçu historique sur le développement de la géométrie algébrique*, Paris: Presses universitaires de France.

Enriques, Federigo and Chisini, Oscar. 1915. *Lezioni sulla teoria geometrica delle equazioni e delle funzioni algebriche*, Vol. 1. Bologna: Nicola Zanichelli.

Guerraggio, Angelo and Nastasi, Pietro, Ed. 1993. *Gentile e i matematici italiani: Lettere 1907–1943*. Turin: Universale Bollati Boringheri.

Hasse, Helmut. 1942. "Zur arithmetischen Theorie der algebraischen Funktionenkörper." *Jahresbericht der Deutschen Mathematiker-Vereinigung* 52, 1-48.

Kneser, Hellmuth. 1935. "Örtliche Uniformisierung der analytischen Funktionen mehrerer Veränderlichen." *Jahresbericht der Deutschen Mathematiker-Vereinigung* 45, 76-77.

Krull, Wolfgang. 1937. "Beiträge zur Arithmetik kommutativer Integritätsbereiche. III: Zum Dimensionsbegriff der Idealtheorie." *Mathematische Zeitschrift* 42, 745-766; Reprinted in Krull, Wolfgang. 1999. *Gesammelte Abhandlungen—Collected Papers*. Ed. Paolo Ribenboim. Berlin/New York: de Gruyter, 1:746-767.

Lang, Serge. 1996. "Comments on Chow's Work." *Notices of the American Mathematical Society* 43, 1119-1123; Reprinted in [Chow, 2002, pp. 493-500].

_____. 2002. "Comments on Non-references in Weil's Works." *Deutsche Mathematiker-Vereinigung Mitteilungen* 1/2002, 49-56.

Lefschetz, Solomon. 1924. *L'analysis situs et la géométrie algébrique*. Paris: Gauthier-Villars.

Noether, Emmy. 1919. "Die arithmetische Theorie der algebraischen Funktionen einer Veränderlichen in ihrer Beziehung zu den übrigen Theorien und zu der Zahlkörpertheorie." *Jahresbericht der Deutschen Mathematiker-Vereinigung* 38, 182-203; reproduced (without the final footnote) in [Noether, 1983, pp. 271-292].

_____. 1921. "Idealtheorie in Ringbereichen." *Mathematische Annalen* 83, 24-66; Reprinted in [Noether, 1983, pp. 254-396].

_____. 1923a. "Bearbeitung von K. Hentzelt: Zur Theorie der Polynomideale und Resultanten." *Mathematische Annalen* 88, 53-79; Reprinted in [Noether, 1983, pp. 409-435].

_____. 1923b. "Eliminationstheorie und allgemeine Idealtheorie." *Mathematische Annalen* 90, 229-261; Reprinted in [Noether, 1983, pp. 444-476].

_____. 1983. *Gesammelte Abhandlungen—Collected Papers.* Ed. Nathan Jacobson. Berlin/Heidelberg/New York: Springer-Verlag.

Parikh, Carol. 1991. *The Unreal Life of Oscar Zariski.* Boston/San Diego/New York: Academic Press, Inc.

Philippon, Patrice. 1991. "Sur des hauteurs alternatives I." *Mathematische Annalen* 289, 255-283.

_____. 1994. "Sur des hauteurs alternatives II." *Annales de l'Institut Fourier* 44, 1043-1065.

_____. 1995. "Sur des hauteurs alternatives III." *Journal des mathématiques pures et appliquées* 74, 345-365.

Rémond, Gaël. 2001. "Élimination multihomogène." In *Introduction to Algebraic Independence Theory.* Ed. Yuri V. Nesterenko and Patrice Philippon. Berlin/Heidelberg/New York: Springer Verlag, pp. 53-81.

Remmert, Volker. 2004. "Die Deutsche Mathematiker-Vereinigung im 'Dritten Reich.' II: Fach- und Parteipolitik." *Mitteilungen der Deutschen Mathematiker-Vereinigung* 12, 223-245.

Roth, Leonard. 1963. "Francesco Severi." *Journal of the London Mathematical Society* 38, 282-307.

Samuel, Pierre. 1951. "La notion de multiplicité en algèbre et en géométrie algébrique." *Thèses présentées à la Faculté des Sciences de l'Université de Paris.* Paris: Gauthier-Villars.

Schubert, Hermann C. H. 1979. *Kalkül der abzählenden Geometrie.* Berlin/Heidelberg/New York: Springer-Verlag (Reprint Ed. of the original 1879 edition with an introduction by Steven L. Kleiman and a bibliography of H. C. H. Schubert by Werner Burau).

Severi, Francesco. 1912. "Sul principio della conservazione del numero." *Rendiconti del Circolo matematico di Palermo* 33, 313-327; Reprinted in [Severi, 1974, pp. 309-323].

_____. 1920. "Una rapida ricostruzione della geometria sopra una curva algebrica." *Atti del Reale Istituto Veneto di scienze, lettere ed arti* 79, 929-938; Reprinted in [Severi, 1977, pp. 12-18].

_____. 1926. *Trattato di geometria algebrica.* Vol. 1. Pt. 1. *Geometria delle serie lineari.* Bologna: Nicola Zanichelli.

_____. 1933. "Über die Grundlagen der algebraischen Geometrie." *Abhandlungen aus dem mathematischen Seminar der Hamburgischen Universität* 9, 335-364; Reproduced, with numerous typographical errors and with added 'Osservazioni complementari' by F. Severi in [Severi, 1980, pp. 102-131].

_____. 1939. "La teoria generale delle corrispondenze fra due varietà algebriche e i sistemi d'equivalenza." *Abhandlungen aus dem mathematischen*

Seminar der Hamburgischen Universität 13, 101-112; Reprinted in [Severi, 1980, pp. 378-389].

_____. 1952. "Una nuova visione della geometria sopra una curva." *Acta pontificia academia scientiarum* 14 (13), 143-152; Reprinted in [Severi, 1989, pp. 225-232].

_____. 1971–1989. *Opere Matematiche, Memorie e Note.* Rome: Accademia Nazionale dei Lincei; 1971. Vol. 1 (1900–1908); 1974. Vol. 2 (1909–1917); 1977. Vol. 3 (1918–1932); 1980. Vol. 4 (1933–1941); 1988. Vol. 5 (1942–1948); 1989. Vol. 6 (1949–1961).

Siegmund-Schultze, Reinhard. 1986. "Faschistische Pläne zur 'Neuordnung' der europäischen Wissenschaft: Das Beispiel Mathematik." *NTM. Schriftenreihe für Geschichte der Naturwissenschaften, Technik und Medizin* 23/2, 1-17.

_____. 2001. *Rockefeller and the Internationalization of Mathematics between the Two World Wars. Documents and Studies for the Social History of Mathematics in the Twentieth Century. Science Networks. Vol. 25.* Basel/Boston/Berlin: Birkhäuser Verlag.

Study, Eduard. 1916. "Das Prinzip der Erhaltung der Anzahl (nebst einer Bemerkung von K. Rohn)." *Berichte der sächsischen Akademie Leipzig* 68, 65-92.

Teichmüller, Oswald. 1982. *Gesammelte Abhandlungen—Collected Papers.* Ed. Lars V. Ahlfors and Frederick W. Gehring. Berlin/Heidelberg/ New York: Springer-Verlag.

_____. 1942. "Skizze einer Begründung der algebraischen Funktionentheorie durch Uniformisierung," Deutsche Mathematik 6, 257-265; Reprinted in [Teichmüller, 1982, pp. 599-607].

_____. 1944. "Beweis der analytischen Abhängigkeit des konformen Moduls einer analytischen Ringflächenschar von den Parametern." Deutsche Mathematik 7, 309-336; Reprinted in [Teichmüller, 1982, pp. 677-704].

van der Waerden, Bartel L. 1926a. "Zur Nullstellentheorie der Polynomideale." *Mathematische Annalen* 96, 183-208; Reprinted in [van der Waerden, 1983, pp. 11-36].

_____. 1926b. "Ein algebraisches Kriterium für die Lösbarkeit von homogenen Gleichungen." *Proceedings of the Royal Academy of Amsterdam* 29, 142-149.

_____. 1927. "Der Multiplizitätsbegriff der algebraischen Geometrie." *Mathematische Annalen* 97, 756-774; Reprinted in [van der Waerden, 1983, pp. 37-55].

_____. 1928a. "Eine Verallgemeinerung des Bézoutschen Theorems." *Mathematische Annalen* 99, 497-541; Reprinted in [van der Waerden, 1983, pp. 56-100].

_____. 1928b. "Die Alternative bei nichtlinearen Gleichungen." *Nachrichten der Gesellschaft der Wissenschaften Göttingen*, 1-11.

_____. 1928c. "On Hilbert's Function, Series of Composition for Ideals and a Generalization of the Theorem of Bezout." *Proceedings of the Royal Academy of Amsterdam* 31, 749-770.

_____. 1929. "Topologische Begründung des Kalküls der abzählenden Geometrie." *Mathematische Annalen* 102, 337-362; Reprinted in [van der Waerden, 1983, pp. 101-126].

_____. 1930–1931. *Moderne Algebra*. 2 Vols. Berlin: Springer-Verlag.

_____. 1931. "Zur Begründung des Restsatzes mit dem Noetherschen Fundamentalsatz." *Mathematische Annalen* 104, 472-3475; Reprinted in [van der Waerden, 1983, pp. 127-130].

_____. 1933. "Zur algebraischen Geometrie I: Gradbestimmung von Schnittmannigfaltigkeiten einer beliebigen Mannigfaltigkeit mit Hyperflächen." *Mathematische Annalen* 108, 113-125; Reprinted in [van der Waerden, 1983, pp. 131-143].

_____. 1934. "Zur algebraischen Geometrie VI: Algebraische Korrespondenzen und rationale Abbildungen." *Mathematische Annalen* 110, 134-160; Reprinted in [van der Waerden, 1983, pp. 168-194].

_____. 1935. "Nachruf auf Emmy Noether." *Mathematische Annalen* 111, 469-476.

_____. 1939. *Einführung in die algebraische Geometrie*. Grundlehren der mathematischen Wissenschaften. Vol. 51. Berlin: Springer-Verlag.

_____. 1941. "Topologie und Uniformisierung der Riemannschen Flächen." *Berichte der sächsischen Akademie Leipzig* 93.

_____. 1942. "Die Bedeutung des Bewertungsbegriffs für die algebraische Geometrie." *Jahresbericht der Deutschen Mathematiker-Vereinigung* 52, 161-172; Reprinted in [van der Waerden, 1983, pp. 308-319].

_____. 1946. "The Foundation of the Invariant Theory of Linear Systems of Curves on an Algebraic Surface." *Proceedings of the Royal Academy of Amsterdam* 49, 223-226.

_____. 1947a. "Divisorenklassen in algebraischen Funktionenkörpern." *Commentarii Mathematici Helvetici* 20, 68-109; Reprinted in [van der Waerden, 1983, pp. 320-361].

_____. 1947b. "Birational Invariants of Algebraic Manifolds." *Acta Salamantica* 2, 1-56.

_____. 1948. "Book Review of [Weil, 1946a]." *Nieuw Archief voor Wiskunde*, 363-366.

_____. 1950a. "Les variétés de chaînes sur une variété abstraite." *Colloque de géométrie algébrique Liège 1949*, 79-85.

_____. 1950b. "Les valuations en géométrie algébrique." *Algèbre et théorie des nombres*, Paris: Services des publications du CNRS, pp. 117-122.

_____. 1956a. "The Invariant Theory of Linear Sets on an Algebraic Variety." *Proceedings of the International Congress of Mathematicians: Amsterdam 1954*. 3 Vols. Nedeln/Lichtenstein: Kraus Reprint Limited, 3:542-544.

_____. 1956b. "On the Definition of Rational Equivalence of Cycles on a Variety." *Proceedings of the International Congress of Mathematicians: Amsterdam 1954*. 3 Vols. Nedeln/Lichtenstein: Kraus Reprint Limited, 3:545-549.

_____. 1958. "Über André Weils Neubegründung der algebraischen Geometrie." *Abhandlungen aus dem Mathematischen Seminar der Universität Hamburg* 22, 158-170.

———. 1959. *Algebra*. Vol. 2. 4th Ed. Berlin/Heidelberg: Springer-Verlag.

———. 1970. "The Theory of Equivalence Systems of Cycles on a Variety." *Symposia Mathematica INdAM* 5, 255-262; Reprinted in [van der Waerden, 1983, pp. 440-447].

———. 1971. "The Foundations of Algebraic Geometry from Severi to André Weil." *Archive for History of Exact Sciences* 7(3), 171-180; Reprinted in [van der Waerden, 1983, pp. 1-10].

———. 1983. *Zur Algebraischen Geometrie: Selected Papers*. Berlin/Heidelberg/New York: Springer-Verlag.

———. 1986. "Francesco Severi and the Foundations of Algebraic Geometry," *Symposia Mathematica INdAM* 27, 239-244.

Walker, Robert J. 1935. "Reduction of the Singularities of an Algebraic Surface." *Annals of Mathematics* 36, 336-365.

Weil, André. 1938b. "Zur algebraischen Theorie der algebraischen Funktionen." *Journal für die reine und angewandte Mathematik* 179, 129-133; Reprinted in [Weil, 1979, pp. 227-231].

———. 1946a. *Foundations of Algebraic Geometry*. American Mathematical Society Colloquium Publications. Vol. 29. New York: American Mathematical Society.

———. 1979. *Œuvres scientifiques—Collected Papers*. Vol. 1. Berlin/Heidelberg/New York: Springer-Verlag.

Weyl, Hermann. 1924. "Randbemerkungen zu Hauptproblemen der Mathematik." *Mathematische Zeitschrift* 20, 131-150; Reprinted in [Weyl, 1968, 2:433-452].

———. 1968. *Gesammelte Abhandlungen*. Ed. Komaravolu Chandrasakharan. 4 Vols. Berlin/New York: Springer-Verlag.

Zariski, Oscar. 1944. "The Compactness of the Riemann Manifold of an Abstract Field of Algebraic Functions." *Bulletin of the American Mathematical Society* 45, 683-691; Reprinted in [Zariski, 1972, 1(2):432-440].

———. 1972. *Collected Papers*. Ed. Heisuke Hironaka and David Mumford. Vol. I (in 2 Pts.). Cambridge, MA and London: The MIT Press.

CHAPTER 12

On the Arithmetization of Algebraic Geometry

Silke Slembek[1]
Zürich, Switzerland

Introduction

The *arithmetization* of algebraic geometry is one of the foundational processes that has shaped algebraic geometry in the twentieth century. One of its defining characteristics is a methodological shift: classical geometrical methods were replaced with concepts from modern algebra and arithmetical ideal theory. Arithmetization went hand in hand with a geographical shift. Until the 1920s, algebraic geometry was largely identified with the work of Italian mathematicians, notably Guido Castelnuovo, Federigo Enriques, and Francesco Severi. In 1927, the area was transferred to North America by Oscar Zariski, a student of Castelnuovo. Ten years later, Zariski began the project of arithmetizing algebraic geometry and, in so doing, completely transformed the discipline. Within another ten years, Zariski's new methods had replaced the old ones, so that by the middle of the twentieth century, Italian algebraic geometry had lost almost all of its influence, while Zariski, who was appointed to a professorship at Harvard University in 1947, "would help to make Harvard into the world center of algebraic geometry that Rome had once been" [Parikh, 1991, p. 115]. The new American school that grew up around Zariski was influential through students such as David Mumford, Joseph Lipman, Heisuke Hironaka, Shreeram Abhyankar, and others.

The origins and conditions of arithmetization have been explained by its protagonists in the light of its success and of later reformulations of algebraic geometry. In order to gain new insight into the historical circumstances of the foundational process, I offer here an analysis of arithmetization through the original texts in order to highlight both the initial scheme underlying the process and the characteristics that differentiate it from earlier formulations.

The chapter opens with preparatory remarks on the methodological variety in algebraic geometry in the nineteenth century by way of establishing the historical context. Among the factors that immediately influenced arithmetization, I focus on the mathematical one, namely, the theory of singularities of algebraic surfaces and, in particular, the theorem on the reduction of singularities. In order to show why this problem became the touchstone of arithmetization, I outline it and discuss how mathematicians between 1897 and 1935 judged its validity. Through Zariski's articles, the design of arithmetization—its issues, its architecture, how it was supposed

---

[1]Due to unforeseen circumstances, the author was unable to do a final proofreading of this chapter. We have tried to assure that her chapter is as accurate as possible.—The Editors.

to work—comes to the fore. This textual analysis reveals a foundational process that differed from nineteenth-century algebraic rewritings of algebraic geometry in its motives, origins, structure, and issues.[2]

Earlier Rewritings of Algebraic Geometry

Algebraic geometry grew in the second half of the nineteenth century from different traditions: complex analysis, geometry, algebra, and number theory. Trying to understand and to use the path-breaking ideas of Bernhard Riemann, mathematicians fostered known and established new relations between algebra, analysis, and geometry. Between 1860 and 1895, a half-dozen approaches to algebraic functions were developed, each with its own methodological focus: algebro-geometric, algebraic, arithmetic, algebro-form-theoretic, geometric, and function-theoretic [Brill and Noether, 1892–1893, p. 287].

The geometric approach was developed mainly by the Italians Corrado Segre, Castelnuovo, and Enriques, inspired by articles of Max Noether and Alexander von Brill. In 1871 and 1873, Noether set out to replace the analytic methods that had been used throughout the nineteenth century in the analysis of singularities of algebraic curves—namely, Puiseux series expansions—with algebraic processes [M. Noether, 1871/1873]. In their 1874 paper "Ueber die algebraischen Functionen und ihre Anwendung in der Geometrie [On Algebraic Functions and Their Application to Geometry]," Brill and Noether rejected transcendental methods and, in particular, the Dirichlet principle. They promoted instead the use of algebraic methods in the application to geometry that Alfred Clebsch had made in 1864 of Riemann's theory of Abelian functions. They explicitly aimed to make the theory of Abelian functions a branch of algebra [Brill and Noether, 1874, p. 269].

At the same time, the work of Segre, Castelnuovo, and Enriques was deeply rooted in the geometric tradition. One of its most influential ideas was that of the use of higher dimensional spaces. Giuseppe Veronese had shown in 1882 that it was often useful to look at geometric objects as the projection of objects in some higher dimensional projective space [Veronese, 1882]. For instance, he showed that a rational plane curve of order m is the projection of a "normal" curve of order m belonging to the m-dimensional projective space. In particular, certain singular curves can be regarded as non-singular curves in a suitable projective space. Segre systematically took up Veronese's idea and proposed to look for families of curves that yield an embedding of a surface in some suitable projective space. Segre, in fact, gave a geometric reformulation of some of the algebraic work by Brill and Noether. In his "Introduzione alla geometria sopra un ente algebrico semplicemente infinito [Introduction to Geometry over a Simply Infinite Algebraic Entity]" of 1894, Segre sought to build the theory of linear series of points solely on geometric considerations. Moreover, he emphasized the birational point of view when he asserted that "[w]e can now ... easily define the object of the *geometry on a variety* M_k. It is *the geometry that studies those properties of the M_k that*

---

[2]A good textbook on the arithmetic theory of algebraic varieties is the three-volume *Methods of Algebraic Geometry* by William V. D. Hodge and Daniel Pedoe [Hodge and Pedoe, 1953–1954]. John Semple and Leonard Roth give numerous geometric examples in their *Introduction to Algebraic Geometry* [Semple and Roth, 1949].

are invariant under birational transformations of the variety" [Segre, 1894/1954–1963, p. 51 or 210].[3] Segre encouraged Castelnuovo and Enriques to take up the study of algebraic surfaces. In their 1895 paper on some then-recent results in the theory of algebraic surfaces, they generalized the ideas found in [Brill and Noether, 1874] to algebraic surfaces, but, in so doing, they minimized algebra and gave the theory a geometric turn [Castelnuovo and Enriques, 1895]. In a series of joint papers, they developed the theory of algebraic surfaces that ultimately became synonymous with algebraic geometry, one of the most outstanding achievements of which was the classification of algebraic surfaces by Castelnuovo and Enriques in 1914 [Castelnuovo and Enriques, 1914].

The central concept in the geometric study of varieties is that of linear systems. They are the tool for manipulating curves and surfaces in projective space. More precisely, to study a surface F in the projective space $\mathbb{P}^n$, choose $r+1$ homogeneous polynomials $G_0(x_0, \ldots, x_n), \ldots, G_r(x_0, \ldots, x_n)$ with the same degree. Then for any complex numbers $\lambda_0, \ldots, \lambda_r$ (not all zero), the hypersurface with equation

$$\lambda_0 G_0 + \lambda_1 G_1 + \cdots + \lambda_r G_r = 0$$

cuts out a curve on F. The set gathering all these curves (for various $\lambda_0, \ldots, \lambda_r$) is called a linear system on F. The importance of linear systems lies in the fact that they yield rational applications of the surface to the r-dimensional projective space $\mathbb{P}^r$ by mapping a point $x_0, \ldots, x_n$ of F to the point

$$(G_0(x_0, \ldots, x_n) : \ldots : G_r(x_0, \ldots, x_n)).$$

Consider the transformation $f : \mathbb{P}^2 \to \mathbb{P}^2$ defined by

$$(x : y : z) \mapsto \left(\frac{1}{x} : \frac{1}{y} : \frac{1}{z}\right);$$

it will arise several times in what follows. This map is defined everywhere in the projective plane, except at the three points $(1:0:0)$, $(0:1:0)$, and $(0:0:1)$ and the three lines $L_0 = \{x = 0\}$, $L_1 = \{y = 0\}$, and $L_2 = \{z = 0\}$, which form the so-called fundamental triangle. Since the map is involutive outside this triangle, it is a birational transformation. Geometrically, it blows up each of the vertices of the fundamental triangle into the opposite side and, conversely, shrinks down each of the sides of the triangle to the opposite vertex.

It is easy to see that this birational transformation is a particular case of the more general construction above. Indeed, it is the birational application afforded by the linear system

$$\{\lambda\, yz + \mu\, xz + \nu\, xy = 0 \mid \lambda, \mu, \nu \in \mathbb{C}\}$$

consisting of all the plane conic sections through the three base points $(1:0:0)$, $(0:1:0)$, and $(0:0:1)$. Moreover, this linear system is a subsystem of the complete linear system

$$\{\rho\, x^2 + \sigma\, y^2 + \tau\, z^3 + \lambda\, yz + \mu\, xz + \nu\, xy = 0 \mid \rho, \sigma, \ldots, \nu \in \mathbb{C}\}$$

of all plane conic sections, which, in turn, yields the embedding

$$(x : y : z) \mapsto (x^2 : y^2 : z^2 : yz : xz : xy)$$

---

[3]"Possiamo ora ... definire l'oggetto della *geometria sopra una varietà* M_k. Essa è la geometria che studia le proprietà della M_k invariabili per trasformazioni birationali della varietà stessa."

of the projective plane $\mathbb{P}^2$ into the 5-dimensional projective space $\mathbb{P}^5$ studied by Veronese in 1882.

Although other approaches to the study of algebraic surfaces developed, algebraic geometry was largely identified with the geometric approach promoted by Italian mathematicians, and from the late nineteenth century, Rome was one of its most active research centers. By 1921 when Oscar Zariski enrolled at the University of Rome, the leading mathematicians of the day—Castelnuovo, Enriques, and Severi—taught in the Eternal City. Zariski subsequently received most of his mathematical training and culture from Italian mathematicians; he was a close disciple of his thesis adviser, Castelnuovo, and also of Enriques, who encouraged the research he published shortly after earning his Ph. D. in 1924. Although many mathematicians from Europe and the United States came to Rome to work with the masters and to learn their methods directly, Zariski actually became part of the Italian school and was regarded as one of the most promising figures of the new generation of mathematicians trained in Italy.[4]

After Zariski left Italy for the the United States in 1927, he was increasingly confronted with other traditions in algebraic geometry. Under the influence of Princeton's Solomon Lefschetz, for example, Zariski not only took a topological turn but he also confronted criticism of the Italian tradition. According to Lefschetz in a letter to Hermann Weyl on 30 November, 1926, "[t]he structure built by Castelnuovo, Enriques, Severi is no doubt magnificent but tremendously special and requires a terrible 'entraînement.' It is significant that since 1909 little has been done in that direction even in Italy. I think a parallel edifice can be built up within the grasp of an average analyst" [ETHZ, HS 91:659].[5] In Lefschetz's view, then, while impressive, the Italian approach was far too specialized; its methods were tailor-made for specific problems and required a huge amount of practice. From his vantage point, the results of the Italians should be obtainable using the more or less elementary methods of analysis.

In 1928, Zariski published his first paper in English in an American journal, and it reflected Lefschetz's critical eye. Zariski stated that after "investigating closer the geometrical proof of Severi[,] certain doubts as to the rigor of the procedure were brought forth" [Zariski, 1928, p. 87 or 1972–1979, 3:81-86 on p. 81]. These doubts were resolved by a methodological shift, namely, by the use of new topological methods.[6] Encouraged by his teachers, by Lefschetz, and by his own result, Zariski then concentrated on topological properties of algebraic varieties.

Ten years later in 1938, however, Zariski began the so-called *arithmetization* of algebraic geometry, which implied a new shift of methods: geometric tools were replaced by methods from modern algebra and arithmetic ideal theory. Arithmetization marked a point of no return in the methodological diversity of algebraic geometry. From this point of view, the algebraic and the geometric approaches that Brill and Noether discussed in 1893 coexist [Brill and Noether, 1892–1893]. Moreover, arithmetization turned out to be irreversible; it definitely abolished the

---

[4]See, for example, [Castelnuovo, 1929, p. 197] and [Severi, 1932/1967, p. 219].

[5]I thank David Rowe for pointing this letter out to me.

[6]The main theorem proved in Zariski's paper is that the general curve of a linear system $|C|$ on a regular surface F has no singular correspondence, provided that $|C|$ is sufficiently non-degenerate. Michael Artin and Barry Mazur gave a modern summary of the paper in the preface to the third volume of Zariski's collected papers [Zariski, 1972–1979, 3:3-4].

geometric argumentation used by Italian mathematicians. Several internal and external factors influenced this process. I focus first on the mathematical problem that prompted arithmetization.

The Mathematical Situation: Why Another Rewriting Seemed Necessary

In the early 1930s, Zariski was asked to write a monograph on the core discipline of Italian algebraic geometry, namely, the theory of algebraic surfaces. Published in 1935 in the series "Ergebnisse der Mathematik und ihrer Grenzgebiete," the book, *Algebraic Surfaces*, aimed to give a systematic exposition of the basics of the theory, but its novelty lay in the fact that it did not primarily display the results. Rather, Zariski insisted that "it is especially true of algebraic geometry that in this domain the methods employed are at least as important as the results" [Zariski, 1935, Preface]. This thus marked a critical shift in what the basics of algebraic geometry were supposed to be: the fundamentals were no longer theorems but methods. It was precisely in the methods and proofs that Zariski found many gaps and errors.

Zariski repeated this warning when he treated the theory of singularities in *Algebraic Surfaces*. "In the theory of singularities," he wrote, "the details of the proofs acquire a special importance and make all the difference between theorems which are rigorously proved and those which are only rendered highly plausible" [Zariski, 1935, p. 18]. One of the central theorems here concerned the resolution of singularities of algebraic surfaces, and the question of whether every given algebraic surface could be transformed with finitely many birational transformations into a surface without singularities had been intensely studied by Italian geometers in the 1890s.[7] All told, between 1897 and 1924, five proofs of that theorem had been given. Consistent with his statement just quoted, Zariski did not just indicate the result in his exposition; he gave a short analysis of each proof. His conclusion? While the Italian geometers had given several proofs, none of them withstood critical examination. The theorem was thus rendered, as Zariski put it, "highly plausible," but not "rigorously proved."[8] Doubts about the proof had, in fact, arisen at least as early as the 1920s [Bliss, 1923], and proofs of the resolution theorem had been analyzed earlier or at around the same time. In *Algebraic Surfaces*, however, its validity was explicitly denied for the first time since 1897.

Zariski's proposal for a new proof of the theorem built on the geometric argument given by Beppo Levi in 1897 [Levi, 1897a]. Consider first, however, the analogous theorem for plane algebraic curves as stated for the first time by Max Noether in 1871: every plane algebraic curve is birationally equivalent to a plane algebraic curve having only ordinary singularities (double points with distinct tangents and cusps) [M. Noether, 1871]. The birational transformation can be carried out with a finite number of suitable quadratic transformations of the projective plane, where "suitable" means a quadratic transformation that has a fundamental point in a singularity of the given curve. The series of quadratic transformations yields a decomposition of the singular point.

---

[7]See [Gario 1988; 1989].

[8]Still, as Brigaglia and Ciliberto have rightly observed and as a closer analysis confirms, although the lack of rigor is sometimes criticized, rigor is not the main issue in *Algebraic Surfaces* [Brigaglia and Ciliberto, 1995, p. 121].

The following example illustrates what happens to a singular curve under this transformation.[9] Let C be the plane quartic curve whose equation in homogeneous coordinates $(x:y:z)$ is $(x^2+y^2)^2 - x^3z + 3xy^2z = 0$. (The picture on the left below represents the curve in the affine plane $z \neq 0$.) Use a quadratic transformation to resolve the singularity at the point $(0:0:1)$ with fundamental triangle with vertices at the points $(0:0:1)$, $(-\frac{1}{2}:\frac{\sqrt{3}}{2}:1)$, and $(-\frac{1}{2}:-\frac{\sqrt{3}}{2}:1)$. In order to use the quadratic transformation defined above, first make the following change of coordinates:

$$x = -\frac{1}{2}(x_1 + y_1), \quad y = \frac{\sqrt{3}}{2}(x_1 - y_1), \quad z = x_1 + y_1 + z_1.$$

In this new system of homogeneous coordinates $(x_1 : y_1 : z_1)$, the equation of the curve is

$$(x_1^2 - x_1 y_1 + y_1^2)^2 = (x_1 + y_1 + z_1)(x_1 + y_1)\left(x_1^2 - \frac{5}{2}x_1 y_1 + y_1^2\right),$$

and the vertices of the fundamental triangle have coordinates $(0:0:1)$, $(1:0:0)$, and $(0:1:0)$. Now the quadratic transformation has become

$$(x_1 : y_1 : z_1) \mapsto (x_2 : y_2 : z_2) \quad \text{with } x_1 = y_2 z_2,\ y_1 = z_2 x_2 \text{ and } z_1 = x_2 y_2,$$

and the equation of the transformed curve is

$$x_2 y_2 z_2^3 \left[(x_2 + y_2)\left(x_2^2 - \frac{5}{2}x_2 y_2 + y_2^2\right) + z_2\left(\frac{3}{2}x_2^2 - 6x_2 y_2 + \frac{3}{2}y_2^2\right)\right] = 0.$$

This transformed curve consists therefore of the union of the sides of the fundamental triangle and of a rational cubic curve with a double point (the figure on the right below).

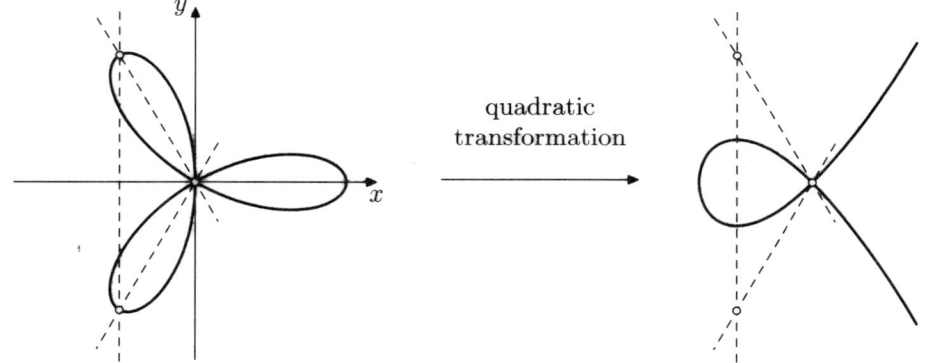

quadratic transformation

With the help of another Cremona plane transformation, one can resolve the double point and define a birational equivalence from the rational cubic to a smooth conic.

Twenty years after Noether's seminal work, Italian mathematicians took up and generalized the question to algebraic surfaces. Segre gave a proof of the assertion that the process of decomposition of an isolated singularity on an algebraic surface is always finite, except in the case when the singular point lies on a multiple component of the surface [Segre, 1894]. Segre's student, Beppo Levi, as noted above, gave his geometric proof of this result in 1897. In that same year, Levi also published a proof of the resolution problem for the singular locus of an algebraic surface, taking

---

[9]For another example, see [Brieskorn and Knörrer, 1981, pp. 604-714].

up (like every successful proof of the resolution theorem) Noether's idea of using Cremona transformations with a fundamental point at a singularity P of the surface F [Levi, 1897b]. What happens to the surface under such a transformation? Levi analyzed the situation this way: to the singular s-fold point P, there correspond on the transformed surface F' either finitely many points with multiplicity s or infinitely many such points all lying on a curve. In the second case, a curve appears that is included in the singular locus of the transformed surface. Depending on which case occurs, Levi applied different types of Cremona transformations. In the first case, he resolved isolated points with multiplicity s. If this gave rise to a finite series of infinitely near points, he was done. During the resolution of an isolated point, however, a curve of multiplicity s could also appear, which returned him to the second case above. In retrospect, Levi's argument that an s-fold curve can be resolved is not un-understandable; it is just not clear. Moreover, new, so-called accidental singularities were introduced at each transformation [Levi, 1897b]. Levi obviously thought that they were easy to get rid of, but again, his argument is unclear.

Today, the status of Levi's proof is undecidable, but his contemporaries did not doubt its correctness and accepted the proof as both valid and rigorous. Other proofs were given in [Severi, 1914], [Albanese, 1924], and [Chisini, 1924], all with the aim of increasing its simplicity or elegance. The rigor of the proof was never doubted, that is, at least not until Zariski's *Algebraic Surfaces* changed the way in which the proof was conceived.

While the proof of the theorem of desingularization marked the beginning of and the touchstone for arithmetization, it was not the only proof that Zariski criticized. In his review of Zariski's book for the *Bulletin of the American Mathematical Society*, Lefschetz cautioned the reader not to get the "unjust impression that Zariski occupied himself chiefly with a critique of things long since done. As a matter of fact this critique takes up only half of the monograph, the rest is being given over to modern questions" [Lefschetz, 1936, p. 14]. Lefschetz's remark might well have reflected his own opinion of Italian algebraic geometry (recall the letter quoted above), but it also captured the image of Italian algebraic geometry one could get from reading *Algebraic Surfaces*. Indeed, from Zariski's choice of the literature, the study of algebraic surfaces seems to have been a vital topic between 1890 and 1910, and during that period it was dominated by Italian mathematicians. After World War I, the scene changed radically. The Italian school seemed to contribute little; research activities moved on to other European countries and to America. The parts of the theory criticized in *Algebraic Surfaces* are precisely the ones "long since done," namely, the results and proofs by Italian geometers. Zariski's critique chiefly centered on methods that he deemed too specific and that made the theory of algebraic surfaces, from his perspective, a collection of case studies. Rigor was a minor issue here. Thus, the image was fostered of Italian algebraic geometry as old-fashioned and not quite meeting contemporary standards of mathematical writing.[10]

In his review, Lefschetz also referred, however, to "modern questions" of algebraic geometry, a hint to the alternatives to the Italian algebro-geometric approach that he and others perceived. Zariski announced in the preface of *Algebraic*

---

[10]On this notion of the "images" of mathematics, see [Corry, 1996] as well as Corry's chapter in the present volume.

Surfaces that his exposition would emphasize "the interrelations between the various aspects of the theory: algebro-geometric, topological, transcendental" [Zariski, 1935, Preface]. In fact, six out of eight chapters deal with the traditional, Italian algebro-geometric approach, one with the transcendental approach, and one with the topological approach. The standpoint of modern algebra had not yet permeated algebraic geometry to the point of being included among the modern approaches, but, among the promising modern questions, were those concerning topological properties of algebraic surfaces. This was exactly the area on which Zariski (influenced by Lefschetz) was focusing in 1935. Nothing in his book indicates that Zariski was about to leave that area or that he wanted to rewrite algebraic geometry arithmetically. He concentrated his research on topological questions (such as the fundamental group) and on topological methods that seemed modern, "clean and new" [Mumford, 1986, p. 891].

Imitating Geometry with Modern Algebra and Arithmetic Ideal Theory

When Zariski decided to undertake another rewriting of algebraic geometry, he turned to modern algebra and arithmetic ideal theory. He first had to acquaint himself with the tools. He knew and had read van der Waerden's *Moderne Algebra* soon after its publication in 1931–1932, and shortly after writing *Algebraic Surfaces*, he was invited to Princeton, where he heard lectures by Emmy Noether. It was, however, in the 1935 book, *Idealtheorie*, of Noether's student Wolfgang Krull that Zariski found concepts that he could interpret geometrically and that offered a basis for the rewriting of algebraic geometry [Krull, 1935]. The arithmetization of the field finally began at the end of the 1930s. As Zariski later put it, "[a]fter spending a couple of years just studying modern algebra, I had to begin somewhere" [Zariski, 1972–1979, 3:xiii]. The starting point, however, was not as arbitrary as the word "somewhere" might suggest. In fact, it should come as no surprise that the starting point was that same problem of the resolution of singularities that had been so important to Italian geometers and that had been singled out by Zariski as particularly insufficiently proved.

As noted above, in his 1935 monograph on *Algebraic Surfaces*, Zariski reviewed all of the geometric proofs that had been given to that point and had rejected them all. He thus showed that there was no valid proof of the theorem of resolution of singularities using geometric methods. There was, however, another proof, an analytic one given by Lefschetz's student Robert Walker in his doctoral dissertation. Walker was perhaps carrying out part of the program Lefschetz had sketched in his 1926 letter to Weyl, namely, to build a theory parallel to the geometric one that was within "the grasp of an average analyst" [Walker, 1935, p. 336]. Walker explicitly motivated his work with the desire to have a "completely rigorous proof for a theorem which is of such fundamental importance in the theory of surfaces." Moreover, he stated that the proof "should make use of as few as possible of the peculiar properties of algebraic surfaces," an echo of Lefschetz's call for less specific methods, and that the methods should extend to higher dimensional varieties [Walker, 1935, p. 336]. According to Zariski, Walker's proof "stands the most critical examination and settles the validity of the theorem beyond any doubt" [Zariski, 1935, p. 23 (note)]. The resolution theorem was thus more than "rendered highly plausible," it was rigorously proved. Yet, it was not proved within a geometric framework, and that was a crucial point. Arithmetization did not have its origins in the need for a proof that would give a rigorous foundation for some result, since such a proof already existed. Its origins lay elsewhere.

Zariski first articulated the arithmetization of algebraic geometry as a research aim in an article "On the Polynomial Ideals Defined by Infinitely Near Base Points" published in 1938 [Zariski, 1938 or 1972–1979, 1:14-67]. There, he developed the fundamental concepts of the new arithmetic theory and defined the structure he planned to give it, inspired by the work of the Italians.

Zariski started from "the foundation of a geometric theory of singularities of algebraic surfaces," namely, the concept of infinitely near points [Zariski, 1935, p. 13]. This concept had been introduced by Bernhard Riemann in the 1850s and had been given a precise meaning by Max Noether in [M. Noether, 1871; 1875]. Noether's main result in these papers was the theorem mentioned earlier: with a finite number of Cremona transformations, every plane algebraic curve can be transformed into a plane algebraic curve having only ordinary singularities. Infinitely near points occur in the reduction process. Recall that Noether used a Cremona transformation that had a fundamental point in the singular point O of a plane curve C. Under the transformation, the point (a vertex of the fundamental triangle) is blown up to a fundamental line (the opposite side of the fundamental triangle). The points of intersection with the fundamental line $O_1, O_2, \ldots$ of $\tilde{C}$ correspond to the singular point O of C. The points $O_1, O_2, \ldots$ with multiplicity smaller than the multiplicity of O constitute the first neighborhood of O and are infinitely near points of the first kind of O. They need not all be simple. If O_1 is not simple, the resolution process is repeated with a quadratic transformation having a fundamental point in O_1 giving rise to the second neighborhood of O formed by the points $O_{1,1}, O_{1,2}, \ldots$. Noether's result states that a finite series of quadratic transformations yields a curve that is birationally equivalent to C and has only ordinary singularities. At each step, the points $O_{i_1, i_2, \ldots i_j}$ are called infinitely near points of kind j forming the jth neighborhood of O. If the process is reversed, the singular point can be seen as composed of a finite series of infinitely near points. Corrado Segre generalized the concept of infinitely near points to algebraic surfaces in 1897. In an article on the decomposition (*"scomposizione"*) of the singular points of algebraic surfaces, he utilized both Noether's notion and the perception of a singular point as composed of a finite number of infinitely near points and gave a rigorous definition of infinitely near points of an algebraic surface.[11]

The other tool Zariski took from Italian algebraic geometry was the notion of a linear system as explained above. These are the geometric concepts on which arithmetization hinged. The link between them was thought of in the following way. A linear system can be defined by base conditions: one can require that the curves of the system pass through a point, through a set of points, or through infinitely near points. In the first case, the arithmetic translation is that the equations of the curves belong to the ideal of the point. In the second, they all belong to the intersection of the ideals of the points. In the third, one must replace the ideals of the points by so-called v-ideals. Zariski observed that the ideals of interest in this context are *complete* ideals, that is, ideals that can be written as the intersection of v-ideals. The terminology was expressly chosen to reflect the analogy with the

---

[11]See [Gario, 1989].

geometric notion of complete linear systems in the sense of linear systems that are uniquely defined by base conditions.

At this point, the process of arithmetization was described as a reformulation of the existing theory. As Zariski put it, "[i]t is the main purpose of the present investigation to develop an arithmetic theory parallel to the geometric theory" [Zariski, 1938, p. 151 or 1972–1979, 1:14]. Zariski, in fact, insisted on the terminological parallel.[12] Still, constructing the arithmetic theory along the lines of the geometric theory meant not only using the same words (although this is an interesting feature). The parallelism manifested itself in other ways. For example, the basic structure was adopted, that is, the geometric theory guided the decisions about what the basic notions of the arithmetic theory should be and about the argumentation that should be used in the presentation of the theory. Furthermore, the limits of the geometric theory were accepted. For example, properties of the simple or complete v-ideals only reflected geometric properties. There were thus no surprises in arithmetization, at least not for someone trained in the Italian tradition.

This parallelism was desirable and possible only because the geometric theory was accepted as a reliable, rigorous, and reasonable basis for the study of algebraic curves and surfaces. Since rigor was not an issue, the new theory did not need to start from a tabula rasa. It could instead rely on some existing foundations.

What then was the aim of arithmetization at this early stage? What could be expected of a theory that imitates an existing one? In 1938, the arithmetization of algebraic geometry was not yet an aim in and of itself; it was a way to recover the results established within the classical geometric framework in an algebraic-arithmetic framework. This would enable generalizations to higher dimensional objects and over arbitrary fields [Zariski, 1938, p. 152 or 1972–1979, 1:15].

Leaving the Beaten Path

Zariski, however, could not realize the resolution solely by following traditional lines. He also had to go beyond the existing theory and invent new concepts and a new operation. He did this in two articles, "Some Results in the Arithmetic Theory of Algebraic Varieties" and "The Resolution of the Singularities of an Algebraic Surface," that appeared in 1939 in the *American Journal of Mathematics* and the *Annals of Mathematics*, respectively [Zariski, 1939a or 1972–1979, 1:73-118; 1939b or 1972–1979, 1:325-375]. The arithmetic theory of algebraic varieties introduced the concept of arithmetically normal varieties and the operation of "deriving" such varieties from an arbitrary variety. These tools were then used to give an arithmetic proof of the resolution of the singularities of an algebraic surface. Zariski still tied the arithmetic theory up with the geometric theory that had been produced in Italy [Zariski, 1939a, p. 251 and pp. 258ff or 1972–1979, 1:75 and 82ff], but he now expressly gave up both the structural and the conceptual parallels between the theories.[13]

The concept of an arithmetically normal variety was inspired by the existing notion of a normal variety. Since the 1880s, normal varieties had been defined as varieties which are not the projection of a variety of the same order belonging to a higher dimensional projective space. Zariski invented an arithmetical concept of normality. To begin with, the very concept of variety adopted by Zariski was

---

[12]See [Zariski 1939a, pp. 151, 170, 193, and 197 or 1972–1979, 1:14, 33, 56, and 60].

[13]See [Zariski, 1939a, p. 288 or 1972–1970, 1:112].

different from the one mainly used before. An affine (projective) algebraic variety was thought of as the locus of zeros of (homogeneous) polynomials. Zariski adopted the point of view that van der Waerden promoted in his works, that is, an irreducible algebraic variety V is given by its generic (or general) point $(\xi_1, \ldots, \xi_n)$, where the notion of "generic point" was inspired by Emmy Noether's conception of the "generic zero [*allgemeine Nullstelle*]" of a prime ideal. If k is a field and if $\Sigma = k(\xi_1, \ldots, \xi_n)$ is a field extension, the polynomials $f(x_1, \ldots, x_n)$ that vanish at $(\xi_1, \ldots, \xi_n)$ form a prime ideal $\mathfrak{p} \subset k[x_1, \ldots, x_n]$. On the other hand, if $\mathfrak{p}'$ is any prime ideal in $k[x_1, \ldots, x_n]$, then there exists a field extension $\Sigma = k(\xi_1, \ldots, \xi_n)$ such that $\mathfrak{p}'$ consists of all polynomials that vanish in $\xi_1, \ldots, \xi_n$. This field is unique up to isomorphism. Noether called $(\xi_1, \ldots, \xi_n)$ a "generic zero" of the ideal $\mathfrak{p}$. Van der Waerden interpreted this as the generic point of the variety V whose ideal is $\mathfrak{p}$ [van der Waerden, 1926, p. 192].[14]

Zariski defined an affine algebraic variety to be (*arithmetically*) *normal*, if the ring $\mathfrak{o} = k[\xi_1, \ldots, \xi_n] = k[x_1, \ldots, x_n]/\mathfrak{p}$ (today termed the ring of affine coordinates) is integrally closed in its quotient field $k(\xi_1, \ldots, \xi_n)$. It took a lot of technical work for Zariski to show that the projective case is analogous: a projective algebraic variety is arithmetically normal, if the ring $\mathfrak{o} = k[\xi_0, \ldots, \xi_n] = k[x_0, \ldots, x_n]/\mathfrak{p}$ is integrally closed in its quotient field $k(\xi_0, \ldots, \xi_n)$ [Zariski, 1939a, pp. 285-294 or 1972–1979, 1:109-118]. What was the point of this definition? Since the manifold of singular points of an arithmetically normal variety of dimension r in the projective space is of dimension at most $r - 2$, the only singularities that an arithmetically normal surface can have are located at isolated singular points. In particular, there are no curves included in the singular locus of the ring.

Thus, it is clear that those surfaces should play an essential role in the resolution process. If one had only to deal with arithmetically normal surfaces, the problem would be considerably simplified. So, what was needed was an operation that transforms a given algebraic surface into an arithmetically normal one. Zariski introduced such an operation in the framework of the arithmetic theory of algebraic varieties to "establish the existence of [arithmetically] normal varieties in any given class of birationally equivalent varieties" [Zariski, 1939a, p. 290 or 1972–1979, 1:114]. This process—what is today called the normalization of varieties, that is, the operation of taking the integral closure of the coordinate ring—produced what Zariski called the "derived normal varieties" [Zariski, 1939a, p. 290 or 1972–1979, 1:114]. The method worked this way. The integral closure $\overline{\mathfrak{o}^*}$ of the homogeneous quotient ring $\mathfrak{o}^* = K[\xi_0^*, \xi_1^*, \ldots, \xi_n^*]$ is a finitely generated $\mathfrak{o}^*$-module, so it can be written as $\overline{\mathfrak{o}^*} = K[\zeta_1^*, \ldots, \zeta_h^*]$, where $\zeta_1^*, \ldots, \zeta_h^*$ have positive degree. Since the elements $\zeta_1^*, \ldots, \zeta_h^*$ are not necessarily of the same degree, however, Zariski could not view them as the homogeneous coordinates of some embedded projective variety. He thus showed the existence of a "character of homogeneity," that is, an integer δ such that the graded ring $\overline{\mathfrak{o}^*}(\delta) = \oplus_{\rho=1}^{\infty} (\overline{\mathfrak{o}^*})_{\delta\rho}$ is generated in degree one, namely, by elements in $\overline{\mathfrak{o}^*}(\delta)_1 = (\overline{\mathfrak{o}^*})_\delta$ [Zariski, 1939a, p. 291 or 1972–1979, 1:115]. This operation had no analog in the geometric theory. Zariski took the concept of the integral closure of a ring in its quotient ring from algebra and arithmetic ideal theory, interpreted it geometrically, and introduced it in the theory of algebraic surfaces to solve the problem of the reduction of singularities.

---

[14]Compare the discussion of these ideas in Norbert Schappacher's chapter above.

The Arithmetic Proof

With the concept of an arithmetically normal surface and with the operation of deriving such surfaces from any algebraic surface at his disposal, Zariski gave a new proof of the reduction theorem.[15] From an arbitrary algebraic surface $\tilde{F}$, he derived an arithmetically normal surface F; the surface F has only isolated singularities. Compared to previous geometric proofs, the operation of deriving an arithmetically normal surface therefore made a decisive difference.

He next took one of the finitely many singular points of F, say P, and applied a quadratic transformation that has a fundamental point at P. To construct this transformation, he considered all quadratic forms that vanish at P. The elements of a basis of these forms (over k) can be specialized at the generic point of F and then regarded as the homogeneous coordinates of the general point of an algebraic surface F'. This transformation has only one fundamental point, namely, P to which corresponds a fundamental curve Γ on F'. A priori F' is not arithmetically normal, so the singular locus of F' does not consist of isolated points and may well contain the whole curve Γ.

After the quadratic transformation (blow up), he next applied what he called an "integral closure transformation"—a normalization, in modern terminology—so that again there were only with finitely many isolated singular points [Zariski, 1939b, p. 688 or 1972–1979, 1:374]. Then followed a quadratic transformation with fundamental point at the singular point P_1 of the arithmetically normal surface F_1 (that was derived from F') and an integral closure transformation. The result was an arithmetically normal surface F_2. Continuing this procedure gave a sequence of birationally equivalent arithmetically normal surfaces $F, F_1, F_2, \ldots$. The reduction theorem as formulated by Zariski then read: "After a finite number of quadratic and integral closure transformations, a normal surface F_i is obtained which is free from singularities" [Zariski, 1939b, p. 688 or 1972–1979, 1:374]. To show that this process really ends with a surface that has no singularities, Zariski used valuation-theoretic arguments. That is, he let $P, P_1, \ldots$ be the sequence of points obtained by the alternating transformations. The union of the quotient rings $Q(P), Q(P_1), \ldots$ (in modern terminology, the local rings at $P_1, P_2, \ldots$) is the valuation ring of a zero-dimensional valuation of the function field Σ. Local uniformization guarantees that for every zero-dimensional valuation of Σ, there exists a projective model F of Σ on which the center of the valuation is a simple point of the surface. On the basis of this result, Zariski proved the "'local reduction' theorem," which states that in the sequence $P, P_1, P_2, \ldots$ a simple point occurs [Zariski, 1939b, p. 688 or 1972–1979, 1:374]. As there were only finitely many points to be resolved, the process ends. With this proof, Zariski accomplished the first important step in the arithmetization process.

Conclusion

From the second half of the nineteenth century, the main research focus of algebraic geometry had been on those results that could be stated about its objects.

---

[15]For a detailed historical reading of the proof as well as a deeper treatment of most of the arguments given in this chapter, see [Slembek, 2003]. For a good modern account of desingularization, see, for example, [Mumford, 1976, pp. 156-180].

With that focus, the use of various methods was welcomed because it showed different aspects of the objects and different ways to obtain results. Oscar Zariski broke with this view and reoriented the issues of algebraic geometry. In his 1935 monograph, *Algebraic Surfaces*, he concentrated on the ontological tools of algebraic geometry, expressing this when he wrote that "[i]t is especially true of algebraic geometry that in this domain the methods employed are at least as important as the results" [Zariski, 1935, Preface]. This shift in research focus owed to the lack of understanding that contemporary mathematicians (namely, Zariski) experienced when confronted with the proofs of the Italian geometers. The methods and tools used in their geometric proofs were difficult to understand; thus, the validity of the proofs was called into question. In the case of the resolution of singularities of algebraic surfaces, Zariski examined five different geometric proofs. The alarming conclusion of his examination was that while all were *plausible*, none was *rigorous*. The only rigorous proof of the desingularization theorem for algebraic surfaces that existed in 1935 was one that employed analytic instead of geometric methods. If the theory of singularities could not be formulated within the geometric framework in a way satisfying to the mathematicians of the 1930s, what were the consequences? Did the theory lose its validity? Did the methods lose their legitimacy? Did they have to be abandoned?

Zariski answered these questions pragmatically by reformulating the theory in a way that was satisfactory in the mathematical climate of the 1930s. He *arithmetized* a key part of algebraic geometry. Initially, this arithmetization imitated the architecture and language of the geometric theory. It was inspired by geometry; it explicitly sought the arithmetic analogs of geometric concepts. The arithmetic proof of the desingularization theorem was, in fact, based on Levi's geometrical ideas. This highlights the continuity in the development of a mathematical theory instead of the changes it involves. Zariski, in fact, hoped to show that "our fathers were right" [Zariski, 1950, p. 88 or 1972–1979, 3:374].[16] Zariski soon realized, however, that rephrasing was not enough. He had to introduce new concepts that had no analog in the geometric framework of his Italian teachers. Although arithmetization aimed to legitimize and to preserve the Italian tradition of algebraic geometry, this goal was achieved only by replacing its methods, by introducing new objects and issues, and thus by reshaping the area altogether.

Zariski's *Collected Papers*, published in the 1970s, display his works under different headings. In this edition, the papers examined in this chapter are grouped under the headings "foundations" ("Polynomial Ideals Defined by Infinitely Near Base Points" [Zariski, 1938 or 1972–1979, 1:14-67] and "Some Results in the Arithmetic Theory of Algebraic Varieties" [Zariski, 1939a or 1972–1979, 1:73-118]) and "resolution of singularities" ("The Reduction of Singularities of Algebraic Surfaces" [Zariski, 1939b or 1972–1979, 1:639-689]). This division suggests that rewriting algebraic geometry took place in two steps: first, the laying down of foundations and, second, the application of them to the resolution of singularities. The interpretation

---

[16]In his 1950 talk on "The Fundamental Ideas of Algebraic Geometry" given at the International Congress of Mathematicians, Zariksi explicitly alluded [Zariski, 1950, p. 88 or 1972–1979, 3:374] to Poincaré's complaint that "in the past, when one invented a new function, it was with some practical goal in mind; today, one invents them expressly to fault the reasonings of our fathers, and nothing more than that [autrefois, quand on inventait une fonction nouvelle, c'était en vue de quelque but pratique; aujourd'hui, on les invente tout exprès pour mettre en défaut les raisonnements de nos pères, et n'en tira jamais que cela]" [Poincaré, 1916–1956, 11:131].

of the arithmetization proposed here abolishes this division. The concrete geometric problem of finding certain birational transformations cannot be separated from the conceptual and methodological concerns at issue.

References

Archival Sources

ETHZ, HS 91:659 = Solomon Lefschetz to Hermann Weyl. 30 November, 1926. Archiv der ETH Zürich.

Printed Sources

Albanese, Giacomo. 1924. "Trasformazione birazionale di una superficie algebrica qualunque in un'altra priva di punti multipli." *Rendiconti del Circolo matematico di Palermo* 48, 321-332.

Baker, Henry Frederick. 1913. "On Some Recent Advances in the Theory of Algebraic Surfaces." *Proceedings of the London Mathematical Society* 12, 1-40.

Bliss, Gilbert A. 1923. "The Reduction of Singularities of Plane Curves." *Bulletin of the American Mathematical Society* 29, 161-183.

Brieskorn, Egbert and Knörrer, Horst. 1981. *Ebene algebraische Kurven*. Basel/Boston: Birkhäuser Verlag.

Brill, Alexander von and Noether, Max. 1874. "Ueber die algebraischen Functionen und ihre Anwendung in der Geometrie." *Mathematische Annalen* 7, 269-310.

_____. 1892–1893. "Bericht über die Entwicklung der Theorie der algebraischen Functionen in älterer und neuerer Zeit." *Jahresbericht der Deutschen Mathematiker-Vereinigung* 3, 107-565 (this actually appeared in 1894).

Brigaglia, Aldo and Ciliberto, Ciro. 1995. *Italian Algebraic Geometry between the Two World Wars*. Vol. 100. Queen's Papers in Pure and Applied Mathematics. Kingston: Queen's University.

Castelnuovo, Guido. 1929. "La geometria algebrica e la scuola italiana." *Atti del Congresso internazionale dei matematici. Bologna 3-10 settembre 1928.* 6 Vols. Bologna: Zanichelli, 1929-1932, 1:191-201.

Castelnuovo, Guido and Enriques, Federigo. 1897. "Sur quelques récents résultats dans la théorie des surfaces algébriques." *Mathematische Annalen* 48, 241-316.

_____. 1914. "Die algebraischen Flächen vom Gesichtspunkte der birationalen Transformationen aus." Vol III.2. 6b. *Encyklopädie der mathematischen Wissenschaften*. Leipzig: B. G. Teubner Verlag.

Chisini, Oscar. 1921. "La risoluzione delle singolarità di una superficie mediante trasformazioni birationali dello spazio." *Memorie della Reale Accademia delle Scienze dell'Istituto di Bologna* 8, 3-42.

Corry, Leo. 1996. *Modern Algebra and the Rise of Mathematical Structures*. Basel/Boston: Birkhäuser Verlag.

Gario, Paola. 1988. "Histoire de la révolution des singularités des surfaces algébriques (une discussion entre C. Segre et P. Del Pezzo)." *Cahiers du Séminaire d'histoire des mathématiques* 9, 123-137.

_____. 1989. "Resolution of Singularities of Surfaces by P. Del Pezzo: A Mathematical Controversy with C. Segre." *Archive for History of Exact Sciences* 40, 247-274.

Hodge, William V. D. and Pedoe, Daniel. 1953–1954. *Methods of Algebraic Geometry*. 3 Vols. Cambridge: Cambridge University Press.

Krull, Wolfgang. 1935. *Idealtheorie*. Berlin: J. Springer Verlag.

Lefschetz, Solomon. 1936. "Review of Oscar Zariski's *Algebraic Surfaces*." *Bulletin of the American Mathematical Society* 42, 13-14.

Levi, Beppo. 1897a. "Sulla risoluzione delle singolarità puntuali delle superficie algebriche dello spazio ordinario per trasformazione quadratiche." *Annali di matematica pura ed applicata* 26, 219-253.

_____. 1897b. "Risoluzione delle singolarità puntuali delle superficie algebriche." *Atti della Reale Accademia delle Scienze di Torino* 33, 56-76.

Mumford, David. 1976. *Algebraic Geometry I: Complex Projective Varieties*. New York: Springer-Verlag.

_____. 1986. "Oscar Zariski 1899–1986." *Notices of the American Mathematical Society* 33, 891-894.

Noether, Max. 1871. "Sulle curve multiple di superficie algebriche." *Annali di matematica pura ed applicata* 5, 163-177.

_____. 1875. "Ueber die algebraischen Functionen und ihre Anwendung in der Geometrie." *Mathematische Annalen* 7, 269-316.

Parikh, Carol. 1991. *The Unreal Life of Oscar Zariski*. San Diego: Academic Press, Inc.

Poincaré, Henri. 1916–1956. *Oeuvres de Henri Poincaré*. 11 Vols. Paris: Gauthier-Villars et Cie.

Segre, Corrado. 1894. "Introduzione alla geometria sopra un ente algebrico semplicemente infinito." *Annali di matematica pura ed applicata* 22, 41-142 or Segre, Corrado. 1957–1963. *Opere*. 4 Vols. Rome: Edizione Cremonese, 1:198-304.

Semple, John and Roth, Leonard. 1949. *Introduction to Algebraic Geometry*. Oxford: Clarendon Press.

Severi, Francesco. 1914. "Trasformazione birazionale di una superficie algebrica qualunque in una priva di punti multipli." *Atti della Accadamia nazionale dei Lincei* 23, 527-539.

_____. 1932/1967. "Le rôle de la géométrie algébrique dans les mathématiques." *Verhandlungen des Internationalen Mathematiker-Kongresses Zürich 1932*. 2 Vols. Reprint Ed. Nendeln/Lichtenstein: Kraus Reprint Limited, 1:209-220.

Slembek, Silke. 2003. "Continuité ou rupture? À propos de l'arithmétisation de la géométrie algébrique selon Oscar Zariski." Unpublished Doctoral Dissertation: Université Louis Pasteur. Preprint. http://www-irma.u-strasbg.fr/irma/publications/2003/titre2003.shtml

van der Waerden, Bartel L. 1926. "Zur Nullstellentheorie der Polynomideale." *Mathematische Annalen* 96, 183-208; Reprinted in [van der Waerden, 1983, pp. 11-36].

_____. 1930–1931. *Moderne Algebra*. 2 Vols. Berlin: Springer-Verlag.

_____. 1983. *Zur algebraischen Geometrie: Selected Papers*. Berlin/Heidelberg/New York: Springer-Verlag.

Veronese, Giuseppe. 1882. "Behandlung der projectivischen Verhältnisse der Räume von verschiedenen Dimensionen durch das Princip des Projicirens und Schneidens." *Mathematische Annalen* 19, 161-234.

Walker, Robert J. 1935. "The Reduction of the Singularities of an Algebraic Surface." *Annals of Mathematics* 36, 336-365.

Zariski, Oscar. 1928. "On a Theorem of Severi." *American Journal of Mathematics* 50, 87-92, or [Zariski, 1972–1979, 3:81-86].

_____. 1935. *Algebraic Surfaces*. Berlin: Springer-Verlag.

_____. 1938. "Polynomial Ideals Defined by Infinitely Near Base Points." *American Journal of Mathematics* 60, 151-204, or [Zariski, 1972–1979, 1:14-67].

_____. 1939a. "Some Results in the Arithmetic Theory of Algebraic Varieties." *American Journal of Mathematics* 61, 249-294, or [Zariski, 1972–1979, 1:73-118].

_____. 1939b. "The Reduction of the Singularities of an Algebraic Surface." *Annals of Mathematics* 40, 639-689 or [Zariski, 1972–1979, 1:325-375].

_____. 1950. "The Fundamental Ideas of Abstract Algebraic Geometry." *Proceedings of the International Congress of Mathematicians, Cambridge 1950*. 2 Vols. Providence: American Mathematical Society, 2:77-89, or [Zariski, 1972–1979, 3:363-375].

_____. 1972–1979. *Oscar Zariski: Collected Papers*. 4 Vols. Cambridge: MIT Press, Vol. 1. 1972. Ed. Heisuke Hironaka and David Mumford; Vol. 2. 1973. Ed. Michael Artin and David Mumford; Vol. 3. 1978. Ed. Michael Artin and Barry Mazur; Vol. 4. 1979. Ed. Joseph Lipman and Bernard Teissier.

CHAPTER 13

The Rising Sea:
Grothendieck on Simplicity and Generality

Colin McLarty
Case Western Reserve University, United States

In 1949, André Weil published striking conjectures linking number theory to topology and a striking strategy for a proof [Weil, 1949]. Around 1953, Jean-Pierre Serre took on the project and soon recruited Alexander Grothendieck. Serre created a series of concise elegant tools which Grothendieck and coworkers simplified into thousands of pages of category theory. Some have complained of this style, but Miles Reid, for example, says "Grothendieck himself can't necessarily be blamed" for the growth of category theory "since his own use of categories was very successful in solving problems" [Reid, 1990, p. 116].[1] This chapter describes the methods Grothendieck made standard in algebraic geometry by 1958, pursuing the Weil conjectures: Abelian categories, derived functor cohomology, and schemes. It touches on Grothendieck topologies, toposes, and étale cohomology, which arose around 1958, to see how Grothendieck himself relates them to the earlier ideas.[2]

Grothendieck describes two styles in mathematics. When he thinks of a theorem to be proved as a nut to be opened, so as to reach "the nourishing flesh protected by the shell," then the *hammer and chisel* principle is: "put the cutting edge of the chisel against the shell and strike hard. If needed, begin again at many different points until the shell cracks—and you are satisfied." He goes on to say:[3]

> I can illustrate the second approach with the same image of a nut to be opened. The first analogy that came to my mind is of immersing the nut in some softening liquid, and why not simply water? From time to time you rub so the liquid penetrates better, and otherwise you let time pass. The shell becomes more flexible through weeks and months—when the time is ripe, hand pressure is enough, the shell opens like a perfectly ripened avocado!

---

[1]Grothendieck began using category theory in functional analysis; see, for example, [Grothendieck, 1952]. Dieudonné punned on the categorical idea of natural transformation to praise Grothendieck's functional analysis as a "constant search for 'natural' definitions and 'functorial' properties" [Dieunonné, 1990, p. 2].

[2]For a recent survey of Grothendieck's life, work, and influence, see [Jackson, 2004].

[3]The abbreviation *ReS* refers to the memoir [Grothendieck, 1985–1987]. Pierre Deligne points out that Grothendieck's mastery of language in *Récoltes et Semailles* parallels the serious responsibility he took for naming new concepts (e-mail to the author, 13 May, 2003). All translations in this paper are my own.

A different image came to me a few weeks ago. The unknown thing to be known appeared to me as some stretch of earth or hard marl, resisting penetration ... the sea advances insensibly in silence, nothing seems to happen, nothing moves, the water is so far off you hardly hear it ... yet it finally surrounds the resistant substance [*ReS*, pp. 552-553].[4]

He writes of a "rising sea [la mer qui monte]," which in French most often means a rising tide but can mean other things.[5] The theorem is "submerged and dissolved by some more or less vast theory, going well beyond the results originally to be established" [*ReS*, p. 555].[6] Grothendieck says this is his approach as well as Bourbaki's, thus comparing his research with Bourbaki writing the *Elements of Mathematics*. He often writes as if research, exposition, and teaching are all the same.

Deligne describes a typical Grothendieck proof as a long series of trivial steps where "nothing seems to happen, and yet at the end a highly non-trivial theorem is there [rien ne semble se passer et pourtant à la fin de l'exposé un théorème clairement non trivial est là]" [Deligne, 1998, p. 12]. Grothendieck makes this simplicity an extreme form of Cantor's stand: "the *essence* of *mathematics* lies precisely in its *freedom* [das *Wesen* der *Mathematik* liegt gerade in ihrer *Freiheit*]" [Cantor, 1932, p. 182]. Grothendieck claims the freedom not only to build a world of set theory for mathematics but also to build an entire world—as large as the universe of all sets—adapted to any single problem such as counting the solutions to a given polynomial equation.

Grothendieck describes himself as creating "new worlds [mondes nouveaux]," but he means what he elsewhere calls "building beautiful houses [belles maisons]," that is, framing theories and methods that become a heritage others can use.[7] He has certainly done that. So has Jean-Pierre Serre in a very different way. Grothendieck says Serre generally uses the hammer and chisel [*ReS*, p. 558]. He calls Serre "Super Yang" against his own "Yin"—but not at all in the sense of being

---

[4]"... la chair nourricière protégée par la coque ...on pose le tranchant du burin contre la coque, et on tape fort. Au besoin, on recommence en plusieurs endroits différents, jusqu'à ce que la coque se casse—et on est content. ... Je pourrais illustrer la deuxième approche, en gardant l'image de la noix qu'il s'agit d'ouvrir. La première parabole qui m'est venue à l'esprit tantôt, c'est qu'on plonge la noix dans un liquide émollient, de l'eau simplement pourquoi pas, de temps en temps on frotte pour qu'elle pénètre mieux, pour le reste on laisse faire le temps. La coque s'assouplit au fil des semaines et des mois—quand le temps est mûr, une pression de la main suffit, la coque s'ouvre comme celle d'un avocat mûr à point! Ou encore, on laisse mûrir la noix sous le soleil et sous la pluie et peut-être aussi sous les gelées de l'hiver. Quand le temps est mûr c'est une pousse délicate sortie de la substantifique chair qui aura percé la coque, comme en se jouant—ou pour mieux dire, la coque se sera ouverte d'elle-même, pour lui laisser passage.

L'image qui m'était venue il y a quelques semaines était différente encore, la chose inconnue qu'il s'agit de connaître m'apparaissait comme quelque étendue de terre ou de marnes compactes, réticente à se laisser pénétrer.... La mer s'avance insensiblement et sans bruit.... Pourtant elle finit par entourer la substance rétive."

[5]It can mean waves rising in a storm, or crashing against rocks, or a generally rising sea level as from global warming. The sometimes psychoanalytic tone of the memoir makes us notice the pun "l'amère qui monte", a rising bitterness. Wordplay on "la mer/l'amère/la mère (the mother)" is familiar in French.

[6]"...submergé et dissous par quelque plus ou moins vaste théorie, allant bien au delà des résultats qu'il était d'abord question d'établir."

[7]See [*ReS*, pp. 554 and P27, resp.].

heavy-handed—rather Serre is the "incarnation of elegance [l'incarnation justement de l'élégance]" [ReS, p. 969]. That is the difference. Serre cuts elegantly to an answer. Grothendieck creates truly massive multi-volume books with numerous coauthors, offering set-theoretically vast yet conceptually simple mathematical systems adapted to express the heart of each matter and to dissolve the problems.[8] This is the sense of world-building I mean.

Across the difference in working style, Grothendieck says that from 1955 to 1970 Serre was at the origin of most of his ideas [ReS, p. 982]. This includes *every* major step towards the Weil conjectures. Their collaboration is comparable to that of Richard Dedekind and Emmy Noether. One difference is that Serre and Grothendieck talked with each other. Another is that Dedekind and Noether shared much the same style. Such collaboration deserves more historical attention. The most important and challenging remark ever made about twentieth-century mathematics was Noether's watchword "it is all already in Dedekind."

The Weil Conjectures

Solving a Diophantine equation, that is, finding integer or rational solutions to an integer polynomial, can be unapproachably difficult. Weil describes one indirect strategy in a letter dated 26 March, 1940 to his sister, the philosopher Simone Weil: first look for solutions in richer fields than the rationals, perhaps fields of rational functions over the complex numbers. But these are quite different from the integers:

> We would be badly blocked if there were no bridge between the two.
> And *voilà* God carries the day against the devil: this bridge exists; it is the theory of algebraic function fields over a finite field of constants [Weil, 1979, 1:252].

A solution modulo 5 to a polynomial equation $P(X, Y, \ldots, Z) = 0$ is a list of integers $X, Y, \ldots, Z$ making the value $P(X, Y, \ldots, Z)$ divisible by 5 or, in other words, equal to $0 \pmod 5$. For example, $X^2 + Y^2 - 3 = 0$ has no integer solutions. That is clear since X and Y would both have to be 0 or ± 1, to keep their squares below 3, and no combination of those works. But it has solutions mod 5 since, among others, $3^2 + 3^2 - 3 = 15$ is divisible by 5. Solutions modulo a given prime p are easier to find than integer solutions, and they amount to the same thing as solutions in the finite field of integers mod p.

A polynomial equation $P(X, Y, \ldots, Z) = 0$ can be exhaustively checked for solutions mod p, by just checking p different values for each variable. Even if p is impractically large, equations are more manageable modulo p. Going farther, we might look at equations mod p, but ask how the number of solutions grows as

---

[8]Deligne emphasizes (in an e-mail to the author, 13 May, 2003) that the set-theoretic size of toposes never fazed Grothendieck but was never the point either; and it is inessential in that the same technical work can be done by small Grothendieck topologies.

Grothendieck posited *universes*, now often called *Grothendieck universes*, as a set-theoretic fix to gain the conceptual unity of toposes over topologies [Artin, Grothendieck, and Verdier, 1972, pp. 185ff.]. Serre suggests Grothendieck got the idea from Dieudonné or Chevalley, who got it from earlier set-theorists (e-mail to the author, 21 June, 2004). But it was only a fix. He later faulted Bourbaki for focusing on set theory instead of simple categorical properties [ReS, p. PU22]. See below for the vision of unity in toposes, and categorical versus set-theoretic properties of schemes. William Lawvere and Myles Tierney's elementary topos axioms sought to formalize the unity directly, with far less set-theoretic strength than universes [Lawvere, 1979].

we allow irrationals of higher degree as solutions—roots of quadratic polynomials, roots of cubic polynomials, and so on. In each case, there are still only finitely many different potential solutions to check. Doing this for all primes p and for all degrees n means looking for solutions in all finite fields, as in Weil's letter.

Answering this question in finite fields does not itself answer those about integer or rational solutions. It might help. It is interesting. And it has surprising applications such as planning efficient networks [Li, 1996]. Building on earlier number theorists, Weil conjectured a penetrating form for the exact answer and some useful approximations. More than that, he conjectured an amazing link with topology.

The key points about finite fields are: for each prime number p, the integers mod p form a field, written $\mathbb{F}_p$. For each natural number $r > 0$, there is (up to isomorphism) just one field with p^r elements, denoted $\mathbb{F}_{p^r}$ or $\mathbb{F}_q$ with $q = p^r$. This comes from $\mathbb{F}_p$ by adjoining the roots of a degree r polynomial. So for any natural number $s > 0$, there is just one field with q^s elements, namely, $\mathbb{F}_{p^{(r+s)}} = \mathbb{F}_{q^s}$. These fields for all prime numbers p are all the finite fields. The union for all r of the $\mathbb{F}_{p^r}$ is the algebraic closure $\overline{\mathbb{F}_p}$. The Frobenius morphism on the algebraic closure takes each $x \in \overline{\mathbb{F}_p}$ to x^p. By Galois theory, $\mathbb{F}_{p^r}$ consists of just the fixed points for the rth iterate of the Frobenius morphism, so $x \in \mathbb{F}_{p^r}$ if and only if $x = x^{p^r}$.[9]

Take any nice enough n-dimensional space defined by polynomials on a finite field $\mathbb{F}_q$.[10] For each $s \in \mathbb{N}$, let N_s be the number of points defined in $\mathbb{F}_{q^s}$. Define the zeta function as an exponential

$$Z(t) = \exp\left(\sum_{s=1}^{\infty} N_s \frac{t^s}{s}\right).$$

The first Weil conjecture says $Z(t)$ is a rational function

$$Z(t) = \frac{P(t)}{Q(t)},$$

for some integer polynomials $P(t)$ and $Q(t)$. This means there are complex algebraic numbers $a_1, \ldots, a_i$ and $b_1, \ldots, b_j$, with each algebraic conjugate of an a (respectively, b) also an a (respectively, b), such that for every s,

$$N_s = (a_1^s + \ldots + a_i^s) - (b_1^s + \ldots + b_j^s).$$

The second conjecture is a functional equation:

$$Z\left(\frac{1}{q^n t}\right) = \pm q^{nE/2} t^E Z(t).$$

This says for each a_i in the list of as (respectively, b_j in the list of bs), the quotient q^n/a_i is also in the list (respectively, q^n/b_j).

The third is a Riemann Hypothesis

$$Z(t) = \frac{P_1(t) P_3(t) \cdots P_{2n-1}(t)}{P_0(t) P_2(t) \cdots P_{2n}(t)},$$

where each P_k is an integer polynomial with all roots of absolute value $q^{-k/2}$. That means each a has absolute value q^k, for some $0 \leq k \leq n$. Each b has absolute value $q^{(2k-1)/2}$, for some $0 \leq k \leq n$. This puts bounds on the numbers N_s.

---

[9]See, for example, [Serre, 1973].

[10]More specifically, take a smooth projective variety over $\mathbb{F}_q$ which lifts to some algebraic number ring.

Over it all is the conjectured link to topology. Since the polynomials are nice, they define a complex manifold M, a continuous space of complex number solutions. Since M is algebraically n-dimensional on the complex numbers, it is topologically $2n$-dimensional. Let $B_0, B_1, \ldots, B_{2n}$ be its Betti numbers so each B_k gives the number of topologically k-dimensional holes or handles on M. Then Weil conjectured that each polynomial P_k in the Riemann hypothesis has degree B_k, and the constant E in the functional equation is the Euler number of the manifold, namely, the alternating sum of the Betti numbers

$$E = \sum_{k=0}^{2n} (-1)^k B_k$$

It is a fascinating, astonishing link between finite arithmetic and the topology of continuous manifolds. The topology of M tells how many as and bs there are with each absolute value. This implies useful approximations to the numbers N_s.

Special cases of these conjectures were known, and Weil proved more. All dealt with curves (algebraically 1-dimensional) or hypersurfaces (defined by a single polynomial). Proofs of these minus the topology make up five chapters in Ireland and Rosen's *A Classical Introduction to Modern Number Theory*. The book never mentions Grothendieck but calls Deligne's completion of the proof "one of the most remarkable achievements of this century" [Ireland and Rosen, 1992, p. 151].

Weil presented the topology as motivating the conjectures for higher dimensional varieties [Weil, 1949, p. 507]. He especially pointed out how the whole series of conjectures would follow quickly if theorems on topological manifolds would apply to the spaces of finite field solutions. The Lefschetz fixed point theorem would reduce the conjectures to textbook exercises in linear algebra—now literally exercises 24 and 25 of chapter fourteen in Serge Lang's *Algebra* [Lang, 1993, p. 570].[11]

The conjectures were the intuition of an encyclopedic mathematician, drawn to classical problems, skilled at calculation, and informed on the latest methods. The topological strategy was powerfully seductive but seriously remote from existing tools. Even Weil did not name Lefschetz or cohomology in his publication. Serre says that "[a]t that time, Weil was explaining things (in conversations) in terms of cohomology and Lefschetz's fixed point formula" yet "did not want to predict the existence" of a cohomology theory that would actually work. "Indeed, in 1949–50, nobody thought that it could be possible to apply topology" to geometry over finite fields.[12] Weil's arithmetic spaces were not even precisely defined.

Abelian Categories

Serre gave a thoroughly *cohomological* turn to the conjectures. As Grothendieck put it, "[a]nyway Serre explained the Weil conjectures to me in cohomological terms around 1955—and it was only in these terms that they could possibly "hook" me .... I am not sure anyone but Serre and I, not even Weil if that is possible, was deeply convinced such [a cohomology] must exist."[13] Serre approached the problem through *sheaves*, a new method in topology. Grothendieck would later describe each

---

[11]See also [Houzel, 1994], [Dieudonné, 1988], [Mumford and Tate, 1978], [Hartshorne, 1997, Appendix C], and [Weil, 1979, 1:568; 2:180-188; 3:279-302].

[12]Jean-Pierre Serre in an e-mail to the author of 1 July, 2004.

[13]"C'est en termes cohomologiques, en tous cas, que Serre m'a expliqué les conjectures de Weil, vers les années 1955—et ce n'est qu'en ces termes qu'elles étaient susceptibles de

sheaf on a space T as a "meter stick" measuring T. The *cohomology* of a sheaf gives a very coarse summary of the information. In the best case, it highlights just what is needed. Certain sheaves on T produce the Betti numbers. If such "meter sticks" applied to Weil's arithmetic spaces, and proved standard topological theorems for them, the conjectures would follow.

By the nuts and bolts definition, a sheaf $\mathcal{F}$ on a topological space T is an assignment of Abelian groups to open subsets of T, plus group homomorphisms among them, all meeting a certain covering condition. Precisely these nuts and bolts were discouraging for the Weil conjectures because the topology of arithmetic spaces in the then-existing sense was pretty clearly hopeless for Weil's proof strategy.

At the École Normale Supérieure, Henri Cartan's seminar spent 1948–1949 and 1950-1951 working on sheaf cohomology. Serre, in his early twenties, ran some sessions. Grothendieck, two years younger, attended, perhaps only as an occasional visitor: "I sat in the 'Séminaire Cartan', as a stupefied witness to his discussions with Serre, loaded with 'Spectral Sequences' (brr!) and drawings (called 'diagrams') full of arrows covering the blackboard. It was the heroic age of the theory of 'sheaves' and 'carapaces' and of a whole arsenal whose sense entirely escaped me" [ReS, p. 19].[14] The seminar was aware of de Rham cohomology on differentiable manifolds, which related topology to differential analysis. It was easily expressed in terms of sheaves. During the time of the seminar, Cartan saw how to define sheaves on a complex analytic variety reflecting not only its topology but also complex analysis on it. He and Serre would develop this over the coming years [Fasanelli, 1981].

Perhaps Weil cohomology could use sheaves reflecting algebra, but the applications to differential analysis and complex analysis used sheaves and cohomology in the usual topological sense. Their innovation was only to find new sheaves capturing analytic or algebraic information. It was amazing how well that worked. It seemed smaller, however, than the innovation needed for the Weil conjectures.

The great challenge to the Séminaire Cartan in 1950–1951 was to relate the cohomology of topological spaces to the cohomology of groups. Instead of sheaves, the cohomology of a group G uses G-modules.[15] This was formally quite different from topology, yet it had grown from topology and was tightly tied to it by applications. Indeed, Samuel Eilenberg and Saunders Mac Lane had created category theory in large part simultaneously with group cohomology in an effort to explain both kinds of cohomology by clarifying the links between them. Eilenberg was in Paris that year at Cartan's invitation and joined the seminar. They aimed to find what was common to the two kinds of cohomology, and they found it in a pattern of functors.

A cohomology theory on a topological space X assigns each sheaf $\mathcal{F}$ on X a series of Abelian groups $H^n\mathcal{F}$, and it assigns each sheaf map $f:\mathcal{F} \to \mathcal{F}'$ a series

---

m'"accrocher'.... je ne suis pas sûr que personne d'autre que Serre et moi, pas même Weil si ça se trouve, avait seulement l'intime conviction que ça devait exister" [ReS, p. 840].

[14] "J'ai été l'hôte... du 'Séminaire Cartan', en témoin ébahi des discussions entre lui et Serre, à grands coups de 'Suites Spectrales' (brr!) et de dessins (appelés 'diagrammes') pleins de flèches recouvrant tout le tableau. C'était l'époque héroïque de la théorie des 'faisceaux', 'carapaces' et de tout un arsenal dont le sens m'échappait totalement." See the spectral sequences below. Carapaces were similar to sheaves [Cartan, 1948–].

[15] A G-module is an Abelian group M (which might also be a module over some ring) plus an action of G on M. See, for example, [Brown, 1982].

of group homomorphisms $H^n f : H^n \mathcal{F} \to H^n \mathcal{F}'$. Each H^n is a functor, the so-called n-dimensional cohomology functor, with properties we need not know in detail. A crucial one is that, for $n > 0$, $H^n \mathcal{F} = 0$, for any *fine* sheaf $\mathcal{F}$, where a sheaf is *fine* if it meets a condition borrowed from differential geometry by way of Cartan's complex analytic geometry.[16]

The cohomology of a group G is a series of functors H^n from G-modules to Abelian groups. These functors have the same categorical properties as topological cohomology except that $H^n M = 0$, for $n > 0$, and for any *injective* module M. A G-module I is *injective* if, for every G-module inclusion $N \rightarrowtail M$ and homomorphism $f : N \to I$, there is at least one $g : M \to I$ making this commute

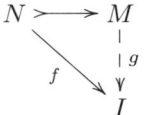

The Séminaire took the analogy no further, and this became a serious problem. One great offshoot was the book *Homological Algebra* usually called "Cartan-Eilenberg." The book never mentions Emmy Noether but opens by declaring victory for her outlook. "During the last decade," it proclaimed, "the methods of algebraic topology have invaded extensively the domain of pure algebra, and initiated a number of internal revolutions. The purpose of this book is to present a unified account of these developments and to lay the foundations for a full-fledged theory" [Cartan and Eilenberg, 1956, p. v]. Yet, this unified account had to exclude its own beginnings.

They could treat the cohomology of several algebraic structures: groups, Lie groups, associative algebras. These all rest on injective resolutions. They could not include topological spaces, the source of the whole, and still one of the main motives for pursuing the others. Topological cohomology rested on completely different resolutions. In the search for Weil cohomology, this left two questions. What would Weil cohomology use in place of topological sheaves or G-modules? And what resolutions would it use for them?

Cartan and Eilenberg defined group cohomology as a *derived functor*, which, in turn, they defined using injective resolutions. So the cohomology of a topological space was not a derived functor in their sense. But a looser sense was current. Grothendieck wrote to Serre on 26 February, 1955:

> I have realized that by formulating the theory of derived functors for categories more general than modules, one gets the cohomology of spaces at the same time at small cost .... The existence follows from a general criterion, and fine sheaves will play the role of *injective* modules.[17] One gets the fundamental spectral sequences as special cases of delectable and useful general spectral sequences. But I am

---

[16] A sheaf is *fine* if it admits partitions of unity in the following sense: for every locally finite cover of X by open subsets U^i, there are endomorphisms ℓ^i of $\mathcal{F}$ such that: (1) for each i, the endomorphism ℓ^i is zero outside of some closed set contained in U^i; and (2) the sum $\sum_i \ell^i$ is the identity. See [Cartan, 1948–, exp. 15].

[17] This is to say that Grothendieck looked at what is now called *effaceability* before injective sheaves. He saw that fine sheaves are acyclic, and each sheaf embeds in one of them [Colmez and Serre, 2001, p. 12]. This answers Serre's question below about which properties of fine sheaves are needed. It also explains why basically every kind of resolution that works at all, works the

not yet sure if it all works as well for non-separated spaces and I recall your doubts on the existence of an exact sequence in cohomology for dimensions ≥ 2. Besides this is probably all more or less explicit in Cartan-Eilenberg's book which I have not yet had the pleasure to see [Colmez and Serre, 2001, pp. 13-14].[18]

Here, he lays out the whole paper commonly called *Tôhoku* for the journal that published it [Grothendieck, 1957]. There are several issues. For one thing, fine resolutions do not work for all topological spaces. They only work for the paracompact spaces—that is, Hausdorff spaces where every open cover has a locally finite refinement. The Séminaire Cartan called these *separated* spaces. The limitation was no problem for differential geometry. All differential manifolds are paracompact. Nor was it a problem for most of analysis. But it was discouraging for the Weil conjectures since non-trivial algebraic varieties are never Hausdorff.

Serre replied on 12 March, 1955 using the same loose sense of derived functor:

> The fact that sheaf cohomology is a special case of derived functors (at least for the paracompact case) is not in Cartan-Sammy. Cartan was aware of it and told Buchsbaum to work on it, but he seems not to have done it.[19] The interest of it would be to show just which properties of fine sheaves we need to use; and so one might be able to figure out whether or not there are enough fine sheaves in the non-separated case (I think the answer is no but I am not at all sure!) [Colmez and Serre, 2001, p. 15].[20]

So Grothendieck began rewriting Cartan-Eilenberg before seeing it. To the Séminaire Bourbaki in 1957, he described his work as a form of Cartan-Eilenberg's homological algebra [Bourbaki, 1949–, p. 149-01].

---

same way: all give the (unique up to isomorphism) universal delta functor over the global section functor.

[18]"Je me suis aperçu qu'en formulant la théorie des foncteurs dérivés pour des catégories plus générales que les modules, on obtient à peu de frais en même temps la cohomologie des espaces à coefficients dans un faisceau .... L'existence résulte d'un critère général, les faisceaux fins joueront le rôle des modules 'injectifs'. On obtient aussi les suites spectrales fondamentales comme cas particuliers de délectables et utiles suites spectrales générales. Mais je ne suis pas encore sûr si tout marche aussi bien dans le cas d'un espace non séparé, et je me rapelle tes doutes sur l'existence d'une suite exact en cohomologie en dimensions ≥ 2. D'ailleurs, probablement tout ça se trouve plus ou moins explicitement dans le bouquin Cartan-Eilenberg, que je n'ai pas encore eu l'heur de voir.

[19]David Buchsbaum's problem was not posed by Cartan (e-mail to the author from Buchsbaum, 1 June, 2003). Buchsbaum had given categorical axioms for derived functors using generalized injectives. Compare Theorems 5.1 in [Buchsbaum, 1955] and [Cartan and Eilenberg, 1956]. Then he tried to show that sheaves on any topological space have enough injectives. He sent Cartan an incomplete proof outline. Cartan encouraged him in it. Buchsbaum dropped it when he too noticed *effaceability*. He later found that a weaker condition suffices to define a cohomology functor. Roughly, each cocycle α of an object A must have some embedding $A \rightarrowtail B$ which kills it.

[20]"Le fait que la cohomologie d'un faisceau soit un cas particulier des foncteurs dérivés (au moins dans le cas paracompact) n'est pas dans le Cartan-Sammy. Cartan en avait conscience, et avait dit à Buchsbaum de s'en occuper, mais il ne semble pas que celui-ci l'ait fait. L'intérêt de ceci serait de voir quelles sont au juste les propriétés des faisceaux fins qu'il faut utiliser; ainsi on pourrait peut-être se rendre compte si, oui or non, il y a suffisamment de faisceaux fins dans le cas non séparé (je pense que la réponse est négative, mais je n'en suis nullement sûr!).

Among other things, he preempted the question of resolutions for Weil cohomology. Before anyone knew what "sheaves" it would use, Grothendieck knew it would use injective resolutions. He did this by asking not what sheaves "are" but how they relate to one another. As he later put it, he set out to

> consider the set[21] of all sheaves on a given topological space or, if you like, the prodigious arsenal of all "meter sticks" that measure it. We consider this "set" or "arsenal" as equipped with its most evident structure, the way it appears so to speak "right in front of your nose"; that is what we call the structure of a "category" .... From here on, this kind of "measuring superstructure" called the "category of sheaves" will be taken as "incarnating" what is most essential to that space [ReS, p. P38].[22]

The Séminaire Cartan had shown that this structure suffices for much of cohomology. Definitions and proofs can be given by commutative diagrams without asking, most of the time, what these are diagrams *of*. Grothendieck and Buchsbaum pursued the idea independently, both extending earlier work in [Mac Lane, 1948].[23]

Grothendieck went the farthest, insisting that the "formal analogy" between sheaf cohomology and group cohomology should become "a common framework including these theories and others" [Grothendieck, 1957, p. 119]. To start with, injectives have a nice categorical sense. An object I in any category is injective if, for every monic $N \rightarrowtail M$ and arrow $f:N \to I$, there is at least one $g:M \to I$ such that

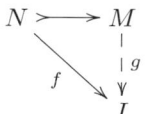

Fine sheaves are not so diagrammatic. Grothendieck saw that Reinhold Baer's original proof that modules have injective resolutions was largely diagrammatic itself.[24] Grothendieck thus gave diagrammatic axioms for the basic properties used

---

[21] Grothendieck knew this "set" is a proper class, the size of the universe of all sets. This is one reason for his Grothendieck universes [Artin *et al.*, 1972, pp. 185ff].

[22] "Considérons l'ensemble formé de tous les faisceaux sur un espace (topologique) donné, ou, si on veut, cet arsenal prodigieux formé de tous ces 'mètres' servant à l'arpenter. Nous considérons cet 'ensemble' ou 'arsenal' comme muni de sa structure la plus évidente, laquelle y apparaît, si on peut dire, 'à vue de nez'; à savoir, une structure dite de 'catégorie' .... C'est cette sorte de 'superstructure d'arpentage', appelée 'catégorie des faisceaux' (sur l'espace envisagé), qui sera dorénavant considérée comme 'incarnant' ce qui est le plus essentiel à l'espace."

[23] Grothendieck has said he did not know Mac Lane's work. Mac Lane has told me that when Grothendieck spoke on Abelian categories in Chicago around 1958 he mentioned no sources. Surely, he did not recall any article by Mac Lane. He was not careful about sources at that time, and he read less than he heard about from friends, notably Serre. But he was traveling in Mac Lane's circles, he was in Kansas in the midwest United States when he did the work, and he used Mac Lane's term "Abelian category," so there was surely an influence. Buchsbaum had seen [Mac Lane, 1948] and rather echoed its title in [Buchsbaum, 1955] but did not use Mac Lane's terminology.

[24] See [Baer, 1940]. Cartan and Eilenberg, and Grothendieck, repair a set-theoretic error. Baer mistakenly says, for any infinite cardinals $\Lambda \leq \Omega$, every function $\Lambda \to \Omega$ is contained in some initial segment of Ω (that is, Λ is not cofinal in Ω). Cartan and Eilenberg correctly say this holds when Ω is the next cardinal above Λ [Cartan and Eilenberg, 1956, p. 10]. Grothendieck gives a proof [Grothendieck, 1957, p. 137]. He was probably reading manuscripts for [Bourbaki, 1958].

in cohomology and called any category that satisfies them an *Abelian category*. He gave further diagrammatic axioms tailored to Baer's proof; every category satisfying these axioms has injective resolutions. Such a category is called an AB5 category, and sometimes around the 1960s a *Grothendieck category*, although that term has been used in several senses.

These axioms are easily verified for sheaf categories on topological spaces, proving that topological cohomology can use injective resolutions. Grothendieck soon learned a "really trivial" proof of that particular claim from Godement.[25] Weibel showed how this proof is easily implicit in methods of the Séminaire Cartan [Weibel, 1999, p. 812]. It is not explicit. The question seems not to have arisen in the Séminaire. Even when Serre wrote to Grothendieck about topological cohomology as a derived functor, he put it in terms of generalizing the definition of derived functor beyond injectives, and not in terms of finding enough injective sheaves [Colmez and Serre, 2001, p. 15 (quoted above)].

Grothendieck showed that sheaves on *any* topological space have injective resolutions and thus have derived functor cohomology in the strict sense. For paracompact spaces, this agrees with cohomology from fine, flabby, or soft resolutions. Those resolutions remain available and useful for many cases. But Grothendieck treats paracompactness as a "restrictive condition" well removed from the basic theory, and he mentions the Weil conjectures [Grothendieck, 1957, p. 120]. His axioms also unify topological cohomology with group cohomology. They are clearly far more general in principle, but there were few, if any, known examples outside that framework.

Grothendieck's axioms also simplified homological algebra by focusing on just the relevant features. Textbooks today rarely use the generality. They rarely even discuss sheaves on topological spaces. Yet, they are generally organized in terms of Abelian categories.

Eisenbud's *Commutative Algebra* [Eisenbud, 1995] takes one common strategy, where Abelian categories are not defined but are referred to, and the definitions and proofs are quite diagrammatic. The proofs are effectively in Abelian category terms, although stated only for categories of modules. Hartshorne's *Algebraic Geometry* gives the Abelian category axioms and relies on several kinds of Abelian categories other than module categories. It does not prove the theorems but describes several ways to do it including directly from the axioms [Hartshorne, 1977, p. 202]. The first and second editions of Lang's *Algebra* famously gave the Abelian category axioms, with an exercise: "Take any book on homological algebra, and prove all the theorems without looking at the proofs given in that book" [Lang, 1993, p. 105]. He dropped that from the third edition because today's homological algebra books are already organized around this axiomatic viewpoint.

Serre's key contribution to the Séminaire Cartan by 1951, taken from work on his dissertation, was to clarify *spectral sequences* and to extend their range and power [Serre, 1951].[26] Spectral sequences were and still are the standard tool for non-trivial calculations in cohomology. At the time, a spectral sequence was an infinite series of infinite two-dimensional arrays of Abelian groups and group

---

[25]Compare Grothendieck's letter to Serre of 16 January, 1956 in [Colmez and Serre, p. 27]. See also [Godement, 1958, p. 260]. Godement cites heavily the Séminaire Cartan and [Grothendieck, 1957].

[26]The dissertation introduced many constructions in topology, especially those for calculating higher homotopy groups, and especially of spheres.

homomorphisms, with each successive array gotten from the homology of the one before. Of course, the groups might also be modules over some ring. No single point about spectral sequences is difficult. They are imposing from sheer mass. Grothendieck would simplify the theory by reconceiving spectral sequences as arrays of objects from any Abelian category.

As Grothendieck wrote to Serre, "I am rid of my horror of spectral sequences [Je suis débarrassé de mon horreur de la suite spectrale]" [Colmez and Serre, 2001, p. 7]. The point of spectral sequences had been to calculate in an orderly way, passing over many details of the Abelian groups involved. He could prove the general spectral sequence theorems while positing no details of the objects in the first place—although applications would depend on using suitably detailed objects. He derived most of the important spectral sequences as special cases of one "delectable and useful general spectral sequence" today called the Grothendieck spectral sequence [Colmez and Serre, 2001, p. 14].[27]

A few pages of definitions of sheaves, resolutions, and spectral sequences from the Séminaire Cartan or from Serre's dissertation were simplified into 102 pages of category theory. Many people found the work completely disproportionate to the problem. It took two years to find a publisher, although this legend may be a bit overstated. Eilenberg was ready to put it in the *Transactions of the American Mathematical Society* in 1956 subject to what Grothendieck called "severe editorial taboos." In a letter to Serre dated 19 September, 1956, Grothendieck said he would do it only if someone else would retype the manuscript [Colmez and Serre, 2001, p. 45]. At any rate, Abelian categories did simplify and extend the theory so that they became and remain the standard setting for (co)homology.

This was a major step. "Grothendieck had shown that, given a category of sheaves, a notion of cohomology groups results" [Deligne, 1998, p. 16].[28] And he radically generalized categories of sheaves to any Abelian category with a generator and enough injectives. Any such category has an intrinsic cohomology theory. It remained to find which Abelian categories give Weil cohomology.

The Larger Vision

Grothendieck never mentions Abelian categories by name in *Récoltes et Semailles*. He focuses on more controversial ideas. He does cite *Tôhoku* in an enlightening way, namely, as an explanation of toposes. He describes a topos as a kind of space. In this sense, the category of sets is a one-point space:

> A "space in the *nouveau style*" (or *topos*), generalizing traditional topological spaces, is given by a "category" which, without necessarily coming from an ordinary space, nonetheless has all the good properties (explicitly designated once and for all, of course) of such a "category of sheaves" ... above all the properties I introduced into category theory under the name "exactness properties" [*ReS*, p. P39].[29]

---

[27]See [Grothendieck, 1957], [Eisenbud, 1995, p. 677], and [Lang, 1993, p. 821].

[28]"Grothendieck avait montré que, une catégorie de faisceaux étant donnée, une notion de groupes de cohomologie en résulte."

[29]"Un 'espace nouveau style' (ou topos), généralisant les espaces topologiques traditionnels, sera décrit tout simplement comme une 'catégorie' qui, sans provenir forcément d'un espace ordinaire, possède néanmoins toutes ces bonnes propriétés (explicitement désignées une fois pour toutes, bien sûr) d'une telle 'catégorie de faisceaux' .... Il s'agit ici surtout de propriétés que j'ai

He introduced those properties in [Grothendieck, 1957]. The specific properties of a topos are very different from those of an Abelian category, but both are expressed in part as exactness properties.

This is the really deep simplification Grothendieck proposed. The way to understand a mathematical problem is to express it in the mathematical world natural to it—that is, in the topos natural to it. Each topos has a natural cohomology, simply taking the category of Abelian groups in that topos as the category of sheaves. The cohomology of that topos may solve the problem. In outline:

(1) Find the natural world for the problem (for example, the étale topos of an arithmetic scheme).
(2) Express the problem cohomologically (state Weil's conjectures as a Lefschetz fixed point theorem).
(3) The cohomology of that world may solve the problem, like a ripe avocado bursts in your hand.

In Grothendieck's own words:

> The crucial thing here, from the viewpoint of the Weil conjectures, is that the new notion [of space] is vast enough, that we can associate to each scheme a "generalized space" or "topos" (called the "étale topos" of the scheme in question). Certain "cohomology invariants" of this topos (as "babyish" as can be!) seemed to have a good chance of offering "what it takes" to give the conjectures their full meaning, and (who knows!) perhaps to give the means of proving them [*ReS*, p. P41].[30]

The unity sought in the Séminaire Cartan was complete. Cohomology gives algebraic invariants of a topos as it used to give invariants of a topological space. Each topological space determines a topos with sheaf cohomology. Each group determines a topos with group cohomology. This would work for cases yet unimagined. Grothendieck notes that his Abelian categories were exactly suited to the cohomology of any topos, although toposes were entirely unforeseen as he wrote *Tôhoku* in 1955. He takes this as one more proof that it is the right idea of cohomology [*ReS*, p. P41n].

For the Weil conjectures, it only remained to find the natural topos for each arithmetic space—recalling that up to 1956 or so, the spaces themselves were not adequately defined. In fact, this conception of "toposes" came to Grothendieck as the way to combine his theory of schemes with Serre's idea of isotrivial covers and produce the cohomology [*ReS*, p. P31 and passim].[31]

---

introduites en théorie des catégories sous le nom de 'propriétés d'exactitude.'" Here "sheaves" are sheaves of sets. Elsewhere, it often means sheaves of groups.

[30]"La chose cruciale ici, dans l'optique des conjectures de Weil, c'est que la nouvelle notion est assez vaste en effet, pour nous permettre d'associer à tout 'schéma' un tel 'espace généralisé' ou 'topos' (appelé le 'topos étale' au schéma envisagé). Certains 'invariants cohomologiques' de ce topos (tout ce qu'il y a de 'bébêtes'!) semblaient alors avoir une bonne chance de fournir 'ce dont on avait besoin' pour donner tout leur sens à ces conjectures, et (qui sait!) de fournir peut-être les moyens de les démontrer."

[31]Deligne's 1972 proof, completing the Weil conjectures, was not as simple as Grothendieck hoped [Deligne, 1974]. Weil's proposed trivial calculation assumed cohomology with ordinary integer coefficients, but étale cohomology gives p-adic integer coefficients which are more general. In fact, the relevant coefficients are ordinary integers. Grothendieck conjectured general theorems on étale cohomology to prove that and more, called the *standard conjectures*. See [Grothendieck,

Anticipations of Schemes

Serre has put it well: no one invented schemes.[32] The brilliant step was to recognize their value. Grothendieck says that "[t]he very idea of scheme is of childish simplicity—so simple, so humble, that no one before me thought of stooping so low. So 'babyish,' in short, that for years, despite all the evidence, for many of my erudite colleagues, it was really 'not serious'!" [ReS, p. P32].[33] The question is, why did Grothendieck believe he should use such an idea to simplify an eighty-page paper by Serre into some 1000 pages of *Éléments de géométrie algébrique*? In fact, others did think of the idea. A look at how they thought of it, and at some of the reasons why they might have dropped it, highlights Grothendieck's achievement.

Basic algebraic geometry studies *varieties* defined by polynomial equations. Varieties have polynomial coordinate functions on them. One stock example is the complex number plane $\mathbb{C}^2$. Its classical points are the pairs (α, β) of complex numbers, subject to no equation. Its coordinate functions are all polynomials $P(x, y)$ with complex number coefficients and variables x, y over the complex numbers. The coordinate ring is then just the ring $\mathbb{C}[x, y]$ of all these polynomials.

Another stock example is the unit circle S^1 in $\mathbb{C}^2$ defined by the equation $x^2 + y^2 = 1$. Its classical points are the pairs of complex numbers (α, β) with $\alpha^2 + \beta^2 = 1$. It is thus the *subvariety* of $\mathbb{C}^2$ defined by that equation. In more systematic terms, it is defined by the ideal in $\mathbb{C}[x, y]$ of all polynomials divisible by the polynomial $x^2 + y^2 - 1$ or, geometrically, the ideal of all polynomials which are 0 all over S^1. The coordinate ring is the quotient of the ring $\mathbb{C}[x, y]$ by that ideal. A coordinate function on S^1 is thus any complex polynomial in x, y, regarding polynomials as equal if they take equal values at each point of the circle. There is a natural ring homomorphism from $\mathbb{C}[x, y]$ to that quotient, which takes each polynomial to its equivalence class. Geometrically, that amounts to restricting each coordinate function on $\mathbb{C}^2$ to a coordinate function on the subvariety $S^1 \subseteq \mathbb{C}^2$.

The subvarieties $V' \subseteq V$ of a variety V correspond to prime ideals in the coordinate ring of V. Since the classical points p of V are the minimal subvarieties, they correspond to the maximal ideals. As a hint of things to come, maximal ideals in any ring are all prime.

The early twentieth-century Italian algebraic geometers made deep and subtle use of generic points of a variety—that is, points with no special properties—so that anything proved of a generic point was true of all except maybe some exceptional points on that variety. Bartel van der Waerden used ideas from Emmy Noether to make this more precise [van der Waerden, 1926].[34]

1969] and [Kleiman, 1994]. They remain unproved. Deligne instead gave a wide-ranging, elegant but difficult geometric argument. See also [Mumford and Tate, 1978] and the review by Nicholas Katz in *Mathematical Reviews* 49, #5013. Deligne, Serre, and others have worked further on Grothendieck's strategy, especially on *motives*. One of the two 2002 Fields Medal winners, Vladimir Voevodsky won for "leading us closer to the world of *motives* that Grothendieck was dreaming about in the sixties" [Soulé, 2003, p. 102]. The other, Laurent Lafforgue, relates his own work to "Grothendieck's conjectural theory of motives" [Lafforgue, 2003, p. 383].

[32]In a conversation with the author in 1995.

[33]"L'idée même de schéma est d'une simplicité enfantine—si simple, si humble, que personne avant moi n'avait songé à se pencher si bas. Si 'bébête' même, pour tout dire, que pendant des années encore et en dépit de l'évidence, pour beaucoup de mes savants collègues, ça faisait vraiment 'pas sérieux'!"

[34]He later learned she had already done it in her lectures [van der Waerden, 1971].

Skipping the details, his version makes generic points not classical points at all but something else. For example, the "generic point" of the unit circle somehow lies over all of the classical points (α, β). David Mumford's famous lecture notes on schemes from the 1960s depict such a point as a blur spread out over the circle [Mumford, 1968]. Anything true of this generic point is true of nearly all the classical points, if stated in the correct form. On this approach, each subvariety of a variety V has a generic point. Since each classical point is a subvariety, it gets a generic point of its own—a silly doubling of points.

For Emmy Noether's school, then, it was natural to look at prime ideals instead of classical and generic points. As we would say today, it was natural to identify all points with prime ideals. Her associate Wolfgang Krull did precisely this. He spoke in Paris before the Second World War on algebraic geometry, taking prime ideals as points and implicitly using a Zariski topology (for which see any current textbook on algebraic geometry). The classical points correspond to the maximal ideas, but maximality plays no role in the basic definitions. In fact, all the basic definitions work for prime ideals in any commutative ring, not only for polynomial rings, so Krull did it in that generality. The audience laughed at him, and he abandoned the idea [Neukirch, 1999, p. 49].

Weil based his *Foundations of Algebraic Geometry* on a version of van der Waerden's generic points [Weil, 1946]. Generic points became a staple of Parisian algebraic geometry. When Serre left them out of his influential rival to Weil foundations, discussed below, people naturally thought about how to add them in. As Pierre Cartier explains:

> [André] Martineau remarked to [Serre] that his arguments remained valid for any commutative ring, provided one takes all prime ideals instead of only maximal ideals [that is, provided one takes generic as well as classical points]. I then proposed a definition of schemes equivalent to the definition of Grothendieck. In my dissertation I confined myself to a framework similar to that of Chevalley, so as to avoid an excessively long exposition of the preliminaries! [Cartier, 2001, p. 398].

Serre probably already knew it, and certainly found it obvious. Grothendieck and Jean Dieudonné soon wrote "Serre himself has remarked that the cohomology theory of algebraic varieties could be transcribed with no difficulty ... to any commutative ring" [Grothendieck and Dieudonné, 1960, p. 7].

Why did Krull abandon his idea, Martineau leave it as an aside, and Cartier judge his excessively long? Since Krull was a foreigner in Paris, perhaps the audience laughed harder at him than Bourbaki would at Grothendieck. Martineau worked in analysis. Perhaps Cartier would have come back to it, if Grothendieck had not taken it over. I will come back to Cartier.

More mathematically, it seems Krull's motive was simply that "it was there." The algebraic definitions of point and subvariety applied over any ring, so he gave them in that generality. Weil later proved hard theorems using generic points, and this seemed to point to an even more penetrating theory than Weil really provided. Perhaps the general theory of schemes could survive only when it had that much work to do.

Also, Krull lacked sheaves. Cartier knew them well but avoided using them in this context. Serre based algebraic geometry on them. Their easy way of pasting varieties together and their facility for cohomology prompted Grothendieck to say that Serre had "the principle of the right definition" of schemes [Grothendieck, 1958, p. 106]. Finally, Krull, Cartier, and Serre all worked without category theory. Grothendieck and Dieudonné remark this on one central issue: "The idea of 'variation' of base ring which we introduce gets easy mathematical expression thanks to the functorial language (whose absence no doubt explains the timidity of earlier attempts)" [Grothendieck and Dieudonné, 1971, p. 6].[35]

Schemes in Paris

A story says that in a Paris café around 1955 Grothendieck asked his friends "what is a scheme?" Compare the story of Hilbert asking John von Neumann "but what is a Hilbert space, really?" Hilbert wanted the idea behind von Neumann's axioms, examples, and theorems. Grothendieck's question was different. At the time, only an undefined idea of "schéma" was current in Paris, meaning more or less whatever would improve on "Weil foundations."[36]

One serious challenge was that Weil wanted "algebraic geometry over the integers" and not over a field at all, following Leopold Kronecker [Weil, 1952]. He wanted each integer polynomial $P(X, Y, \ldots, Z)$ to define a space X over the integers, which would *specialize* to other spaces defined by the same polynomial over the field $\mathbb{Q}$ of rational numbers and all finite prime fields $\mathbb{F}_p$.[37] When Grothendieck described schemes in his memoir years later, this is all he said about them: A single scheme can be a "magical fan [éventail magique]"[38] combining varieties over all those fields [*ReS*, p. P32].

There were two leading contenders by 1956. One, which did not use the word "scheme," was Serre's paper "Coherent Algebraic Sheaves" or, in the original French, "Faisceaux algébriques cohérents" generally cited as FAC [Serre, 1955]. The other was "the Chevalley-Nagata theory of schemes" with a variant of it by Pierre Cartier which he says "closely follows the exposition in Serre [1955] only avoiding the use of sheaves" [Cartier, 1956, p. 1-01].[39]

---

[35]"L'idée de 'variation' de l'anneau de base que nous venons d'introduire s'exprime mathématiquement sans peine grâce au langage fonctoriel (dont l'absence explique sans doute la timidité des tentatives antérieures."

[36]Otto Schilling's enthusiastic but awestruck review of Weil's 1946 book will dispel any thought that the ideas were more accessible and naturally geometric in those days. See *Mathematical Reviews* 9, #303c in 1946.

[37]In effect, Weil wanted geometry over any commutative ring. He later called this "the natural evolution of the subject," largely achieved by Goro Shimura and "above all by the theory of schemes as created by Grothendieck and developed by his students and successors" [Weil, 1979, 1:576].

[38]The image is of an oriental hand fan that collapses to a rod. In stories, such a fan works as a magic wand or can extinguish fires or set them by fanning. When a stage magician fans out a deck of cards, this is also an "éventail magique."

[39]"On a suivi de près l'exposition de Serre (*Ann. de Math.*, **61**, 1955, p. 197-278) en évitant seulement l'emploi des faisceaux." Cartier cites neither Chevalley nor Nagata. Grothendieck mentions them and Cartier in [Grothendieck, 1957, p. 161] and [Grothendieck, 1962, p. 190-01].

Cartier defines a spectrum Ω_A for each finite type algebra A over a field k, with a Zariski topology.[40] Elements of A are construed as "functions" from the spectrum to a field extension of k (as sections of the structure sheaf of a scheme over a field k are construed as "functions" to k). Cartier defines "algebraic sets" by pasting together spectra. He proves various theorems of current scheme theory for spectra of finite type over a field. His axiom "EA1" for an algebraic set requires a finite cover by spectra, while "EA2" is the current definition of a so-called separated scheme [Cartier, 1956, p. 1-12]. The Chevalley-Nagata theory is a slight variant [Cartier, 1956, p. 2-18]. The definitions easily generalize to all commutative rings, supporting Cartier's claim of anticipating Grothendieck's schemes [Cartier, 2001, p. 398].

The key feature of FAC is missing in all of that. It is the idea of structure sheaves. An affine variety in FAC does not just have a ring of coordinate functions, but a sheaf of rings of functions. For later comparison, note that all of these functions are functions in the set-theoretic sense. The structure sheaf on an affine variety V gives no more information than a coordinate ring, but it serves two purposes.[41]

First, Weil could only deal with more general, non-affine varieties in terms of their parts. He had to speak of several affine varieties plus a pasting relation, yet the parts and pasting would not be unique. There would be other ways to assemble the same "abstract variety," but the abstract variety did not exist in itself, only the many ways to assemble it actually exist. Serre could paste several affine varieties into a single space with structure sheaf. That space may also have a natural description without pasting.

Second, the structure sheaf is directly suited to cohomology. Serre's FAC produced a cohomology for varieties, now standard in algebraic geometry called coherent cohomology. Serre suggested this might give the right Betti numbers of varieties for the Weil conjectures [Serre, 1955, p. 233], but he also knew it could not be the Weil cohomology because it could not give an adequate Lefschetz fixed point theorem.[42] Today, coherent cohomology is generally given in Grothendieck's form as a derived functor cohomology on any scheme, as in [Hartshorne, 1977].[43] Serre varieties are much less general than schemes. The ring of coordinate functions

---

[40]In Exposé 1, Ω_A is the set of homomorphisms from A to an algebraically closed extension K of k. If K is the algebraic closure, then (modulo the Galois group of K over k) these amount to the maximal ideals of A as used in [Serre, 1955]. If K has infinite transcendence degree over k, then they amount (modulo the Galois group) to the prime ideals. Exposé 2 uses prime ideals of A rather than homomorphisms, so the Galois group disappears, leaving spectra in Grothendieck's sense.

[41]The sheaf assigns each open subset $U \subseteq V$ a coordinate ring. For all $p \in U$, $g(p) \neq 0$, it assigns U the ring of fractions f/g, where f and g are coordinate functions on V. For example, $1/x$ is defined on the open subset $\{x \in \mathbb{R} |\ x \neq 0\}$. All is determined by the coordinate ring of V.

[42]Applied to a variety over any $\mathbb{F}_{p^r}$ or its algebraic closure, this cohomology gives coefficients modulo p. It could thus count the fixed points modulo p at best.

[43]Serre used a version of Čech cohomology. A key theorem shows that this cohomology is a derived functor: affine varieties have vanishing cohomology [Serre, 1955, p. 239 (corollary 1)]. Yet, Grothendieck says that "this should be considered an accidental phenomenon" and that "it is important for technical reasons not to take as *definition* of cohomology the Čech cohomology" [Grothendieck, 1958, p. 108]. The derived functor definition trivializes the broad generalities before any substantial theorems.

on a Serre variety has to be an integral algebra of finite type over an algebraically closed field.

Schemes in Grothendieck

Grothendieck made it simple. As the correct definition of cohomology applied to every topological space, so every commutative ring should be the coordinate ring of a space. Every such space would have cohomology in Serre's sense. The space of a commutative ring R is its spectrum or $\mathrm{Spec}(R)$ with the prime ideals p as points and the Zariski topology. Every ideal I of R gives a closed set, where a point p lies in the closed set of I, if $I \subseteq p$ as ideals. Schemes in general come from pasting together spectra. At the 1958 International Congress of Mathematicians, Grothendieck called this topology on the prime ideals of commutative rings "classical" [Grothendieck, 1958, p. 106]. The next year in the Séminaire Bourbaki he called it "well known [bien connu]" [Grothendieck, 1962, p. 182-01].

Grothendieck's originality, according to Serre, was that only he saw it as geometry. Before Grothendieck convinced him, Serre thought geometry would require the rings to "meet some conditions, at least be Noetherian."[44] Indeed, many classical geometric results need further assumptions, but uncanny amounts of geometric intuition are directly expressed in all generality for all commutative rings. The later assumptions are visibly irrelevant to the first steps—that is, now visibly, but visible to few in the 1950s.

Famously, Grothendieck wrote the work in collaboration with Jean Dieudonné. Their original edition published in the series of the Institut des Hautes Études Scientifiques (IHES) is wary of history in two senses. They find any history of schemes "beyond our competence [hors de notre compétence]" [Grothendieck and Dieudonné, 1960, p. 7]. They also warn that prior knowledge of algebraic geometry, "despite its obvious advantages, can sometimes (by the too exclusive habituation to the birational viewpoint implied in it) cause problems for those who wish familiarity with the viewpoint and techniques given here" [Grothendieck and Dieudonné, 1960, p. 5].[45]

Yet, their introduction closes with an historic perspective:

> To conclude, we believe it helpful to warn readers that, like the authors themselves, they will no doubt have some trouble before they are accustomed to the language of schemes and before they convince themselves that the usual constructions suggested by geometric intuition can be transcribed in essentially just one reasonable way into this language. As in many parts of modern mathematics the initial intuition seemingly draws farther and farther away from the language suited to expressing it in all the desired precision and generality. In the present case the psychological difficulty is in transporting notions familiar from sets into the objects of rather different categories (that is, the category of schemes[46] or of schemes over a given scheme): cartesian products, the laws of a group or a ring or module, fiber bundles,

---

[44]In a conversation with the author in 1995.

[45]"... malgré ses avantages évidents, pouvait parfois (par l'habitude trop exclusive du point de vue birationnel qu'elle implique) être nuisible à celui qui désire se familiariser avec le point de vue et les techniques exposés ici."

[46]I translate "préschémas" as "schemes" to follow current usage.

principal homogeneous fiber bundles, etc. No doubt it will be difficult for future mathematicians to do without this new effort at abstraction, which is perhaps quite small, compared to that our fathers faced, familiarizing themselves with Set Theory [Grothendieck and Dieudonné, 1960, p. 9].[47]

In the Springer-Verlag edition, they claim their own historical heritage and trace the basic idea back a hundred years to Dedekind and Heinrich Weber [Grothendieck and Dieudonné, 1971, p. 11].

They also signal evolution in thinking about categories. They drop the warning and add a section on transporting structures into categories.[48] Twenty-five years after that Deligne wrote: "if the decision to let every commutative ring define a scheme gives standing to bizarre *schemes*, allowing it gives a *category of* schemes with nice properties" [Deligne, 1998, p. 13 (his emphasis)].[49] The nice properties are those used to transport the various structures listed by Grothendieck and Dieudonné.

The 1960 list of structures highlights how these categorically nice properties are a mess set-theoretically. The product $X \times Y$ of schemes X and Y is a scheme with projection morphisms p_1 and p_2 with the familiar categorical property:

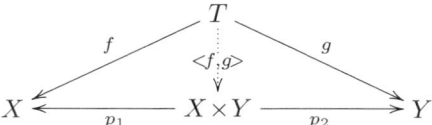

A morphism from any scheme T to $X \times Y$ is given by a pair $<f,g>$ of morphisms to X and Y. The set of points of $X \times Y$, however, is nothing like the set-theoretic product of the sets of points of X and Y. It can happen that X and Y each have points, while $X \times Y$ is empty.[50] Or X and Y may each have a single point, while $X \times Y$ has many.[51]

An ordinary group is a set G with a binary operation $G \times G \rightarrow G$ and a group inverse function $G \rightarrow G$ which satisfy certain equations, that is, they make certain

---

[47]"Pour terminer, nous croyons utile de prévenir les lecteurs que, tout comme les auteurs eux-mêmes, ils auront sans doute quelque difficulté avant de s'accoutumer au langage des schémas, et de se convaincre que les constructions habituelles que suggère l'intuition géométrique peuvent se transcrire, essentiellement d'une seule façon raisonnable, dans ce langage. Comme dans beaucoup de parties de la Mathématique moderne, l'intuition première s'éloigne de plus en plus, en apparence, du langage propre à l'exprimer avec toute la précision et la généralité voulues. En l'occurrence, la difficulté psychologique tient à la nécessité de transporter aux objets d'une catégorie déjà assez différente de la catégorie des ensembles (à savoir la catégorie des préschémas, ou la catégorie des préschémas sur un préschéma donné) des notions familières pour les ensembles: produits cartésiens, lois de groupe, d'anneau, de module, fibrés, fibrés principaux homogènes, etc. Il sera sans doute difficile au mathématicien, dans l'avenir, de se dérober à ce nouvel effort d'abstraction, peut-être assez minime, somme toute, en comparaison de celui fourni par nos pères, se familiarisait avec la Théorie des Ensembles."

[48]See the first section of [Grothendieck and Dieudonné, 1971].

[49]"Si permettre que tout anneau commutatif définisse un schéma affine donne droit de cité à des schémas bizarres, le permettre fournit une catégorie de schémas ayant de bonnes propriétés."

[50]This happens, for example, if X and Y are schemes over two fields with different characteristics.

[51]If $X = Y$ is the spectrum of a field k, then $X \times Y$ has as many points as the Galois group of k over its prime field.

diagrams commute. This transfers directly to schemes. A group scheme X has morphisms $X \times X \to X$ and $X \to X$ making the same diagrams commute. Set-theoretically, however, $X \times X$ does not have the elements of a product, and $X \times X \to X$ is not a binary operation. A group scheme is not a group from the set-theoretic viewpoint, although it is from the categorical viewpoint. Thus, the slogan "a group scheme is just a group object in the category of schemes."

The greatest objection to schemes was a certain non-set-theoretic feature.[52] The elements of any commutative ring R appear as coordinate functions on the spectrum $\mathrm{Spec}(R)$. Of course, these are generally not functions in the set-theoretic sense. The scheme context makes them act rather like set-theoretic functions. Each one can be evaluated at any point p of the scheme (taking values in the fiber of the scheme at that point as in any standard text on schemes). Yet, a "function" $g \in R$ may have $g(p) = 0$ at every point p of the scheme and not be the zero function! In geometric terms, this happens when the scheme has infinitesimal fringe around it, and g is 0 at each point but has nonzero derivative in some directions through the fringe.[53]

A "function" in this sense is not determined by its values. According to David Mumford, "[i]t is this aspect of schemes which was most scandalous when Grothendieck defined them" [Mumford, 1966, p. 12].[54] It is nevertheless tremendously helpful, for example, in describing a singular point x of a scheme X. Looking at "arbitrarily small neighborhoods" of x is not helpful in the very coarse Zariski topology where no neighborhood is small, but there are subschemes of X containing just the point x and infinitesimal fringe around it. The contortions of X around x are retained in this fringe with no other complexities of the larger space X.

Indeed, the set of points of a scheme is rarely the best handle on it:

> The audacity of Grothendieck's definition is to accept that *every* commutative ring A (with unit) has a scheme $\mathrm{Spec}(A)$. ... This has a price. The points of $\mathrm{Spec}(A)$ (prime ideals of A) have no ready to hand geometric sense .... When one needs to construct a scheme one generally does not begin by constructing the set of points [Deligne, 1998, p. 12].[55]

Rather, one begins with geometric relations to other schemes. This holds as well for some older notions of space and many newer ones [Cartier, 2001]. None, however, is as important as schemes.

---

[52] This is not about foundations of mathematics. On any foundation, geometers will treat sections of the structure sheaf as functions (analogous to polynomial functions on classical varieties), while they are not set-theoretic functions.

[53] Algebraically, this means that $g \neq 0$ has some power $g^n = 0$ and so belongs to every prime ideal. For a picture, consider a polynomial $f(x)$ with $f(0) = 0$ but derivative $f'(0) \neq 0$; f is not the zero function on a first-order infinitesimal fringe around 0, although 0 is the only point within that fringe. In this case, $f^2(x)$ has value $f^2(0) = 0$ and first derivative $2f(0) \cdot f'(0) = 0$. So f^2 is the zero function all over that fringe.

[54] Mumford has said Oscar Zariski was particularly put off by this, but Mumford made him waver by describing a scheme-theoretic proof of Zariski's Main Theorem.

[55] "L'audace de la définition de Grothendieck est d'accepter que tout anneau commutatif (à unité) A définisse un schéma affine $\mathrm{Spec}(A)$.... Ceci a un prix. Les points de $\mathrm{Spec}(A)$ (idéaux premiers de A) n'ont pas un sens géométrique maniable.... Quand on a à construire un schéma, on ne commence pas en général par construire l'ensemble de ses points."

Another issue was decisive in the success of schemes. Classically, there were two different ways a variety V could be "over" something. It could be *defined over* a field k, meaning roughly that it is defined by polynomials with coefficients in k, and the coordinates of its points lie in k. Or it could *vary over* a parameter space P. For example, a complex polynomial $x^2 + \alpha xy + \beta y^2$ in two variables x, y with parameters α, β defines a conic section varying over the complex plane $\mathbb{C}^2$. For parameter values $(\alpha, \beta) = (1, 1) \in \mathbb{C}^2$, it is the conic $x^2 + xy + y^2 = 0$. For values $(\alpha, \beta) = (2, 1) \in \mathbb{C}^2$, it is the degenerate conic

$$x^2 + 2xy + y^2 = (x+y)^2 = 0.$$

Scheme theory expresses both in the same way. To say a scheme X varies over the complex plane $\mathbb{C}^2$ is just to say it has a scheme morphism, $X \to \mathbb{C}^2$, to $\mathbb{C}^2$. To say X is defined over a field k is just to say it has a scheme morphism, $X \to \text{Spec}(k)$, to the spectrum of k.

Grothendieck treats a scheme morphism $X \to S$ as a single scheme or, more precisely, as a relative scheme over the base S. A morphism of relative schemes over S is a commuting triangle of scheme morphisms:

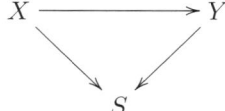

This could be a morphism of schemes over a field k, preserving the coefficients in k. It could be a morphism between families X, Y of schemes over, say, the complex plane $S = \mathbb{C}^2$, so the scheme in X with given parameters (α, β) is mapped to the scheme in Y with the same parameters.

Given $X \to S$, Grothendieck could largely ignore the coefficients or the parameters and let them take care of themselves. The category of schemes over any base scheme S is very much like the category of schemes *per se*, although with specific differences reflecting the algebraic geometry of S. Demazure and Grothendieck note the advantages of this, and of infinitesimal fringe, to group schemes [Demazure and Grothendieck, 1970, 1:viii]. They can treat a parameterized family of group schemes as a single group; a group scheme over a base scheme S with just one point plus infinitesimal fringe is an infinitesimal deformation of one group scheme.

Relative schemes produce the simple and general functorial account of base change that Grothendieck and Dieudonné mentioned [Grothendieck and Dieudonné, 1971, p. 6]. For example, given a scheme $X \to \text{Spec}(\mathbb{R})$ over the real numbers, to focus on its complex points means extending it to a scheme $X' \to \text{Spec}(\mathbb{C})$ over the complex numbers. This is just a pullback in the category of schemes

$$\begin{array}{ccc} X' & \longrightarrow & X \\ \downarrow & & \downarrow \\ \text{Spec}(\mathbb{C}) & \longrightarrow & \text{Spec}(\mathbb{R}) \end{array}$$

The natural morphism $\text{Spec}(\mathbb{C}) \to \text{Spec}(\mathbb{R})$ corresponds to the field extension $\mathbb{R} \subseteq \mathbb{C}$. The same works for any extension field $k \subseteq K$.

Alternatively, take a family of schemes $X \to \mathbb{C}^2$ varying over the complex plane $\mathbb{C}^2$. To look at just the part lying over the unit circle S^1 in that plane, take the pullback along the natural inclusion $S^1 \to \mathbb{C}^2$.

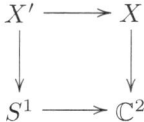

The formalism makes good, general theorems trivial so as to clear the way to the particulars in any given case. Of course, the particulars of a given case may remain difficult themselves because they can give the solution to some genuine problem in geometry or algebra. In any case, the set-theoretic details are rebarbative and uninteresting.

Toward the Séminaire de Géométrie Algébrique

We have seen "the sea advance insensibly in silence ... so far off you hardly hear it [La mer s'avance insensiblement et sans bruit ... si loin on l'entend à peine]" towards the Weil conjectures, up to about 1958 [ReS, p. 552]. Along the way, in 1957, Grothendieck found the Grothendieck-Riemann-Roch Theorem, but left it to Armand Borel and Serre to publish a proof [Borel and Serre]. Raoul Bott's review in the *Mathematical Reviews*[56] notes that "Grothendieck has generalized the theorem to the point where not only is it more generally applicable than Hirzebruch's version, but it depends on a simpler and more natural proof." Grothendieck says this theorem "made me a 'big star' overnight," dispelled Bourbaki's doubts about him, and made him "somewhat feared" by other mathematicians at the 1958 International Congress of Mathematicians. He says he proved it by "the rising sea," even though it was not a question of his own making. Serre put him onto it. We have seen not one step specifically on the Weil conjectures.[57]

Serre made a crucial step in 1958, actually finding the 1-dimensional Weil cohomology groups, using isotrivial coverings.[58] Grothendieck was at the talk and immediately told Serre this would produce the cohomology in all dimensions. Serre was "absolutely unconvinced," since he felt he had "brutally forced" the bundles to yield the H^1s. "But Grothendieck was always an optimist in those days."[59] Serre thought higher dimensional Weil cohomology would need to use higher homotopy groups [Serre, 2001, p. 255]. Those seemed far out of reach. The expanded print version of Serre's talk appeared a few months later. By then, Grothendieck had shown him that it indeed gives cohomology in all dimensions, and convinced him that this was likely the "true cohomology needed to prove the Weil conjectures."[60]

Grothendieck's optimism grew from his method. Cohomology is uniquely determined, once we know *what* we want the cohomology *of*. Serre had found that the Weil conjectures need the cohomology of isotrivial covers (soon modified to étale covers). The job was thus finished in principle—from Grothendieck's viewpoint—but he did not rush to work it all out. That would have been striking at the chisel. Rather, he conceived a larger framework to embrace at once spaces, their sheaves, and cohomology. Technically, this framework is all about "covering." Conceptually,

---

[56]See 22, #6817.

[57]Quotes from [ReS, pp. P23, 705, 32, 554-555].

[58]He used locally isotrivial fiber bundles, but the relation to coverings was obvious. See [Serre, 1958], which cites [Grothendieck, 1958] for scheme theory.

[59]Quotes from a conversation with the author in the fall of 1995.

[60]See [Serre, 1958, p. 125] and compare [Grothendieck, 1958, p. 104].

it is about transporting geometric ideas into new categories. It first appeared as Grothendieck topology: "the technical, provisional form of the crucial notion of *topos*" [ReS, p. P24 (his emphasis)].

Each sheaf lives in a world as big as the universe of all sets: the category of all sheaves of that type—"the way it appears so to speak 'right in front of your nose'" [ReS, p. P38]. To find the right sheaves for a new problem, find a new world: the right topos. The sheaves will be the Abelian groups in that world. To work with a scheme, look at a suitable category of relative schemes. Into each of these worlds transport familiar geometric constructions. The ideas would grow through the 1960s at Grothendieck's Séminaire de Géometrie Algébrique at the IHES along with his proofs of the first and second Weil conjectures.

Categorical world-building is young. Abelian categories are standard in research, although they are not very common in textbooks. Grothendieck topology is entirely accepted in research and rare in textbooks. Toposes are still widely avoided by geometers, although the theory continues to grow.[61] Lafforgue's and Voevodsky's Fields Medals bear testament to the fact that Grothendieck's largest vision is still progressing in algebraic geometry.

As to schemes, Grothendieck and Dieudonné focused on the finally decisive point: "It is fitting to give algebraic geometry all desirable generality and suppleness by resting it on the notion of *scheme*" [Grothendieck and Dieudonné, 1971, p. 1].[62] When Serre spoke at the Stockholm International Congress of Mathematicians in 1962 on algebraic geometry, he said "I must specify that I take this term in the sense it has had for several years now: the theory of schemes" [Serre, 1963, p. 190].[63]

Acknowledgments

My debt to Jean-Pierre Serre, Pierre Deligne, and David Buchsbaum is obvious. Nor could this paper exist without the help of many others. Vladimir Berkovich, Pierre Cartier, David Corfield, Mic Detlefsen, David Eisenbud, Peter Freyd, Jeremy Gray, Alain Herreman, William Lawvere, Pierre Lochak, Barry Mazur, David Mumford, Frédéric Patras, Leila Schneps, Norbert Schappacher, and anonymous referees contributed information, orientation, and critique.

References

Artin, Michael; Grothendieck, Alexander; and Verdier, Jean-Louis. 1972. *Théorie des topos et cohomologie étale des schémas I*. Séminaire de géométrie algébrique du Bois-Marie. Vol. 4. Berlin: Springer-Verlag (generally cited as SGA4).

Baer, Reinhold. 1940. "Abelian Groups That Are Direct Summands of Every Containing Abelian Group." *Bulletin of the American Mathematical Society* 46, 800-806.

Borel, Armand and Serre, Jean-Pierre. 1958. "Le théorème de Riemann-Roch." *Bulletin de la Société mathématique de France* 86, 97-136.

---

[61]For the current state of the general theory, see [Johnstone, 2002].

[62]"Il convient de donner à la Géométrie algébrique toute la souplesse et la généralité désirables, en la faisant reposer sur la notion de *schéma*.".

[63]"Je dois préciser que je prends ce dernier terme au sens qui est devenu le sien depuis quelques années: celui de *théorie des schémas*."

Bourbaki, Nicolas. 1949–. *Séminaire Bourbaki*. Secrétariat mathématique. Paris: École Normale Supérieure.

──────────. 1958. *Théorie des ensembles*. 3rd Ed. Paris: Hermann.

Brown, Kenneth S. 1982. *Cohomology of Groups*. New York: Springer-Verlag.

Buchsbaum, David A. 1955. "Exact Categories and Duality." *Transactions of the American Mathematical Society* 80, 1-34.

Cantor, Georg. 1932. *Gesammelte Abhandlungen mathematischen und philosophischen Inhalts*. Berlin: Julius Springer Verlag.

Cartan, Henri. 1948–. *Séminaire Henri Cartan*. Secrétariat mathématique. Paris: École Normale Supérieure.

Cartan, Henri and Eilenberg, Samuel. 1956. *Homological Algebra*. Princeton: Princeton University Press.

Cartier, Pierre. 1956. "Définition des variétés algébriques"; and "Schémas des variétés algébriques." In *Séminaire Chevalley*. Secrétariat mathématique. Paris: Institut Henri Poincaré.

Cartier, Pierre. 2001. "A Mad Day's Work: From Grothendieck to Connes and Kontsevich: The Evolution of Concepts of Space and Symmetry." *Bulletin of the American Mathematical Society* 38, 389-408.

Colmez, Pierre and Serre, Jean-Pierre, Ed. 2001. *Correspondance Grothendieck-Serre*. Société Mathématique de France; Recently expanded to: *Grothendieck-Serre Correspondence: Bilingual Edition*. Providence: American Mathematical Society and Paris: Société Mathématique de France, 2004.

Deligne, Pierre. 1974. "La conjecture de Weil I." In *Publications mathématiques*. No. 43. Paris: Institut des Hautes Études Scientifiques, 273-307.

──────────. 1998. "Quelques idées maîtresses de l'œuvre de A. Grothendieck." In *Matériaux pour l'histoire des mathématiques au XXe siècle (Nice, 1996)*. Paris: Société mathématique de France, pp. 11-19.

Demazure, Michel and Grothendieck, Alexander. 1970. *Schémas en groupes* [SGA 3]. Berlin: Springer-Verlag.

Dieudonné, Jean. 1988. "On the History of the Weil Conjectures." In *Étale Cohomology and the Weil Conjectures*. Ed. Eberhard Freitag and Reinhard Kiehl. Berlin: Springer-Verlag, pp. ix-xviii.

──────────. 1990. "De l'analyse fonctionnelle aux fondements de la géométrie algébrique." In *The Grothendieck Festschrift*. Ed. Pierre Cartier, Luc Illusie, *et al.* Vol. 1. Basel: Birkhäuser Verlag, pp. 1-14.

Eisenbud, David. 1995. *Commutative Algebra*. New York: Springer-Verlag.

Fasanelli, Florence. 1981. "The Creation of Sheaf Theory." Unpublished Doctoral Dissertation: American University.

Godement, Roger. 1958. *Topologie algébrique et théorie des faisceaux*. Paris: Hermann.

Grothendieck, Alexander. 1952. "Résumé des résultats essentiels dans la théorie des produits tensoriels topologiques et des espaces nucléaires." *Annales de l'Institut Fourier, Grenoble* 4, 73-112.

──────────. 1957. "Sur quelques points d'algèbre homologique." *Tôhoku Mathematical Journal* 9, 119-221.

──────────. 1958. "The Cohomology Theory of Abstract Algebraic Varieties." In *Proceedings of the International Congress of Mathematicians, 1958*. Cambridge: Cambridge University Press, pp. 103-118.

———. 1962. *Fondements de la géométrie algébrique; Extraits du Séminaire Bourbaki 1957–1962*. Secrétariat mathématique. Paris: Institut Henri Poincaré.

———. 1969. "Standard Conjectures on Algebraic Cycles." In *International Colloquium on Algebraic Geometry 1968: Bombay*. Ed. Shreeram Abhyankar. Oxford: Oxford University Press, pp. 193-199.

———. 1985–1987. *Récoltes et Semailles*. Montpelier: Université des Sciences et Techniques du Languedoc (published in several successive volumes).

Grothendieck, Alexander and Dieudonné, Jean. 1960. *Éléments de géométrie algébrique I: Le langage des schémas*. No. 4. Publications mathématiques. Paris: Institut des Hautes Études Scientifiques.

———. 1961. *Éléments de géométrie algébrique III: Étude cohomologique des faisceaux cohérents*. No. 11. Publications mathématiques. Paris: Institut des Hautes Études Scientifiques.

———. 1971. *Éléments de géométrie algébrique I*. Berlin: Springer-Verlag.

Hartshorne, Robin. 1977. *Algebraic Geometry*. New York: Springer-Verlag.

Houzel, Christian. 1994. "La préhistoire des conjectures de Weil." In *Development of Mathematics 1900–1950*. Ed. Jean-Paul Pier. Basel: Birkhäuser Verlag, pp. 385-414.

Ireland, Kenneth and Rosen, Michael. 1992. *A Classical Introduction to Modern Number Theory*. New York: Springer-Verlag.

Jackson, Allyn. 2004. "Comme appelé du néant—As If Summoned from the Void: The Life of Alexandre Grothendieck." *Notices of the American Matheamtical Society* 51, 1038-1056.

Johnstone, Peter. 2002. *Sketches of an Elephant: A Topos Theory Compendium*. Vol. 1. Oxford: Oxford University Press (to be finished as three volumes).

Kleiman, Steven. 1994. "The Standard Conjectures." In *Motives*. Providence: American Mathematical Society, pp. 3-20.

Lafforgue, Laurent. 2003. "Chtoucas de Drinfeld, formule des traces d'Arthur-Selberg et correspondance de Langlands." In *Proceedings of the International Congress of Mathematicians, Beijing 2002*. Ed. L. Tatsien. 2 Vols. Singapore: World Scientific Publishers, 1:383-400.

Lang, Serge. 1993. *Algebra*. New York: Addison-Wesley.

Lawvere, F. William. 1979. "Categorical Dynamics." In *Topos Theoretic Methods in Geometry*. No. 30. Aarhus: Aarhus University, pp. 1-28.

Li, Winnie. 1996. *Number Theory with Applications*. Singapore: World Scientific Publishing.

Mac Lane, Saunders. 1948. "Groups, Categories and Duality." *Proceedings of the National Academy of Sciences of the United States* 34, 263-267.

Mumford, David. 1966. *Lectures on Curves on an Algebraic Surface*. Princeton: Princeton University Press.

———. 1988. *The Red Book of Varieties and Schemes*. New York: Springer-Verlag.

Mumford, David and Tate, John. 1978. "Fields Medals IV: An Instinct for the Key Idea." *Science* 202, 737-739.

Neukirch, Jürgen. 1999. "Erinnerungen an Wolfgang Krull." In *Wolfgang Krull: Gesammelte Abhandlungen*. Ed. Paolo Ribenboim. Berlin: Walter de Gruyter, pp. 47-52.

Reid, Miles. 1990. *Undergraduate Algebraic Geometry*. Cambridge: Cambridge University Press.

Serre, Jean-Pierre. 1951. "Homologie singulière des espaces fibrés: Applications." *Annals of Mathematics* 54, 425-505.

—————. 1955. "Faisceaux algébriques cohérents." *Annals of Mathematics* 61, 197-277.

—————. 1958. "Espaces fibrés algébriques." In *Séminaire Chevalley*. No. 1. Secrétariat mathématique. Paris: Institut Henri Poincaré.

—————. 1963. "Géométrie algébrique." In *Proceedings of the International Congress of Mathematicians (Stockholm, 1962)*. Djursholm: Institut Mittag-Leffler, pp. 190-196.

—————. 1973. *A Course in Arithmetic*. New York: Springer-Verlag.

—————. 2001. *Exposés de Séminaires 1950–1999*. Paris: Société mathématique de France.

Soulé, Christophe. 2003. "The Work of Vladimir Voevodsky." In *Proceedings of the International Congress of Mathematicians, Beijing 2002*. Ed. L. Tatsien. 2 Vols. Singapore: World Scientific Publishers, 1:99-104.

van der Waerden, Bartel L. 1926. "Zur Nullstellentheorie der Polynomideale." *Mathematische Annalen* 96, 183-208.

—————. 1971. "The Foundation of Algebraic Geometry from Severi to André Weil." *Archive for History of Exact Sciences* 7, 171-180.

Weibel, Charles. 1999. "History of Homological Algebra." In *History of Topology*. Ed. Ioan M. James. Amsterdam: North-Holland, pp. 797-836.

Weil, André. 1946. *Foundations of Algebraic Geometry*. New York: American Mathematical Society.

—————. 1949. "Number of Solutions of Equations in Finite Fields." *Bulletin of the American Mathematical Society* 55, 487-495.

—————. 1952. "Number Theory and Algebraic Geometry." In *Proceedings of the International Congress of Mathematicians, Cambridge 1950*. New York: American Mathematical Society, pp. 90-100.

—————. 1979. *Oeuvres scientifiques*. New York: Springer-Verlag.

Index

Abel, Niels Henrik, 4, 76
  as father of the theory of algebraic functions, 87
  on the solution of algebraic equations, 111
Abelian categories, 305–311
Abelian functions
  Riemann's work on, 87–88
  theory of, 87–88
al-Khwārizmī, 3
Albert, A. Adrian
  as an adviser, 190
  broader community interests of, 189–190
  career of, 188–192
  correspondence of with Hasse, 184–187
  early education of, 181–183
  joint paper of with Hasse, 186–187, 193–194
  mathematical style of, 189
  work of on cyclic algebras, 183–184, 206–210
  work of on non-associative algebras, 188–189
  work of on Riemann matrices, 188
  work of on splitting fields, 132
  work of on the Brauer-Hasse-Noether Theorem, 7–8, 185–186, 194
  working style of, 187–188
Albert-Brauer-Hasse-Noether Theorem, *see also* Brauer-Hasse-Noether Theorem, 209–210
Algebra
  as a discipline in the early twentieth century, 8, 222–225
  as a discipline in the nineteenth century, 8, 221–222
  as language of science, 15, 39–40
  Condillac's grammar of, 18–19
  De Gérando's views on, 34–35
  Destutt de Tracy on, 38
  Dickson's work in, 181
  emergence of as a branch of mathematics, 3–5
  Euler's views on, 73–74
  Gregory's definition of, 58
  internationalization of, 192
  Kronecker's views on the foundations of, 107–109
  origin of the term, 3
  Peacock's views on, 58–59
  research in around 1800, 74
  the role of the calculus of operations in the development of in Britain, 68
Algebraic functions
  Abel as father of, 87
  Dedekind's work on, 88–89
  E. Noether's work on, 89
  Kronecker's work on, 88
  theory of, 87–89
  Weber's work on, 88–89
Algebraic geometry
  American development of, 285
  and the infusion of modern algebra, 247–248
  and van der Waerden's series "Zur algebraischen Geometrie" [ZAG], 263–272
  and van der Waerden's *Einführung in die algebraische Geometrie* (1939), 270–271
  and Zariski's *Algebraic Surfaces* (1935), 289–292, 296–297
  arithmetization of, 292–296
  Cartier's work in, 314–315
  Castelnuovo and Enriques's work in, 287
  definitions of, 245–246
  five practices of as of 1947, 276
  Grothendieck's work in, 269–270, 321–322
  historical context of van der Waerden's work in, 272–276
  Italian development of, 246–249, 285–288, 291, 313
  Krull's work in, 314–315
  Levi's work in, 289–291
  M. Noether and Brill's work in, 286
  M. Noether's work in, 289–291, 293
  Mumford's views on Italian contributions to, 248–249
  Segre's work in, 286–287
  Serre's work in, 314–315, 321
  Severi's views on, 266–267

van der Waerden's contributions to, 8–9, 246–276, 313–314
van der Waerden's early algebraization of, 250–255
Veronese's work in, 286, 288
Weil's contributions to, 249, 258–259, 315–316
Zariski's arithmetization of, 9, 275, 285–286, 288–289, 297–298
Algebraic number fields
Hensel's work on, 138
Hilbert's work on, 137–138
Algebraic varieties
Zariski's use of in the arithmetization of algebraic geometry, 294–296
Algebras
beginnings of the structure theory of, 118–121
Benjamin Peirce's work on, 121
Brauer's work on central simple, 201–205
classification of as a branch of algebra, 230–231
Clifford's work on, 121
cyclic, *see also* Cyclic algebras
Dickson's work on the arithmetic of, 7, 117
discovery of noncommutativity of, 119–121
Du Pasquier's work on the arithmetic of, 124
E. Noether's views on the importance of in the commutative setting, 200–202, 207, 210–212
E. Noether's work on central simple, 201, 203
Grassmann's work on, 121
Molien's work on, 121
Scheffers's work on, 121
semisimple, *see also* Semisimple algebras
Wedderburn's structure theory of, 121–122
Ampère, André-Marie, 30
Analytical Society (Cambridge), 20–21, 53–54
Memoirs of the, 54
Argand, Robert
work of on complex numbers, 118–119
Arithmetization
and algebraic geometry, 285–286, 288–289, 292–298
Artin, Emil, 170
seminar of on class field theory, 214–215
work of on class field theory, 128, 142
work of on noncommutative algebra, 117–118

Babbage, Charles, 5, 13–14, 30, 34–35, 54
calculus of functions of, 19–23, 25, 40

definition of function of, 21
on language, 19–21, 39–41
on the theory of signs, 36–37
work of on functional equations, 23–28
Bernstein, Benjamin Abram, 234
Bertholet, Claude-Louis, 35
Bezout's theorem, 93–94
van der Waerden's work on, 259–260
Bezout, Étienne, 75
Biquaternions
Hamilton's discovery of, 120
Birkhoff, Garrett
work of in lattice theory, 237
Bombelli, Rafael, 3
Boole, George, 4
and the *Cambridge Mathematical Journal*, 65–66
work of on the calculus of operations, 65–66
Brauer group, 133
discovery of the, 203–204
Brauer, Richard, 170
and the Brauer-Hasse-Noether Theorem, 130, 133–134, 170
and the construction of factor sets, 203–205
emigration of to the United States, 215–217
work of on central simple algebras, 201–205
work of on splitting fields, 132
Brauer-Hasse-Noether Theorem, *see also* Albert-Brauer-Hasse-Noether Theorem, 133–134, 142, 170, 186, 201, 206–210
Albert's contributions to the, 7–8, 185–186, 194, 201
and cyclic algebras, 130
Brill, Alexander von, 246
work of in algebraic geometry, 286
Buchsbaum, David, 308, 309
Burckhardt, Johann Jakob, 117, 127

Cabanis, Pierre, 14, 29–30
Calculus
foundations of, 34
Calculus of functions
Babbage's work on the, 19–23, 25, 40
Calculus of operations, 50, 52–66
and the *Cambridge Mathematical Journal*, 54–66
and the development of algebra in Britain, 68
Boole's work on the, 65–66
Cauchy's views on the, 53
Craufurd's work on the, 60–61
Ellis's work on the, 61–63
examples in the, 52–53
Greatheed's work on the, 55

INDEX

Gregory's appeal for applicability of the, 57–58
Gregory's appeal for use of the as a tool in pure mathematics, 58–60
Gregory's work on the, 55–60
Herschel's work on the, 54, 60
Murphy's work on the, 59
role of Smith's Prizes in disseminating the, 64–65
role of Tripos moderators in disseminating the, 64
Cambridge Mathematical Journal, 5–6, 49–50
and the calculus of operations, 54–66
founding of the, 51–52
goals of the, 51–52
role of the in British mathematics, 67–68
Cardano, Girolamo, 3
Carnot, Lazare, 34
Cartan, Henri, 308
and the definition of the cohomology of groups, 307–308
Cartier, Pierre
early work of on schemes, 314–316
Castelnuovo, Guido, 246
work of in algebraic geometry, 287
Category theory
Eilenberg's work on, 306
Mac Lane's work on, 306
Cauchy, Augustin-Louis
and roots of polynomials, 78
on the calculus of operations, 53
Cayley, Arthur, 4, 64–65
and the development of matrices, 120–121
and the discovery of the octonions, 119–120
Chevalley, Claude
work of on class field theory, 214
Chomsky, Noam, 5
Chow, Wei-Liang, 269–270
Class field theory
and the Reciprocity Law, 135–141
Artin's work on, 128
E. Noether's work on, 210–212
Hasse's work on, 128, 135–141
Takagi's work on, 128
Clebsch, Alfred, 246
Clifford, William Kingdon
work of on algebras, 121
Cohomology of groups, 128, 306–307
as analogous to the cohomology of sheaves, 309–310
Cartan and Eilenberg's definition of the, 307–308
E. Noether's role in the creation of the, 118, 139, 215
Cohomology of sheaves, 306–307
and the Weil conjectures, 305–306

as analogous to the cohomology of groups, 309–310
Serre's use of the, 305–306
Commutative algebra
definition of, 74
Commutativity (of multiplication)
Sylvester's views on, 73
Complex analytic functions
theory of, 87–89
Complex numbers
and issues of factorization, 74–75
Argand's work on, 118–119
Euler's work on, 118
Gauss's work on, 118
Wessel's work on, 118–119
Condillac, Abbé de, Étienne Bonnot, 5, 13, 28–33
conception of language of, 17–18, 39
De Gérando's critique of, 31–33
grammar of algebra of, 18–19
idea of human statue of, 16
on language and algebra, 15
on the theory of signs, 16–17, 37
Congruences
Dedekind's work on higher, 77–78
Eisenstein's work on higher, 77–78
Gauss's definitions of, 77
Mertens's work on higher, 86
Schönemann's work on higher, 77–78
theory of, 78–80
Zolotarev's work on higher, 84–86
Craufurd, Alexander
work of on the calculus of operations, 60–61
Crossed products
Chevalley's application of, 214
E. Noether's work on, 131–132, 204–205
Hasse's work on, 131–133, 204
Cyclic algebras
Albert's work on, 183–184, 206
and the Brauer-Hasse-Noether Theorem, 130
Dickson's work on, 129–130, 182–183, 206
E. Noether's extension of to crossed products, 131–132
Hasse's work on, 129–131, 135–141, 183–184
Wedderburn's work on, 182–183, 206

D'Alembert, Jean Le Rond, 25, 40
De Gérando, Joseph Marie, 13–14, 30
critique of Condillac of, 31–33
on algebra, 34–35
on the theory of signs, 33–37
De Morgan, Augustus, 49, 66
de Vries, Hendrik, 257
Dedekind, Richard, 83, 85, 87
and binary quadratic forms, 81–82

and higher congruences, 77–78
and the definition of a field, 122
and the definition of modules, 79–80
as founder of lattice theory, 79
early work of on the theory of ideals, 122–123
work of on algebraic functions, 88–89
work of on divisibility, 82, 84–85
work of on primary ideals, 96–97
Derived functor
  early understanding of, 307–308
Descartes, René, 3, 17
Destutt de Tracy, Antoine-Louis-Claude, 14, 24, 28–30
  and the *Élémens d'idéologie*, 37–38
  critique of Condillac of, 39–40
  on algebra, 38
  on language, 39–40
Determinants, 75
Deuring, Max, 215
Dickson, Leonard Eugene, 230–231, 235
  as a formative influence on Albert, 181–182
  as an adviser, 190
  comparative views of on mathematics in the United States and in Europe, 180
  early career of, 179–181
  foreign study tour of, 180
  influence of on American mathematics, 181
  work of in algebra, 181
  work of on algebras and their arithmetics, 7, 117, 127–128, 142
  work of on cyclic algebras, 129–130, 182–183, 206
  work of on skew fields, 124–127, 129–130
  work of on the history of number theory, 155
Dieudonné, Jean
  and the *Éléments de géométrie algébrique*, 317–318
Diophantine analysis
  and integral binary quadratic forms, 155–158
Diophantus of Alexandria, 3
Dirichlet, Gustav Peter Lejeune
  and binary quadratic forms, 81
  and divisibility, 82
Divisibility
  Dedekind's work on, 82, 84–85
  definitions of, 73
  development of theories of, 6
  Dirichlet's theory of, 82
  Gauss's use of, 83
  Kronecker's work on, 83–87, 91–93
  Kummer's work on, 85
  Zolotarev's work on, 84–86
Du Pasquier, Louis-Gustave
  work of on the arithmetic of algebras, 124

Écoles Centrales, 37
Eilenberg, Samuel
  and category theory, 306
  and the definition of the cohomology of groups, 307–308
Eisenbud, David
  treatment of Abelian categories of, 310
Eisenstein, Gotthold
  and higher congruences, 77–78
Elimination theory
  algebraic theory of, 75–76
  Kronecker's use of in modular systems, 91–94
  van der Waerden's use of in algebraic geometry, 252
Ellis, Robert Leslie, 67
  work of on the calculus of operations, 61–63
Encyklopädie der mathematischen Wissenschaften
  classificatory scheme of the, 228–229
Enriques, Federigo, 246
  notion of generic points of, 254
  work of in algebraic geometry, 287
Equations
  solvability of, 76
Euclid of Alexandria, 3
Euler, Leonhard, 4
  and complex numbers, 118
  and the arithmetic theory of binary quadratic forms, 76–77
  views of on algebra, 74
  work of on quadratic forms, 156–157

Factor sets
  Brauer's construction of, 203–205
Fermat, Pierre de, 3
Field theory, 236
Fourcroy, Antoine de, 35
Fraenkel, Abraham
  and ring theory, 234–235
Frobenius, Georg
  results of on skew fields, 122
Function
  Babbage's definition of, 21
Functional equations
  Babbage's work on, 23–28
  definition of, 19–20
Fundamental Theorem of Algebra, 74–75
  Gauss's work on the, 109–110
  in relation to Kronecker's work, 114–115
  Kronecker's views on the, 107–108
Funding
  role of in mathematics, 193
Furtwängler, Philipp, 168

Galois, Évariste, 4, 76

    and the use of symmetric polynomials, 110–113
Gauss's lemma, 83
    Kronecker's extension of, 83–84
Gauss, Carl Friedrich
    and binary quadratic forms, 80–81, 157–158
    and complex numbers, 118
    and congruence, 77
    and divisibility, 83–84
    cyclotomy theory of, 76
    proofs of the Reciprocity Law of, 141
    work of on polynomials, 82–83
    work of on the Fundamental Theorem of Algebra, 109–110
Generic points
    Enriques's notion of, 254
    van der Waerden's notion of, 253–255
Genus
    Gauss's definition of, 157–158
Geometry of numbers
    Minkowski's early ideas on the, 160
Gergonne, Joseph, 30
Gordan, Paul, 4, 246
Grassmann, Hermann, 73–74
    work of on algebras, 121
Graves, John Thomas
    and the discovery of the octonions, 119–120
Greatheed, Samuel, 51
    work of on the calculus of operations, 55
Gregory, Duncan, 5–6, 49–51, 66–67
    appeal of for applicability of calculus of operations, 57–58
    appeal of for use of calculus of operations as a tool in pure mathematics, 58–60
    definition of algebra of, 58
    views of on Ellis's work on the calculus of operations, 62–63
    work of on the calculus of operations, 55–60, 63–64
Grothendieck, Alexandre, 246
    and the cohomology of groups, 311
    and the development of category theory, 9–10, 301
    and the Weil conjectures, 301, 303
    and the *Éléments de géométrie algébrique*, 317–318
    philosophy of mathematics of, 301–303, 312, 322
    Tôhoku paper of, 308–310
    views of on spectral sequences, 311
    work of in algebraic geometry, 269–270, 321–322
    work of on schemes, 317–321
    work of on sheaves, 308–309, 311
    work of on the cohomology of groups, 309–310
    work of on the cohomology of sheaves, 309–310
    work of on topos, 311–312
    working style of, 302–303
Group theory, 233–235
Grunwald, Wilhelm
    Grunwald-Wang Theorem, 207, 209–210

Hamilton, William Rowan, 68, 73–74, 232
    and the discovery of the biquaternions, 120
    and the discovery of the quaternions, 119
Harriot, Thomas, 3
Hartshorne, Robin
    treatment of Abelian categories of, 310
Hasse, Helmut, 155
    and Hasse invariants, 212–214
    and the application of p-adic numbers to number theory, 153–155
    and the Brauer-Hasse-Noether Theorem, 130, 133–134, 170
    and the definition of the norm residue symbol, 135–142, 214
    and the Local-Global Principle, 133–134, 154, 168–169
    and the Strong Hasse Principle, 167–168
    joint paper of with Albert, 186–187
    letter of to Hermann Weyl of 15 December, 1931, 171–173
    mathematical education of, 165–167
    noncommutative proof of Artin's reciprocity law of, 212–214
    on developments in the theory of cyclic algebras as of 1931, 184–185
    paper of on cyclic algebras, 204–205
    relations of van der Waerden with, 273–275
    relations of with Severi, 273, 276–278
    review by of Dickson's *Algebren und ihre Zahlentheorie*, 127–128
    work of on algebraic number fields, 138–141
    work of on class field theory, 128, 135–141
    work of on crossed products, 131–133
    work of on cyclic algebras, 129–131, 135–141, 183–184
    work of on noncommutative algebra, 7, 117–118
    work of on quadratic forms, 167–170
    work of on skew fields, 126
    working style of, 187–188
Hecke, Erich, 166
Heine, Eduard, 108
Hensel, Kurt, 84, 86, 155, 206–207
    influence of on Hasse, 166
    work of on p-adic numbers, 153, 162–165
    work of on algebraic number fields, 138
Hermite, Charles, 4

Herschel, John F. W., 19–20, 23, 49, 54
    work of on the calculus of operations, 54, 60
Hey, Käte, 215
Hilbert's Basis Theorem, 94
Hilbert, David, 4, 160, 165–166, 168
    views of on Kronecker, 109
    work of on algebraic number fields, 137–138
    work of on modular systems, 94–95
    work of on the norm residue symbol, 136, 138, 141
Hurwitz, Adolf, 83, 87, 160
    work of on the arithmetic of quaternions, 122–124, 143
Hypercomplex numbers, *see also* Algebras

Idéologie, 28–31, 33–35, 37
    Destutt de Tracy's views on, 37–38
Ideal numbers
    Kummer's work on, 84
    theory of, 78
Ideal theory
    and Zariski's rewriting of algebraic geometry, 292–296
    Dedekind's early work on, 122–123
Institut national de France
    Second Class of, 29
Internationalization
    of algebra, 192
Intersection theory
    van der Waerden's development of, 256–264
Invariant theory, 4–5

Jahrbuch über die Fortschritte der Mathematik
    classificatory scheme 1905–1915 of the, 229–231
    classificatory scheme after 1916 of the, 233–237
    classificatory scheme in 1900 of the, 227–228
    founding of the, 225–226
Joly, Charles J., 230
Jordan, Camille
    and the *Traité des substitutions et des équations algébriques*, 221

König, Julius, 83
Königsberg seminar for mathematics and physics
    Minkowski at the, 159–160
Kürschák, Josef, 163
Kamke, Erich, 27–28
Kant, Immanuel, 30
Klein, Felix, 81
Kneser, Hellmuth, 266, 274–275
Kolchin, Ellis R., 99
Kronecker, Leopold, 78–79, 85
    and computation with roots of polynomials, 112–113
    and the Fundamental Theorem of General Arithmetic, 6, 107–109, 113–114
    definition of modular system of, 79
    extension of Gauss's lemma of, 83–84
    views of on infinite series, 108
    views of on real numbers, 108–109
    views of on the foundations of algebra, 107–109
    views of on the Fundamental Theorem of Algebra, 107–108
    work of on algebraic functions, 88
    work of on divisibility, 83–87
    work of on modular systems, 89–94
    work of on primary ideals, 97
Krull, Wolfgang, 99, 235, 255
    early conception of schemes of, 314–315
    influence of on Zariski, 292
    work of in ring theory, 235
    work of on algebraic geometry, 314–315
Kummer, Ernst Eduard
    and ideal numbers, 78, 84
    work of on divisibility, 85

Lacroix, Silvestre, 25–26
Lagrange, Joseph-Louis, 4, 34
    and the arithmetic theory of binary quadratic forms, 76–77
    work of on quadratic forms, 156
Lampe, Emil, 226
Lang, Serge
    treatment of Abelian categories of, 310
Language of science
    algebra as the, 15, 39
    Babbage's construction of the, 21
    Babbage's views on the, 40
    chemistry as the, 35, 39
    Condillac's views on the, 40
    construction of the, 24
    De Gérando's views on the, 41
    limitations of the, 14–15
    mathematics as the, 15
Laplace, Pierre Simon, 19–20
Lasker, Emanuel, 93, 231
    work of on modular systems, 95–96
Lattice theory
    Øre's work in, 237
    Dedekind's role in founding, 79
    work of Garrett Birkhoff in, 237
Lavoisier, Antoine, 24, 35
Law of inertia, 158–159
Lefschetz, Solomon
    influences of on Zariski, 288
    review by of Zariski's *Algebraic Surfaces* (1935), 291
    topological ideas of applied by van der Waerden to algebraic geometry, 261–264

Legendre, Adrien-Marie, 169
  and the arithmetic theory of binary quadratic forms, 76–77
Leibniz, Gottfried Wilhelm, 31
  and the calculus of operations, 52
Levi, Beppo
  work of in algebraic geometry, 289–291
Lichtenstein, Leon, 233
Lindemann, Ferdinand, 160
Lipschitz, Rudolph
  work of on the arithmetic of quaternions, 142–143
Local-Global Principle, 7, 170–171, 206–209
  attributed by Hasse to Minkowski, 170
  Hasse's work on the, 133–134, 154, 168–169
Locke, John, 31
Loewy, Alfred, 234

Macaulay, Francis, 93, 235
MacFarlane, Alexander, 230
Mac Lane, Saunders
  and category theory, 306
Maine de Biran, François-Pierre-Gonthier, 25–26, 31
Marburg Mathematical Colloquium (26-28 February, 1931), 199–200
Mathematical and Philosophical Repository, 51
Mathematical Reviews
  classificatory scheme of the, 238–239
Mathematical Tripos (Cambridge)
  role of moderators of the in disseminating the calculus of operations, 64
Matrices
  Cayley's development of, 120–121
  Sylvester's coining of the term, 120–121
Mertens, Franz, 86
Minkowski, Hermann, 155
  and the origins of the geometry of numbers, 160
  and the *Grand Prix des Sciences Mathématiques* of 1882, 159–160
  mathematical education of, 159–161
  work of on quadratic forms, 154, 161–162
Modern algebra
  and Zariski's rewriting of algebraic geometry, 292–296
  evolution of the image of, 221–239
  influence of van der Waerden in shaping the image of, 222–225
Modular systems
  Hilbert's work on, 94–95
  Kronecker's definition of, 79
  Kronecker's use of elimination in, 91
  Kronecker's work on, 89–94
  Lasker's work on, 95–96
Module
  and Dedekind's theory of quadratic forms, 82
  Dedekind's definition of, 79–80
Molien, Theodor
  work of on algebras, 121
Molk, Jules, 94
Monge, Gaspard, 19
Morveau, Louis Guyton de, 35
Mumford, David
  views of on Italian algebraic geometry, 248–249
Murphy, Robert
  work of on the calculus of operations, 59

Neugebauer, Otto, 237, 238
Newton-Puiseux series, 76, 88
Noether, Emmy, 93, 235
  and the Brauer-Hasse-Noether Theorem, 130, 133–134, 170
  and the creation of the cohomology of groups, 118, 139, 205, 215
  and the definition of crossed products, 131–132, 204
  and the Principal Genus Theorem, 210–212
  and the recasting of algebraic geometry in terms of modern algebra, 247–248
  as an adviser, 251–252
  at Göttingen University, 166, 170
  emigration of to the United States, 215–217
  ICM plenary lecture of in 1932, 8, 199–201, 210–212
  influence of on Zariski, 292
  lectures of on noncommutative algebra at Göttingen, 199–200, 204
  relationship of with van der Waerden, 251–252
  views of on the importance of algebras in the commutative setting, 200–202, 207, 210–212
  work of on algebraic functions, 88–89
  work of on central simple algebras, 201, 203
  work of on crossed products, 131–132, 204–205
  work of on noncommutative algebra, 117–118
  work of on primary ideals, 97–99
  work of on splitting fields, 132
Noether, Max, 246
  work of in algebraic geometry, 286, 289–291, 293
Non-associative algebras
  Albert's work on, 188–189
Norm residue symbol
  Hasse's definition of the, 135–141
  Hilbert's work on the, 136, 138, 141

Notation
  role of in mathematics, 23
Nullstellensatz, 95

Octonions
  Cayley's discovery of the, 119–120
  Graves's discovery of the, 119–120
Øre, Oystein, 85
  work of in lattice theory, 237

p-adic numbers
  Hasse's application of to number theory, 153–155
  Hensel's work on, 153, 162–165
Pascal, Ernesto, 228
Peacock, George, 36, 54
  views of on algebra, 58–59
Peirce, Benjamin, 232
  work of on algebras, 121
Peirce, Charles S.
  results of on skew fields, 122
Philosophical Magazine
  obstacles to publishing mathematics in the, 50
Phlogiston theory, 39
Poincaré, Henri, 81
Polynomials
  Gauss's work on, 82–83
  roots of, 78–79
  Weber's work on roots of, 78
Postulational analysis, 232, 234
Primary ideals
  Dedekind's work on, 96–97
  E. Noether's work on, 97–99
  Kronecker's work on, 97
Principal Genus Theorem
  E. Noether's proof of the, 210–212
Pseudonyms
  use of in British scientific journals, 57

Quadratic forms
  Dedekind's work on, 81–82
  Dirichlet's work on, 81
  Euler's work on, 76–77, 156–157
  Gauss's work on, 80–81
  Hasse's work on, 167–170
  Lagrange's work on, 76–77, 156
  Legendre's work on, 76–77
  Minkowski's work on, 154, 161–162
  Smith's work on, 81
  theory of, 76–77, 158–159
Quaternions
  considered as a branch of algebra, 230
  Hamilton's discovery of, 119
  Hurwitz's work on the arithmetic of, 122–124

Reciprocity Law
  and class field theory, 135, 141
  Gauss's proofs of the, 141
Reinhold, Baer, 309
Resolution of singularities
  early geometric proofs of the, 289–291
  Walker's geometric proof of the, 292–293
  Zariski's arithmetic proof of the, 296
  Zariski's critique of the, 289–291
Resolvent, 76
Resultant, 75–76
Riemann hypothesis
  and the Weil conjectures, 304–305
Riemann, Bernhard, 245
  role of in the development of algebraic geometry, 286
  work of on Abelian functions, 87–88
Ring theory, 236
  Fraenkel's work in, 234–235
  Krull's work in, 235
  Schur's work in, 235
Ritt, Joseph, 99
Rosati, Carlo, 246
Roth, Leonard, 265

Schönemann, Theodor
  and higher congruences, 77–78
Scheffers, Georg
  work of on algebras, 121
Schemes
  Cartier's early work on, 314–316
  Grothendieck's work on, 317–321
  Krull's early work on, 314–315
  Serre's early work on, 314–315
Schröder, Ernst, 233
Schubert, Hermann C. H.
  and the development of Schubert calculus, 256–257
Schur, Issai, 202, 231, 235
  work of in ring theory, 235
Segre, Beniamino, 246
Segre, Corrado
  work of in algebraic geometry, 286–287
Selling, Eduard, 78
Semisimple algebras
  arithmetic theory of, 127–133
  Wedderburn's definition of, 121
Serre, Jean-Pierre
  and the cohomology of sheaves, 305–306
  and the Weil conjectures, 301, 303, 316–317
  early conception of schemes of, 314–315
  early work of on schemes, 315
  work of in algebraic geometry, 314–315, 321
  work of on spectral sequences, 310–311
  working style of, 302–303
Serret, Joseph
  and the *Cours d'algèbre supérieure*, 221
Severi, Francesco, 246

influence of on van der Waerden, 264–272
   personality of, 265–266
   relations of with Zariski, 275
   views of on algebraic geometry, 266–267
Sheaves
   definition of, 306
   Grothendieck's work on, 308–309, 311
Siegel, Carl Ludwig, 169–170
Silberstein, Ludwick, 27
Skew fields
   Charles Peirce's results on, 122
   Dickson's work on, 124–127, 129–130
   Frobenius's results on, 122
   Hasses's work on, 126
   structure theory of, 124–127
   Wedderburn's work on, 126, 129–130
Smith's Prizes (Cambridge)
   role of the in disseminating the calculus of operations, 64–65
Smith, Archibald, 51
Smith, Henry J. S.
   and binary quadratic forms, 81
   and the *Grand Prix des Sciences Mathématiques* of 1882, 160
Specialization
   van der Waerden's notion of, 257–259
   Weil's notion of, 258
Spectral sequences
   Grothendieck's views on, 311
   Serre's work on, 310–311
Speiser, Andreas, 117, 127, 143
Splitting fields
   Albert's work on, 132
   Brauer's work on, 132
   E. Noether's work on, 132
Steinitz, Ernst
   work of on fields, 231
Strong Hasse Principle, 167–168
Sylow, Ludvig, 76
Sylvester, James Joseph, 4, 232
   coins the word matrix, 120–121
   definition of resultant of, 75–76
   law of inertia of, 158–159
   views of on commutativity of multiplication, 73
Symmetric polynomials
   and the computation of roots of polynomials, 110–112

Takagi, Teiji
   work of on class field theory, 118, 128, 142
Tartaglia, Niccolò, 3
Taylor's Theorem
   and the calculus of operations, 52–53
Thomson, William (Lord Kelvin), 67
Topology
   and its connection with the Weil conjectures, 304–305

Topos
   Grothendieck's work on, 311–312
Transactions of the American Mathematical Society
   classificatory scheme of the, 232–233
Transactions of the Royal Society of London
   obstacles to publishing mathematics in the, 50

University of Chicago
   research tradition in algebra at the, 191

van der Waerden, Bartel, 4
   and his series "Zur algebraischen Geometrie" [ZAG], 263–272
   and the algebraization of algebraic geometry, 250–255
   and the application of Lefschetz's topological ideas to algebraic geometry, 261–264
   and *Einführung in die algebraische Geometrie* (1939), 270–271
   and *Moderne Algebra*, 8, 143, 222–225
   development of intersection theory of, 256–264
   first paper of on algebraic geometry (1926), 250–255
   influence of on Weil, 275–276
   notion of generic points of, 253–255
   notion of specialization of, 257–259
   relations of with Hasse and his school, 273–275
   relations of with the Italian school of algebraic geometry, 272–273
   relationship of with E. Noether, 251–252
   second paper of on algebraic geometry (1927), 257–259
   Severi's influence on, 264–272
   work of in algebraic geometry, 8–9, 252, 272–276, 313–314
   work of on Bezout's theorem, 259–260
Vandermonde, Alexandre, 4
Varieties
   examples of, 313
Vaugelas, Claude Favre de, 17
Veronese, Giuseppe
   work of in algebraic geometry, 286, 288
Viète, François, 3
Voigt, Woldemar, 159

Walker, Robert
   geometric proof of the resolution of singularities of, 292–293
Wang, Shianghaw, 209–210
Weber, Heinrich, 84, 159, 168
   and roots of polynomials, 78
   and the *Lehrbuch der Algebra*, 221–222
   work of on algebraic functions, 88–89
   work of on Galois theory, 227

Wedderburn, Joseph H. M., 230–231
  and the definition of semisimple algebras, 122
  and the structure theory of algebras, 121–122, 233
  work of on cyclic algebras, 182–183, 206
  work of on skew fields, 126, 129–130
Weil conjectures, 308
  and the cohomology of sheaves, 305–306
  and the Riemann hypothesis, 304–305
  connection of the with topology, 304–305
  Grothendieck's work on the, 301, 303, 312
  Serre's work on the, 301, 303, 316–317
  Weil's formulation of the, 303–305
Weil, André, 246
  and the Weil conjectures, 301, 303–305
  influence of van der Waerden on, 275–276
  notion of specialization of, 258
  work of in algebraic geometry, 249, 259, 315–316
Wessel, Caspar
  work of on complex numbers, 118–119
Weyl, Hermann, 170–171
Whewell, William, 64

Zariski, Oscar, 246, 261
  and critique of proofs of the resolution of algebraic singularities, 289–291
  and the arithmetization of algebraic geometry, 285–286, 288–289, 297–298
  and *Algebraic Surfaces* (1935), 289–292, 296–297
  arithmetic proof of the resolution of singularities of, 296
  contributions of to algebraic geometry, 9, 275
  E. Noether's influence on, 292
  influence of Krull on, 292
  Italian influences on, 288
  Lefschetz's influence on, 288
  relations of with Severi, 275
  use of algebraic varieties in the arithmetization of algebraic geometry of, 294–296
Zentralblatt für Mathematik und ihre Grenzgebiete
  classificatory scheme of the, 237–238
Zolotarev (Zolotareff), Egor Ivanovič, 78
  work of on divisibility, 84–86
  work of on higher congruences, 84–86